Applied Calculus
4th Edition

Hughes-Hallett • Gleason • Lock • Flath • et al.

Mathematics
An Applied Approach
8th Edition

Sullivan • Mizrahi

WILEY Custom
LEARNING SOLUTIONS

To order books or for customer service, please call 1(800)-CALL-WILEY (225-5945).

ISBN 978-0-470-89179-7

Printed and Bound by Lightning Source, Inc.

10 9 8 7 6 5 4 3 2 1

Contents

APPLIED CALCULUS

Fourth Edition

Produced by the Calculus Consortium and initially funded by a National Science Foundation Grant.

Deborah Hughes-Hallett
University of Arizona

William G. McCallum
University of Arizona

Andrew M. Gleason
Harvard University

Brad G. Osgood
Stanford University

Patti Frazer Lock
St. Lawrence University

Douglas Quinney
University of Keele

Daniel E. Flath
Macalester College

Karen Rhea
University of Michigan

David O. Lomen
University of Arizona

Jeff Tecosky-Feldman
Haverford College

David Lovelock
University of Arizona

Thomas W. Tucker
Colgate University

with the assistance of

Otto K. Bretscher
Colby College

Eric Connally
Harvard Extension School

Richard D. Porter
Northeastern University

Sheldon P. Gordon
SUNY at Farmingdale

Andrew Pasquale
Chelmsford High School

Joe B. Thrash
University of Southern Mississippi

Coordinated by
Elliot J. Marks

WILEY

John Wiley & Sons, Inc.

PUBLISHER	Laurie Rosatone
ACQUISITIONS EDITOR	David Dietz
ASSOCIATE EDITOR	Shannon Corliss
EDITORIAL ASSISTANT	Pamela Lashbrook
DEVELOPMENTAL EDITOR	Anne Scanlan-Rohrer/Two Ravens Editorial
MARKETING MANAGER	Sarah Davis
MEDIA EDITOR	Melissa Edwards
SENIOR PRODUCTION EDITOR	Ken Santor
COVER DESIGNER	Madelyn Lesure
COVER AND CHAPTER OPENING PHOTO	©Patrick Zephyr/Patrick Zephyr Nature Photography Images

Problems from *Calculus: The Analysis of Functions*, by Peter D. Taylor (Toronto: Wall & Emerson, Inc., 1992). Reprinted with permission of the publisher.

This book was set in Times Roman by the Consortium using TeX, Mathematica, and the package AsTeX, which was written by Alex Kasman.

This book is printed on acid-free paper.

This material is based upon work supported by the National Science Foundation under Grant No. DUE-9352905. Opinions expressed are those of the authors and not necessarily those of the Foundation.

ISBN: 978-0-470-17052-6
Binder Ready: 978-0-470-55662-7

10 9 8 7 6 5 4 3 2 1

We dedicate this book to Andrew M. Gleason.

His brilliance and the extraordinary kindness and dignity with which he treated others made an enormous difference to us, and to many, many people. Andy brought out the best in everyone.

Deb Hughes Hallett
for the Calculus Consortium

PREFACE

Calculus is one of the greatest achievements of the human intellect. Inspired by problems in astronomy, Newton and Leibniz developed the ideas of calculus 300 years ago. Since then, each century has demonstrated the power of calculus to illuminate questions in mathematics, the physical sciences, engineering, and the social and biological sciences.

Calculus has been so successful because of its extraordinary power to reduce complicated problems to simple rules and procedures. Therein lies the danger in teaching calculus: it is possible to teach the subject as nothing but the rules and procedures—thereby losing sight of both the mathematics and of its practical value. This edition of *Applied Calculus* continues our effort to promote courses in which understanding reinforces computation.

Origin of the Text: A Community of Instructors

This text, like others we write, draws on the experience of a diverse group of authors and users. We have benefitted enormously from input from a broad spectrum of instructors—at research universities, four-year colleges, community colleges, and secondary schools. For *Applied Calculus*, the contributions of colleagues in biology, economics, medicine, business, and other life and social sciences have been equally central to the development of the text. It is the collective wisdom of this community of mathematicians, teachers, natural and social scientists that forms the basis for the new edition.

A Balance Between Skills and Concepts

The first edition of our text struck a new balance between concepts and skills. As instructors ourselves, we know that the balance we choose depends on the students we have: sometimes a focus on conceptual understanding is best; sometimes more drill is appropriate. The flexibility of this new fourth edition allows instructors to tailor the course to their students.

Since 1992, we have continued to find new ways to help students learn. Under our approach, which we call the "Rule of Four," ideas are presented graphically, numerically, symbolically, and verbally, thereby encouraging students with a variety of learning styles to expand their knowledge. Our problems probe student understanding in areas often taken for granted. The influence of these problems, praised for their creativity and variety, has extended far beyond the users of our textbook.

Mathematical Thinking: A Balance Between Theory and Modeling

The first stage in the development of mathematical thinking is the acquisition of a clear intuitive picture of the central ideas. In the next stage, the student learns to reason with the intuitive ideas and explain the reasoning clearly in plain English. After this foundation has been laid, there is a choice of direction. All students benefit from both theory and modeling, but the balance may differ for different groups. In our experience as instructors, students of this book are motivated both by understanding the concepts and by seeing the power of mathematics applied to their fields of interest—be it the spread of a disease or the analysis of a company. Some instructors may choose a more theoretical approach; others may choose to ground the mathematics in applied examples. This text is flexible enough to support both approaches.

Mathematical Skills: A Balance Between Symbolic Manipulation and Technology

To use calculus effectively, students need skill in both symbolic manipulation and the use of technology. The balance between them may vary, depending on the needs of the students and the wishes of the instructor. The book is adaptable to many different combinations.

The book does not require any specific software or technology. It has been used with graphing calculators, graphing software, and computer algebra systems. Any technology with the ability to graph functions and perform numerical integration will suffice. Students are expected to use their own judgment to determine where technology is useful.

What Student Background is Expected?

This book is intended for students in business, the social sciences, and the life sciences. We have found the material to be thought-provoking for well-prepared students while still accessible to students with weak algebra backgrounds. Providing numerical and graphical approaches as well as the algebraic gives students several ways of mastering the material. This approach encourages students to persist, thereby lowering failure rate; a pre-test over background material is available in the appendix to the book; An algebra refresher is avalable at the student book companion site at www.wiley.com/college/hughes-hallett.

The Fourth Edition: Expanded Options

Because different users often choose very different topics to cover in a one-semester applied calculus course, we have designed this book for either a one-semester course (with much flexibility in choosing topics) or a two-semester course. Sample syllabi are provided in the Instructor's Manual.

The fourth edition has the same vision as the first three editions. In preparing this edition, we solicited comments from a large number of mathematicians who had used the text. We continued to discuss with our colleagues in client disciplines the mathematical needs of their students. We were offered many valuable suggestions, which we have tried to incorporate, while maintaining our original commitment to a focused treatment of a limited number of topics. The changes we have made include:

- **Updated data** and **fresh applications** throughout the book.
- Many **new problems** have been added. These have been designed to build student confidence with basic concepts and to reinforce skills.
- **Chapter 1 has been shortened**; the material on polynomials has been streamlined to focus on quadratics. Focus sections have been moved to the appendix or removed.
- Material on **relative change** and **relative rates of change** has been added in Sections 1.3, 2.3, and 3.3, thereby allowing instructors to reinforce the distinction between rates of change and relative rates of change.
- A new **project on relative growth rates in economics** has been added in Chapter 3.
- A new section on **integration by parts** has been added to Chapter 7.
- True-False **Check Your Understanding** questions have been added at the end of every chapter, enabling students to check their progress.
- As in the previous edition, a **Pre-test** is included for students whose skills may need a refresher prior to taking the course.
- **Pointers** have been inserted in chapters referring students to Appendices or Focus sections.

Content

This content represents our vision of how applied calculus can be taught. It is flexible enough to accommodate individual course needs and requirements. Topics can easily be added or deleted, or the order changed.

Chapter 1: Functions and Change

Chapter 1 introduces the concept of a function and the idea of change, including the distinction between total change, rate of change, and relative change. All elementary functions are introduced here. Although the functions are probably familiar, the graphical, numerical, verbal, and modeling approach to them is likely to be new. We introduce exponential functions early, since they are fundamental to the understanding of real-world processes.

Chapter 2: Rate of Change: The Derivative

Chapter 2 presents the key concept of the derivative according to the Rule of Four. The purpose of this chapter is to give the student a practical understanding of the meaning of the derivative and its interpretation as an instantaneous rate of change. Students will learn how the derivative can be used to represent relative rates of change. After finishing this chapter, a student will be able to approximate derivatives numerically by taking difference quotients, visualize derivatives graphically as the slope of the graph, and interpret the meaning of first and second derivatives in various applications. The student will also understand the concept of marginality and recognize the derivative as a function in its own right.

Focus on Theory: This section discusses limits and continuity and presents the symbolic definition of the derivative.

Chapter 3: Short-Cuts to Differentiation

The derivatives of all the functions in Chapter 1 are introduced, as well as the rules for differentiating products, quotients, and composite functions. Students learn how to find relative rates of change using logarithms.

Focus on Theory: This section uses the definition of the derivative to obtain the differentiation rules.

Focus on Practice: This section provides a collection of differentiation problems for skill-building.

Chapter 4: Using the Derivative

The aim of this chapter is to enable the student to use the derivative in solving problems, including optimization and graphing. It is not necessary to cover all the sections.

Chapter 5: Accumulated Change: The Definite Integral

Chapter 5 presents the key concept of the definite integral, in the same spirit as Chapter 2.

The purpose of this chapter is to give the student a practical understanding of the definite integral as a limit of Riemann sums, and to bring out the connection between the derivative and the definite integral in the Fundamental Theorem of Calculus. We use the same method as in Chapter 2, introducing the fundamental concept in depth without going into technique. The student will finish the chapter with a good grasp of the definite integral as a limit of Riemann sums, and the ability to approximate a definite integral numerically and interpret it graphically. The chapter includes applications of definite iintegrals in a variety of contexts.

Chapter 5 can be covered immediately after Chapter 2 without difficulty.

Focus on Theory: This section presents the Second Fundamental Theorem of Calculus and the properties of the definite integral.

Chapter 6: Using the Definite Integral

This chapter presents applications of the definite integral. It is not meant to be comprehensive and it is not necessary to cover all the sections.

Chapter 7: Antiderivatives

This chapter covers antiderivatives from a graphical, numerical, and algebraic point of view. Sections on integration by substitution and integration by parts are included. The Fundamental Theorem of Calculus is used to evaluate definite integrals and to analyze antiderivatives.

Focus on Practice: This section provides a collection of integration problems for skill-building.

Chapter 8: Probability

This chapter covers probability density functions, cumulative distribution functions, the median, and the mean.

Chapter 9: Functions of Several Variables

Chapter 9 introduces functions of two variables from several points of view, using contour diagrams, formulas, and tables. It gives students the skills to read contour diagrams and think graphically, to read tables and think numerically, and to apply these skills, along with their algebraic skills, to modeling. The idea of the partial derivative is introduced from graphical, numerical, and symbolic viewpoints. Partial derivatives are then applied to optimization problems, ending with a discussion of constrained optimization using Lagrange multipliers.

Focus on Theory: This section uses optimization to derive the formula for the regression line.

Chapter 10: Mathematical Modeling Using Differential Equations

This chapter introduces differential equations. The emphasis is on modeling, qualitative solutions, and interpretation. This chapter includes applications of systems of differential equations to population models, the spread of disease, and predator-prey interactions.

Focus on Theory: This section explains the technique of separation of variables.

Chapter 11: Geometric Series

This chapter covers geometric series and their applications to business, economics, and the life sciences.

Appendices

The first appendix introduces the student to fitting formulas to data; the second appendix provides further discussion of compound interest and the definition of the number e. The third appendix contains a selection of spreadsheet projects.

Supplementary Materials

Supplements for the instructor can be obtained by sending a request on your institutional letterhead to Mathematics Marketing Manager, John Wiley & Sons Inc., 111 River Street, Hoboken, NJ 07030-5774, or by contacting your local Wiley representative. The following supplementary materials are available.

- **Instructor's Manual** (ISBN 978-0-470-60189-1) containing teaching tips, sample syllabii, calculator programs, and overhead transparency masters.
- **Instructor's Solution Manual** (ISBN 978-0-470-60190-7) with complete solutions to all problems.
- **Student's Solution Manual** (ISBN 978-0-470-17053-3) with complete solutions to half the odd-numbered problems.
- **Student Study Guide** (ISBN 978-0-470-45820-4) with additional study aids for students that are tied directly to the book.
- **Additional Material for Instructors**, elaborating specially marked points in the text, as well as password protected electronic versions of the instructor ancillaries, can be found on the web at www.wiley.com/college/hughes-hallett.
- **Additional Material for Students**, at the student book companion site at www.wiley.com/college/hughes-hallett, includes an algebra refresher and web quizzes.

Getting Started Technology Manual Series:
- **Getting Started with Mathematica**, 3rd edn, by C-K. Cheung, G.E. Keough, Robert H. Gross, and Charles Landraitis of Boston College (ISBN 978-0-470-45687-3)
- **Getting Started with Maple**, 3rd edn, by C-K. Cheung, G.E. Keough, both of Boston College, and Michael May of St. Louis University (ISBN 978-0-470-45554-8)

ConcepTests

ConcepTests, (ISBN 978-0-470-60191-4) modeled on the pioneering work of Harvard physicist Eric Mazur, are questions designed to promote active learning during class, particularly (but not exclusively) in large lectures. Evaluation data shows that students taught with ConcepTests outperformed students taught by traditional lecture methods 73% versus 17% on conceptual questions, and 63% versus 54% on computational problems.[1] A new supplement to *Applied Calculus*, 4th edn, containing ConcepTests by section, is available from your Wiley representative.

Wiley Faculty Network

The Wiley Faculty Network is a peer-to-peer network of academic faculty dedicated to the effective use of technology in the classroom. This group can help you apply innovative classroom techniques and implement specific software packages. Visit www.wherefacultyconnect.com or ask your Wiley representative for details.

WileyPLUS

WileyPLUS, Wiley's digital learning environment, is loaded with all of the supplements above, and also features:
- E-book, which is an exact version of the print text, but also features hyperlinks to questions, definitions, and supplements for quicker and easier support.
- Homework management tools, which easily enable the instructor to assign and automatically grade questions, using a rich set of options and controls.
- QuickStart pre-designed reading and homework assignments. Use them as-is or customize them to fit the needs of your classroom.
- Guided Online (GO) Exercises, which prompt students to build solutions step-by-step. Rather than simply grading an exercise answer as wrong, GO problems show students precisely where they are making a mistake.
- Student Study Guide, providing key ideas, additional worked examples with corresponding exercises, and study skills.
- Algebra & Trigonometry Refresher quizzes, which provide students with an opportunity to brush-up on material necessary to master calculus, as well as to determine areas that require further review.
- Graphing Calculator Manual, to help students get the most out of their graphing calculator, and to show how they can apply the numerical and graphing functions of their calculators to their study of calculus.

Acknowledgements

First and foremost, we want to express our appreciation to the National Science Foundation for their faith in our ability to produce a revitalized calculus curriculum and, in particular, to Louise Raphael, John Kenelly, John Bradley, Bill Haver, and James Lightbourne. We also want to thank the members of our Advisory Board, Benita Albert, Lida Barrett, Bob Davis, Lovenia DeConge-Watson, John Dossey, Ron Douglas, Don Lewis, Seymour Parter, John Prados, and Steve Rodi for their ongoing guidance and advice.

In addition, we want to thank all the people across the country who encouraged us to write this book and who offered so many helpful comments. We would like to thank the following people, for all that they have done to help our project succeed: Lauren Akers, Ruth Baruth, Jeffery Bergen, Ted Bick, Graeme Bird, Kelly Boyle, Kelly Brooks, Lucille Buonocore, R.B. Burckel, J. Curtis Chipman, Dipa Choudhury, Scott Clark, Larry Crone, Jane Devoe, Jeff Edmunds, Gail Ferrell, Joe Fiedler, Holland Filgo, Sally Fischbeck, Ron Frazer, Lynn Garner, David Graser, Ole Hald, Jenny Harrison, John Hennessey, Yvette Hester, David Hornung, Richard Iltis, Adrian Iovita, Jerry Johnson, Thomas Judson, Selin Kalaycioglu, Bonnie Kelly, Mary Kittell, Donna Krawczyk, Theodore Laetsch, T.-Y. Lam, Sylvain Laroche, Kurt Lemmert, Suzanne Lenhart,

[1]"Peer Instruction in Physics and Mathematics" by Scott Pilzer in *Primus*, Vol XI, No 2, June 2001. At the start of Calculus II, students earned 73% on conceptual questions and 63% on computational questions if they were taught with ConcepTests in Calculus I; 17% and 54% otherwise.

Madelyn Lesure, Ben Levitt, Thomas Lucas, Alfred Manaster, Peter McClure, Georgia Kamvosoulis Mederer, Kurt Mederer, David Meredith, Nolan Miller, Mohammad Moazzam, Saadat Moussavi, Patricia Oakley, Mary Ellen O'Leary, Jim Osterburg, Mary Parker, Ruth Parsons, Greg Peters, Kim Presser, Sarah Richardson, Laurie Rosatone, Daniel Rovey, Harry Row, Kenneth Santor, Anne Scanlan-Rohrer, Alfred Schipke, Adam Spiegler, Virginia Stallings, Brian Stanley, Mary Jane Sterling, Robert Styer, "Suds" Sudholz, Ralph Teixeira, Thomas Timchek, Jake Thomas, J. Jerry Uhl, Tilaka Vijithakumara, Alan Weinstein, Rachel Deyette Werkema, Aaron Wootton, Hung-Hsi Wu, and Sam Xu.

Reports from the following reviewers were most helpful in shaping the third edition:

Victor Akatsa, Carol Blumberg, Mary Ann Collier, Murray Eisenberg, Donna Fatheree, Dan Fuller, Ken Hannsgen, Marek Kossowski, Sheri Lehavi, Deborah Lurie, Jan Mays, Jeffery Meyer, Bobra Palmer, Barry Peratt, Russ Potter, Ken Price, Maijian Qian, Emily Roth, Lorenzo Traldi, Joan Weiss, Christos Xenophontos.

Reports from the following reviewers were most helpful in shaping the second edition:

Victor Akatsa, Carol Blumberg, Jennifer Fowler, Helen Hancock, Ken Hannsgen, John Haverhals, Mako E. Haruta, Linda Hill, Thom Kline, Jill Messer Lamping, Dennis Lewandowski, Lige Li, William O. Martin, Ted Marsden, Michael Mocciola, Maijian Qian, Joyce Quella, Peter Penner, Barry Peratt, Emily Roth, Jerry Schuur, Barbara Shabell, Peter Sternberg, Virginia Stover, Bruce Yoshiwara, Katherine Yoshiwara.

Deborah Hughes-Hallett	David O. Lomen	Douglas Quinney
Andrew M. Gleason	David Lovelock	Karen Rhea
Patti Frazer Lock	William G. McCallum	Jeff Tecosky-Feldman
Daniel E. Flath	Brad G. Osgood	Thomas W. Tucker

APPLICATIONS INDEX

Life Sciences and Ecology

To Students: How to Learn from this Book

- This book may be different from other math textbooks that you have used, so it may be helpful to know about some of the differences in advance. At every stage, this book emphasizes the *meaning* (in practical, graphical or numerical terms) of the symbols you are using. There is much less emphasis on "plug-and-chug" and using formulas, and much more emphasis on the interpretation of these formulas than you may expect. You will often be asked to explain your ideas in words or to explain an answer using graphs.

- The book contains the main ideas of calculus in plain English. Success in using this book will depend on reading, questioning, and thinking hard about the ideas presented. It will be helpful to read the text in detail, not just the worked examples.

- There are few examples in the text that are exactly like the homework problems, so homework problems can't be done by searching for similar–looking "worked out" examples. Success with the homework will come by grappling with the ideas of calculus.

- For many problems in the book, there is more than one correct approach and more than one correct solution. Sometimes, solving a problem relies on common sense ideas that are not stated in the problem explicitly but which you know from everyday life.

- Some problems in this book assume that you have access to a graphing calculator or computer. There are many situations where you may not be able to find an exact solution to a problem, but you can use a calculator or computer to get a reasonable approximation.

- This book attempts to give equal weight to four methods for describing functions: graphical (a picture), numerical (a table of values), algebraic (a formula), and verbal (words). Sometimes it's easier to translate a problem given in one form into another. For example, you might replace the graph of a parabola with its equation, or plot a table of values to see its behavior. It is important to be flexible about your approach: if one way of looking at a problem doesn't work, try another.

- Students using this book have found discussing these problems in small groups helpful. There are a great many problems which are not cut-and-dried; it can help to attack them with the other perspectives your colleagues can provide. If group work is not feasible, see if your instructor can organize a discussion session in which additional problems can be worked on.

- You are probably wondering what you'll get from the book. The answer is, if you put in a solid effort, you will get a real understanding of one of the crowning achievements of human creativity—calculus— as well as a real sense of the power of mathematics in the age of technology.

Chapter One

FUNCTIONS AND CHANGE

Contents

1.1   WHAT IS A FUNCTION?

In mathematics, a *function* is used to represent the dependence of one quantity upon another.

Let's look at an example. In December 2008, International Falls, Minnesota, was very cold over winter vacation. The daily low temperatures for December 17–26 are given in Table 1.1.[1]

Table 1.1   *Daily low temperature in International Falls, Minnesota, December 17–26, 2008*

Date	17	18	19	20	21	22	23	24	25	26
Low temperature (°F)	−14	−15	−12	−4	−15	−18	1	−9	−10	16

Although you may not have thought of something so unpredictable as temperature as being a function, the temperature *is* a function of date, because each day gives rise to one and only one low temperature. There is no formula for temperature (otherwise we would not need the weather bureau), but nevertheless the temperature does satisfy the definition of a function: Each date, t, has a unique low temperature, L, associated with it.

We define a function as follows:

> A **function** is a rule that takes certain numbers as inputs and assigns to each a definite output number. The set of all input numbers is called the **domain** of the function and the set of resulting output numbers is called the **range** of the function.

The input is called the *independent variable* and the output is called the *dependent variable*. In the temperature example, the set of dates $\{17, 18, 19, 20, 21, 22, 23, 24, 25, 26\}$ is the domain and the set of temperatures $\{-18, -15, -14, -12, -10, -9, -4, 1, 16\}$ is the range. We call the function f and write $L = f(t)$. Notice that a function may have identical outputs for different inputs (December 18 and 21, for example).

Some quantities, such as date, are *discrete*, meaning they take only certain isolated values (dates must be integers). Other quantities, such as time, are *continuous* as they can be any number. For a continuous variable, domains and ranges are often written using interval notation:

The set of numbers t such that $a \leq t \leq b$ is written $[a, b]$.

The set of numbers t such that $a < t < b$ is written (a, b).

The Rule of Four: Tables, Graphs, Formulas, and Words

Functions can be represented by tables, graphs, formulas, and descriptions in words. For example, the function giving the daily low temperatures in International Falls can be represented by the graph in Figure 1.1, as well as by Table 1.1.

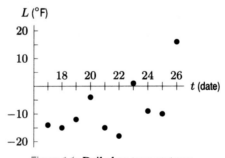

Figure 1.1: Daily low temperatures, International Falls, December 2008

[1] www.wunderground.com.

Other functions arise naturally as graphs. Figure 1.2 contains electrocardiogram (EKG) pictures showing the heartbeat patterns of two patients, one normal and one not. Although it is possible to construct a formula to approximate an EKG function, this is seldom done. The pattern of repetitions is what a doctor needs to know, and these are more easily seen from a graph than from a formula. However, each EKG gives electrical activity as a function of time.

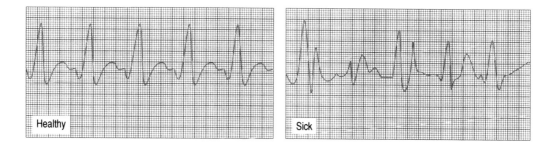

Figure 1.2: EKG readings on two patients

Consider the snow tree cricket. Surprisingly enough, all such crickets chirp at essentially the same rate if they are at the same temperature. That means that the chirp rate is a function of temperature. In other words, if we know the temperature, we can determine the chirp rate. Even more surprisingly, the chirp rate, C, in chirps per minute, increases steadily with the temperature, T, in degrees Fahrenheit, and can be computed, to a fair degree of accuracy, using the formula

$$C = f(T) = 4T - 160.$$

The graph of this function is in Figure 1.3.

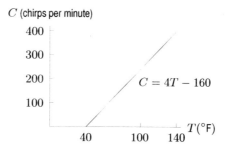

Figure 1.3: Cricket chirp rate as a function of temperature

Mathematical Modeling

A *mathematical model* is a mathematical description of a real situation. In this book we consider models that are functions.

Modeling almost always involves some simplification of reality. We choose which variables to include and which to ignore—for example, we consider the dependence of chirp rate on temperature, but not on other variables. The choice of variables is based on knowledge of the context (the biology of crickets, for example), not on mathematics. To test the model, we compare its predictions with observations.

In this book, we often model a situation that has a discrete domain with a continuous function whose domain is an interval of numbers. For example, the annual US gross domestic product (GDP) has a value for each year, $t = 0, 1, 2, 3, \ldots$ etc. We may model it by a function of the form $G = f(t)$, with values for t in a continuous interval. In doing this, we expect that the values of $f(t)$ match the values of the GDP at the points $t = 0, 1, 2, 3, \ldots$ etc., and that information obtained from $f(t)$ closely matches observed values.

Used judiciously, a mathematical model captures trends in the data to enable us to analyze and make predictions. A common way of finding a model is described in Appendix A.

Function Notation and Intercepts

We write $y = f(t)$ to express the fact that y is a function of t. The independent variable is t, the dependent variable is y, and f is the name of the function. The graph of a function has an *intercept* where it crosses the horizontal or vertical axis. Horizontal intercepts are also called the *zeros* of the function.

Example 1   The value of a car, V, is a function of the age of the car, a, so $V = g(a)$, where g is the name we are giving to this function.

(a) Interpret the statement $g(5) = 9$ in terms of the value of a car if V is in thousands of dollars and a is in years.

(b) In the same units, the value of a Honda[2] is approximated by $g(a) = 13.78 - 0.8a$. Find and interpret the vertical and horizontal intercepts of the graph of this depreciation function g.

Solution   (a) Since $V = g(a)$, the statement $g(5) = 9$ means $V = 9$ when $a = 5$. This tells us that the car is worth $9000 when it is 5 years old.

(b) Since $V = g(a)$, a graph of the function g has the value of the car on the vertical axis and the age of the car on the horizontal axis. The vertical intercept is the value of V when $a = 0$. It is $V = g(0) = 13.78$, so the Honda was valued at $13,780 when new. The horizontal intercept is the value of a such that $g(a) = 0$, so

$$13.78 - 0.8a = 0$$
$$a = \frac{13.78}{0.8} = 17.2.$$

At age 17 years, the Honda has no value.

Increasing and Decreasing Functions

In the previous examples, the chirp rate increases with temperature, while the value of the Honda decreases with age. We express these facts saying that f is an increasing function, while g is decreasing. In general:

> A function f is **increasing** if the values of $f(x)$ increase as x increases.
> A function f is **decreasing** if the values of $f(x)$ decrease as x increases.
>
> The graph of an *increasing* function *climbs* as we move from left to right.
> The graph of a *decreasing* function *descends* as we move from left to right.

Figure 1.4: Increasing and decreasing functions

[2]From data obtained from the Kelley Blue Book, www.kbb.com.

Problems for Section 1.1

1. Which graph in Figure 1.5 best matches each of the following stories?[3] Write a story for the remaining graph.

   (a) I had just left home when I realized I had forgotten my books, and so I went back to pick them up.

   (b) Things went fine until I had a flat tire.

   (c) I started out calmly but sped up when I realized I was going to be late.

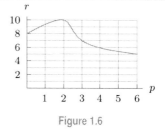

Figure 1.6

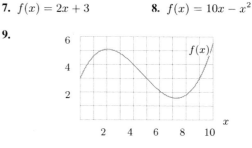

Figure 1.5

2. The population of a city, P, in millions, is a function of t, the number of years since 1970, so $P = f(t)$. Explain the meaning of the statement $f(35) = 12$ in terms of the population of this city.

3. Let $W = f(t)$ represent wheat production in Argentina,[4] in millions of metric tons, where t is years since 1990. Interpret the statement $f(12) = 9$ in terms of wheat production.

4. The concentration of carbon dioxide, $C = f(t)$, in the atmosphere, in parts per million (ppm), is a function of years, t, since 1960.

   (a) Interpret $f(40) = 370$ in terms of carbon dioxide.[5]

   (b) What is the meaning of $f(50)$?

5. The population of Washington DC grew from 1900 to 1950, stayed approximately constant during the 1950s, and decreased from about 1960 to 2005. Graph the population as a function of years since 1900.

6. **(a)** The graph of $r = f(p)$ is in Figure 1.6. What is the value of r when p is 0? When p is 3?

   (b) What is $f(2)$?

For the functions in Problems 7–11, find $f(5)$.

7. $f(x) = 2x + 3$ **8.** $f(x) = 10x - x^2$

9.

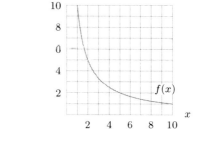

10.

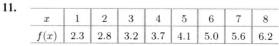

11.

x	1	2	3	4	5	6	7	8
$f(x)$	2.3	2.8	3.2	3.7	4.1	5.0	5.6	6.2

12. Let $y = f(x) = x^2 + 2$.

   (a) Find the value of y when x is zero.

   (b) What is $f(3)$?

   (c) What values of x give y a value of 11?

   (d) Are there any values of x that give y a value of 1?

13. The use of CFCs (chlorofluorocarbons) has declined since the 1987 Montreal Protocol came into force to reduce the use of substances that deplete the ozone layer. World annual CFC consumption, $C = f(t)$, in million tons, is a function of time, t, in years since 1987. (CFCs are measured by the weight of ozone that they could destroy.)

   (a) Interpret $f(10) = 0.2$ in terms of CFCs.[6]

---

[3] Adapted from Jan Terwel, "Real Math in Cooperative Groups in Secondary Education." *Cooperative Learning in Mathematics*, ed. Neal Davidson, p. 234 (Reading: Addison Wesley, 1990).

[4] www.usda.gov.oce/weather/pubs/Other/MWCACP/Graphs/Argentina/ArgentinaWheat.pdf.

[5] *Vital Signs 2007-2008*, The Worldwatch Institute, W.W. Norton & Company, 2007, p. 43.

[6] *Vital Signs 2007-2008*, The Worldwatch Institute, W.W. Norton & Company, 2007, p. 47.

(b) Interpret the vertical intercept of the graph of this function in terms of CFCs.

(c) Interpret the horizontal intercept of the graph of this function in terms of CFCs.

14. Figure 1.7 shows the amount of nicotine, $N = f(t)$, in mg, in a person's bloodstream as a function of the time, t, in hours, since the person finished smoking a cigarette.

   (a) Estimate $f(3)$ and interpret it in terms of nicotine.
   (b) About how many hours have passed before the nicotine level is down to 0.1 mg?
   (c) What is the vertical intercept? What does it represent in terms of nicotine?
   (d) If this function had a horizontal intercept, what would it represent?

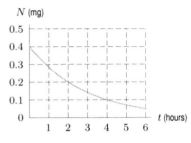

Figure 1.7

15. The number of sales per month, S, is a function of the amount, a (in dollars), spent on advertising that month, so $S = f(a)$.

   (a) Interpret the statement $f(1000) = 3500$.
   (b) Which of the graphs in Figure 1.8 is more likely to represent this function?
   (c) What does the vertical intercept of the graph of this function represent, in terms of sales and advertising?

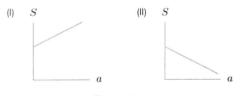

Figure 1.8

16. A deposit is made into an interest-bearing account. Figure 1.9 shows the balance, B, in the account t years later.

   (a) What was the original deposit?
   (b) Estimate $f(10)$ and interpret it.
   (c) When does the balance reach $5000?

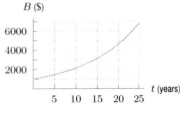

Figure 1.9

17. When a patient with a rapid heart rate takes a drug, the heart rate plunges dramatically and then slowly rises again as the drug wears off. Sketch the heart rate against time from the moment the drug is administered.

18. In the Andes mountains in Peru, the number, N, of species of bats is a function of the elevation, h, in feet above sea level, so $N = f(h)$.

   (a) Interpret the statement $f(500) = 100$ in terms of bat species.
   (b) What are the meanings of the vertical intercept, k, and horizontal intercept, c, in Figure 1.10?

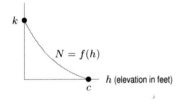

Figure 1.10

19. After an injection, the concentration of a drug in a patient's body increases rapidly to a peak and then slowly decreases. Graph the concentration of the drug in the body as a function of the time since the injection was given. Assume that the patient has none of the drug in the body before the injection. Label the peak concentration and the time it takes to reach that concentration.

20. An object is put outside on a cold day at time $t = 0$. Its temperature, $H = f(t)$, in °C, is graphed in Figure 1.11.

   (a) What does the statement $f(30) = 10$ mean in terms of temperature? Include units for 30 and for 10 in your answer.
   (b) Explain what the vertical intercept, a, and the horizontal intercept, b, represent in terms of temperature of the object and time outside.

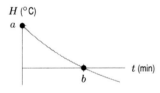

Figure 1.11

21. Financial investors know that, in general, the higher the expected rate of return on an investment, the higher the corresponding risk.

   (a) Graph this relationship, showing expected return as a function of risk.
   (b) On the figure from part (a), mark a point with high expected return and low risk. (Investors hope to find such opportunities.)

22. In tide pools on the New England coast, snails eat algae. Describe what Figure 1.12 tells you about the effect of snails on the diversity of algae.[7] Does the graph support the statement that diversity peaks at intermediate predation levels?

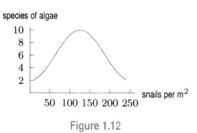

Figure 1.12

23. (a) A potato is put in an oven to bake at time $t = 0$. Which of the graphs in Figure 1.13 could represent the potato's temperature as a function of time?
    (b) What does the vertical intercept represent in terms of the potato's temperature?

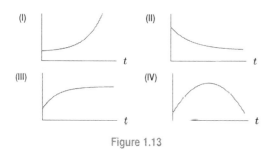

Figure 1.13

24. Figure 1.14 shows fifty years of fertilizer use in the US, India, and the former Soviet Union.[8]

    (a) Estimate fertilizer use in 1970 in the US, India, and the former Soviet Union.
    (b) Write a sentence for each of the three graphs describing how fertilizer use has changed in each region over this 50-year period.

Figure 1.14

25. The gas mileage of a car (in miles per gallon) is highest when the car is going about 45 miles per hour and is lower when the car is going faster or slower than 45 mph. Graph gas mileage as a function of speed of the car.

26. The six graphs in Figure 1.15 show frequently observed patterns of age-specific cancer incidence rates, in number of cases per 1000 people, as a function of age.[9] The scales on the vertical axes are equal.

    (a) For each of the six graphs, write a sentence explaining the effect of age on the cancer rate.
    (b) Which graph shows a relatively high incidence rate for children? Suggest a type of cancer that behaves this way.
    (c) Which graph shows a brief decrease in the incidence rate at around age 50? Suggest a type of cancer that might behave this way.
    (d) Which graph or graphs might represent a cancer that is caused by toxins which build up in the body over time? (For example, lung cancer.) Explain.

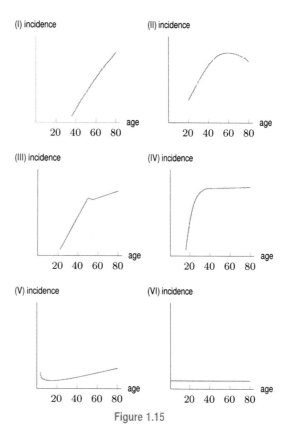

Figure 1.15

---

[7] Rosenzweig, M.L., *Species Diversity in Space and Time*, p. 343 (Cambridge: Cambridge University Press, 1995).
[8] The Worldwatch Institute, *Vital Signs 2001*, p. 32 (New York: W.W. Norton, 2001).
[9] Abraham M. Lilienfeld, *Foundations of Epidemiology*, p. 155 (New York: Oxford University Press, 1976).

1.2 LINEAR FUNCTIONS

Probably the most commonly used functions are the *linear functions*, whose graphs are straight lines. The chirp-rate and the Honda depreciation functions in the previous section are both linear. We now look at more examples of linear functions.

Olympic and World Records

During the early years of the Olympics, the height of the men's winning pole vault increased approximately 8 inches every four years. Table 1.2 shows that the height started at 130 inches in 1900, and increased by the equivalent of 2 inches a year between 1900 and 1912. So the height was a linear function of time.

Table 1.2   *Winning height (approximate) for Men's Olympic pole vault*

Year	1900	1904	1908	1912
Height (inches)	130	138	146	154

If y is the winning height in inches and t is the number of years since 1900, we can write

$$y = f(t) = 130 + 2t.$$

Since $y = f(t)$ increases with t, we see that f is an increasing function. The coefficient 2 tells us the rate, in inches per year, at which the height increases. This rate is the *slope* of the line in Figure 1.16. The slope is given by the ratio

$$\text{Slope} = \frac{\text{Rise}}{\text{Run}} = \frac{146 - 138}{8 - 4} = \frac{8}{4} = 2 \text{ inches/year.}$$

Calculating the slope (rise/run) using any other two points on the line gives the same value.

What about the constant 130? This represents the initial height in 1900, when $t = 0$. Geometrically, 130 is the intercept on the vertical axis.

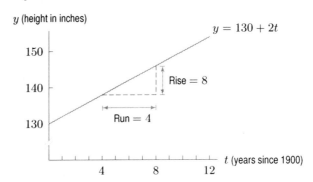

Figure 1.16: Olympic pole vault records

You may wonder whether the linear trend continues beyond 1912. Not surprisingly, it does not exactly. The formula $y = 130 + 2t$ predicts that the height in the 2008 Olympics would be 346 inches or 28 feet 10 inches, which is considerably higher than the actual value of 19 feet 4.25 inches.[10] There is clearly a danger in *extrapolating* too far from the given data. You should also observe that the data in Table 1.2 is *discrete*, because it is given only at specific points (every four years). However, we have treated the variable t as though it were *continuous*, because the function $y = 130 + 2t$ makes sense for all values of t. The graph in Figure 1.16 is of the continuous function because it is a solid line, rather than four separate points representing the years in which the Olympics were held.

---

[10]http://sports.espn.go.com/olympics/summer08.

Example 1   If y is the world record time to run the mile, in seconds, and t is the number of years since 1900, then records show that, approximately,

$$y = g(t) = 260 - 0.4t.$$

Explain the meaning of the intercept, 260, and the slope, -0.4, in terms of the world record time to run the mile and sketch the graph.

Solution   The intercept, 260, tells us that the world record was 260 seconds in 1900 (at $t = 0$). The slope, -0.4, tells us that the world record decreased at a rate of about 0.4 seconds per year. See Figure 1.17.

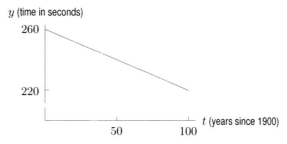

Figure 1.17: World record time to run the mile

Slope and Rate of Change

We use the symbol Λ (the Greek letter capital delta) to mean "change in," so Δx means change in x and Δy means change in y.

The slope of a linear function $y = f(x)$ can be calculated from values of the function at two points, given by x_1 and x_2, using the formula

$$\text{Slope} = \frac{\text{Rise}}{\text{Run}} = \frac{\Delta y}{\Delta x} = \frac{f(x_2) - f(x_1)}{x_2 - x_1}.$$

The quantity $(f(x_2) - f(x_1))/(x_2 - x_1)$ is called a *difference quotient* because it is the quotient of two differences. (See Figure 1.18.) Since slope $= \Delta y/\Delta x$, the slope represents the *rate of change* of y with respect to x. The units of the slope are y-units over x-units.

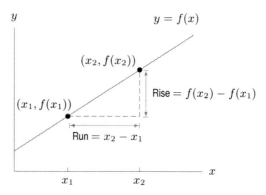

Figure 1.18: Difference quotient $= \dfrac{f(x_2) - f(x_1)}{x_2 - x_1}$

Linear Functions in General

> **A linear function** has the form
>
> $$y = f(x) = b + mx.$$
>
> Its graph is a line such that
> - m is the **slope**, or rate of change of y with respect to x.
> - b is the **vertical intercept** or value of y when x is zero.

If the slope, m, is positive, then f is an increasing function. If m is negative, then f is decreasing.

Notice that if the slope, m, is zero, we have $y = b$, a horizontal line. For a line of slope m through the point (x_0, y_0), we have

$$\text{Slope} = m = \frac{y - y_0}{x - x_0}.$$

Therefore we can write the equation of the line in the *point-slope form*:

> The equation of a line of slope m through the point (x_0, y_0) is
>
> $$y - y_0 = m(x - x_0).$$

Example 2   The solid waste generated each year in the cities of the US is increasing. The solid waste generated,[11] in millions of tons, was 238.3 in 2000 and 251.3 in 2006.

(a) Assuming that the amount of solid waste generated by US cities is a linear function of time, find a formula for this function by finding the equation of the line through these two points.

(b) Use this formula to predict the amount of solid waste generated in the year 2020.

Solution   (a) We think of the amount of solid waste, W, as a function of year, t, and the two points are $(2000, 238.3)$ and $(2006, 251.3)$. The slope of the line is

$$m = \frac{\Delta W}{\Delta t} = \frac{251.3 - 238.3}{2006 - 2000} = \frac{13}{6} = 2.167 \text{ million tons/year.}$$

We use the point-slope form to find the equation of the line. We substitute the point $(2000, 238.3)$ and the slope $m = 2.167$ into the equation:

$$W - W_0 = m(t - t_0)$$
$$W - 238.3 = 2.167(t - 2000)$$
$$W - 238.3 = 2.167t - 4334$$
$$W = 2.167t - 4095.7.$$

The equation of the line is $W = 2.167t - 4095.7$. Alternatively, we could use the slope-intercept form of a line to find the vertical intercept.

(b) To calculate solid waste predicted for the year 2020, we substitute $t = 2020$ into the equation of the line, $W = -4095.7 + 2.167t$, and calculate W:

$$W = -4095.7 + 2.167(2020) = 281.64.$$

The formula predicts that in the year 2020, there will be 281.64 million tons of solid waste.

[11] *Statistical Abstracts of the US*, 2009, Table 361.

> **Recognizing Data from a Linear Function:** Values of x and y in a table could come from a linear function $y = b + mx$ if differences in y-values are constant for equal differences in x.

Example 3   Which of the following tables of values could represent a linear function?

x	0	1	2	3
$f(x)$	25	30	35	40

x	0	2	4	6
$g(x)$	10	16	26	40

t	20	30	40	50
$h(t)$	2.4	2.2	2.0	1.8

Solution   Since $f(x)$ increases by 5 for every increase of 1 in x, the values of $f(x)$ could be from a linear function with slope $= 5/1 = 5$.

Between $x = 0$ and $x = 2$, the value of $g(x)$ increases by 6 as x increases by 2. Between $x = 2$ and $x = 4$, the value of y increases by 10 as x increases by 2. Since the slope is not constant, $g(x)$ could not be a linear function.

Since $h(t)$ decreases by 0.2 for every increase of 10 in t, the values of $h(t)$ could be from a linear function with slope $= -0.2/10 = -0.02$.

Example 4   The data in the following table lie on a line. Find formulas for each of the following functions, and give units for the slope in each case:

(a) q as a function of p

(b) p as a function of q

p(dollars)	5	10	15	20
q(tons)	100	90	80	70

Solution   (a) If we think of q as a linear function of p, then q is the dependent variable and p is the independent variable. We can use any two points to find the slope. The first two points give

$$\text{Slope} = m = \frac{\Delta q}{\Delta p} = \frac{90 - 100}{10 - 5} = \frac{-10}{5} = -2.$$

The units are the units of q over the units of p, or tons per dollar.

To write q as a linear function of p, we use the equation $q = b + mp$. We know that $m = -2$, and we can use any of the points in the table to find b. Substituting $p = 10, q = 90$ gives

$$q = b + mp$$
$$90 = b + (-2)(10)$$
$$90 = b - 20$$
$$110 = b.$$

Thus, the equation of the line is

$$q = 110 - 2p.$$

(b) If we now consider p as a linear function of q, then p is the dependent variable and q is the independent variable. We have

$$\text{Slope} = m = \frac{\Delta p}{\Delta q} = \frac{10 - 5}{90 - 100} = \frac{5}{-10} = -0.5.$$

The units of the slope are dollars per ton.

Since p is a linear function of q, we have $p = b + mq$ and $m = -0.5$. To find b, we substitute any point from the table, such as $p = 10$, $q = 90$, into this equation:

$$p = b + mq$$
$$10 = b + (-0.5)(90)$$
$$10 = b - 45$$
$$55 = b.$$

Thus, the equation of the line is

$$p = 55 - 0.5q.$$

Alternatively, we could take our answer to part (a), that is $q = 110 - 2p$, and solve for p. Appendix A shows how to fit a linear function to data that is not exactly linear.

Families of Linear Functions

Formulas such as $f(x) = b + mx$, in which the constants m and b can take on various values, represent a *family of functions*. All the functions in a family share certain properties—in this case, the graphs are lines. The constants m and b are called *parameters*. Figures 1.19 and 1.20 show graphs with several values of m and b. Notice the greater the magnitude of m, the steeper the line.

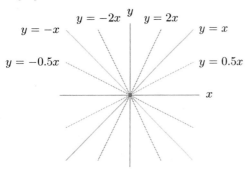

Figure 1.19: The family $y = mx$ (with $b = 0$)

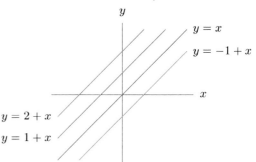

Figure 1.20: The family $y = b + x$ (with $m = 1$)

Problems for Section 1.2

For Problems 1–4, determine the slope and the y-intercept of the line whose equation is given.

1. $7y + 12x - 2 = 0$

2. $3x + 2y = 8$

3. $12x = 6y + 4$

4. $-4y + 2x + 8 = 0$

For Problems 5–8, find an equation for the line that passes through the given points.

5. $(0, 2)$ and $(2, 3)$

6. $(0, 0)$ and $(1, 1)$

7. $(-2, 1)$ and $(2, 3)$

8. $(4, 5)$ and $(2, -1)$

9. Figure 1.21 shows four lines given by equation $y = b + mx$. Match the lines to the conditions on the parameters m and b.

(a) $m > 0, b > 0$

(b) $m < 0, b > 0$

(c) $m > 0, b < 0$

(d) $m < 0, b < 0$

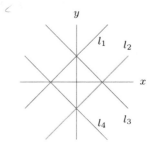

Figure 1.21

10. **(a)** Which two lines in Figure 1.22 have the same slope? Of these two lines, which has the larger y-intercept?
   (b) Which two lines have the same y-intercept? Of these two lines, which has the larger slope?

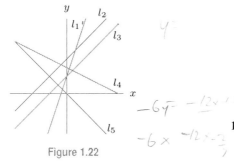

Figure 1.22

11. A city's population was 30,700 in the year 2000 and is growing by 850 people a year.

   (a) Give a formula for the city's population, P, as a function of the number of years, t, since 2000.
   (b) What is the population predicted to be in 2010?
   (c) When is the population expected to reach 45,000?

12. A cell phone company charges a monthly fee of $25 plus $0.05 per minute. Find a formula for the monthly charge, C, in dollars, as a function of the number of minutes, m, the phone is used during the month.

13. A company rents cars at $40 a day and 15 cents a mile. Its competitor's cars are $50 a day and 10 cents a mile.

   (a) For each company, give a formula for the cost of renting a car for a day as a function of the distance traveled.
   (b) On the same axes, graph both functions.
   (c) How should you decide which company is cheaper?

14. World milk production rose at an approximately constant rate between 1997 and 2003.[12] See Figure 1.23.

   (a) Estimate the vertical intercept and interpret it in terms of milk production.
   (b) Estimate the slope and interpret it in terms of milk production.
   (c) Give an approximate formula for milk production, M, as a function of t.

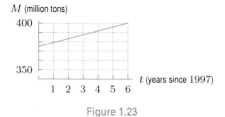

Figure 1.23

15. Annual revenue R from McDonald's restaurants worldwide can be estimated by $R = 19.1 + 1.8t$, where R is in billion dollars and t is in years since January 1, 2005.[13]

   (a) What is the slope of this function? Include units. Interpret the slope in terms of McDonald's revenue.
   (b) What is the vertical intercept of this function? Include units. Interpret the vertical intercept in terms of McDonald's revenue.
   (c) What annual revenue does the function predict for 2010?
   (d) When is annual revenue predicted to hit 30 billion dollars?

16. Let y be the percent increase in annual US national production during a year when the unemployment rate changes by u percent. (For example, $u = 2$ if unemployment increases from 4% to 6%.) Okun's law states that

$$y = 3.5 - 2u.$$

   (a) What is the meaning of the number 3.5 in Okun's law?
   (b) What is the effect on national production of a year when unemployment rises from 5% to 8%?
   (c) What change in the unemployment rate corresponds to a year when production is the same as the year before?
   (d) What is the meaning of the coefficient -2 in Okun's law?

17. Which of the following tables could represent linear functions?

   (a)

x	0	1	2	3
y	27	25	23	21

   (b)

t	15	20	25	30
s	62	72	82	92

   (c)

u	1	2	3	4
w	5	10	18	28

18. For each table in Problem 17 that could represent a linear function, find a formula for that function.

19. A company's pricing schedule in Table 1.3 is designed to encourage large orders. (A gross is 12 dozen.) Find a formula for:

   (a) q as a linear function of p.
   (b) p as a linear function of q.

Table 1.3

q (order size, gross)	3	4	5	6
p (price/dozen)	15	12	9	6

[12]*Statistical Abstracts of the US 2004–2005*, Table 1355.
[13]Based on McDonald's Annual Report 2007, accessed at www.mcdonalds.com.

20. Figure 1.24 shows the distance from home, in miles, of a person on a 5-hour trip.

   (a) Estimate the vertical intercept. Give units and interpret it in terms of distance from home.
   (b) Estimate the slope of this linear function. Give units, and interpret it in terms of distance from home.
   (c) Give a formula for distance, D, from home as a function of time, t in hours.

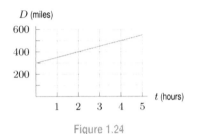

Figure 1.24

21. The percentage of people, P, below the poverty level in the US[14] is given in Table 1.4.

   (a) Find a formula for the percentage in poverty as a linear function of time in years since 2000.
   (b) Use the formula to predict the percentage in poverty in 2006.
   (c) What is the difference between the prediction and the actual percentage, 12.3%?

Table 1.4

Year (since 2000)	0	1	2	3
P (percentage)	11.3	11.7	12.1	12.5

22. World grain production was 1241 million tons in 1975 and 2048 million tons in 2005, and has been increasing at an approximately constant rate.[15]

   (a) Find a linear function for world grain production, P, in million tons, as a function of t, the number of years since 1975.
   (b) Using units, interpret the slope in terms of grain production.
   (c) Using units, interpret the vertical intercept in terms of grain production.
   (d) According to the linear model, what is the predicted world grain production in 2015?
   (e) According to the linear model, when is grain production predicted to reach 2500 million tons?

23. Search and rescue teams work to find lost hikers. Members of the search team separate and walk parallel to one another through the area to be searched. Table 1.5 shows the percent, P, of lost individuals found for various separation distances, d, of the searchers.[16]

Table 1.5

Separation distance d (ft)	20	40	60	80	100
Approximate percent found, P	90	80	70	60	50

   (a) Explain how you know that the percent found, P, could be a linear function of separation distance, d.
   (b) Find P as a linear function of d.
   (c) What is the slope of the function? Give units and interpret the answer.
   (d) What are the vertical and horizontal intercepts of the function? Give units and interpret the answers.

24. Annual sales of music compact discs (CDs) have declined since 2000. Sales were 942.5 million in 2000 and 384.7 million in 2008.[17]

   (a) Find a formula for annual sales, S, in millions of music CDs, as a linear function of the number of years, t, since 2000.
   (b) Give units for and interpret the slope and the vertical intercept of this function.
   (c) Use the formula to predict music CD sales in 2012.

25. In a California town, the monthly charge for waste collection is $8 for 32 gallons of waste and $12.32 for 68 gallons of waste.

   (a) Find a linear formula for the cost, C, of waste collection as a function of the number of gallons of waste, w.
   (b) What is the slope of the line found in part (a)? Give units and interpret your answer in terms of the cost of waste collection.
   (c) What is the vertical intercept of the line found in part (a)? Give units and interpret your answer in terms of the cost of waste collection.

26. The number of species of coastal dune plants in Australia decreases as the latitude, in °S, increases. There are 34 species at 11°S and 26 species at 44°S.[18]

   (a) Find a formula for the number, N, of species of coastal dune plants in Australia as a linear function of the latitude, l, in °S.
   (b) Give units for and interpret the slope and the vertical intercept of this function.
   (c) Graph this function between $l = 11°S$ and $l = 44°S$. (Australia lies entirely within these latitudes.)

[14] www.census.gov/hhes/www/poverty/histpov/hstpov2.html
[15] *Vital Signs 2007-2008*, The Worldwatch Institute, W.W. Norton & Company, 2007, p. 21
[16] From *An Experimental Analysis of Grid Sweep Searching*, by J. Wartes (Explorer Search and Rescue, Western Region, 1974).
[17] *The World Almanac and Book of Facts 2008*, (New York).
[18] Rosenzweig, M.L., *Species Diversity in Space and Time*, p. 292, (Cambridge: Cambridge University Press, 1995).

27. Table 1.6 gives the average weight, w, in pounds, of American men in their sixties for height, h, in inches.[19]

   (a) How do you know that the data in this table could represent a linear function?

   (b) Find weight, w, as a linear function of height, h. What is the slope of the line? What are the units for the slope?

   (c) Find height, h, as a linear function of weight, w. What is the slope of the line? What are the units for the slope?

Table 1.6

h (inches)	68	69	70	71	72	73	74	75
w (pounds)	166	171	176	181	186	191	196	201

Problems 28–33 concern the maximum heart rate (MHR), which is the maximum number of times a person's heart can safely beat in one minute. If MHR is in beats per minute and a is age in years, the formulas used to estimate MHR, are

$$\text{For females: MHR} = 226 - a,$$

$$\text{For males: MHR} = 220 - a.$$

28. Which of the following is the correct statement?

   (a) As you age, your maximum heart rate decreases by one beat per year.

   (b) As you age, your maximum heart rate decreases by one beat per minute.

   (c) As you age, your maximum heart rate decreases by one beat per minute per year.

29. Which of the following is the correct statement for a male and female of the same age?

   (a) Their maximum heart rates are the same.

   (b) The male's maximum heart rate exceeds the female's.

   (c) The female's maximum heart rate exceeds the male's.

30. What can be said about the ages of a male and a female with the same maximum heart rate?

31. Recently[20] it has been suggested that a more accurate predictor of MHR for both males and females is given by

$$\text{MHR} = 208 - 0.7a.$$

   (a) At what age do the old and new formulas give the same MHR for females? For males?

   (b) Which of the following is true?

      (i) The new formula predicts a higher MHR for young people and a lower MHR for older people than the old formula.

      (ii) The new formula predicts a lower MHR for young people and a higher MHR for older people than the old formula.

   (c) When testing for heart disease, doctors ask patients to walk on a treadmill while the speed and incline are gradually increased until their heart rates reach 85 percent of the MHR. For a 65-year-old male, what is the difference in beats per minute between the heart rate reached if the old formula is used and the heart rate reached if the new formula is used?

32. Experiments[21] suggest that the female MHR decreases by 12 beats per minute by age 21, and by 19 beats per minute by age 33. Is this consistent with MHR being approximately linear with age?

33. Experiments[22] suggest that the male MHR decreases by 9 beats per minute by age 21, and by 26 beats per minute by age 33. Is this consistent with MHR being approximately linear with age?

34. An Australian[23] study found that, if other factors are constant (education, experience, etc.), taller people receive higher wages for the same work. The study reported a "height premium" for men of 3% of the hourly wage for a 10 cm increase in height; for women the height premium reported was 2%. We assume that hourly wages are a linear function of height, with slope given by the height premium at the average hourly wage for that gender.

   (a) The average hourly wage[24] for a 178 cm Australian man is AU$29.40. Express the average hourly wage of an Australian man as a function of his height, x cm.

   (b) The average hourly wage for a 164 cm Australian woman is AU$24.78. Express the average hourly wage of an Australian woman as a function of her height, y cm.

   (c) What is the difference in average hourly wages between men and women of height 178 cm?

   (d) Is there a height for which men and women are predicted to have the same wage? If so, what is it?

---

[19] Adapted from "Average Weight of Americans by Height and Age," *The World Almanac* (New Jersey: Funk and Wagnalls, 1992), p. 956.

[20] www.physsportsmed.com/issues/2001/07_01/jul01news.htm, accessed January 4, 2005.

[21] www.css.edu/users/tboone2/asep/May2002JEPonline.html, accessed January 4, 2005.

[22] www.css.edu/users/tboone2/asep/May2002JEPonline.html, accessed January 4, 2005.

[23] "Study finds tall people at top of wages ladder", Yahoo News, May 17, 2009.

[24] Australian Fair Pay Commission, August 2007.

1.3 AVERAGE RATE OF CHANGE AND RELATIVE CHANGE

Average Rate of Change

In the previous section, we saw that the height of the winning Olympic pole vault increased at an approximately constant rate of 2 inches/year between 1900 and 1912. Similarly, the world record for the mile decreased at an approximately constant rate of 0.4 seconds/year. We now see how to calculate rates of change when they are not constant.

Example 1 Table 1.7 shows the height of the winning pole vault at the Olympics[25] during the 1960s and 1990s. Find the rate of change of the winning height between 1960 and 1968, and between 1992 and 2000. In which of these two periods did the height increase faster than during the period 1900–1912?

Table 1.7 *Winning height in men's Olympic pole vault (approximate)*

Year	1960	1964	1968	$\cdots$	1992	1996	2000
Height (inches)	185	201	213	$\cdots$	228	233	232

Solution From 1900 to 1912, the height increased by 2 inches/year. To compare the 1960s and 1990s, we calculate

$$\begin{matrix} \text{Average rate of change of height} \\ \text{1960 to 1968} \end{matrix} = \frac{\text{Change in height}}{\text{Change in time}} = \frac{213 - 185}{1968 - 1960} = 3.5 \text{ inches/year.}$$

$$\begin{matrix} \text{Average rate of change of height} \\ \text{1992 to 2000} \end{matrix} = \frac{\text{Change in height}}{\text{Change in time}} = \frac{232 - 228}{2000 - 1992} = 0.5 \text{ inches/year.}$$

Thus, on average, the height was increasing more quickly during the 1960s than from 1900 to 1912. During the 1990s, the height was increasing more slowly than from 1900 to 1912.

In Example 1, the function does not have a constant rate of change (it is not linear). However, we can compute an *average rate of change* over any interval. The word average is used because the rate of change may vary within the interval. We have the following general formula.

If y is a function of t, so $y = f(t)$, then

$$\begin{matrix} \textbf{Average rate of change} \text{ of } y \\ \text{between } t = a \text{ and } t = b \end{matrix} = \frac{\Delta y}{\Delta t} = \frac{f(b) - f(a)}{b - a}.$$

The units of average rate of change of a function are units of y per unit of t.

The average rate of change of a linear function is the slope, and a function is linear if the average rate of change is the same on all intervals.

---

[25] *The World Almanac and Book of Facts, 2005*, p. 866 (New York).

Example 2   Using Figure 1.25, estimate the average rate of change of the number of farms[26] in the US between 1950 and 1970.

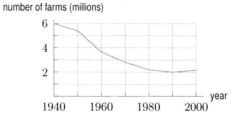

Figure 1.25: Number of farms in the US (in millions)

Solution   Figure 1.25 shows that the number, N, of farms in the US was approximately 5.4 million in 1950 and approximately 2.8 million in 1970. If time, t, is in years, we have

$$\text{Average rate of change} = \frac{\Delta N}{\Delta t} = \frac{2.8 - 5.4}{1970 - 1950} = -0.13 \text{ million farms per year.}$$

The average rate of change is negative because the number of farms is decreasing. During this period, the number of farms decreased at an average rate of 0.13 million, or 130,000, per year.

We have looked at how an Olympic record and the number of farms change over time. In the next example, we look at average rate of change with respect to a quantity other than time.

Example 3   High levels of PCB (polychlorinated biphenyl, an industrial pollutant) in the environment affect pelicans' eggs. Table 1.8 shows that as the concentration of PCB in the eggshells increases, the thickness of the eggshell decreases, making the eggs more likely to break.[27]

Find the average rate of change in the thickness of the shell as the PCB concentration changes from 87 ppm to 452 ppm. Give units and explain why your answer is negative.

Table 1.8   *Thickness of pelican eggshells and PCB concentration in the eggshells*

Concentration, c, in parts per million (ppm)	87	147	204	289	356	452
Thickness, h, in millimeters (mm)	0.44	0.39	0.28	0.23	0.22	0.14

Solution   Since we are looking for the average rate of change of thickness with respect to change in PCB concentration, we have

$$\text{Average rate of change of thickness} = \frac{\text{Change in the thickness}}{\text{Change in the PCB level}} = \frac{\Delta h}{\Delta c} = \frac{0.14 - 0.44}{452 - 87}$$
$$= -0.00082 \frac{\text{mm}}{\text{ppm}}.$$

The units are thickness units (mm) over PCB concentration units (ppm), or millimeters over parts per million. The average rate of change is negative because the thickness of the eggshell decreases as the PCB concentration increases. The thickness of pelican eggs decreases by an average of 0.00082 mm for every additional part per million of PCB in the eggshell.

---

[26]*The World Almanac and Book of Facts, 2005*, p. 136 (New York).

[27]Risebrough, R. W., "Effects of environmental pollutants upon animals other than man." *Proceedings of the 6th Berkeley Symposium on Mathematics and Statistics, VI*, p. 443–463 (Berkeley: University of California Press, 1972).

Visualizing Rate of Change

For a function $y = f(x)$, the change in the value of the function between $x = a$ and $x = c$ is $\Delta y = f(c) - f(a)$. Since Δy is a difference of two y-values, it is represented by the vertical distance in Figure 1.26. The average rate of change of f between $x = a$ and $x = c$ is represented by the slope of the line joining the points A and C in Figure 1.27. This line is called the *secant line* between $x = a$ and $x = c$.

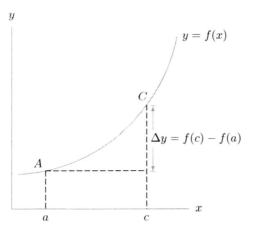

Figure 1.26: The change in a function is represented by a vertical distance

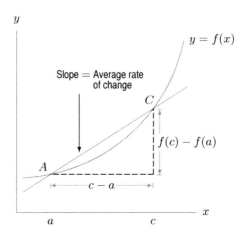

Figure 1.27: The average rate of change is represented by the slope of the line

Example 4

(a) Find the average rate of change of $y = f(x) = \sqrt{x}$ between $x = 1$ and $x = 4$.

(b) Graph $f(x)$ and represent this average rate of change as the slope of a line.

(c) Which is larger, the average rate of change of the function between $x = 1$ and $x = 4$ or the average rate of change between $x = 4$ and $x = 5$? What does this tell us about the graph of the function?

Solution

(a) Since $f(1) = \sqrt{1} = 1$ and $f(4) = \sqrt{4} = 2$, between $x = 1$ and $x = 4$, we have

$$\text{Average rate of change} = \frac{\Delta y}{\Delta x} = \frac{f(4) - f(1)}{4 - 1} = \frac{2 - 1}{3} = \frac{1}{3}.$$

(b) A graph of $f(x) = \sqrt{x}$ is given in Figure 1.28. The average rate of change of f between 1 and 4 is the slope of the secant line between $x = 1$ and $x = 4$.

(c) Since the secant line between $x = 1$ and $x = 4$ is steeper than the secant line between $x = 4$ and $x = 5$, the average rate of change between $x = 1$ and $x = 4$ is larger than it is between $x = 4$ and $x = 5$. The rate of change is decreasing. This tells us that the graph of this function is bending downward.

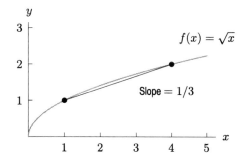

Figure 1.28: Average rate of change = Slope of secant line

Concavity

We now look at the graphs of functions whose rates of change are increasing throughout an interval or decreasing throughout an interval.

Figure 1.28 shows a graph that is bending downward because the rate of change is decreasing. The graph in Figure 1.26 bends upward because the rate of change of the function is increasing. We make the following definitions.

> The graph of a function is **concave up** if it bends upward as we move left to right; the graph is **concave down** if it bends downward. (See Figure 1.29.) A line is neither concave up nor concave down.

Figure 1.29: Concavity of a graph

Example 5   Using Figure 1.30, estimate the intervals over which:
(a) The function is increasing; decreasing.   (b) The graph is concave up; concave down.

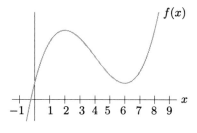

Figure 1.30

Solution   (a) The graph suggests that the function is increasing for $x < 2$ and for $x > 6$. It appears to be decreasing for $2 < x < 6$.
(b) The graph is concave down on the left and concave up on the right. It is difficult to tell exactly where the graph changes concavity, although it appears to be about $x = 4$. Approximately, the graph is concave down for $x < 4$ and concave up for $x > 4$.

Example 6   From the following values of $f(t)$, does f appear to be increasing or decreasing? Do you think its graph is concave up or concave down?

t	0	5	10	15	20	25	30
$f(t)$	12.6	13.1	14.1	16.2	20.0	29.6	42.7

Solution   Since the given values of $f(t)$ increase as t increases, f appears to be increasing. As we read from left to right, the change in $f(t)$ starts small and gets larger (for constant change in t), so the graph is climbing faster. Thus, the graph appears to be concave up. Alternatively, plot the points and notice that a curve through these points bends up.

Distance, Velocity, and Speed

A grapefruit is thrown up in the air. The height of the grapefruit above the ground first increases and then decreases. See Table 1.9.

Table 1.9   *Height, y, of the grapefruit above the ground t seconds after it is thrown*

t (sec)	0	1	2	3	4	5	6
y (feet)	6	90	142	162	150	106	30

Example 7   Find the change and average rate of change of the height of the grapefruit during the first 3 seconds. Give units and interpret your answers.

Solution   The change in height during the first 3 seconds is $\Delta y = 162 - 6 = 156$ ft. This means that the grapefruit goes up a total of 156 meters during the first 3 seconds. The average rate of change during this 3 second interval is $156/3 = 52$ ft/sec. During the first 3 seconds, the grapefruit is rising at an average rate of 52 ft/sec.

The average rate of change of height with respect to time is *velocity*. You may recognize the units (feet per second) as units of velocity.

$$\text{Average velocity} \quad = \frac{\text{Change in distance}}{\text{Change in time}} = \frac{\text{Average rate of change of distance}}{\text{with respect to time}}$$

There is a distinction between *velocity* and *speed*. Suppose an object moves along a line. If we pick one direction to be positive, the velocity is positive if the object is moving in that direction and negative if it is moving in the opposite direction. For the grapefruit, upward is positive and downward is negative. Speed is the magnitude of velocity, so it is always positive or zero.

Example 8   Find the average velocity of the grapefruit over the interval $t = 4$ to $t = 6$. Explain the sign of your answer.

Solution   Since the height is $y = 150$ feet at $t = 4$ and $y = 30$ feet at $t = 6$, we have

$$\text{Average velocity} = \frac{\text{Change in distance}}{\text{Change in time}} = \frac{\Delta y}{\Delta t} = \frac{30 - 150}{6 - 4} = -60 \text{ ft/sec.}$$

The negative sign means the height is decreasing and the grapefruit is moving downward.

Example 9   A car travels away from home on a straight road. Its distance from home at time t is shown in Figure 1.31. Is the car's average velocity greater during the first hour or during the second hour?

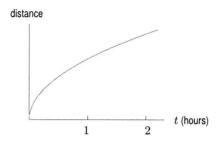

Figure 1.31: Distance of car from home   Figure 1.32: Average velocities of the car

Solution   Average velocity is represented by the slope of a secant line. Figure 1.32 shows that the secant line between $t = 0$ and $t = 1$ is steeper than the secant line between $t = 1$ and $t = 2$. Thus, the average velocity is greater during the first hour.

Relative Change

Is a population increase of 1000 a significant change? It depends on the original size of the community. If the town of Coyote, NM, population 1559, increases by 1000 people, the townspeople would definitely notice. On the other hand, if New York City, population 8.25 million, increases by 1000 people, almost no one will notice. To visualize the impact of the increase on the two different communities, we look at the change, 1000, as a fraction, or percentage, of the initial population. This percent change is called the *relative change*.

Example 10   If the population increases by 1000 people, find the relative change in the population for

(a) Coyote, NM (population 1559)
(b) New York City (population 8,250,000)

Solution   (a) The population increases by 1000 from 1559 so

$$\text{Relative change} = \frac{\text{Change in population}}{\text{Initial population}} = \frac{1000}{1559} = 0.641.$$

The population has increased by 64.1%, a significant increase.
(b) The population increases by 1000 from 8,250,000 so

$$\text{Relative change} = \frac{\text{Change in population}}{\text{Initial population}} = \frac{1000}{8,250,000} = 0.00012.$$

The population has increased by 0.012%, or less than one-tenth of one percent.

In general, when a quantity P changes from P_0 to P_1, we define

$$\text{Relative change in } P = \frac{\text{Change in } P}{P_0} = \frac{P_1 - P_0}{P_0}.$$

The relative change is a number, without units. It is often expressed as a percentage.

Example 11   A price increase can be significant or inconsequential depending on the item. In each of the following cases, find the relative change in price of a $2 price increase; give your answer as a percent.

(a) A gallon of gas costing $2.25   (b) A cell phone costing $180

Solution   (a) The change in the price is $2 so we have

$$\text{Relative change in price of gas } = \frac{\text{Change in price}}{\text{Initial price}} = \frac{2}{2.25} = 0.889.$$

The price of gas has gone up 88.9%.
(b) We have

$$\text{Relative change in price of cell phone } = \frac{\text{Change in price}}{\text{Initial price}} = \frac{2}{180} = 0.011.$$

The price of the cell phone has gone up only 1.1%.

Relative change can be positive or negative, as we see in the following example.

Example 12   Find the relative change in the price of a $75.99 pair of jeans if the sale price is $52.99.

Solution   The price has dropped from $75.99 to $52.99. We have

$$\text{Relative change } = \frac{52.99 - 75.99}{75.99} = \frac{-23}{75.99} = -0.303.$$

The price has been reduced by 30.3% for the sale.

Problems for Section 1.3

In Problems 1–4, decide whether the graph is concave up, concave down, or neither.

In Problems 5–8, find the relative, or percent, change.

5. S changes from 400 to 450

6. B changes from 12,000 to 15,000

7. R changes from 50 to 47

8. W changes from 0.3 to 0.05

1.

2.

3.

4.

9. Table 1.10 gives values of a function $w = f(t)$. Is this function increasing or decreasing? Is the graph of this function concave up or concave down?

Table 1.10

t	0	4	8	12	16	20	24
w	100	58	32	24	20	18	17

10. For which pairs of consecutive points in Figure 1.33 is the function graphed:

    (a) Increasing and concave up
    (b) Increasing and concave down
    (c) Decreasing and concave up
    (d) Decreasing and concave down

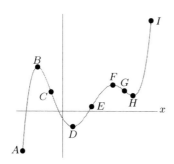

Figure 1.33

11. Graph a function $f(x)$ which is increasing everywhere and concave up for negative x and concave down for positive x.

12. Find the average rate of change of $f(x) - 2x^2$ between $x = 1$ and $x = 3$.

13. Find the average rate of change of $f(x) = 3x^2 + 4$ between $x = -2$ and $x = 1$. Illustrate your answer graphically.

14. When a deposit of $1000 is made into an account paying 8% interest, compounded annually, the balance, B, in the account after t years is given by $B = 1000(1.08)^t$. Find the average rate of change in the balance over the interval $t = 0$ to $t = 5$. Give units and interpret your answer in terms of the balance in the account.

15. Table 1.11 shows world bicycle production.[28]

    (a) Find the change in bicycle production between 1950 and 2000. Give units.
    (b) Find the average rate of change in bicycle production between 1950 and 2000. Give units and interpret your answer in terms of bicycle production.

Table 1.11   *World bicycle production, in millions*

Year	1950	1960	1970	1980	1990	2000
Bicycles	11	20	36	62	92	101

16. Table 1.12 gives the net sales of The Gap, Inc, which operates nearly 3000 clothing stores.[29]

    (a) Find the change in net sales between 2005 and 2008.
    (b) Find the average rate of change in net sales between 2005 and 2008. Give units and interpret your answer.
    (c) From 2003 to 2008, were there any one-year intervals during which the average rate of change was positive? If so, when?

Table 1.12   *Gap net sales, in millions of dollars*

Year	2003	2004	2005	2006	2007	2008
Sales	15,854	16,267	16,019	15,923	15,763	14,526

17. Table 1.13 shows attendance at NFL football games.[30]

    (a) Find the average rate of change in the attendance from 2003 to 2007. Give units.
    (b) Find the annual increase in the attendance for each year from 2003 to 2007. (Your answer should be four numbers.)
    (c) Show that the average rate of change found in part (a) is the average of the four yearly changes found in part (b).

Table 1.13   *Attendance at NFL football games, in millions of fans*

Year	2003	2004	2005	2006	2007
Attendance	21.64	21.71	21.79	22.20	22.26

18. Figure 1.34 shows the total value of US imports, in billions of dollars.[31]

    (a) Was the value of the imports higher in 1985 or in 2003? Approximately how much higher?
    (b) Estimate the average rate of change of US imports between 1985 and 2003. Give units and interpret your answer in terms of imports.

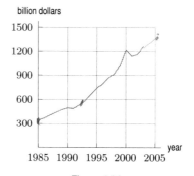

Figure 1.34

---

[28]www.earth-policy.org/Indicators/indicator11_data1.htm, accessed April 19, 2005.
[29]www.gapinc.com/public/investors/inv_financials.shtml/ accessed May 24, 2009.
[30]*Statistical Abstracts of the United States 2009, Table 1204.*
[31]www.ita.doc.gov/td/industry/otea/usfth/aggregate/H03t26.pdf, accessed April 19, 2005.

19. Table 1.14 gives sales of Pepsico, which operates two major businesses: beverages (including Pepsi) and snack foods.[32]

  (a) Find the change in sales between 2003 and 2008.

  (b) Find the average rate of change in sales between 2003 and 2008. Give units and interpret your answer.

Table 1.14 *Pepsico sales, in millions of dollars*

Year	2003	2004	2005	2006	2007	2008
Sales	26,971	29,261	32,562	35,137	39,474	45,251

20. Table 1.15 shows world population, P, in billions of people, world passenger automobile production, A, in millions of cars, and world cell phone subscribers, C, in millions of subscribers.[33]

  (a) Find the average rate of change, with units, for each of P, A, and C between 1995 and 2005.

  (b) Between 1995 and 2005, which increased faster:

    (i) Population or the number of automobiles?

    (ii) Population or the number of cell phone subscribers?

Table 1.15

Year	1995	2000	2005
P (billions)	5.68	6.07	6.45
A (millions)	36.1	41.3	45.9
C (millions)	91	740	2168

21. Figure 1.35 shows a particle's distance from a point. What is the particle's average velocity from $t = 0$ to $t = 3$?

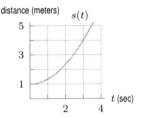

Figure 1.35

22. Figure 1.36 shows a particle's distance from a point. What is the particle's average velocity from $t = 1$ to $t = 3$?

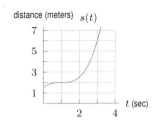

Figure 1.36

23. At time t in seconds, a particle's distance $s(t)$, in cm, from a point is given in the table. What is the average velocity of the particle from $t = 3$ to $t = 10$?

t	0	3	6	10	13
$s(t)$	0	72	92	144	180

24. Table 1.16 shows the production of tobacco in the US.[34]

  (a) What is the average rate of change in tobacco production between 1996 and 2003? Give units and interpret your answer in terms of tobacco production.

  (b) During this seven-year period, is there any interval during which the average rate of change was positive? If so, when?

Table 1.16 *Tobacco production, in millions of pounds*

Year	1996	1997	1998	1999	2000	2001	2002	2003
Production	1517	1787	1480	1293	1053	991	879	831

25. Do you expect the average rate of change (in units per year) of each of the following to be positive or negative? Explain your reasoning.

  (a) Number of acres of rain forest in the world.

  (b) Population of the world.

  (c) Number of polio cases each year in the US, since 1950.

  (d) Height of a sand dune that is being eroded.

  (e) Cost of living in the US.

---

[32]www.pepsico.com, accessed May 23, 2009.

[33]*Vital Signs 2007-2008*, The Worldwatch Institute, W.W. Norton & Company, 2007, p. 51 and 67.

[34]*The World Almanac and Book of Facts 2005*, pp. 138–139 (New York).

26. Figure 1.37 shows the length, L, in cm, of a sturgeon (a type of fish) as a function of the time, t, in years.[35]

   (a) Is the function increasing or decreasing? Is the graph concave up or concave down?

   (b) Estimate the average rate of growth of the sturgeon between $t = 5$ and $t = 15$. Give units and interpret your answer in terms of the sturgeon.

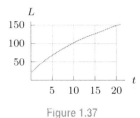

Figure 1.37

27. Table 1.17 shows the total US labor force, L. Find the average rate of change between 1940 and 2000; between 1940 and 1960; between 1980 and 2000. Give units and interpret your answers in terms of the labor force.[36]

Table 1.17   *US labor force, in thousands of workers*

Year	1940	1960	1980	2000
L	47,520	65,778	99,303	136,891

28. The total world marine catch[37] of fish, in metric tons, was 17 million in 1950 and 99 million in 2001. What was the average rate of change in the marine catch during this period? Give units and interpret your answer.

29. Table 1.18 gives the revenues, R, of General Motors, formerly the world's largest auto manufacturer.[38]

   (a) Find the change in revenues between 2003 and 2008.

   (b) Find the average rate of change in revenues between 2003 and 2008. Give units and interpret your answer.

   (c) From 2003 to 2008, were there any one-year intervals during which the average rate of change was negative? If so, which?

Table 1.18   *GM revenues, billions of dollars*

Year	2003	2004	2005	2006	2007	2008
R	184.0	192.9	193.1	205.6	181.1	149.0

30. The number of US households with cable television[39] was 12,168,450 in 1977 and 73,365,880 in 2003. Estimate the average rate of change in the number of US households with cable television during this 26-year period. Give units and interpret your answer.

31. Figure 1.7 in Problem 14 of Section 1.1 shows the amount of nicotine $N = f(t)$, in mg, in a person's bloodstream as a function of the time, t, in hours, since the last cigarette.

   (a) Is the average rate of change in nicotine level positive or negative? Explain.

   (b) Find the average rate of change in the nicotine level between $t = 0$ and $t = 3$. Give units and interpret your answer in terms of nicotine.

32. Table 1.19 shows the concentration, c, of creatinine in the bloodstream of a dog.[40]

   (a) Including units, find the average rate at which the concentration is changing between the

   (i) 6th and 8th minutes. (ii) 8th and 10th minutes.

   (b) Explain the sign and relative magnitudes of your results in terms of creatinine.

Table 1.19

t (minutes)	2	4	6	8	10
c (mg/ml)	0.439	0.383	0.336	0.298	0.266

33. The population of the world reached 1 billion in 1804, 2 billion in 1927, 3 billion in 1960, 4 billion in 1974, 5 billion in 1987 and 6 billion in 1999. Find the average rate of change of the population of the world, in people per minute, during each of these intervals (that is, from 1804 to 1927, 1927 to 1960, etc.).

Problems 34–35 refer to Figure 1.38 which shows the contraction velocity of a muscle as a function of the load it pulls against.

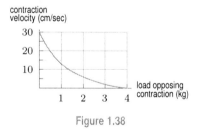

Figure 1.38

---

[35]Data from von Bertalanffy, L., *General System Theory*, p. 177 (New York: Braziller, 1968).

[36]*The World Almanac and Book of Facts 2005*, p. 144 (New York).

[37]*The World Almanac and Book of Facts 2005*, p. 143 (New York).

[38]www.gm.com/company/investor_information/earnings/hist_earnings/index.html, accessed May 23, 2009.

[39]*The World Almanac and Book of Facts 2005*, p. 310 (New York).

[40]From Cullen, M.R., *Linear Models in Biology* (Chichester: Ellis Horwood, 1985).

34. In terms of the muscle, interpret the

    (a) Vertical intercept   **(b)** Horizontal intercept

35. (a) Find the change in muscle contraction velocity when the load changes from 1 kg to 3 kg. Give units.

    (b) Find the average rate of change in the contraction velocity between 1 kg and 3 kg. Give units.

36. Table 1.20 gives the sales, S, of Intel Corporation, a leading manufacturer of integrated circuits.[41]

    (a) Find the change in sales between 2003 and 2008.

    (b) Find the average rate of change in sales between 2003 and 2008. Give units and interpret your answer.

Table 1.20   *Intel sales, in millions of dollars*

Year	2003	2004	2005	2006	2007	2008
S	30,100	34,200	38,800	35,400	38,300	37,600

37. In an experiment, a lizard is encouraged to run as fast as possible. Figure 1.39 shows the distance run in meters as a function of the time in seconds.[42]

    (a) If the lizard were running faster and faster, what would be the concavity of the graph? Does this match what you see?

    (b) Estimate the average velocity of the lizard during this 0.8 second experiment.

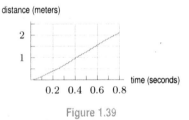

Figure 1.39

38. Values of $F(t)$, $G(t)$, and $H(t)$ are in Table 1.21. Which graph is concave up and which is concave down? Which function is linear?

Table 1.21

t	$F(t)$	$G(t)$	$H(t)$
10	15	15	15
20	22	18	17
30	28	21	20
40	33	24	24
50	37	27	29
60	40	30	35

39. Experiments suggest that the male maximum heart rate (the most times a male's heart can safely beat in a minute) decreases by 9 beats per minute during the first 21 years of his life, and by 26 beats per minute during the first 33 years.[43] If you model the maximum heart rate as a function of age, should you use a function that is increasing or decreasing, concave up or concave down?

40. A car starts slowly and then speeds up. Eventually the car slows down and stops. Graph the distance that the car has traveled against time.

41. Figure 1.40 shows the position of an object at time t.

    (a) Draw a line on the graph whose slope represents the average velocity between $t = 2$ and $t = 8$.

    (b) Is average velocity greater between $t = 0$ and $t = 3$ or between $t = 3$ and $t = 6$?

    (c) Is average velocity positive or negative between $t = 6$ and $t = 9$?

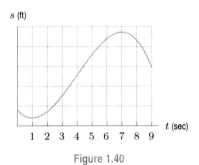

Figure 1.40

In Problems 42–45, which relative change is bigger in magnitude? Justify your answer.

42. The change in the Dow Jones average from 164.6 to 77.9 in 1931; the change in the Dow Jones average from 13261.8 to 8776.4 in 2008.

43. The change in the US population from 5.2 million to 7.2 million from 1800 to 1810; the change in the US population from 151.3 to 179.3 from 1950 to 1960.

44. An increase in class size from 5 to 10; an increase in class size from 30 to 50.

45. An increase in sales from \$100,000 to \$500,000; an increase in sales from \$20,000,000 to \$20,500,000.

46. Find the relative change of a population if it changes

    (a) From 1000 to 2000   **(b)** From 2000 to 1000

    (c) From 1,000,000 to 1,001,000

---

[41] Intel 2008 Annual Report, www.intel.com/intel/finance, accessed June 13, 2009.

[42] Data from Huey, R.B. and Hertz, P.E., "Effects of Body Size and Slope on the Acceleration of a Lizard," *J. Exp. Biol.*, Volume 110, 1984, p. 113-123.

[43] www.css.edu/users/tboone2/asep/May2002JEPonline.html, accessed January 4, 2005.

47. On Black Monday, October 28, 1929, the stock market on Wall Street crashed. The Dow Jones average dropped from 298.94 to 260.64 in one day. What was the relative change in the index?

48. On May 11, 2009, the cost to mail a letter in the US was raised from 42 cents to 44 cents. Find the relative change in the cost.

49. The US Consumer Price Index (CPI) is a measure of the cost of living. The inflation rate is the annual relative rate of change of the CPI. Use the January data in Table 1.22[44] to estimate the inflation rate for each of years 2005–2008.

Table 1.22

Year	2005	2006	2007	2008	2009
CPI	190.7	198.3	202.416	211.08	211.143

50. During 2008 the US economy stopped growing and began to shrink. Table 1.23[45] gives quarterly data on the US Gross Domestic Product (GDP) which measures the size of the economy.

(a) Estimate the relative growth rate (percent per year) at the first four times in the table.

(b) Economists often say an economy is in recession if the GDP decreases for two quarters in a row. Was the US in recession in 2008?

Table 1.23

t (years since 2008)	0	0.25	0.5	0.75	1.0
GDP (trillion dollars)	14.15	14.29	14.41	14.2	14.09

1.4 APPLICATIONS OF FUNCTIONS TO ECONOMICS

In this section, we look at some of the functions of interest to decision-makers in a firm or industry.

The Cost Function

> The **cost function**, $C(q)$, gives the total cost of producing a quantity q of some good.

What sort of function do you expect C to be? The more goods that are made, the higher the total cost, so C is an increasing function. Costs of production can be separated into two parts: the *fixed costs*, which are incurred even if nothing is produced, and the *variable costs*, which depend on how many units are produced.

An Example: Manufacturing Costs

Let's consider a company that makes radios. The factory and machinery needed to begin production are fixed costs, which are incurred even if no radios are made. The costs of labor and raw materials are variable costs since these quantities depend on how many radios are made. The fixed costs for this company are $24,000 and the variable costs are $7 per radio. Then,

$$\text{Total costs for the company} = \text{Fixed costs} + \text{Variable costs}$$
$$= 24{,}000 + 7 \cdot \text{Number of radios},$$

so, if q is the number of radios produced,

$$C(q) = 24{,}000 + 7q.$$

This is the equation of a line with slope 7 and vertical intercept 24,000.

The variable cost for one additional unit is called the *marginal cost*. For a linear cost function, the marginal cost is the rate of change, or slope, of the cost function.

---

[44]www.bls.gov/cpi, accessed June 7, 2009.
[45]www.bea.gov/national/nipaweb, Table 1.1.5.

Example 1   Graph the cost function $C(q) = 24{,}000 + 7q$. Label the fixed costs and marginal cost.

Solution   The graph of $C(q)$ is the line in Figure 1.41. The fixed costs are represented by the vertical intercept of 24,000. The marginal cost is represented by the slope of 7, which is the change in cost corresponding to a unit change in production.

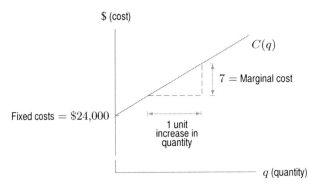

Figure 1.41: Cost function for the radio manufacturer

If $C(q)$ is a linear cost function,

- Fixed costs are represented by the vertical intercept.
- Marginal cost is represented by the slope.

Example 2   In each case, draw a graph of a linear cost function satisfying the given conditions:

(a) Fixed costs are large but marginal cost is small.

(b) There are no fixed costs but marginal cost is high.

Solution   (a) The graph is a line with a large vertical intercept and a small slope. See Figure 1.42.

(b) The graph is a line with a vertical intercept of zero (so the line goes through the origin) and a large positive slope. See Figure 1.43. Figures 1.42 and 1.43 have the same scales.

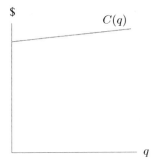

Figure 1.42: Large fixed costs, small marginal cost

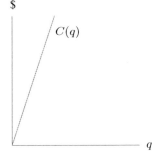

Figure 1.43: No fixed costs, high marginal cost

The Revenue Function

> The **revenue function**, $R(q)$, gives the total revenue received by a firm from selling a quantity, q, of some good.

If the good sells for a price of p per unit, and the quantity sold is q, then

$$\text{Revenue} = \text{Price} \cdot \text{Quantity}, \quad \text{so} \quad R = pq.$$

If the price does not depend on the quantity sold, so p is a constant, the graph of revenue as a function of q is a line through the origin, with slope equal to the price p.

Example 3   If radios sell for \$15 each, sketch the manufacturer's revenue function. Show the price of a radio on the graph.

Solution   Since $R(q) = pq = 15q$, the revenue graph is a line through the origin with a slope of 15. See Figure 1.44. The price is the slope of the line.

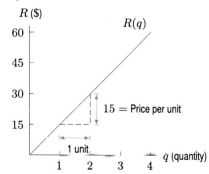

Figure 1.44: Revenue function for the radio manufacturer

Example 4   Graph the cost function $C(q) = 24{,}000 + 7q$ and the revenue function $R(q) = 15q$ on the same axes. For what values of q does the company make money?

Solution   The company makes money whenever revenues are greater than costs, so we find the values of q for which the graph of $R(q)$ lies above the graph of $C(q)$. See Figure 1.45.
We find the point at which the graphs of $R(q)$ and $C(q)$ cross:

$$\text{Revenue} = \text{Cost}$$
$$15q = 24{,}000 + 7q$$
$$8q = 24{,}000$$
$$q = 3000.$$

The company makes a profit if it produces and sells more than 3000 radios. The company loses money if it produces and sells fewer than 3000 radios.

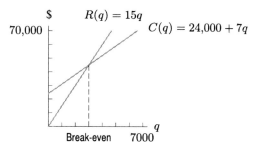

Figure 1.45: Cost and revenue functions for the radio manufacturer: What values of q generate a profit?

The Profit Function

Decisions are often made by considering the profit, usually written[46] as π to distinguish it from the price, p. We have

$$\text{Profit} = \text{Revenue} - \text{Cost} \quad \text{so} \quad \pi = R - C.$$

The *break-even point* for a company is the point where the profit is zero and revenue equals cost. See Figure 1.45.

Example 5   Find a formula for the profit function of the radio manufacturer. Graph it, marking the break-even point.

Solution   Since $R(q) = 15q$ and $C(q) = 24,000 + 7q$, we have

$$\pi(q) = R(q) - C(q) = 15q - (24,000 + 7q) = -24,000 + 8q.$$

Notice that the negative of the fixed costs is the vertical intercept and the break-even point is the horizontal intercept. See Figure 1.46.

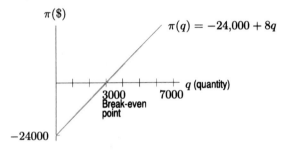

Figure 1.46: **Profit for radio manufacturer**

Example 6   (a) Using Table 1.24, estimate the break-even point for this company.
(b) Find the company's profit if 1000 units are produced.
(c) What price do you think the company is charging for its product?

Table 1.24   *Company's estimates of cost and revenue for a product*

q	500	600	700	800	900	1000	1100
$C(q)$, in \$	5000	5500	6000	6500	7000	7500	8000
$R(q)$, in \$	4000	4800	5600	6400	7200	8000	8800

Solution   (a) The break-even point is the value of q for which revenue equals cost. Since revenue is below cost at $q = 800$ and revenue is greater than cost at $q = 900$, the break-even point is between 800 and 900. The values in the table suggest that the break-even point is closer to 800, as the cost and revenue are closer there. A reasonable estimate for the break-even point is $q = 830$.
(b) If the company produces 1000 units, the cost is \$7500 and the revenue is \$8000, so the profit is $8000 - 7500 = 500$ dollars.
(c) From the table, it appears that $R(q) = 8q$. This indicates the company is selling the product for \$8 each.

---

[46]This π has nothing to do with the area of a circle, and merely stands for the Greek equivalent of the letter "p."

The Marginal Cost, Marginal Revenue, and Marginal Profit

Just as we used the term marginal cost to mean the rate of change, or slope, of a linear cost function, we use the terms *marginal revenue* and *marginal profit* to mean the rate of change, or slope, of linear revenue and profit functions, respectively. The term *marginal* is used because we are looking at how the cost, revenue, or profit change "at the margin," that is, by the addition of one more unit. For example, for the radio manufacturer, the marginal cost is 7 dollars/item (the additional cost of producing one more item is $7), the marginal revenue is 15 dollars/item (the additional revenue from selling one more item is $15), and the marginal profit is 8 dollars/item (the additional profit from selling one more item is $8).

The Depreciation Function

Suppose that the radio manufacturer has a machine that costs $20,000 and is sold ten years later for $3000. We say the value of the machine *depreciates* from $20,000 today to a resale value of $3000 in ten years. The depreciation formula gives the value, $V(t)$, in dollars, of the machine as a function of the number of years, t, since the machine was purchased. We assume that the value of the machine depreciates linearly.

The value of the machine when it is new ($t = 0$) is $20,000, so $V(0) = 20,000$. The resale value at time $t = 10$ is $3000, so $V(10) = 3000$. We have

$$\text{Slope} = m = \frac{3000 - 20{,}000}{10 - 0} = \frac{-17{,}000}{10} = -1700 \text{ dollars per year.}$$

This slope tells us that the value of the machine is decreasing at a rate of $1700 per year. Since $V(0) = 20,000$, the vertical intercept is 20,000, so

$$V(t) = 20{,}000 - 1700t \text{ dollars.}$$

Supply and Demand Curves

The quantity, q, of an item that is manufactured and sold depends on its price, p. As the price increases, manufacturers are usually willing to supply more of the product, whereas the quantity demanded by consumers falls.

> The **supply curve**, for a given item, relates the quantity, q, of the item that manufacturers are willing to make per unit time to the price, p, for which the item can be sold.
> The **demand curve** relates the quantity, q, of an item demanded by consumers per unit time to the price, p, of the item.

Economists often think of the quantities supplied and demanded as functions of price. However, for historical reasons, the economists put price (the independent variable) on the vertical axis and quantity (the dependent variable) on the horizontal axis. (The reason for this state of affairs is that economists originally took price to be the dependent variable and put it on the vertical axis. Later, when the point of view changed, the axes did not.) Thus, typical supply and demand curves look like those shown in Figure 1.47.

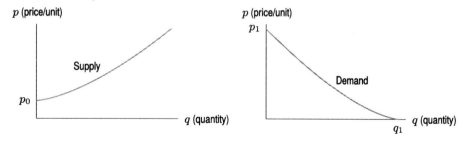

Figure 1.47: Supply and demand curves

Example 7   What is the economic meaning of the prices p_0 and p_1 and the quantity q_1 in Figure 1.47?

Solution   The vertical axis corresponds to a quantity of zero. Since the price p_0 is the vertical intercept on the supply curve, p_0 is the price at which the quantity supplied is zero. In other words, for prices below p_0, the suppliers will not produce anything. The price p_1 is the vertical intercept on the demand curve, so it corresponds to the price at which the quantity demanded is zero. In other words, for prices above p_1, consumers buy none of the product.

The horizontal axis corresponds to a price of zero, so the quantity q_1 on the demand curve is the quantity demanded if the price were zero—the quantity that could be given away free.

Equilibrium Price and Quantity

If we plot the supply and demand curves on the same axes, as in Figure 1.48, the graphs cross at the *equilibrium point*. The values p^* and q^* at this point are called the *equilibrium price* and *equilibrium quantity*, respectively. It is assumed that the market naturally settles to this equilibrium point. (See Problem 23.)

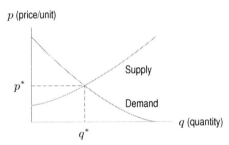

Figure 1.48: The equilibrium price and quantity

Example 8   Find the equilibrium price and quantity if

$$\text{Quantity supplied} = 3p - 50 \quad \text{and} \quad \text{Quantity demanded} = 100 - 2p.$$

Solution   To find the equilibrium price and quantity, we find the point at which

$$\text{Supply} = \text{Demand}$$
$$3p - 50 = 100 - 2p$$
$$5p = 150$$
$$p = 30.$$

The equilibrium price is $30. To find the equilibrium quantity, we use either the demand curve or the supply curve. At a price of $30, the quantity produced is $100 - 2 \cdot 30 = 40$ items. The equilibrium quantity is 40 items. In Figure 1.49, the demand and supply curves intersect at $p^* = 30$ and $q^* = 40$.

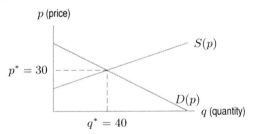

Figure 1.49: Equilibrium: $p^* = 30$, $q^* = 40$

The Effect of Taxes on Equilibrium

What effect do taxes have on the equilibrium price and quantity for a product? We distinguish between two types of taxes.[47] A *specific tax* is a fixed amount per unit of a product sold regardless of the selling price. This is the case with such items as gasoline, alcohol, and cigarettes. A specific tax is usually imposed on the producer. A *sales tax* is a fixed percentage of the selling price. Many cities and states collect sales tax on a wide variety of items. A sales tax is usually imposed on the consumer. We consider a specific tax now; a sales tax is considered in Problems 38 and 39.

Example 9   A specific tax of $5 per unit is now imposed upon suppliers in Example 8. What are the new equilibrium price and quantity?

Solution   The consumers pay p dollars per unit, but the suppliers receive only $p - 5$ dollars per unit because $5 goes to the government as taxes. Since

$$\text{Quantity supplied} = 3(\text{Amount per unit received by suppliers}) - 50,$$

the new supply equation is

$$\text{Quantity supplied} = 3(p - 5) - 50 = 3p - 65;$$

the demand equation is unchanged:

$$\text{Quantity demanded} = 100 - 2p.$$

At the equilibrium price, we have

$$\text{Demand} = \text{Supply}$$
$$100 - 2p = 3p - 65$$
$$165 = 5p$$
$$p = 33.$$

The equilibrium price is $33. The equilibrium quantity is 34 units, since the quantity demanded is $q = 100 - 2 \cdot 33 = 34$.

In Example 8, the equilibrium price was $30; with the imposition of a $5 tax in Example 9, the equilibrium price is $33. Thus the equilibrium price increases by $3 as a result of the tax. Notice that this is less than the amount of the tax. The consumer ends up paying $3 more than if the tax did not exist. However the government receives $5 per item. The producer pays the other $2 of the tax, retaining $28 of the price paid per item. Although the tax was imposed on the producer, some of the tax is passed on to the consumer in terms of higher prices. The tax has increased the price and reduced the number of items sold. See Figure 1.50. Notice that the taxes have the effect of moving the supply curve up by $5 because suppliers have to be paid $5 more to produce the same quantity.

---

[47]Adapted from Barry Bressler, *A Unified Approach to Mathematical Economics*, p. 81–88 (New York: Harper & Row, 1975).

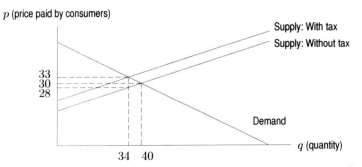

Figure 1.50: Specific tax shifts the supply curve, altering the equilibrium price and quantity

A Budget Constraint

An ongoing debate in the federal government concerns the allocation of money between defense and social programs. In general, the more that is spent on defense, the less that is available for social programs, and vice versa. Let's simplify the example to guns and butter. Assuming a constant budget, we show that the relationship between the number of guns and the quantity of butter is linear. Suppose that there is $12,000 to be spent and that it is to be divided between guns, costing $400 each, and butter, costing $2000 a ton. Suppose the number of guns bought is g, and the number of tons of butter is b. Then the amount of money spent on guns is $400g$, and the amount spent on butter is $2000b$. Assuming all the money is spent,

$$\text{Amount spent on guns } + \text{ Amount spent on butter } = \$12,000$$

or

$$400g + 2000b = 12,000.$$

Thus, dividing both sides by 400,

$$g + 5b = 30.$$

This equation is the budget constraint. Since the budget constraint can be written as

$$g = 30 - 5b,$$

the graph of the budget constraint is a line. See Figure 1.51.

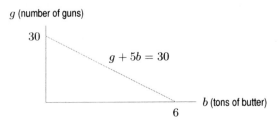

Figure 1.51: Budget constraint

Problems for Section 1.4

1. Figure 1.52 shows cost and revenue for a company.

    (a) Approximately what quantity does this company have to produce to make a profit?

    (b) Estimate the profit generated by 600 units.

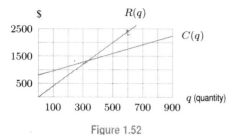

Figure 1.52

2. In Figure 1.53, which shows the cost and revenue functions for a product, label each of the following:

    (a) Fixed costs   **(b)** Break-even quantity

    (c) Quantities at which the company:

        (i) Makes a profit   (ii) Loses money

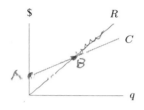

Figure 1.53

3. (a) Estimate the fixed costs and the marginal cost for the cost function in Figure 1.54.

    (b) Estimate $C(10)$ and interpret it in terms of cost.

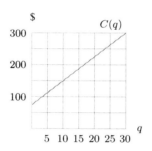

Figure 1.54

4. A company has cost and revenue functions, in dollars, given by $C(q) = 6000 + 10q$ and $R(q) = 12q$.

    (a) Find the cost and revenue if the company produces 500 units. Does the company make a profit? What about 5000 units?

    (b) Find the break-even point and illustrate it graphically.

5. Suppose that $q = f(p)$ is the demand curve for a product, where p is the selling price in dollars and q is the quantity sold at that price.

    (a) What does the statement $f(12) = 60$ tell you about demand for this product?

    (b) Do you expect this function to be increasing or decreasing? Why?

6. The demand curve for a quantity q of a product is $q = 5500 - 100p$ where p is price in dollars. Interpret the 5500 and the 100 in terms of demand. Give units.

7. A demand curve is given by $75p + 50q = 300$, where p is the price of the product, in dollars, and q is the quantity demanded at that price. Find p- and q-intercepts and interpret them in terms of consumer demand.

8. An amusement park charges an admission fee of $7 per person as well as an additional $1.50 for each ride.

    (a) For one visitor, find the park's total revenue $R(n)$ as a function of the number of rides, n, taken.

    (b) Find $R(2)$ and $R(8)$ and interpret your answers in terms of amusement park fees.

9. A company that makes Adirondack chairs has fixed costs of $5000 and variable costs of $30 per chair. The company sells the chairs for $50 each.

    (a) Find formulas for the cost and revenue functions.

    (b) Find the marginal cost and marginal revenue.

    (c) Graph the cost and the revenue functions on the same axes.

    (d) Find the break-even point.

10. A photocopying company has two different price lists. The first price list is $100 plus 3 cents per copy; the second price list is $200 plus 2 cents per copy.

    (a) For each price list, find the total cost as a function of the number of copies needed.

    (b) Determine which price list is cheaper for 5000 copies.

    (c) For what number of copies do both price lists charge the same amount?

11. A company has cost function $C(q) = 4000 + 2q$ dollars and revenue function $R(q) = 10q$ dollars.

    (a) What are the fixed costs for the company?

    (b) What is the marginal cost?

    (c) What price is the company charging for its product?

    (d) Graph $C(q)$ and $R(q)$ on the same axes and label the break-even point, q_0. Explain how you know the company makes a profit if the quantity produced is greater than q_0.

    (e) Find the break-even point q_0.

12. Values of a linear cost function are in Table 1.25. What are the fixed costs and the marginal cost? Find a formula for the cost function.

Table 1.25

q	0	5	10	15	20
$C(q)$	5000	5020	5040	5060	5080

13. A movie theater has fixed costs of $5000 per day and variable costs averaging $2 per customer. The theater charges $7 per ticket.

(a) How many customers per day does the theater need in order to make a profit?

(b) Find the cost and revenue functions and graph them on the same axes. Mark the break-even point.

14. A company producing jigsaw puzzles has fixed costs of $6000 and variable costs of $2 per puzzle. The company sells the puzzles for $5 each.

(a) Find formulas for the cost function, the revenue function, and the profit function.

(b) Sketch a graph of $R(q)$ and $C(q)$ on the same axes. What is the break-even point, q_0, for the company?

15. Production costs for manufacturing running shoes consist of a fixed overhead of $650,000 plus variable costs of $20 per pair of shoes. Each pair of shoes sells for $70.

(a) Find the total cost, $C(q)$, the total revenue, $R(q)$, and the total profit, $\pi(q)$, as a function of the number of pairs of shoes produced, q.

(b) Find the marginal cost, marginal revenue, and marginal profit.

(c) How many pairs of shoes must be produced and sold for the company to make a profit?

16. The cost C, in millions of dollars, of producing q items is given by $C = 5.7 + 0.002q$. Interpret the 5.7 and the 0.002 in terms of production. Give units.

17. The table shows the cost of manufacturing various quantities of an item and the revenue obtained from their sale.

Quantity	0	10	20	30	40	50	60	70	80
Cost ($)	120	400	600	780	1000	1320	1800	2500	3400
Revenue ($)	0	300	600	900	1200	1500	1800	2100	2400

(a) What range of production levels appears to be profitable?

(b) Calculate the profit or loss for each of the quantities shown. Estimate the most profitable production level.

18. (a) Give an example of a possible company where the fixed costs are zero (or very small).

(b) Give an example of a possible company where the marginal cost is zero (or very small).

19. A $15,000 robot depreciates linearly to zero in 10 years.

(a) Find a formula for its value as a function of time.

(b) How much is the robot worth three years after it is purchased?

20. A new bus worth $100,000 in 2010 depreciates linearly to $20,000 in 2030.

(a) Find a formula for the value of the bus, V, as a function of time, t, in years since 2010.

(b) What is the value of the bus in 2015?

(c) Find and interpret the vertical and horizontal intercepts of the graph of the function.

(d) What is the domain of the function?

21. A $50,000 tractor has a resale value of $10,000 twenty years after it was purchased. Assume that the value of the tractor depreciates linearly from the time of purchase.

(a) Find a formula for the value of the tractor as a function of the time since it was purchased.

(b) Graph the value of the tractor against time.

(c) Find the horizontal and vertical intercepts, give units, and interpret them.

22. A corporate office provides the demand curve in Figure 1.55 to its ice cream shop franchises. At a price of $1.00 per scoop, 240 scoops per day can be sold.

(a) Estimate how many scoops could be sold per day at a price of 50¢ per scoop. Explain.

(b) Estimate how many scoops per day could be sold at a price of $1.50 per scoop. Explain.

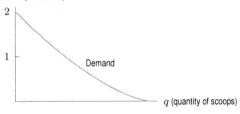

p (price per scoop in dollars)

Figure 1.55

23. Figure 1.56 shows supply and demand for a product.

(a) What is the equilibrium price for this product? At this price, what quantity is produced?

(b) Choose a price above the equilibrium price—for example, $p = 12$. At this price, how many items are suppliers willing to produce? How many items do consumers want to buy? Use your answers to these questions to explain why, if prices are above the equilibrium price, the market tends to push prices lower (toward the equilibrium).

(c) Now choose a price below the equilibrium price—for example, $p = 8$. At this price, how many items are suppliers willing to produce? How many items do consumers want to buy? Use your answers to these questions to explain why, if prices are below the equilibrium price, the market tends to push prices higher (toward the equilibrium).

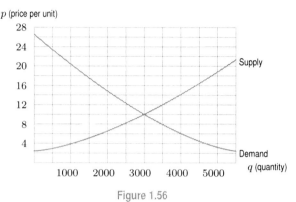

Figure 1.56

24. One of Tables 1.26 and 1.27 represents a supply curve; the other represents a demand curve.

(a) Which table represents which curve? Why?

(b) At a price of $155, approximately how many items would consumers purchase?

(c) At a price of $155, approximately how many items would manufacturers supply?

(d) Will the market push prices higher or lower than $155?

(e) What would the price have to be if you wanted consumers to buy at least 20 items?

(f) What would the price have to be if you wanted manufacturers to supply at least 20 items?

Table 1.26

p ($/unit)	182	167	153	143	133	125	118
q (quantity)	5	10	15	20	25	30	35

Table 1.27

p ($/unit)	6	35	66	110	166	235	316
q (quantity)	5	10	15	20	25	30	35

25. A company produces and sells shirts. The fixed costs are $7000 and the variable costs are $5 per shirt.

(a) Shirts are sold for $12 each. Find cost and revenue as functions of the quantity of shirts, q.

(b) The company is considering changing the selling price of the shirts. Demand is $q = 2000 - 40p$, where p is price in dollars and q is the number of shirts. What quantity is sold at the current price of $12? What profit is realized at this price?

(c) Use the demand equation to write cost and revenue as functions of the price, p. Then write profit as a function of price.

(d) Graph profit against price. Find the price that maximizes profits. What is this profit?

26. When the price, p, charged for a boat tour was $25, the average number of passengers per week, N, was 500. When the price was reduced to $20, the average number of passengers per week increased to 650. Find a formula for the demand curve, assuming that it is linear.

27. Table 1.28 gives data for the linear demand curve for a product, where p is the price of the product and q is the quantity sold every month at that price. Find formulas for the following functions. Interpret their slopes in terms of demand.

(a) q as a function of p.   (b) p as a function of q.

Table 1.28

p (dollars)	16	18	20	22	24
q (tons)	500	460	420	380	340

28. The demand curve for a product is given by $q = 120,000 - 500p$ and the supply curve is given by $q = 1000p$ for $0 \le q \le 120,000$, where price is in dollars.

(a) At a price of $100, what quantity are consumers willing to buy and what quantity are producers willing to supply? Will the market push prices up or down?

(b) Find the equilibrium price and quantity. Does your answer to part (a) support the observation that market forces tend to push prices closer to the equilibrium price?

29. World production, Q, of zinc in thousands of metric tons and the value, P, in dollars per metric ton are given[48] in Table 1.29. Plot the value as a function of production. Sketch a possible supply curve.

Table 1.29   *World zinc production*

Year	2003	2004	2005	2006	2007
Q	9520	9590	9930	10,000	10,900
P	896	1160	1480	3500	3400

[48] http://minerals.usgs.gov/ds/2005/140/, accessed May 24, 2009.

30. A taxi company has an annual budget of $720,000 to spend on drivers and car replacement. Drivers cost the company $30,000 each and car replacements cost $20,000 each.

   (a) What is the company's budget constraint equation? Let d be the number of drivers paid and c be the number of cars replaced.

   (b) Find and interpret both intercepts of the graph of the equation.

31. You have a budget of $1000 for the year to cover your books and social outings. Books cost (on average) $40 each and social outings cost (on average) $10 each. Let b denote the number of books purchased per year and s denote the number of social outings in a year.

   (a) What is the equation of your budget constraint?

   (b) Graph the budget constraint. (It does not matter which variable you put on which axis.)

   (c) Find the vertical and horizontal intercepts, and give a financial interpretation for each.

32. A company has a total budget of $500,000 and spends this budget on raw materials and personnel. The company uses m units of raw materials, at a cost of $100 per unit, and hires r employees, at a cost of $25,000 each.

   (a) What is the equation of the company's budget constraint?

   (b) Solve for m as a function of r.

   (c) Solve for r as a function of m.

33. Linear supply and demand curves are shown in Figure 1.57, with price on the vertical axis.

   (a) Label the equilibrium price p_0 and the equilibrium quantity q_0 on the axes.

   (b) Explain the effect on equilibrium price and quantity if the slope, $\Delta p / \Delta q$, of the supply curve increases. Illustrate your answer graphically.

   (c) Explain the effect on equilibrium price and quantity if the slope, $\Delta p / \Delta q$, of the demand curve becomes more negative. Illustrate your answer graphically.

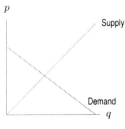

Figure 1.57

34. A demand curve has equation $q = 100 - 5p$, where p is price in dollars. A $2 tax is imposed on consumers. Find the equation of the new demand curve. Sketch both curves.

35. A supply curve has equation $q = 4p - 20$, where p is price in dollars. A $2 tax is imposed on suppliers. Find the equation of the new supply curve. Sketch both curves.

36. A tax of $8 per unit is imposed on the supplier of an item. The original supply curve is $q = 0.5p - 25$ and the demand curve is $q = 165 - 0.5p$, where p is price in dollars. Find the equilibrium price and quantity before and after the tax is imposed.

37. The demand and supply curves for a product are given in terms of price, p, by

$$q = 2500 - 20p \quad \text{and} \quad q = 10p - 500.$$

   (a) Find the equilibrium price and quantity. Represent your answers on a graph.

   (b) A specific tax of $6 per unit is imposed on suppliers. Find the new equilibrium price and quantity. Represent your answers on the graph.

   (c) How much of the $6 tax is paid by consumers and how much by producers?

   (d) What is the total tax revenue received by the government?

38. In Example 8, the demand and supply curves are given by $q = 100 - 2p$ and $q = 3p - 50$, respectively; the equilibrium price is $30 and the equilibrium quantity is 40 units. A sales tax of 5% is imposed on the consumer.

   (a) Find the equation of the new demand and supply curves.

   (b) Find the new equilibrium price and quantity.

   (c) How much is paid in taxes on each unit? How much of this is paid by the consumer and how much by the producer?

   (d) How much tax does the government collect?

39. Answer the questions in Problem 38, assuming that the 5% sales tax is imposed on the supplier instead of the consumer.

1.5   EXPONENTIAL FUNCTIONS

The function $f(x) = 2^x$, where the power is variable, is an *exponential function*. The number 2 is called the base. Exponential functions of the form $f(x) = k \cdot a^x$, where a is a positive constant, are used to represent many phenomena in the natural and social sciences.

Population Growth

The population of Nevada[49] from 2000 to 2006 is given in Table 1.30. To see how the population is growing, we look at the increase in population in the third column of Table 1.30. If the population had been growing linearly, all the numbers in the third column would be the same. But populations usually grow much faster as they get bigger, so it is not surprising that the numbers in the third column increase.

Table 1.30   *Population of Nevada (estimated), 2000–2006*

Year	Population (millions)	Change in population (millions)
2000	2.020	
		0.073
2001	2.093	
		0.075
2002	2.168	
		0.078
2003	2.246	
		0.081
2004	2.327	
		0.084
2005	2.411	
		0.087
2006	2.498	

Suppose we divide each year's population by the previous year's population. We get, approximately,

$$\frac{\text{Population in 2001}}{\text{Population in 2000}} = \frac{2.093 \text{ million}}{2.020 \text{ million}} = 1.036$$

$$\frac{\text{Population in 2002}}{\text{Population in 2001}} = \frac{2.168 \text{ million}}{2.093 \text{ million}} = 1.036.$$

$$\frac{\text{Population in 2003}}{\text{Population in 2002}} = \frac{2.246 \text{ million}}{2.168 \text{ million}} = 1.036.$$

The fact that all calculations are near 1.036 shows the population grew by about 3.6% between 2000 and 2001, between 2001 and 2002, and between 2002 and 2003. Whenever we have a constant percent increase (here 3.6%), we have *exponential growth*. If t is the number of years since 2000 and population is in millions,

When $t = 0$,   population $= 2.020 = 2.020(1.036)^0$.

When $t = 1$,   population $= 2.093 = 2.020(1.036)^1$.

When $t = 2$,   population $= 2.168 = 2.093(1.036) = 2.020(1.036)^2$.

When $t = 3$,   population $= 2.246 = 2.168(1.036) = 2.020(1.036)^3$.

So P, the population in millions t years after 2000, is given by

$$P = 2.020(1.036)^t \text{ million}.$$

Since the variable t is in the exponent, this is an exponential function. The base, 1.036, represents the factor by which the population grows each year and is called the *growth factor*. Assuming that the formula holds for 50 years, the population graph has the shape in Figure 1.58. The population is growing, so the function is increasing. Since the population grows faster as time passes, the graph is concave up. This behavior is typical of an exponential function. Even exponential functions that climb slowly at first, such as this one, eventually climb extremely quickly.

---

[49]www.census.gov, accessed May 14, 2007.

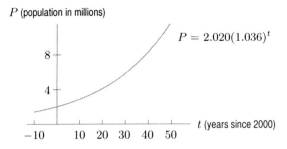

Figure 1.58: Population of Nevada (estimated): Exponential growth

Elimination of a Drug from the Body

Now we look at a quantity that is decreasing instead of increasing. When a patient is given medication, the drug enters the bloodstream. The rate at which the drug is metabolized and eliminated depends on the particular drug. For the antibiotic ampicillin, approximately 40% of the drug is eliminated every hour. A typical dose of ampicillin is 250 mg. Suppose $Q = f(t)$, where Q is the quantity of ampicillin, in mg, in the bloodstream at time t hours since the drug was given. At $t = 0$, we have $Q = 250$. Since the quantity remaining at the end of each hour is 60% of the quantity remaining the hour before, we have

$$f(0) = 250$$
$$f(1) = 250(0.6)$$
$$f(2) = 250(0.6)(0.6) = 250(0.6)^2$$
$$f(3) = 250(0.6)^2(0.6) = 250(0.6)^3.$$

So, after t hours,

$$Q = f(t) = 250(0.6)^t.$$

This function is called an *exponential decay* function. As t increases, the function values get arbitrarily close to zero. The t-axis is a *horizontal asymptote* for this function.

Notice the way the values in Table 1.31 are decreasing. Each additional hour a smaller quantity of drug is removed than the previous hour (100 mg the first hour, 60 mg the second, and so on). This is because as time passes, there is less of the drug in the body to be removed. Thus, the graph in Figure 1.59 bends upward. Compare this to the exponential growth in Figure 1.58, where each step upward is larger than the previous one. Notice that both graphs are concave up.

Table 1.31 *Value of decay function*

t (hours)	Q (mg)
0	250
1	150
2	90
3	54
4	32.4
5	19.4

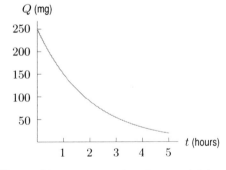

Figure 1.59: Drug elimination: Exponential decay

The General Exponential Function

Exponential growth is often described in terms of percent growth rates. The population of Nevada is growing at 3.6% per year, so it increases by a factor of $a = 1 + 0.036 = 1.036$ every year. Similarly, 40% of the ampicillin is removed every hour, so the quantity remaining decays by a factor of $a = 1 - 0.40 = 0.6$ each hour. We have the following general formulas.

> We say that P is an **exponential function** of t with base a if
>
> $$P = P_0 a^t,$$
>
> where P_0 is the initial quantity (when $t = 0$) and a is the factor by which P changes when t increases by 1. If $a > 1$, we have **exponential growth**; if $0 < a < 1$, we have **exponential decay**. The factor a is given by
>
> $$a = 1 + r$$
>
> where r is the decimal representation of the percent rate of change; r may be positive (for growth) or negative (for decay).

The largest possible domain for the exponential function is all real numbers,[50] provided $a > 0$.

Comparison Between Linear and Exponential Functions

Every exponential function changes at a constant percent, or *relative*, rate. For example, the population of Nevada increased approximately 3.6% per year. Every linear function changes at a constant absolute rate. For example, the Olympic pole vault record increased by 2 inches per year.

> A **linear** function has a constant rate of change.
> An **exponential** function has a constant percent, or relative, rate of change.

Example 1

The amount of adrenaline in the body can change rapidly. Suppose the initial amount is 15 mg. Find a formula for A, the amount in mg, at a time t minutes later if A is:

(a) Increasing by 0.4 mg per minute.   (b) Decreasing by 0.4 mg per minute.
(c) Increasing by 3% per minute.   (d) Decreasing by 3% per minute.

Solution

(a) This is a linear function with initial quantity 15 and slope 0.4, so

$$A = 15 + 0.4t.$$

(b) This is a linear function with initial quantity 15 and slope -0.4, so

$$A = 15 - 0.4t.$$

(c) This is an exponential function with initial quantity 15 and base $1 + 0.03 = 1.03$, so

$$A = 15(1.03)^t.$$

(d) This is an exponential function with initial quantity 15 and base $1 - 0.03 = 0.97$, so

$$A = 15(0.97)^t.$$

---

[50]The reason we do not want $a \leq 0$ is that, for example, we cannot define $a^{1/2}$ if $a < 0$. Also, we do not usually have $a = 1$, since $P = P_0 a^t = P_0 1^t = P_0$ is then a constant function.

Example 2 Sales[51] at Borders Books and Music stores increased from \$2503 million in 1997 to \$3699 million in 2003. Assuming that sales have been increasing exponentially, find an equation of the form $P = P_0 a^t$, where P is Borders sales in millions of dollars and t is the number of years since 1997. What is the percent growth rate?

Solution We know that $P = 2503$ when $t = 0$, so $P_0 = 2503$. To find a, we use the fact that $P = 3699$ when $t = 6$. Substituting gives

$$P = P_0 a^t$$
$$3699 = 2503a^6.$$

Dividing both sides by 2503, we get

$$\frac{3699}{2503} = a^6$$
$$1.478 = a^6.$$

Taking the sixth root of both sides gives

$$a = (1.478)^{1/6} = 1.07.$$

Since $a = 1.07$, the equation for Borders sales as a function of the number of years since 1997 is

$$P = 2503(1.07)^t.$$

During this period, sales increased by 7% per year.

Recognizing Data from an Exponential Function: Values of t and P in a table could come from an exponential function $P = P_0 a^t$ if ratios of P values are constant for equally spaced t values.

Example 3 Which of the following tables of values could correspond to an exponential function, a linear function, or neither? For those which could correspond to an exponential or linear function, find a formula for the function.

(a)

x	$f(x)$
0	16
1	24
2	36
3	54
4	81

(b)

x	$g(x)$
0	14
1	20
2	24
3	29
4	35

(c)

x	$h(x)$
0	5.3
1	6.5
2	7.7
3	8.9
4	10.1

Solution (a) We see that f cannot be a linear function, since $f(x)$ increases by different amounts ($24-16=8$ and $36-24=12$) as x increases by one. Could f be an exponential function? We look at the ratios of successive $f(x)$ values:

$$\frac{24}{16} = 1.5 \quad \frac{36}{24} = 1.5 \quad \frac{54}{36} = 1.5 \quad \frac{81}{54} = 1.5.$$

Since the ratios are all equal to 1.5, this table of values could correspond to an exponential function with a base of 1.5. Since $f(0) = 16$, a formula for $f(x)$ is

$$f(x) = 16(1.5)^x.$$

Check by substituting $x = 0, 1, 2, 3, 4$ into this formula; you get the values given for $f(x)$.

(b) As x increases by one, $g(x)$ increases by 6 (from 14 to 20), then 4 (from 20 to 24), so g is not linear. We check to see if g could be exponential:

[51] http://phx.corporate-ir.net/phoenix.zhtml?c=65380&p=irol-annualreports, accessed April 14, 2005.

$$\frac{20}{14} = 1.43 \quad \text{and} \quad \frac{24}{20} = 1.2.$$

Since these ratios (1.43 and 1.2) are different, g is not exponential.

(c) For h, notice that as x increases by one, the value of $h(x)$ increases by 1.2 each time. So h could be a linear function with a slope of 1.2. Since $h(0) = 5.3$, a formula for $h(x)$ is

$$h(x) = 5.3 + 1.2x.$$

The Family of Exponential Functions and the Number e

The formula $P = P_0 a^t$ gives a family of exponential functions with parameters P_0 (the initial quantity) and a (the base). The base tells us whether the function is increasing ($a > 1$) or decreasing ($0 < a < 1$). Since a is the factor by which P changes when t is increased by 1, large values of a mean fast growth; values of a near 0 mean fast decay. (See Figures 1.60 and 1.61.) All members of the family $P = P_0 a^t$ are concave up if $P_0 > 0$.

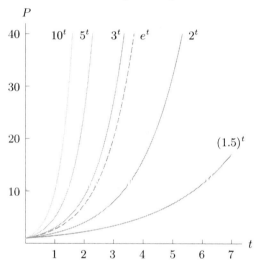

Figure 1.60: Exponential growth: $P = a^t$, for $a > 1$ Figure 1.61: Exponential decay: $P = a^t$, for $0 < a < 1$

In practice the most commonly used base is the number $e = 2.71828\ldots$. The fact that most calculators have an e^x button is an indication of how important e is. Since e is between 2 and 3, the graph of $y = e^t$ in Figure 1.60 is between the graphs of $y = 2^t$ and $y = 3^t$.

The base e is used so often that it is called the natural base. At first glance, this is somewhat mysterious: What could be natural about using 2.71828... as a base? The full answer to this question must wait until Chapter 3, where you will see that many calculus formulas come out more neatly when e is used as the base. (See Appendix B for the relation to compound interest.)

Problems for Section 1.5

1. The following functions give the populations of four towns with time t in years.

(i) $P = 600(1.12)^t$ (ii) $P - 1,000(1.03)^t$
(iii) $P = 200(1.08)^t$ (iv) $P = 900(0.90)^t$

(a) Which town has the largest percent growth rate? What is the percent growth rate?
(b) Which town has the largest initial population? What is that initial population?
(c) Are any of the towns decreasing in size? If so, which one(s)?

2. Each of the following functions gives the amount of a substance present at time t. In each case, give the amount present initially (at $t = 0$), state whether the function represents exponential growth or decay, and give the percent growth or decay rate.

(a) $A = 100(1.07)^t$ (b) $A = 5.3(1.054)^t$
(c) $A = 3500(0.93)^t$ (d) $A = 12(0.88)^t$

3. Figure 1.62 shows graphs of several cities' populations against time. Match each of the following descriptions to a graph and write a description to match each of the remaining graphs.

  (a) The population increased at 5% per year.
  (b) The population increased at 8% per year.
  (c) The population increased by 5000 people per year.
  (d) The population was stable.

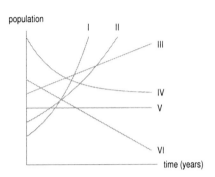

Figure 1.62

4. A town has a population of 1000 people at time $t = 0$. In each of the following cases, write a formula for the population, P, of the town as a function of year t.

  (a) The population increases by 50 people a year.
  (b) The population increases by 5% a year.

5. The gross domestic product, G, of Switzerland was 310 billion dollars in 2007. Give a formula for G (in billions of dollars) t years after 2007 if G increases by
  (a)   3% per year   **(b)**   8 billion dollars per year

6. A product costs $80 today. How much will the product cost in t days if the price is reduced by

  (a) $4 a day   **(b)**   5% a day

7. An air-freshener starts with 30 grams and evaporates. In each of the following cases, write a formula for the quantity, Q grams, of air-freshener remaining t days after the start and sketch a graph of the function. The decrease is:

  (a)   2 grams a day   **(b)**   12% a day

8. World population is approximately $P = 6.4(1.0126)^t$, with P in billions and t in years since 2004.

  (a) What is the yearly percent rate of growth of the world population?
  (b) What was the world population in 2004? What does this model predict for the world population in 2010?
  (c) Use part (b) to find the average rate of change of the world population between 2004 and 2010.

9. A 50 mg dose of quinine is given to a patient to prevent malaria. Quinine leaves the body at a rate of 6% per hour.

  (a) Find a formula for the amount, A (in mg), of quinine in the body t hours after the dose is given.
  (b) How much quinine is in the body after 24 hours?
  (c) Graph A as a function of t.
  (d) Use the graph to estimate when 5 mg of quinine remains.

10. The Hershey Company is the largest US producer of chocolate. In 2008, annual net sales were 5.1 billion dollars and were increasing at a continuous rate of 4.3% per year.[52]

  (a) Write a formula for annual net sales, S, as a function of time, t, in years since 2008.
  (b) Estimate annual net sales in 2015.
  (c) Use a graph or trial and error to estimate the year in which annual net sales are expected to pass 8 billion dollars.

11. The consumer price index (CPI) for a given year is the amount of money in that year that has the same purchasing power as $100 in 1983. At the start of 2009, the CPI was 211. Write a formula for the CPI as a function of t, years after 2009, assuming that the CPI increases by 2.8% every year.

12. During the 1980s, Costa Rica had the highest deforestation rate in the world, at 2.9% per year. (This is the rate at which land covered by forests is shrinking.) Assuming the rate continues, what percent of the land in Costa Rica covered by forests in 1980 will be forested in 2015?

13. (a) Make a table of values for $y = e^x$ using $x = 0, 1, 2, 3$.
  (b) Plot the points found in part (a). Does the graph look like an exponential growth or decay function?
  (c) Make a table of values for $y = e^{-x}$ using $x = 0, 1, 2, 3$.
  (d) Plot the points found in part (c). Does the graph look like an exponential growth or decay function?

14. Graph $y = 100e^{-0.4x}$. Describe what you see.

For Problems 15–16, find a possible formula for the function represented by the data.

15.

x	0	1	2	3
$f(x)$	4.30	6.02	8.43	11.80

16.

t	0	1	2	3
$g(t)$	5.50	4.40	3.52	2.82

[52] 2008 Annual Report to Stockholders, accessed at www.thehersheycompany.com.

Give a possible formula for the functions in Problems 17–18.

17.

18.

x	$f(x)$	$g(x)$	$h(x)$
-2	12	16	37
-1	17	24	34
0	20	36	31
1	21	54	28
2	18	81	25

19. If the world's population increased exponentially from 4.453 billion in 1980 to 5.937 billion in 1998 and continued to increase at the same percentage rate between 1998 and 2008, calculate what the world's population would have been in 2008. How does this compare to the actual population of 6.771 billion?

20. The number of passengers using a railway fell from 190,205 to 174,989 during a 5-year period. Find the annual percentage decrease over this period.

21. The company that produces Cliffs Notes (abridged versions of classic literature) was started in 1958 with $4000 and sold in 1998 for $14,000,000. Find the annual percent increase in the value of this company over the 40 years.

22. Find a formula for the number of zebra mussels in a bay as a function of the number of years since 2003, given that there were 2700 at the start of 2003 and 3186 at the start of 2004.

(a) Assume that the number of zebra mussels is growing linearly. Give units for the slope of the line and interpret it in terms of zebra mussels.

(b) Assume that the number of zebra mussels is growing exponentially. What is the percent rate of growth of the zebra mussel population?

23. Worldwide, wind energy[53] generating capacity, W, was 40,300 megawatts in 2003 and 121,188 megawatts in 2008.

(a) Use the values given to write W, in megawatts, as a linear function of t, the number of years since 2003.

(b) Use the values given to write W as an exponential function of t.

(c) Graph the functions found in parts (a) and (b) on the same axes. Label the given values.

24. (a) Which (if any) of the functions in the following table could be linear? Find formulas for those functions.

(b) Which (if any) of these functions could be exponential? Find formulas for those functions.

25. Determine whether each of the following tables of values could correspond to a linear function, an exponential function, or neither. For each table of values that could correspond to a linear or an exponential function, find a formula for the function.

(a)
x	$f(x)$
0	10.5
1	12.7
2	18.9
3	36.7

(b)
t	$s(t)$
-1	50.2
0	30.12
1	18.072
2	10.8432

(c)
u	$g(u)$
0	27
2	24
4	21
6	18

26. (a) Could the data on annual world soybean production[54] in Table 1.32 correspond to a linear function or an exponential function? If so, which?

(b) Find a formula for P, world soybean production in millions of tons, as a function of time, t, in years since 2000.

(c) What is the annual percent increase in soybean production?

Table 1.32 *Soybean production, in millions of tons*

Year	2000	2001	2002	2003	2004	2005
Production	161.0	170.3	180.2	190.7	201.8	213.5

27. The 2004 US presidential debates questioned whether the minimum wage has kept pace with inflation. Decide the question using the following information:[55] In 1938, the minimum wage was 25¢; in 2004, it was $5.15. During the same period, inflation averaged 4.3%.

28. (a) Niki invested $10,000 in the stock market. The investment was a loser, declining in value 10% per year each year for 10 years. How much was the investment worth after 10 years?

(b) After 10 years, the stock began to gain value at 10% per year. After how long will the investment regain its initial value ($10,000)?

[53]World Wind Energy Report 2008, WWEA.
[54]*Vital Signs 2007-2008*, The Worldwatch Institute, W.W. Norton & Company, 2007, p. 23.
[55]http://www.dol.gov/esa/minwage/chart.htm#5.

29. A photocopy machine can reduce copies to 80% of their original size. By copying an already reduced copy, further reductions can be made.

(a) If a page is reduced to 80%, what percent enlargement is needed to return it to its original size?

(b) Estimate the number of times in succession that a page must be copied to make the final copy less than 15% of the size of the original.

30. Whooping cough was thought to have been almost wiped out by vaccinations. It is now known that the vaccination wears off, leading to an increase in the number of cases, w, from 1248 in 1981 to 18,957 in 2004.

(a) With t in years since 1980, find an exponential function that fits this data.

(b) What does your answer to part (a) give as the average annual percent growth rate of the number of cases?

(c) On May 4, 2005, the *Arizona Daily Star* reported (correctly) that the number of cases had more than doubled between 2000 and 2004. Does your model confirm this report? Explain.

31. Aircraft require longer takeoff distances, called takeoff rolls, at high altitude airports because of diminished air density. The table shows how the takeoff roll for a certain light airplane depends on the airport elevation. (Takeoff rolls are also strongly influenced by air temperature; the data shown assume a temperature of $0°$ C.) Determine a formula for this particular aircraft that gives the takeoff roll as an exponential function of airport elevation.

Elevation (ft)	Sea level	1000	2000	3000	4000
Takeoff roll (ft)	670	734	805	882	967

1.6 THE NATURAL LOGARITHM

In Section 1.5, we projected the population of Nevada (in millions) by the function

$$P = f(t) = 2.020(1.036)^t,$$

where t is the number of years since 2000. Now suppose that instead of calculating the population at time t, we ask when the population will reach 4 million. We want to find the value of t for which

$$4 = f(t) = 2.020(1.036)^t.$$

We use logarithms to solve for a variable in an exponent.

Definition and Properties of the Natural Logarithm

We define the natural logarithm of x, written $\ln x$, as follows:

The **natural logarithm** of x, written $\ln x$, is the power of e needed to get x. In other words,

$$\ln x = c \quad \text{means} \quad e^c = x.$$

The natural logarithm is sometimes written $\log_e x$.

For example, $\ln e^3 = 3$ since 3 is the power of e needed to give e^3. Similarly, $\ln(1/e) = \ln e^{-1} = -1$. A calculator gives $\ln 5 = 1.6094$, because $e^{1.6094} = 5$. However if we try to find $\ln(-7)$ on a calculator, we get an error message because e to any power is never negative or 0. In general

---

$\ln x$ is not defined if x is negative or 0.

To work with logarithms, we use the following properties:

> ## Properties of the Natural Logarithm
> 1. $\ln(AB) = \ln A + \ln B$
> 2. $\ln\left(\dfrac{A}{B}\right) = \ln A - \ln B$
> 3. $\ln(A^p) = p\ln A$
> 4. $\ln e^x = x$
> 5. $e^{\ln x} = x$
>
> In addition, $\ln 1 = 0$ because $e^0 = 1$, and $\ln e = 1$ because $e^1 = e$.

Using the $\boxed{\text{LN}}$ button on a calculator, we get the graph of $f(x) = \ln x$ in Figure 1.63. Observe that, for large x, the graph of $y = \ln x$ climbs very slowly as x increases. The x-intercept is $x = 1$, since $\ln 1 = 0$. For $x > 1$, the value of $\ln x$ is positive; for $0 < x < 1$, the value of $\ln x$ is negative.

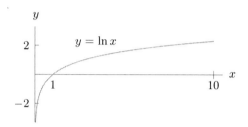

Figure 1.63: The natural logarithm function climbs very slowly

Solving Equations Using Logarithms

Natural logs can be used to solve for unknown exponents.

Example 1   Find t such that $3^t = 10$.

Solution   First, notice that we expect t to be between 2 and 3, because $3^2 = 9$ and $3^3 = 27$. To find t exactly, we take the natural logarithm of both sides and solve for t:

$$\ln(3^t) = \ln 10.$$

The third property of logarithms tells us that $\ln(3^t) = t\ln 3$, so we have

$$t\ln 3 = \ln 10$$
$$t = \frac{\ln 10}{\ln 3}.$$

Using a calculator to find the natural logs gives

$$t = 2.096.$$

Example 2   We return to the question of when the population of Nevada reaches 4 million. To get an answer, we solve $4 = 2.020(1.036)^t$ for t, using logs.

Solution   Dividing both sides of the equation by 2.020, we get

$$\frac{4}{2.020} = (1.036)^t.$$

Now take natural logs of both sides:

$$\ln\left(\frac{4}{2.020}\right) = \ln(1.036^t).$$

Using the fact that $\ln(1.036^t) = t\ln 1.036$, we get

$$\ln\left(\frac{4}{2.020}\right) = t\ln(1.036).$$

Solving this equation using a calculator to find the logs, we get

$$t = \frac{\ln(4/2.020)}{\ln(1.036)} = 19.317 \text{ years}.$$

Since $t = 0$ in 2000, this value of t corresponds to the year 2019.

Example 3   Find t such that $12 = 5e^{3t}$.

Solution   It is easiest to begin by isolating the exponential, so we divide both sides of the equation by 5:

$$2.4 = e^{3t}.$$

Now take the natural logarithm of both sides:

$$\ln 2.4 = \ln(e^{3t}).$$

Since $\ln(e^x) = x$, we have

$$\ln 2.4 = 3t,$$

so, using a calculator, we get

$$t = \frac{\ln 2.4}{3} = 0.2918.$$

Exponential Functions with Base e

An exponential function with base a has formula

$$P = P_0 a^t.$$

For any positive number a, we can write $a = e^k$ where $k = \ln a$. Thus, the exponential function can be rewritten as

$$P = P_0 a^t = P_0(e^k)^t = P_0 e^{kt}.$$

If $a > 1$, then k is positive, and if $0 < a < 1$, then k is negative. We conclude:

Writing $a = e^k$, so $k = \ln a$, any exponential function can be written in two forms

$$P = P_0 a^t \quad \text{or} \quad P = P_0 e^{kt}.$$

- If $a > 1$, we have exponential growth; if $0 < a < 1$, we have exponential decay.
- If $k > 0$, we have exponential growth; if $k < 0$, we have exponential decay.
- k is called the *continuous* growth or decay rate.

The word continuous in continuous growth rate is used in the same way to describe continuous compounding of interest earned on money. (See Appendix A.)

Example 4   (a) Convert the function $P = 1000e^{0.05t}$ to the form $P = P_0 a^t$.
(b) Convert the function $P = 500(1.06)^t$ to the form $P = P_0 e^{kt}$.

Solution   (a) Since $P = 1000e^{0.05t}$, we have $P_0 = 1000$. We want to find a so that

$$1000a^t = 1000e^{0.05t} = 1000(e^{0.05})^t.$$

We take $a = e^{0.05} = 1.0513$, so the following two functions give the same values:

$$P = 1000e^{0.05t} \quad \text{and} \quad P = 1000(1.0513)^t.$$

So a continuous growth rate of 5% is equivalent to a growth rate of 5.13% per unit time.
(b) We have $P_0 = 500$ and we want to find k with

$$500(1.06)^t = 500(e^k)^t,$$

so we take

$$1.06 = e^k$$
$$k = \ln(1.06) = 0.0583.$$

The following two functions give the same values:

$$P = 500(1.06)^t \quad \text{and} \quad P = 500e^{0.0583t}.$$

So a growth rate of 6% per unit time is equivalent to a continuous growth rate of 5.83%.

Example 5   Sketch graphs of $P = e^{0.5t}$, a continuous growth rate of 50%, and $Q = 5e^{-0.2t}$, a continuous decay rate of 20%.

Solution   The graph of $P = e^{0.5t}$ is in Figure 1.64. Notice that the graph is the same shape as the previous exponential growth curves: increasing and concave up. The graph of $Q = 5e^{-0.2t}$ is in Figure 1.65; it has the same shape as other exponential decay functions.

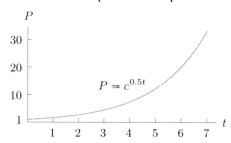

Figure 1.64: Continuous exponential growth function

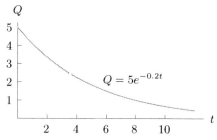

Figure 1.65: Continuous exponential decay function

49

Problems for Section 1.6

For Problems 1–16, solve for t using natural logarithms.

1. $5^t = 7$

2. $10 = 2^t$

3. $2 = (1.02)^t$

4. $130 = 10^t$

5. $50 = 10 \cdot 3^t$

6. $100 = 25(1.5)^t$

7. $10 = e^t$

8. $5 = 2e^t$

9. $e^{3t} = 100$

10. $10 = 6e^{0.5t}$

11. $40 = 100e^{-0.03t}$

12. $a = b^t$

13. $B = Pe^{rt}$

14. $2P = Pe^{0.3t}$

15. $5e^{3t} = 8e^{2t}$

16. $7 \cdot 3^t = 5 \cdot 2^t$

The functions in Problems 17–20 represent exponential growth or decay. What is the initial quantity? What is the growth rate? State if the growth rate is continuous.

17. $P = 5(1.07)^t$

18. $P = 7.7(0.92)^t$

19. $P = 15e^{-0.06t}$

20. $P = 3.2e^{0.03t}$

21. The following formulas give the populations of four different towns, A, B, C, and D, with t in years from now.

$$P_A = 600e^{0.08t} \qquad P_B = 1000e^{-0.02t}$$
$$P_C = 1200e^{0.03t} \qquad P_D = 900e^{0.12t}$$

   (a) Which town is growing fastest (that is, has the largest percentage growth rate)?

   (b) Which town is the largest now?

   (c) Are any of the towns decreasing in size? If so, which one(s)?

22. A city's population is 1000 and growing at 5% a year.

   (a) Find a formula for the population at time t years from now assuming that the 5% per year is an:

   (i)   Annual rate   (ii)   Continuous annual rate

   (b) In each case in part (a), estimate the population of the city in 10 years.

Write the functions in Problems 23–26 in the form $P = P_0 a^t$. Which represent exponential growth and which represent exponential decay?

23. $P = 15e^{0.25t}$

24. $P = 2e^{-0.5t}$

25. $P = P_0 e^{0.2t}$

26. $P = 7e^{-\pi t}$

In Problems 27–30, put the functions in the form $P = P_0 e^{kt}$.

27. $P = 15(1.5)^t$

28. $P = 10(1.7)^t$

29. $P = 174(0.9)^t$

30. $P = 4(0.55)^t$

31. The population of the world can be represented by $P = 6.4(1.0126)^t$, where P is in billions of people and t is years since 2004. Find a formula for the population of the world using a continuous growth rate.

32. A fishery stocks a pond with 1000 young trout. The number of trout t years later is given by $P(t) = 1000e^{-0.5t}$.

   (a) How many trout are left after six months? After 1 year?

   (b) Find $P(3)$ and interpret it in terms of trout.

   (c) At what time are there 100 trout left?

   (d) Graph the number of trout against time, and describe how the population is changing. What might be causing this?

33. During a recession a firm's revenue declines continuously so that the revenue, R (measured in millions of dollars), in t years' time is given by $R = 5e^{-0.15t}$.

   (a) Calculate the current revenue and the revenue in two years' time.

   (b) After how many years will the revenue decline to $2.7 million?

34. **(a)** What is the continuous percent growth rate for $P = 100e^{0.06t}$, with time, t, in years?

   (b) Write this function in the form $P = P_0 a^t$. What is the annual percent growth rate?

35. **(a)** What is the annual percent decay rate for $P = 25(0.88)^t$, with time, t, in years?

   (b) Write this function in the form $P = P_0 e^{kt}$. What is the continuous percent decay rate?

36. The gross world product is $W = 32.4(1.036)^t$, where W is in trillions of dollars and t is years since 2001. Find a formula for gross world product using a continuous growth rate.

37. The population, P, in millions, of Nicaragua was 5.4 million in 2004 and growing at an annual rate of 3.4%. Let t be time in years since 2004.

   (a) Express P as a function in the form $P = P_0 a^t$.

   (b) Express P as an exponential function using base e.

   (c) Compare the annual and continuous growth rates.

38. What annual percent growth rate is equivalent to a continuous percent growth rate of 8%?

39. What continuous percent growth rate is equivalent to an annual percent growth rate of 10%?

40. The population of a city is 50,000 in 2008 and is growing at a continuous yearly rate of 4.5%.

   (a) Give the population of the city as a function of the number of years since 2008. Sketch a graph of the population against time.

   (b) What will be the city's population in the year 2018?

   (c) Calculate the time for the population of the city to reach 100,000. This is called the doubling time of the population.

41. In 2000, there were about 213 million vehicles (cars and trucks) and about 281 million people in the US. The number of vehicles has been growing at 4% a year, while the population has been growing at 1% a year. If the growth rates remain constant, when is there, on average, one vehicle per person?

1.7   EXPONENTIAL GROWTH AND DECAY

Many quantities in nature change according to an exponential growth or decay function of the form $P = P_0 e^{kt}$, where P_0 is the initial quantity and k is the continuous growth or decay rate.

Example 1   The Environmental Protection Agency (EPA) recently investigated a spill of radioactive iodine. The radiation level at the site was about 2.4 millirems/hour (four times the maximum acceptable limit of 0.6 millirems/hour), so the EPA ordered an evacuation of the surrounding area. The level of radiation from an iodine source decays at a continuous hourly rate of $k = -0.004$.

(a) What was the level of radiation 24 hours later?

(b) Find the number of hours until the level of radiation reached the maximum acceptable limit, and the inhabitants could return.

Solution   (a) The level of radiation, R, in millirems/hour, at time t, in hours since the initial measurement, is given by

$$R = 2.4e^{-0.004t},$$

so the level of radiation 24 hours later was

$$R = 2.4e^{(-0.004)(24)} = 2.18 \text{ millirems per hour.}$$

(b) A graph of $R = 2.4e^{-0.004t}$ is in Figure 1.66. The maximum acceptable value of R is 0.6 millirems per hour, which occurs at approximately $t = 350$. Using logarithms, we have

$$0.6 = 2.4e^{-0.004t}$$
$$0.25 = e^{-0.004t}$$
$$\ln 0.25 = -0.004t$$
$$t = \frac{\ln 0.25}{-0.004} = 346.57$$

The inhabitants will not be able to return for 346.57 hours, or about 15 days.

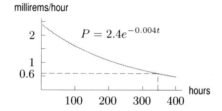

millirems/hour

$$P = 2.4e^{-0.004t}$$

Figure 1.66: The level of radiation from radioactive iodine

Example 2   The population of Kenya[56] was 19.5 million in 1984 and 39.0 million in 2009. Assuming the population increases exponentially, find a formula for the population of Kenya as a function of time.

Solution   We measure the population, P, in millions and time, t, in years since 1984. We can express P in terms of t using the continuous growth rate k by

$$P = P_0 e^{kt} = 19.5e^{kt},$$

where $P_0 = 19.5$ is the initial value of P. We find k using the fact that $P = 39.0$ when $t = 25$:

$$39.0 = 19.5e^{k \cdot 25}.$$

Divide both sides by 19.5, giving

$$\frac{39.0}{19.5} = e^{25k}.$$

[56] www.cia.gov/library/publications/the-world-factbook.

Take natural logs of both sides:

$$\ln\left(\frac{39.0}{19.5}\right) = \ln(e^{25k}).$$

Since $\ln(e^{25k}) = 25k$, this becomes

$$\ln\left(\frac{39.0}{19.5}\right) = 25k.$$

We get

$$k = \frac{1}{25}\ln\left(\frac{39.0}{19.5}\right) = 0.028,$$

and therefore

$$P = 19.5e^{0.028t}.$$

Since $k = 0.028 = 2.8\%$, the population of Kenya was growing at a continuous rate of 2.8% per year.

Doubling Time and Half-Life

Every exponential growth function has a constant doubling time and every exponential decay function has a constant half-life.

> The **doubling time** of an exponentially increasing quantity is the time required for the quantity to double.
> The **half-life** of an exponentially decaying quantity is the time required for the quantity to be reduced by a factor of one half.

Example 3   Show algebraically that every exponentially growing function has a fixed doubling time.

Solution   Consider the exponential function $P = P_0 a^t$. For any base a with $a > 1$, there is a positive number d such that $a^d = 2$. We show that d is the doubling time. If the population is P at time t, then at time $t + d$, the population is

$$P_0 a^{t+d} = P_0 a^t a^d = (P_0 a^t)(2) = 2P.$$

So, no matter what the initial quantity and no matter what the initial time, the size of the population is doubled d time units later.

Example 4   The release of chlorofluorocarbons used in air conditioners and household sprays (hair spray, shaving cream, etc.) destroys the ozone in the upper atmosphere. The quantity of ozone, Q, is decaying exponentially at a continuous rate of 0.25% per year. What is the half-life of ozone? In other words, at this rate, how long will it take for half the ozone to disappear?

Solution   If Q_0 is the initial quantity of ozone and t is in years, then

$$Q = Q_0 e^{-0.0025t}.$$

We want to find the value of t making $Q = Q_0/2$, so

$$\frac{Q_0}{2} = Q_0 e^{-0.0025t}.$$

Dividing both sides by Q_0 and taking natural logs gives

$$\ln\left(\frac{1}{2}\right) = -0.0025t,$$

so

$$t = \frac{\ln(1/2)}{-0.0025} = 277 \text{ years.}$$

Half the present atmospheric ozone will be gone in 277 years.

Financial Applications: Compound Interest

We deposit $100 in a bank paying interest at a rate of 8% per year. How much is in the account at the end of the year? This depends on how often the interest is compounded. If the interest is paid into the account *annually*, that is, only at the end of the year, then the balance in the account after one year is $108. However, if the interest is paid twice a year, then 4% is paid at the end of the first six months and 4% at the end of the year. Slightly more money is earned this way, since the interest paid early in the year will earn interest during the rest of the year. This effect is called *compounding*.

In general, the more often interest is compounded, the more money is earned (although the increase may not be large). What happens if interest is compounded more frequently, such as every minute or every second? The benefit of increasing the frequency of compounding becomes negligible beyond a certain point. When that point is reached, we find the balance using the number e and we say that the interest per year is *compounded continuously*. If we have deposited $100 in an account paying 8% interest per year compounded continuously, the balance after one year is $100e^{0.08} = 108.33. Compounding is discussed further in Appendix B. In general:

> An amount P_0 is deposited in an account paying interest at a rate of r per year. Let P be the balance in the account after t years.
> - If interest is compounded annually, then $P = P_0(1 + r)^t$.
> - If interest is compounded continuously, then $P = P_0 e^{rt}$, where $e = 2.71828....$

We write P_0 for the initial deposit because it is the value of P when $t = 0$. Note that for a 7% interest rate, $r = 0.07$. If a rate is continuous, we will say so explicitly.

Example 5   A bank advertises an interest rate of 8% per year. If you deposit $5000, how much is in the account 3 years later if the interest is compounded (a) Annually?   (b) Continuously?

Solution   (a) For annual compounding, $P = P_0(1 + r)^t = 5000(1.08)^3 = 6298.56.
(b) For continuous compounding, $P = P_0 e^{rt} = 5000 e^{0.08 \cdot 3} = 6356.25. As expected, the amount in the account 3 years later is larger if the interest is compounded continuously ($6356.25) than if the interest is compounded annually ($6298.56).

Example 6   If $10,000 is deposited in an account paying interest at a rate of 5% per year, compounded continuously, how long does it take for the balance in the account to reach $15,000?

Solution   Since interest is compounded continuously, we use $P = P_0 e^{rt}$ with $r = 0.05$ and $P_0 = 10,000$. We want to find the value of t for which $P = 15,000$. The equation is

$$15,000 = 10,000 e^{0.05t}.$$

Now divide both sides by 10,000, then take logarithms and solve for t:

$$1.5 = e^{0.05t}$$
$$\ln(1.5) = \ln(e^{0.05t})$$
$$\ln(1.5) = 0.05t$$
$$t = \frac{\ln(1.5)}{0.05} = 8.1093.$$

It takes about 8.1 years for the balance in the account to reach $15,000.

Example 7   (a) Calculate the doubling time, D, for interest rates of 2%, 3%, 4%, and 5% per year, compounded annually.

(b) Use your answers to part (a) to check that an interest rate of i% gives a doubling time approximated for small values of i by

$$D \approx \frac{70}{i} \text{ years.}$$

This is the "Rule of 70" used by bankers: To compute the approximate doubling time of an investment, divide 70 by the percent annual interest rate.

Solution   (a) We find the doubling time for an interest rate of 2% per year using the formula $P = P_0(1.02)^t$ with t in years. To find the value of t for which $P = 2P_0$, we solve

$$2P_0 = P_0(1.02)^t$$
$$2 = (1.02)^t$$
$$\ln 2 = \ln(1.02)^t$$
$$\ln 2 = t \ln(1.02) \qquad \text{(using the third property of logarithms)}$$
$$t = \frac{\ln 2}{\ln 1.02} = 35.003 \text{ years.}$$

With an annual interest rate of 2%, it takes about 35 years for an investment to double in value. Similarly, we find the doubling times for 3%, 4%, and 5% in Table 1.33.

Table 1.33   *Doubling time as a function of interest rate*

i (% annual growth rate)	2	3	4	5
D (doubling time in years)	35.003	23.450	17.673	14.207

(b) We compute $(70/i)$ for $i = 2, 3, 4, 5$. The results are shown in Table 1.34.

Table 1.34   *Approximate doubling time as a function of interest rate: Rule of 70*

i (% annual growth rate)	2	3	4	5
$(70/i)$ (Approximate doubling time in years)	35.000	23.333	17.500	14.000

Comparing Tables 1.33 and 1.34, we see that the quantity $(70/i)$ gives a reasonably accurate approximation to the doubling time, D, for the small interest rates we considered.

Present and Future Value

Many business deals involve payments in the future. For example, when a car is bought on credit, payments are made over a period of time. Being paid $100 in the future is clearly worse than being paid $100 today for many reasons. If we are given the money today, we can do something else with it—for example, put it in the bank, invest it somewhere, or spend it. Thus, even without considering inflation, if we are to accept payment in the future, we would expect to be paid more to compensate for this loss of potential earnings.[57] The question we consider now is, how much more?

To simplify matters, we consider only what we would lose by not earning interest; we do not consider the effect of inflation. Let's look at some specific numbers. Suppose we deposit $100 in an account that earns 7% interest per year compounded annually, so that in a year's time we have $107. Thus, $100 today is worth $107 a year from now. We say that the $107 is the *future value* of the $100, and that the $100 is the *present value* of the $107. In general, we say the following:

[57] This is referred to as the time value of money.

- The **future value**, B, of a payment, P, is the amount to which the P would have grown if deposited today in an interest-bearing bank account.
- The **present value**, P, of a future payment, B, is the amount that would have to be deposited in a bank account today to produce exactly B in the account at the relevant time in the future.

Due to the interest earned, the future value is larger than the present value. The relation between the present and future values depends on the interest rate, as follows.

Suppose B is the *future value* of P and P is the *present value* of B.
If interest is compounded annually at a rate r for t years, then

$$B = P(1 + r)^t, \quad \text{or equivalently,} \quad P = \frac{B}{(1 + r)^t}.$$

If interest is compounded continuously at a rate r for t years, then

$$B = Pe^{rt}, \quad \text{or equivalently,} \quad P = \frac{B}{e^{rt}} = Be^{-rt}.$$

The rate, r, is sometimes called the *discount rate*. The present value is often denoted by PV and the future value by FV.

Example 8   You win the lottery and are offered the choice between \$1 million in four yearly installments of \$250,000 each, starting now, and a lump-sum payment of \$920,000 now. Assuming a 6% interest rate per year, compounded continuously, and ignoring taxes, which should you choose?

Solution   We assume that you pick the option with the largest present value. The first of the four \$250,000 payments is made now, so

$$\text{Present value of first payment} = \$250,000.$$

The second payment is made one year from now and so

$$\text{Present value of second payment} = \$250,000e^{-0.06(1)}.$$

Calculating the present value of the third and fourth payments similarly, we find:

$$\begin{aligned}
\text{Total present value} &= \$250,000 + \$250,000e^{-0.06(1)} + \$250,000e^{-0.06(2)} + \$250,000e^{-0.06(3)} \\
&= \$250,000 + \$235,441 + \$221,730 + \$208,818 \\
&= \$915,989.
\end{aligned}$$

Since the present value of the four payments is less than \$920,000, you are better off taking the \$920,000 now.

Alternatively, we can compare the future values of the two pay schemes. We calculate the future value of both schemes three years from now, on the date of the last \$250,000 payment. At that time,

$$\text{Future value of the lump-sum payment} = \$920,000e^{0.06(3)} = \$1,101,440.$$

The future value of the first \$250,000 payment is $\$250,000e^{0.06(3)}$. Calculating the future value of the other payments similarly, we find:

$$\begin{aligned}
\text{Total future value} &= \$250,000e^{0.06(3)} + \$250,000e^{0.06(2)} + \$250,000e^{0.06(1)} + \$250,000 \\
&= \$299,304 + \$281,874 + \$265,459 + \$250,000 \\
&= \$1,096,637.
\end{aligned}$$

As we expect, the future value of the $920,000 payment is greater, so you are better off taking the $920,000 now.[58]

Problems for Section 1.7

1. World wind energy generating[59] capacity, W, was 18,000 megawatts in 2000 and has been increasing at a continuous rate of approximately 27% per year. Assume this rate continues.

   (a) Give a formula for W, in megawatts, as a function of time, t, in years since 2000.
   (b) When is wind capacity predicted to pass 250,000 megawatts?

2. The half-life of nicotine in the blood is 2 hours. A person absorbs 0.4 mg of nicotine by smoking a cigarette. Fill in the following table with the amount of nicotine remaining in the blood after t hours. Estimate the length of time until the amount of nicotine is reduced to 0.04 mg.

t (hours)	0	2	4	6	8	10
Nicotine (mg)	0.4					

3. If you deposit $10,000 in an account earning interest at an 8% annual rate compounded continuously, how much money is in the account after five years?

4. If you need $20,000 in your bank account in 6 years, how much must be deposited now? The interest rate is 10%, compounded continuously.

5. If a bank pays 6% per year interest compounded continuously, how long does it take for the balance in an account to double?

6. Suppose $1000 is invested in an account paying interest at a rate of 5.5% per year. How much is in the account after 8 years if the interest is compounded

   (a) Annually?   **(b)** Continuously?

7. Find the doubling time of a quantity that is increasing by 7% per year.

8. If $12,000 is deposited in an account paying 8% interest per year, compounded continuously, how long will it take for the balance to reach $20,000?

9. You want to invest money for your child's education in a certificate of deposit (CD). You want it to be worth $12,000 in 10 years. How much should you invest if the CD pays interest at a 9% annual rate compounded
   (a) Annually?   **(b)** Continuously?

10. From October 2002 to October 2006 the number $N(t)$ of Wikipedia articles was approximated by $N(t) =$

$N_0 e^{t/500}$, where t is the number of days after October 1, 2002. Find the doubling time for the number of Wikipedia articles during this period.

11. A cup of coffee contains 100 mg of caffeine, which leaves the body at a continuous rate of 17% per hour.

   (a) Write a formula for the amount, A mg, of caffeine in the body t hours after drinking a cup of coffee.
   (b) Graph the function from part (a). Use the graph to estimate the half-life of caffeine.
   (c) Use logarithms to find the half-life of caffeine.

12. A population, currently 200, is growing at 5% per year.

   (a) Write a formula for the population, P, as a function of time, t, years in the future.
   (b) Graph P against t.
   (c) Estimate the population 10 years from now.
   (d) Use the graph to estimate the doubling time of the population.

13. Figure 1.67 shows the balances in two bank accounts. Both accounts pay the same interest rate, but one compounds continuously and the other compounds annually. Which curve corresponds to which compounding method? What is the initial deposit in each case?

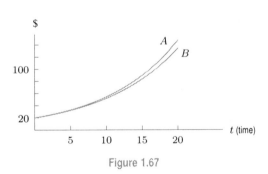

Figure 1.67

14. The exponential function $y(x) = Ce^{\alpha x}$ satisfies the conditions $y(0) = 2$ and $y(1) = 1$. Find the constants C and α. What is $y(2)$?

---

[58] If you read the fine print, you will find that many lotteries do not make their payments right away, but often spread them out, sometimes far into the future. This is to reduce the present value of the payments made, so that the value of the prizes is less than it might first appear!

[59] World Wind Energy Association, www.wwindea.org, accessed September 6, 2009.

15. Air pressure, P, decreases exponentially with the height, h, in meters above sea level:

$$P = P_0 e^{-0.00012h}$$

where P_0 is the air pressure at sea level.

(a) At the top of Mount McKinley, height 6194 meters (about 20,320 feet), what is the air pressure, as a percent of the pressure at sea level?

(b) The maximum cruising altitude of an ordinary commercial jet is around 12,000 meters (about 39,000 feet). At that height, what is the air pressure, as a percent of the sea level value?

16. The antidepressant fluoxetine (or Prozac) has a half-life of about 3 days. What percentage of a dose remains in the body after one day? After one week?

17. A firm decides to increase output at a constant relative rate from its current level of 20,000 to 30,000 units during the next five years. Calculate the annual percent rate of increase required to achieve this growth.

18. The half-life of a radioactive substance is 12 days. There are 10.32 grams initially.

(a) Write an equation for the amount, A, of the substance as a function of time.

(b) When is the substance reduced to 1 gram?

19. One of the main contaminants of a nuclear accident, such as that at Chernobyl, is strontium-90, which decays exponentially at a rate of approximately 2.5% per year.

(a) Write the percent of strontium-90 remaining, P, as a function of years, t, since the nuclear accident. [Hint: 100% of the contaminant remains at $t = 0$.]

(b) Graph P against t.

(c) Estimate the half-life of strontium-90.

(d) After the Chernobyl disaster, it was predicted that the region would not be safe for human habitation for 100 years. Estimate the percent of original strontium-90 remaining at this time.

20. The number of people living with HIV infections increased worldwide approximately exponentially from 2.5 million in 1985 to 37.8 million in 2003.[60] (HIV is the virus that causes AIDS.)

(a) Give a formula for the number of HIV infections, H, (in millions) as a function of years, t, since 1985. Use the form $H = H_0 e^{kt}$. Graph this function.

(b) What was the yearly continuous percent change in the number of HIV infections between 1985 and 2003?

21. (a) Figure 1.68 shows exponential growth. Starting at $t = 0$, estimate the time for the population to double.

(b) Repeat part (a), but this time start at $t = 3$.

(c) Pick any other value of t for the starting point, and notice that the doubling time is the same no matter where you start.

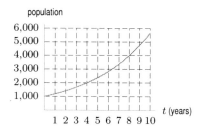

Figure 1.68

22. An exponentially growing animal population numbers 500 at time $t = 0$; two years later, it is 1500. Find a formula for the size of the population in t years and find the size of the population at $t = 5$.

23. If the quantity of a substance decreases by 4% in 10 hours, find its half-life.

24. Pregnant women metabolize some drugs at a slower rate than the rest of the population. The half-life of caffeine is about 4 hours for most people. In pregnant women, it is 10 hours.[61] (This is important because caffeine, like all psychoactive drugs, crosses the placenta to the fetus.) If a pregnant woman and her husband each have a cup of coffee containing 100 mg of caffeine at 8 am, how much caffeine does each have left in the body at 10 pm?

25. The half-life of radioactive strontium-90 is 29 years. In 1960, radioactive strontium-90 was released into the atmosphere during testing of nuclear weapons, and was absorbed into people's bones. How many years does it take until only 10% of the original amount absorbed remains?

26. In 1923, koalas were introduced on Kangaroo Island off the coast of Australia. In 1996, the population was 5000. By 2005, the population had grown to 27,000, prompting a debate on how to control their growth and avoid koalas dying of starvation.[62] Assuming exponential growth, find the (continuous) rate of growth of the koala population between 1996 and 2005. Find a formula for the population as a function of the number of years since 1996, and estimate the population in the year 2020.

27. The total world marine catch in 1950 was 17 million tons and in 2001 was 99 million tons.[63] If the marine catch is increasing exponentially, find the (continuous) rate of increase. Use it to predict the total world marine catch in the year 2020.

[60] *The World Almanac and Book of Facts 2005*, p. 89 (New York).
[61] From Robert M. Julien, *A Primer of Drug Action*, 7th ed., p. 159 (New York: W. H. Freeman, 1995).
[62] news.yahoo.com/s/afp/australiaanimalskoalas, accessed June 1, 2005.
[63] *The World Almanac and Book of Facts 2005*, p. 143 (New York).

28. **(a)** Use the Rule of 70 to predict the doubling time of an investment which is earning 8% interest per year.
(b) Find the doubling time exactly, and compare your answer to part (a).

29. The island of Manhattan was sold for $24 in 1626. Suppose the money had been invested in an account which compounded interest continuously.

(a) How much money would be in the account in the year 2005 if the yearly interest rate was

  (i) 5%?   (ii) 7%?

(b) If the yearly interest rate was 6%, in what year would the account be worth one million dollars?

30. In 2004, the world's population was 6.4 billion, and the population was projected to reach 8.5 billion by the year 2030. What annual growth rate is projected?

31. A picture supposedly painted by Vermeer (1632–1675) contains 99.5% of its carbon-14 (half-life 5730 years). From this information decide whether the picture is a fake. Explain your reasoning.

32. Find the future value in 8 years of a $10,000 payment today, if the interest rate is 3% per year compounded continuously.

33. Find the future value in 15 years of a $20,000 payment today, if the interest rate is 3.8% per year compounded continuously.

34. Find the present value of an $8000 payment to be made in 5 years. The interest rate is 4% per year compounded continuously.

35. Find the present value of a $20,000 payment to be made in 10 years. Assume an interest rate of 3.2% per year compounded continuously.

36. Interest is compounded annually. Consider the following choices of payments to you:

Choice 1: $1500 now and $3000 one year from now

Choice 2: $1900 now and $2500 one year from now

(a) If the interest rate on savings were 5% per year, which would you prefer?
(b) Is there an interest rate that would lead you to make a different choice? Explain.

37. A person is to be paid $2000 for work done over a year. Three payment options are being considered. Option 1 is to pay the $2000 in full now. Option 2 is to pay $1000 now and $1000 in a year. Option 3 is to pay the full $2000 in a year. Assume an annual interest rate of 5% a year, compounded continuously.

(a) Without doing any calculations, which option is the best option financially for the worker? Explain.
(b) Find the future value, in one year's time, of all three options.
(c) Find the present value of all three options.

38. A business associate who owes you $3000 offers to pay you $2800 now, or else pay you three yearly installments of $1000 each, with the first installment paid now. If you use only financial reasons to make your decision, which option should you choose? Justify your answer, assuming a 6% interest rate per year, compounded continuously.

39. Big Tree McGee is negotiating his rookie contract with a professional basketball team. They have agreed to a three-year deal which will pay Big Tree a fixed amount at the end of each of the three years, plus a signing bonus at the beginning of his first year. They are still haggling about the amounts and Big Tree must decide between a big signing bonus and fixed payments per year, or a smaller bonus with payments increasing each year. The two options are summarized in the table. All values are payments in millions of dollars.

	Signing bonus	Year 1	Year 2	Year 3
Option #1	6.0	2.0	2.0	2.0
Option #2	1.0	2.0	4.0	6.0

(a) Big Tree decides to invest all income in stock funds which he expects to grow at a rate of 10% per year, compounded continuously. He would like to choose the contract option which gives him the greater future value at the end of the three years when the last payment is made. Which option should he choose?
(b) Calculate the present value of each contract offer.

40. A company is considering whether to buy a new machine, which costs $97,000. The cash flows (adjusted for taxes and depreciation) that would be generated by the new machine are given in the following table:

Year	1	2	3	4
Cash flow	$50,000	$40,000	$25,000	$20,000

(a) Find the total present value of the cash flows. Treat each year's cash flow as a lump sum at the end of the year and use an interest rate of 7.5% per year, compounded annually.
(b) Based on a comparison of the cost of the machine and the present value of the cash flows, would you recommend purchasing the machine?

41. You win $38,000 in the state lottery to be paid in two installments—$19,000 now and $19,000 one year from now. A friend offers you $36,000 in return for your two lottery payments. Instead of accepting your friend's offer, you take out a one-year loan at an interest rate of 8.25% per year, compounded annually. The loan will be paid back by a single payment of $19,000 (your second lottery check) at the end of the year. Which is better, your friend's offer or the loan?

42. You are considering whether to buy or lease a machine whose purchase price is $12,000. Taxes on the machine will be $580 due in one year, $464 due in two years, and $290 due in three years. If you buy the machine, you expect to be able to sell it after three years for $5,000. If you lease the machine for three years, you make an initial payment of $2650 and then three payments of $2650 at the end of each of the next three years. The leasing company will pay the taxes. The interest rate is 7.75% per year, compounded annually. Should you buy or lease the machine? Explain.

43. You are buying a car that comes with a one-year warranty and are considering whether to purchase an extended warranty for $375. The extended warranty covers the two years immediately after the one-year warranty expires. You estimate that the yearly expenses that would have been covered by the extended warranty are $150 at the end of the first year of the extension and $250 at the end of the second year of the extension. The interest rate is 5% per year, compounded annually. Should you buy the extended warranty? Explain.

1.8 NEW FUNCTIONS FROM OLD

We have studied linear and exponential functions, and the logarithm function. In this section, we learn how to create new functions by composing, stretching, and shifting functions we already know.

Composite Functions

A drop of water falls onto a paper towel. The area, A of the circular damp spot is a function of r, its radius, which is a function of time, t. We know $A = f(r) = \pi r^2$; suppose $r = g(t) = t + 1$. By substitution, we express A as a function of t:

$$A = f(g(t)) = \pi(t + 1)^2.$$

The function $f(g(t))$ is a "function of a function," or a *composite function*, in which there is an *inside function* and an *outside function*. To find $f(g(2))$, we first add one ($g(2) = 2 + 1 = 3$) and then square and multiply by π. We have

$$f(g(2)) = \pi(2 + 1)^2 \quad = \quad \pi 3^2 \quad = \quad 9\pi.$$

First calculation   Second calculation

The inside function is $t + 1$ and the outside function is squaring and multiplying by π. In general, the inside function represents the calculation that is done first and the outside function represents the calculation done second.

Example 1 If $f(t) = t^2$ and $g(t) = t + 2$, find

(a) $f(t + 1)$ (b) $f(t) + 3$ (c) $f(t + h)$ (d) $f(g(t))$ (e) $g(f(t))$

Solution (a) Since $t + 1$ is the inside function, $f(t + 1) = (t + 1)^2$.
(b) Here 3 is added to $f(t)$, so $f(t) + 3 = t^2 + 3$.
(c) Since $t + h$ is the inside function, $f(t + h) = (t + h)^2$.
(d) Since $g(t) = t + 2$, substituting $t + 2$ into f gives $f(g(t)) = f(t + 2) = (t + 2)^2$.
(e) Since $f(t) = t^2$, substituting t^2 into g gives $g(f(t)) = g(t^2) = t^2 + 2$.

Example 2 If $f(x) = e^x$ and $g(x) = 5x + 1$, find (a) $f(g(x))$ (b) $g(f(x))$

Solution (a) Substituting $g(x) = 5x + 1$ into f gives $f(g(x)) = f(5x + 1) = e^{5x+1}$.
(b) Substituting $f(x) = e^x$ into g gives $g(f(x)) = g(e^x) = 5e^x + 1$.

Example 3 Using the following table, find $g(f(0))$, $f(g(0))$, $f(g(1))$, and $g(f(1))$.

x	0	1	2	3
$f(x)$	3	1	−1	−3
$g(x)$	0	2	4	6

Solution   To find $g(f(0))$, we first find $f(0) = 3$ from the table. Then we have $g(f(0)) = g(3) = 6$. For $f(g(0))$, we must find $g(0)$ first. Since $g(0) = 0$, we have $f(g(0)) = f(0) = 3$. Similar reasoning leads to $f(g(1)) = f(2) = -1$ and $g(f(1)) = g(1) = 2$.

We can write a composite function using a new variable u to represent the value of the inside function. For example

$$y = (t+1)^4 \quad \text{is the same as} \quad y = u^4 \quad \text{with} \quad u = t+1.$$

Other expressions for u, such as $u = (t+1)^2$, with $y = u^2$, are also possible.

Example 4   Use a new variable u for the inside function to express each of the following as a composite function:
(a) $y = \ln(3t)$ (b) $w = 5(2r+3)^2$ (c) $P = e^{-0.03t}$

Solution   (a) We take the inside function to be $3t$, so $y = \ln u$ with $u = 3t$.
(b) We take the inside function to be $2r + 3$, so $w = 5u^2$ with $u = 2r + 3$.
(c) We take the inside function to be $-0.03t$, so $P = e^u$ with $u = -0.03t$.

Stretches of Graphs

If the demand function is linear, the graph of a possible revenue function $R = f(p)$ is in Figure 1.69. What does the graph of $R = 3f(p)$ look like? The factor 3 in the function $R = 3f(p)$ stretches each $f(p)$ revenue value by multiplying it by 3. See Figure 1.70. If c is positive, the graph of $R = cf(p)$ is the graph of $R = f(p)$ stretched or shrunk vertically by c units. If c is negative, the function no longer makes sense as a revenue function, but we can still draw the graph. What does the graph of $R = -2f(p)$ look like? The factor -2 in the function $R = -2f(p)$ stretches $f(p)$ by multiplying by 2 and reflecting it about the x-axis. See Figure 1.70.

> Multiplying a function by a constant, c, stretches the graph vertically (if $c > 1$) or shrinks the graph vertically (if $0 < c < 1$). A negative sign (if $c < 0$) reflects the graph about the x-axis, in addition to shrinking or stretching.

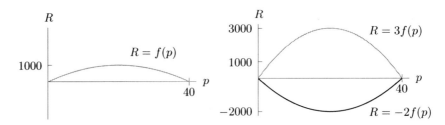

Figure 1.69: Graph of $f(p)$ Figure 1.70: Multiples of the function $f(p)$

Shifted Graphs

Consider the function $y = x^2 + 4$. The y-coordinates for this function are exactly 4 units larger than the corresponding y-coordinates of the function $y = x^2$. So the graph of $y = x^2 + 4$ is obtained from the graph of $y = x^2$ by adding 4 to the y-coordinate of each point; that is, by moving the graph of $y = x^2$ up 4 units. (See Figure 1.71.)

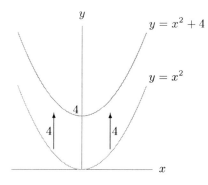

Figure 1.71: Vertical shift: Graphs of $y = x^2$ and $y = x^2 + 4$

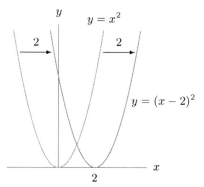

Figure 1.72: Horizontal shift: Graphs of $y = x^2$ and $y = (x - 2)^2$

A graph can also be shifted to the left or to the right. In Figure 1.72, we see that the graph of $y = (x - 2)^2$ is the graph of $y = x^2$ shifted to the right 2 units. In general,

> - The graph of $y = f(x) + k$ is the graph of $y = f(x)$ moved up k units (down if k is negative).
> - The graph of $y = f(x - k)$ is the graph of $y = f(x)$ moved to the right k units (to the left if k is negative).

Example 5 (a) A cost function, $C(q)$, for a company is shown in Figure 1.73. The fixed cost increases by $1000. Sketch a graph of the new cost function.

(b) A supply curve, S, for a product is given in Figure 1.74. A new factory opens and produces 100 units of the product no matter what the price. Sketch a graph of the new supply curve.

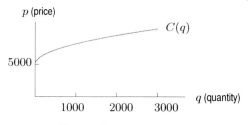

Figure 1.73: A cost function

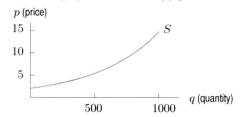

Figure 1.74: A supply function

Solution (a) For each quantity, the new cost is $1000 more than the old cost. The new cost function is $C(q) + 1000$, whose graph is the graph of $C(q)$ shifted vertically up 1000 units. (See Figure 1.75.)

(b) To see the effect of the new factory, look at an example. At a price of 10 dollars, approximately 800 units are currently produced. With the new factory, this amount increases by 100 units, so the new amount produced is 900 units. At each price, the quantity produced increases by 100, so the new supply curve is S shifted horizontally to the right by 100 units. (See Figure 1.76.)

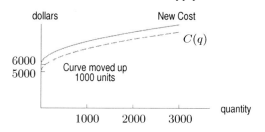

Figure 1.75: New cost function (original curve dashed)

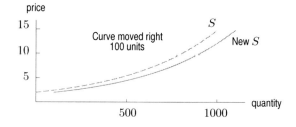

Figure 1.76: New supply curve (original curve dashed)

61

Problems for Section 1.8

1. For $g(x) = x^2 + 2x + 3$, find and simplify:

   (a) $g(2 + h)$ (b) $g(2)$

   (c) $g(2 + h) - g(2)$

2. If $f(x) = x^2 + 1$, find and simplify:

   (a) $f(t + 1)$ (b) $f(t^2 + 1)$ (c) $f(2)$

   (d) $2f(t)$ (e) $[f(t)]^2 + 1$

For the functions f and g in Problems 3–6, find

(a) $f(g(1))$ (b) $g(f(1))$ (c) $f(g(x))$

(d) $g(f(x))$ (e) $f(t)g(t)$

3. $f(x) = x^2, g(x) = x + 1$

4. $f(x) = \sqrt{x + 4}, g(x) = x^2$

5. $f(x) = e^x, g(x) = x^2$

6. $f(x) = 1/x, g(x) = 3x + 4$

7. Let $f(x) = x^2$ and $g(x) = 3x - 1$. Find the following:

   (a) $f(2) + g(2)$ (b) $f(2) \cdot g(2)$

   (c) $f(g(2))$ (d) $g(f(2))$

In Problems 8–10, find the following:

(a) $f(g(x))$ (b) $g(f(x))$ (c) $f(f(x))$

8. $f(x) = 2x^2$ and $g(x) = x + 3$

9. $f(x) = 2x + 3$ and $g(x) = 5x^2$

10. $f(x) = x^2 + 1$ and $g(x) = \ln x$

11. Use Table 1.35 to find:

   (a) $f(g(0))$ (b) $f(g(1))$ (c) $f(g(2))$

   (d) $g(f(2))$ (e) $g(f(3))$

Table 1.35

x	0	1	2	3	4	5
$f(x)$	10	6	3	4	7	11
$g(x)$	2	3	5	8	12	15

12. Make a table of values for each of the following functions using Table 1.35:

   (a) $f(x) + 3$ (b) $f(x - 2)$ (c) $5g(x)$

   (d) $-f(x) + 2$ (e) $g(x - 3)$ (f) $f(x) + g(x)$

13. Use the variable u for the inside function to express each of the following as a composite function:

   (a) $y = 2^{3x - 1}$ (b) $P = \sqrt{5t^2 + 10}$

   (c) $w = 2\ln(3r + 4)$

14. Use the variable u for the inside function to express each of the following as a composite function:

   (a) $y = (5t^2 - 2)^6$ (b) $P = 12e^{-0.6t}$

   (c) $C = 12\ln(q^3 + 1)$

Simplify the quantities in Problems 15–18 using $m(z) = z^2$.

15. $m(z + 1) - m(z)$ 16. $m(z + h) - m(z)$

17. $m(z) - m(z - h)$ 18. $m(z + h) - m(z - h)$

For Problems 19–21, use the graphs in Figure 1.77.

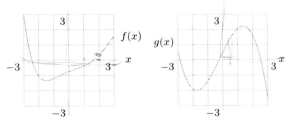

Figure 1.77

19. Estimate $f(g(1))$.   20. Estimate $g(f(2))$.

21. Estimate $f(f(1))$.

22. Using Table 1.36, create a table of values for $f(g(x))$ and for $g(f(x))$.

Table 1.36

x	-3	-2	-1	0	1	2	3
$f(x)$	0	1	2	3	2	1	0
$g(x)$	3	2	2	0	-2	-2	-3

23. A tree of height y meters has, on average, B branches, where $B = y - 1$. Each branch has, on average, n leaves where $n = 2B^2 - B$. Find the average number of leaves of a tree as a function of height.

In Problems 24–27, use Figure 1.78 to estimate the function value or explain why it cannot be done.

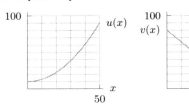

Figure 1.78

24. $u(v(10))$ 25. $u(v(40))$

26. $v(u(10))$ 27. $v(u(40))$

28. The Heaviside step function, H, is graphed in Figure 1.79. Graph the following functions.

   (a) $2H(x)$ **(b)** $H(x) + 1$ **(c)** $H(x + 1)$
   (d) $-H(x)$ **(e)** $H(-x)$

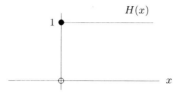

Figure 1.79

For the functions f in Exercises 29–31, graph:

  (a) $f(x + 2)$ **(b)** $f(x - 1)$ **(c)** $f(x) - 4$
  (d) $f(x + 1) + 3$ **(e)** $3f(x)$ **(f)** $-f(x) + 1$

29. 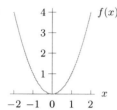   **30.**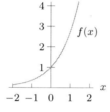

31.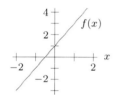

Graph the functions in Problems 32–37 using Figure 1.80.

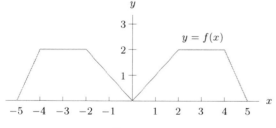

Figure 1.80

32. $y = f(x) + 2$ **33.** $y = 2f(x)$

34. $y = f(x - 1)$ **35.** $y = -3f(x)$

36. $y = 2f(x) - 1$ **37.** $y = 2 - f(x)$

For the functions $f(x)$ in Problems 38–41, graph:

  (a) $y = f(x) + 2$ **(b)** $y = f(x - 1)$
  (c) $y = 3f(x)$ **(d)** $y = -f(x)$

38. 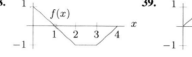   **39.**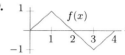

40. 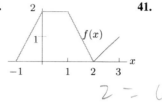   **41.**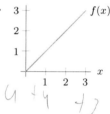

42. Morphine, a pain-relieving drug, is administered to a patient intravenously starting at 8 am. The drug saturation curve $Q = f(t)$ in Figure 1.81 gives the quantity, Q, of morphine in the blood t hours after 8 am.

  (a) Draw the drug saturation curve $Q = g(t)$ if the IV line is started at noon instead of 8 am.
  (b) Is g one of the following transformations of f: vertical shift, vertical stretch, horizontal shift, horizontal stretch? If so, which?
  (c) Write $g(t)$ in terms of the function f.

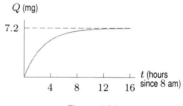

Figure 1.81

43. (a) Write an equation for a graph obtained by vertically stretching the graph of $y = x^2$ by a factor of 2, followed by a vertical upward shift of 1 unit. Sketch it.
  (b) What is the equation if the order of the transformations (stretching and shifting) in part (a) is interchanged?
  (c) Are the two graphs the same? Explain the effect of reversing the order of transformations.

1.9 PROPORTIONALITY AND POWER FUNCTIONS

Proportionality

A common functional relationship occurs when one quantity is *proportional* to another. For example, if apples are $1.40 a pound, we say the price you pay, p dollars, is proportional to the weight you buy, w pounds, because

$$p = f(w) = 1.40w.$$

As another example, the area, A, of a circle is proportional to the square of the radius, r:

$$A = f(r) = \pi r^2.$$

> We say y is (directly) **proportional** to x if there is a nonzero constant k such that
>
> $$y = kx.$$
>
> This k is called the constant of proportionality.

We also say that one quantity is *inversely proportional* to another if one is proportional to the reciprocal of the other. For example, the speed, v, at which you make a 50-mile trip is inversely proportional to the time, t, taken, because v is proportional to $1/t$:

$$v = 50\left(\frac{1}{t}\right) = \frac{50}{t}.$$

Notice that if y is directly proportional to x, then the magnitude of one variable increases (decreases) when the magnitude of the other increases (decreases). If, however, y is inversely proportional to x, then the magnitude of one variable increases when the magnitude of the other decreases.

Example 1 The heart mass of a mammal is proportional to its body mass.[64]

(a) Write a formula for heart mass, H, as a function of body mass, B.

(b) A human with a body mass of 70 kilograms has a heart mass of 0.42 kilograms. Use this information to find the constant of proportionality.

(c) Estimate the heart mass of a horse with a body mass of 650 kg.

Solution (a) Since H is proportional to B, for some constant k, we have

$$H = kB.$$

(b) We use the fact that $H = 0.42$ when $B = 70$ to solve for k:

$$H = kB$$
$$0.42 = k(70)$$
$$k = \frac{0.42}{70} = 0.006.$$

(c) Since $k = 0.006$, we have $H = 0.006B$, so the heart mass of the horse is given by

$$H = 0.006(650) = 3.9 \text{ kilograms.}$$

---

[64]K. Schmidt-Nielson: *Scaling—Why is Animal Size So Important?* (Cambridge: CUP, 1984).

Example 2   The period of a pendulum, T, is the amount of time required for the pendulum to make one complete swing. For small swings, the period, T, is approximately proportional to the square root of l, the pendulum's length. So

$$T = k\sqrt{l} \quad \text{where } k \text{ is a constant.}$$

Notice that T is not directly proportional to l, but T is proportional to $\sqrt{l}$.

Example 3   An object's weight, w, is inversely proportional to the square of its distance, r, from the earth's center. So, for some constant k,

$$w = \frac{k}{r^2}.$$

Here w is not inversely proportional to r, but to r^2.

Power Functions

In each of the previous examples, one quantity is proportional to the power of another quantity. We make the following definition:

> We say that $Q(x)$ is a **power function** of x if $Q(x)$ is proportional to a constant power of x. If k is the constant of proportionality, and if p is the power, then
>
> $$Q(x) = k \cdot x^p.$$

For example, the function $H = 0.006B$ is a power function with $p = 1$. The function $T = k\sqrt{l} = kl^{1/2}$ is a power function with $p = 1/2$, and the function $w = k/r^2 = kr^{-2}$ is a power function with $p = -2$.

Example 4   Which of the following are power functions? For those which are, write the function in the form $y = kx^p$, and give the coefficient k and the exponent p.

(a) $y = \dfrac{5}{x^3}$ (b) $y = \dfrac{2}{3x}$ (c) $y = \dfrac{5x^2}{2}$

(d) $y = 5 \cdot 2^x$ (e) $y = 3\sqrt{x}$ (f) $y = (3x^2)^3$

Solution

(a) Since $y = 5x^{-3}$, this is a power function with $k = 5$ and $p = -3$.
(b) Since $y = (2/3)x^{-1}$, this is a power function with $k = 2/3$ and $p = -1$.
(c) Since $y = (5/2)x^2$, this is a power function with $k = 5/2 = 2.5$ and $p = 2$.
(d) This is not a power function. It is an exponential function.
(e) Since $y = 3x^{1/2}$, this is a power function with $k = 3$ and $p = 1/2$.
(f) Since $y = 3^3 \cdot (x^2)^3 = 27x^6$, this is a power function with $k = 27$ and $p = 6$.

Graphs of Power Functions

The graph of $y = x^2$ is shown in Figure 1.82. It is decreasing for negative x and increasing for positive x. Notice that it is bending upward, or concave up, for all x. The graph of $y = x^3$ is shown in Figure 1.83. Notice that it is bending downward, or concave down for negative x and bending upward, or concave up for positive x. The graph of $y = \sqrt{x} = x^{1/2}$ is shown in Figure 1.84. Notice that the graph is increasing and concave down.

Since x^2 increases without bound as x increases, we often say that it tends to infinity as x approaches infinity, which we write in symbols as

$$x^2 \to \infty \quad \text{as} \quad x \to \infty.$$

Since x^3 decreases without bound as x decreases, we write

$$x^3 \to -\infty \quad \text{as} \quad x \to -\infty.$$

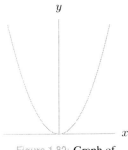

Figure 1.82: Graph of
$y = x^2$

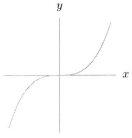

Figure 1.83: Graph of
$y = x^3$

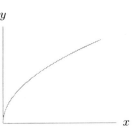

Figure 1.84: Graph of
$y = x^{1/2}$

Example 5   If N is the average number of species found on an island and A is the area of the island, observations have shown[65] that N is approximately proportional to the cube root of A. Write a formula for N as a function of A and describe the shape of the graph of this function.

Solution   For some positive constant k, we have

$$N = k \sqrt[3]{A} = kA^{1/3}.$$

It turns out that the value of k depends on the region of the world in which the island is found. The graph of N against A (for $A > 0$) has a shape similar to the graph in Figure 1.84. It is increasing and concave down. Thus, larger islands have more species on them (as we would expect), but the increase slows as the island gets larger.

The function $y = x^0 = 1$ has a graph that is a horizontal line. For negative powers, rewriting

$$y = x^{-1} = \frac{1}{x} \quad \text{and} \quad y = x^{-2} = \frac{1}{x^2}$$

makes it clear that as $x > 0$ increases, the denominators increase and the functions decrease. The graphs of $y = x^{-1}$ and $y = x^{-2}$ have both the x- and y-axes as asymptotes. (See Figure 1.85.)

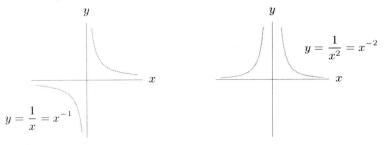

Figure 1.85: Graphs of x^0 and of negative powers of x

Quadratic Functions and Polynomials

Sums of power functions with nonnegative integer exponents are called *polynomials*, which are functions of the form

$$y = p(x) = a_n x^n + a_{n-1} x^{n-1} + \cdots + a_1 x + a_0.$$

Here, n is a nonnegative integer, called the *degree* of the polynomial, and a_n is a nonzero number called the *leading coefficient*. We call $a_n x^n$ the *leading term*.

If $n = 2$, the polynomial is called *quadratic* and has the form $ax^2 + bx + c$ with $a \neq 0$. The graph of a quadratic polynomial is a parabola. It opens up if the leading coefficient a is positive and opens down if a is negative.

[65] *Scientific American*, p. 112 (September, 1989).

Example 6   A company finds that the average number of people attending a concert is 75 if the price is $50 per person. At a price of $35 per person, the average number of people in attendance is 120.

(a) Assume that the demand curve is a line. Write the demand, q, as a function of price, p.

(b) Use your answer to part (a) to write the revenue, R, as a function of price, p.

(c) Use a graph of the revenue function to determine what price should be charged to obtain the greatest revenue.

Solution   (a) Two points on the line are $(p, q) = (50, 75)$ and $(p, q) = (35, 120)$. The slope of the line is

$$m = \frac{120 - 75}{35 - 50} = \frac{45}{-15} = -3 \text{ people/dollar.}$$

To find the vertical intercept of the line, we use the slope and one of the points:

$$75 = b + (-3)(50)$$
$$225 = b$$

The demand function is $q = 225 - 3p$.

(b) Since $R = pq$ and $q = 225 - 3p$, we see that $R = p(225 - 3p) = 225p - 3p^2$.

(c) The revenue function is the quadratic polynomial graphed in Figure 1.86. The maximum revenue occurs at $p = 37.5$. Thus, the company maximizes revenue by charging $37.50 per person.

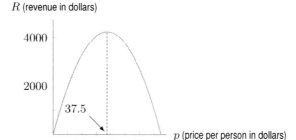

Figure 1.86: Revenue function for concert ticket sales

Problems for Section 1.9

In Problems 1–12, determine whether or not the function is a power function. If it is a power function, write it in the form $y = kx^p$ and give the values of k and p.

1. $y = 5\sqrt{x}$ **2.** $y = \dfrac{3}{x^2}$ **3.** $y = 2^x$

4. $y = \dfrac{3}{8x}$ **5.** $y = (3x^5)^2$ **6.** $y = \dfrac{5}{2\sqrt{x}}$

7. $y = 3 \cdot 5^x$ **8.** $y = \dfrac{2x^2}{10}$ **9.** $y = \dfrac{8}{x}$

10. $y = (5x)^3$ **11.** $y = 3x^2 + 4$ **12.** $y = \dfrac{x}{5}$

In Problems 13–16, write a formula representing the function.

13. The strength, S, of a beam is proportional to the square of its thickness, h.

14. The energy, E, expended by a swimming dolphin is proportional to the cube of the speed, v, of the dolphin.

15. The average velocity, v, for a trip over a fixed distance, d, is inversely proportional to the time of travel, t.

16. The gravitational force, F, between two bodies is inversely proportional to the square of the distance d between them.

17. The number of species of lizards, N, found on an island off Baja California is proportional to the fourth root of the area, A, of the island.[66] Write a formula for N as a function of A. Graph this function. Is it increasing or decreasing? Is the graph concave up or concave down? What does this tell you about lizards and island area?

18. The surface area of a mammal, S, satisfies the equation $S = kM^{2/3}$, where M is the body mass, and the constant of proportionality k depends on the body shape of the mammal. A human of body mass 70 kilograms has surface area 18,600 cm^2. Find the constant of proportionality for humans. Find the surface area of a human with body mass 60 kilograms.

19. Use shifts of power functions to find a possible formula for each of the graphs:

(a)   **(b)**

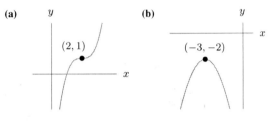

20. Kleiber's Law states that the metabolic needs (such as calorie requirements) of a mammal are proportional to its body weight raised to the 0.75 power.[67] Surprisingly, the daily diets of mammals conform to this relation well. Assuming Kleiber's Law holds:

   (a) Write a formula for C, daily calorie consumption, as a function of body weight, W.

   (b) Sketch a graph of this function. (You do not need scales on the axes.)

   (c) If a human weighing 150 pounds needs to consume 1800 calories a day, estimate the daily calorie requirement of a horse weighing 700 lbs and of a rabbit weighing 9 lbs.

   (d) On a per-pound basis, which animal requires more calories: a mouse or an elephant?

21. Allometry is the study of the relative size of different parts of a body as a consequence of growth. In this problem, you will check the accuracy of an allometric equation: the weight of a fish is proportional to the cube of its length.[68] Table 1.37 relates the weight, y, in gm, of plaice (a type of fish) to its length, x, in cm. Does this data support the hypothesis that (approximately) $y = kx^3$? If so, estimate the constant of proportionality, k.

Table 1.37

x	y	x	y	x	y
33.5	332	37.5	455	41.5	623
34.5	363	38.5	500	42.5	674
35.5	391	39.5	538	43.5	724
36.5	419	40.5	574		

22. The blood mass of a mammal is proportional to its body mass. A rhinoceros with body mass 3000 kilograms has blood mass of 150 kilograms. Find a formula for the blood mass of a mammal as a function of the body mass and estimate the blood mass of a human with body mass 70 kilograms.

23. Biologists estimate that the number of animal species of a certain body length is inversely proportional to the square of the body length.[69] Write a formula for the number of animal species, N, of a certain body length as a function of the length, L. Are there more species at large lengths or at small lengths? Explain.

$$N^2 = L^{-1}$$

24. The specific heat, s, of an element is the number of calories of heat required to raise the temperature of one gram of the element by one degree Celsius. Use the following table to decide if s is proportional or inversely proportional to the atomic weight, w, of the element. If so, find the constant of proportionality.

Element	Li	Mg	Al	Fe	Ag	Pb	Hg
w	6.9	24.3	27.0	55.8	107.9	207.2	200.6
s	0.92	0.25	0.21	0.11	0.056	0.031	.033

25. The circulation time of a mammal (that is, the average time it takes for all the blood in the body to circulate once and return to the heart) is proportional to the fourth root of the body mass of the mammal.

   (a) Write a formula for the circulation time, T, in terms of the body mass, B.

   (b) If an elephant of body mass 5230 kilograms has a circulation time of 148 seconds, find the constant of proportionality.

   (c) What is the circulation time of a human with body mass 70 kilograms?

[66]Rosenzweig, M.L., *Species Diversity in Space and Time*, p. 143 (Cambridge: Cambridge University Press, 1995).

[67]Strogatz, S., "Math and the City", The New York Times, May 20, 2009. Kleiber originally estimated the exponent as 0.74; it is now believed to be 0.75.

[68]Adapted from "On the Dynamics of Exploited Fish Populations" by R. J. H. Beverton and S. J. Holt, *Fishery Investigations*, Series II, 19, 1957.

[69]*US News & World Report*, August 18, 1997, p. 79.

26. The DuBois formula relates a person's surface area s, in m^2, to weight w, in kg, and height h, in cm, by

$$s = 0.01w^{0.25}h^{0.75}.$$

(a) What is the surface area of a person who weighs 65 kg and is 160 cm tall?

(b) What is the weight of a person whose height is 180 cm and who has a surface area of 1.5 m^2?

(c) For people of fixed weight 70 kg, solve for h as a function of s. Simplify your answer.

27. The infrastructure needs of a region (for example, the number of miles of electrical cable, the number of miles of roads, the number of gas stations) depend on its population. Cities enjoy economies of scale.[70] For example, the number of gas stations is proportional to the population raised to the power of 0.77.

(a) Write a formula for the number, N, of gas stations in a city as a function of the population, P, of the city.

(b) If city A is 10 times bigger than city B, how do their number of gas stations compare?

(c) Which is expected to have more gas stations per person, a town of 10,000 people or a city of 500,000 people?

28. According to the National Association of Realtors,[71] the minimum annual gross income, m, in thousands of dollars, needed to obtain a 30-year home loan of A thousand dollars at 9% is given in Table 1.38. Note that the larger the loan, the greater the income needed. Of course, not every mortgage is financed at 9%. In fact, excepting for slight variations, mortgage interest rates are generally determined not by individual banks but by the economy as a whole. The minimum annual gross income, m, in thousands of dollars, needed for a home loan of $100,000 at various interest rates, r, is given in Table 1.39. Note that obtaining a loan at a time when interest rates are high requires a larger income.

(a) Is the size of the loan, A, proportional to the minimum annual gross income, m?

(b) Is the percentage rate, r, proportional to the minimum annual gross income, m?

Table 1.38

A	50	75	100	150	200
m	17.242	25.863	34.484	51.726	68.968

Table 1.39

r	8	9	10	11	12
m	31.447	34.484	37.611	40.814	44.084

29. A standard tone of 20,000 dynes/cm^2 (about the loudness of a rock band) is assigned a value of 10. A subject listened to other sounds, such as a light whisper, normal conversation, thunder, a jet plane at takeoff, and so on. In each case, the subject was asked to judge the loudness and assign it a number relative to 10, the value of the standard tone. This is a "judgment of magnitude" experiment. The power law $J = al^{0.3}$ was found to model the situation well, where l is the actual loudness (measured in dynes/cm^2) and J is the judged loudness.

(a) What is the value of a?

(b) What is the judged loudness if the actual loudness is 0.2 dynes/cm^2 (normal conversation)?

(c) What is the actual loudness if judged loudness is 20?

30. A sporting goods wholesaler finds that when the price of a product is $25, the company sells 500 units per week. When the price is $30, the number sold per week decreases to 460 units.

(a) Find the demand, q, as a function of price, p, assuming that the demand curve is linear.

(b) Use your answer to part (a) to write revenue as a function of price.

(c) Graph the revenue function in part (b). Find the price that maximizes revenue. What is the revenue at this price?

31. A health club has cost and revenue functions given by $C = 10,000 + 35q$ and $R = pq$, where q is the number of annual club members and p is the price of a one-year membership. The demand function for the club is $q = 3000 - 20p$.

(a) Use the demand function to write cost and revenue as functions of p.

(b) Graph cost and revenue as a function of p, on the same axes. (Note that price does not go above $170 and that the annual costs of running the club reach $120,000.)

(c) Explain why the graph of the revenue function has the shape it does.

(d) For what prices does the club make a profit?

(e) Estimate the annual membership fee that maximizes profit. Mark this point on your graph.

---

[70] Strogatz, S., "Math and the City", *The New York Times*, May 20, 2009.
[71] "Income Needed to Get a Mortgage," *The World Almanac 1992*, p. 720.

1.10 PERIODIC FUNCTIONS

What Are Periodic Functions?

Many functions have graphs that oscillate, resembling a wave. Figure 1.87 shows the number of new housing construction starts (one-family units) in the US, 2002–2005, where t is time in quarter-years.[72] Notice that few new homes begin construction during the first quarter of a year (January, February, and March), whereas many new homes are begun in the second quarter (April, May, and June). Since 2008, as the economy has slowed, the pattern of oscillations has been replaced by a sharp drop in housing construction.

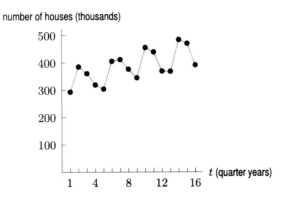

number of houses (thousands)

Figure 1.87: New housing construction starts, 2002–2005

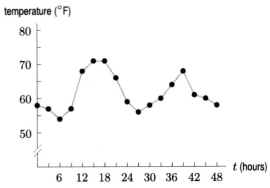

temperature (°F)

Figure 1.88: Temperature in Phoenix after midnight February 17, 2005

Let's look at another example. Figure 1.88 is a graph of the temperature (in °F) in Phoenix, AZ, in hours after midnight, February 17, 2005. Notice that the maximum is in the afternoon and the minimum is in the early morning.[73] Again, the graph looks like a wave.

Functions whose values repeat at regular intervals are called *periodic*. Many processes, such as the number of housing starts or the temperature, are approximately periodic. The water level in a tidal basin, the blood pressure in a heart, retail sales in the US, and the position of air molecules transmitting a musical note are also all periodic functions of time.

Amplitude and Period

Periodic functions repeat exactly the same cycle forever. If we know one cycle of the graph, we know the entire graph.

> For any periodic function of time:
> - The **amplitude** is half the difference between its maximum and minimum values.
> - The **period** is the time for the function to execute one complete cycle.

Example 1   Estimate the amplitude and period of the new housing starts function shown in Figure 1.87.

Solution   Figure 1.87 is not exactly periodic, since the maximum and minimum are not the same for each cycle. Nonetheless, the minimum is about 300, and the maximum is about 450. The difference between them is 150, so the amplitude is about $\frac{1}{2}(150) = 75$ thousand houses.

The wave completes a cycle between $t = 1$ and $t = 5$, so the period is $t = 4$ quarter-years, or one year. The business cycle for new housing construction is one year.

[72]http://www.census.gov/const/www/quarterly_starts_completions.pdf, accessed May 25, 2009.
[73]http://www.weather.com, accessed February 20, 2005.

Example 2   Figure 1.89 shows the temperature in an unopened freezer. Estimate the temperature in the freezer at 12:30 and at 2:45.

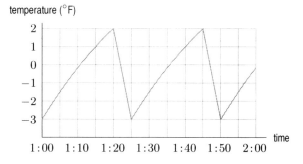

Figure 1.89: Oscillating freezer temperature. Estimate the temperature at 12:30 and 2:45

Solution   The maximum and minimum values each occur every 25 minutes, so the period is 25 minutes. The temperature at 12:30 should be the same as at 12:55 and at 1:20, namely, 2°F. Similarly, the temperature at 2:45 should be the same as at 2:20 and 1:55, or about −1.5°F.

The Sine and Cosine

Many periodic functions are represented using the functions called *sine* and *cosine*. The keys for the sine and cosine on a calculator are usually labeled as sin and cos.

Warning: Your calculator can be in either "degree" mode or "radian" mode. For this book, always use "radian" mode.

Graphs of the Sine and Cosine

The graphs of the sine and the cosine functions are periodic; see Figures 1.90 and 1.91. Notice that the graph of the cosine function is the graph of the sine function, shifted $\pi/2$ to the left.

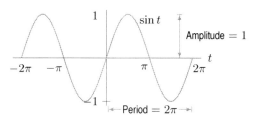

Figure 1.90: Graph of $\sin t$

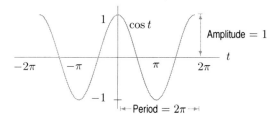

Figure 1.91: Graph of $\cos t$

The maximum and minimum values of $\sin t$ are $+1$ and -1, so the amplitude of the sine function is 1. The graph of $y = \sin t$ completes a cycle between $t = 0$ and $t = 2\pi$; the rest of the graph repeats this portion. The period of the sine function is 2π.

Example 3   Use a graph of $y = 3 \sin 2t$ to estimate the amplitude and period of this function.

Solution   In Figure 1.92, the waves have a maximum of $+3$ and a minimum of -3, so the amplitude is 3. The graph completes one complete cycle between $t = 0$ and $t = \pi$, so the period is π.

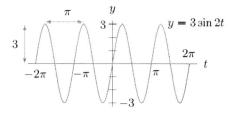

Figure 1.92: The amplitude is 3 and the period is π

Example 4   Explain how the graphs of each of the following functions differ from the graph of $y = \sin t$.

(a) $y = 6\sin t$ (b) $y = 5 + \sin t$ (c) $y = \sin(t + \frac{\pi}{2})$

Solution   (a) The graph of $y = 6\sin t$ is in Figure 1.93. The maximum and minimum values are $+6$ and -6, so the amplitude is 6. This is the graph of $y = \sin t$ stretched vertically by a factor of 6.

(b) The graph of $y = 5 + \sin t$ is in Figure 1.94. The maximum and minimum values of this function are 6 and 4, so the amplitude is $(6 - 4)/2 = 1$. The amplitude (or size of the wave) is the same as for $y = \sin t$, since this is a graph of $y = \sin t$ shifted up 5 units.

(c) The graph of $y = \sin(t + \frac{\pi}{2})$ is in Figure 1.95. This has the same amplitude, namely 1, and period, namely 2π, as the graph of $y = \sin t$. It is the graph of $y = \sin t$ shifted $\pi/2$ units to the left. (In fact, this is the graph of $y = \cos t$.)

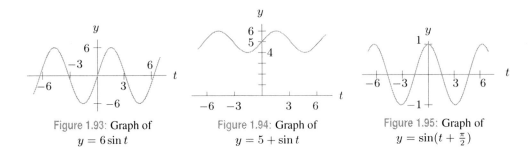

Figure 1.93: Graph of   Figure 1.94: Graph of   Figure 1.95: Graph of
$y = 6\sin t$ $y = 5 + \sin t$ $y = \sin(t + \frac{\pi}{2})$

Families of Curves: The Graph of $y = A\sin(Bt)$

The constants A and B in the expression $y = A\sin(Bt)$ are called *parameters*. We can study families of curves by varying one parameter at a time and studying the result.

Example 5   (a) Graph $y = A\sin t$ for several positive values of A. Describe the effect of A on the graph.
(b) Graph $y = \sin(Bt)$ for several positive values of B. Describe the effect of B on the graph.

Solution   (a) From the graphs of $y = A\sin t$ for $A = 1, 2, 3$ in Figure 1.96, we see that A is the amplitude.

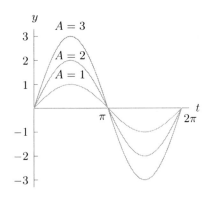

Figure 1.96: Graphs of $y = A\sin t$ with $A = 1, 2, 3$

(b) The graphs of $y = \sin(Bt)$ for $B = \frac{1}{2}, B = 1$, and $B = 2$ are shown in Figure 1.97. When $B = 1$, the period is 2π; when $B = 2$, the period is π; and when $B = \frac{1}{2}$, the period is 4π.

The parameter B affects the period of the function. The graphs suggest that the larger B is, the shorter the period. In fact, the period is $2\pi/B$.

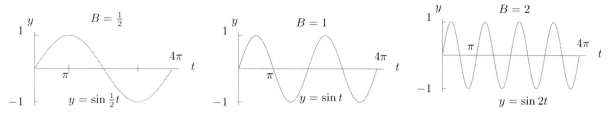

Figure 1.97: Graphs of $y = \sin(Bt)$ with $B = \frac{1}{2}, 1, 2$

In Example 5, the amplitude of $y = A\sin(Bt)$ was determined by the parameter A, and the period was determined by the parameter B. In general, we have

The functions $y = A\sin(Bt) + C$ and $y = A\cos(Bt) + C$ are periodic with

$$\text{Amplitude} = |A|, \qquad \text{Period} = \frac{2\pi}{|B|}, \qquad \text{Vertical shift} = C$$

Example 6 Find possible formulas for the following periodic functions.

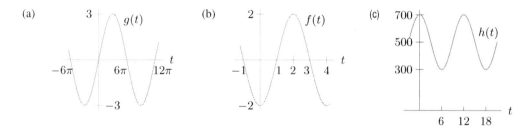

Solution (a) This function looks like a sine function of amplitude 3, so $g(t) = 3\sin(Bt)$. Since the function executes one full oscillation between $t = 0$ and $t = 12\pi$, when t changes by 12π, the quantity Bt changes by 2π. This means $B \cdot 12\pi = 2\pi$, so $B = 1/6$. Therefore, $g(t) = 3\sin(t/6)$ has the graph shown.

(b) This function looks like an upside-down cosine function with amplitude 2, so $f(t) = -2\cos(Bt)$. The function completes one oscillation between $t = 0$ and $t = 4$. Thus, when t changes by 4, the quantity Bt changes by 2π, so $B \cdot 4 = 2\pi$, or $B = \pi/2$. Therefore, $f(t) = -2\cos(\pi t/2)$ has the graph shown.

(c) This function looks like a cosine function. The maximum is 700 and the minimum is 300, so the amplitude is $\frac{1}{2}(700 - 300) = 200$. The height halfway between the maximum and minimum is 500, so the cosine curve has been shifted up 500 units, so $h(t) = 500 + 200\cos(Bt)$. The period is 12, so $B \cdot 12 = 2\pi$. Thus, $B = \pi/6$. The function $h(t) = 500 + 200\cos(\pi t/6)$ has the graph shown.

Example 7   On June 23, 2009, high tide in Portland, Maine was at midnight.[74] The height of the water in the harbor is a periodic function, since it oscillates between high and low tide. If t is in hours since midnight, the height (in feet) is approximated by the formula

$$y = 4.9 + 4.4 \cos\left(\frac{\pi}{6}t\right).$$

(a) Graph this function from $t = 0$ to $t = 24$.
(b) What was the water level at high tide?
(c) When was low tide, and what was the water level at that time?
(d) What is the period of this function, and what does it represent in terms of tides?
(e) What is the amplitude of this function, and what does it represent in terms of tides?

Solution   (a) See Figure 1.98.
(b) The water level at high tide was 9.3 feet (given by the y-intercept on the graph).
(c) Low tide occurs at $t = 6$ (6 am) and at $t = 18$ (6 pm). The water level at this time is 0.5 feet.
(d) The period is 12 hours and represents the interval between successive high tides or successive low tides. Of course, there is something wrong with the assumption in the model that the period is 12 hours. If so, the high tide would always be at noon or midnight, instead of progressing slowly through the day, as it in fact does. The interval between successive high tides actually averages about 12 hours 39 minutes, which could be taken into account in a more precise mathematical model.
(e) The maximum is 9.3, and the minimum is 0.5, so the amplitude is $(9.3 - 0.5)/2$, which is 4.4 feet. This represents half the difference between the depths at high and low tide.

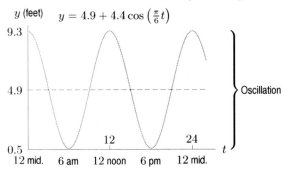

Figure 1.98: Graph of the function approximating the depth of the water in Portland, Maine on June 23, 2009

Problems for Section 1.10

1. A graduate student in environmental science studied the temperature fluctuations of a river. Figure 1.99 shows the temperature of the river (in °C) every hour, with hour 0 being midnight of the first day.

   (a) Explain why a periodic function could be used to model these data.
   (b) Approximately when does the maximum occur? The minimum? Why does this make sense?
   (c) What is the period for these data? What is the amplitude?

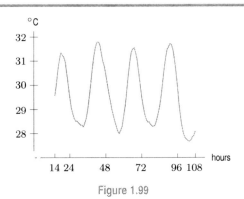

Figure 1.99

---

[74]www.maineboats.com/tidecharts/tides?T=junpt09.

2. Figure 1.100 shows quarterly beer production during the period 1997 to 1999. Quarter 1 reflects production during the first three months of the year, etc.[75]

   (a) Explain why a periodic function should be used to model these data.

   (b) Approximately when does the maximum occur? The minimum? Why does this make sense?

   (c) What are the period and amplitude for these data?

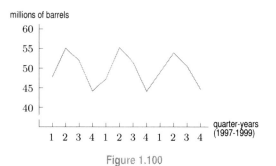

Figure 1.100

3. Sketch a possible graph of sales of sunscreen in the northeastern US over a 3-year period, as a function of months since January 1 of the first year. Explain why your graph should be periodic. What is the period?

For Problems 4–9, sketch graphs of the functions. What are their amplitudes and periods?

4. $y = 3\sin x$

5. $y = 3\sin 2x$

6. $y = -3\sin 2\theta$

7. $y = 4\cos 2x$

8. $y = 4\cos(\frac{1}{2}t)$

9. $y = 5 - \sin 2t$

10. Values of a function are given in the following table. Explain why this function appears to be periodic. Approximately what are the period and amplitude of the function? Assuming that the function is periodic, estimate its value at $t = 15$, at $t = 75$, and at $t = 135$.

t	20	25	30	35	40	45	50	55	60
$f(t)$	1.8	1.4	1.7	2.3	2.0	1.8	1.4	1.7	2.3

11. The following table shows values of a periodic function $f(x)$. The maximum value attained by the function is 5.

   (a) What is the amplitude of this function?

   (b) What is the period of this function?

   (c) Find a formula for this periodic function.

x	0	2	4	6	8	10	12
$f(x)$	5	0	-5	0	5	0	-5

12. Figure 1.101 shows the levels of the hormones estrogen and progesterone during the monthly ovarian cycles in females.[76] Is the level of both hormones periodic? What is the period in each case? Approximately when in the monthly cycle is estrogen at a peak? Approximately when in the monthly cycle is progesterone at a peak?

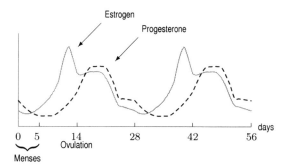

Figure 1.101

13. Average daily high temperatures in Ottawa, the capital of Canada, range from a low of $-6°$ Celsius on January 1 to a high of $26°$ Celsius on July 1 six months later. See Figure 1.102. Find a formula for H, the average daily high temperature in Ottawa in, $°C$, as a function of t, the number of months since January 1.

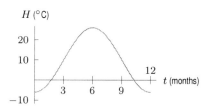

Figure 1.102

14. A person breathes in and out every three seconds. The volume of air in the person's lungs varies between a minimum of 2 liters and a maximum of 4 liters. Which of the following is the best formula for the volume of air in the person's lungs as a function of time?

   (a) $y = 2 + 2\sin\left(\frac{\pi}{3}t\right)$ **(b)** $y = 3 + \sin\left(\frac{2\pi}{3}t\right)$

   (c) $y = 2 + 2\sin\left(\frac{2\pi}{3}t\right)$ **(d)** $y = 3 + \sin\left(\frac{\pi}{3}t\right)$

[75] www.beerinstitute.org/pdfs/PRODUCTION_AND_WITHDRAWALS_OF_MALT_BEVERAGES_1997_1999.pdf, accessed May 7, 2005.

[76] Robert M. Julien, *A Primer of Drug Action*, Seventh Edition, p. 360 (W. H. Freeman and Co., New York: 1995).

15. Delta Cephei is one of the most visible stars in the night sky. Its brightness has periods of 5.4 days, the average brightness is 4.0 and its brightness varies by ± 0.35. Find a formula that models the brightness of Delta Cephei as a function of time, t, with $t = 0$ at peak brightness.

16. Most breeding birds in the northeast US migrate elsewhere during the winter. The number of bird species in an Ohio forest preserve oscillates between a high of 28 in June and a low of 10 in December.[77]

(a) Graph the number of bird species in this preserve as a function of t, the number of months since June. Include at least three years on your graph.
(b) What are the amplitude and period of this function?
(c) Find a formula for the number of bird species, B, as a function of the number of months, t since June.

For Problems 17–28, find a possible formula for each graph.

17.
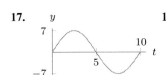

18.
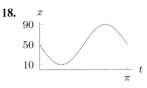

19.
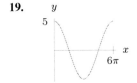

20.
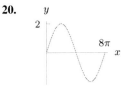

21.
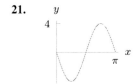

22.
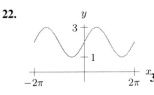

23.
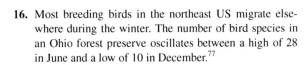

24.

25.
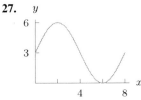

26.
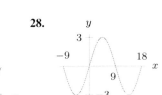

27.

28.
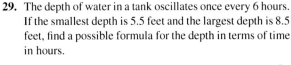

29. The depth of water in a tank oscillates once every 6 hours. If the smallest depth is 5.5 feet and the largest depth is 8.5 feet, find a possible formula for the depth in terms of time in hours.

30. The desert temperature, H, oscillates daily between $40°F$ at 5 am and $80°F$ at 5 pm. Write a possible formula for H in terms of t, measured in hours from 5 am.

31. Table 1.40 gives values for $g(t)$, a periodic function.

(a) Estimate the period and amplitude for this function.
(b) Estimate $g(34)$ and $g(60)$.

Table 1.40

t	0	2	4	6	8	10	12	14
$g(t)$	14	19	17	15	13	11	14	19
t	16	18	20	22	24	26	28	
$g(t)$	17	15	13	11	14	19	17	

32. The Bay of Fundy in Canada has the largest tides in the world. The difference between low and high water levels is 15 meters (nearly 50 feet). At a particular point the depth of the water, y meters, is given as a function of time, t, in hours since midnight by

$$y = D + A\cos\left(B(t - C)\right).$$

(a) What is the physical meaning of D?
(b) What is the value of A?
(c) What is the value of B? Assume the time between successive high tides is 12.4 hours.
(d) What is the physical meaning of C?

---

[77]Rosenzweig, M.L., *Species Diversity in Space and Time*, p. 71 (Cambridge University Press, 1995).

CHAPTER SUMMARY

- **Function terminology**
  Domain/range, increasing/decreasing, concavity, intercepts.
- **Linear functions**
  Slope, y-intercept. Grow by equal amounts in equal times.
- **Economic applications**
  Cost, revenue, and profit functions, break-even point. Supply and demand curves, equilibrium point. Depreciation function. Budget constraint. Present and future value.
- **Change, average rate of change, relative change**

- **Exponential functions**
  Exponential growth and decay, growth rate, the number e, continuous growth rate, doubling time, half-life, compound interest. Grow by equal percentages in equal times.
- **The natural logarithm function**
- **New functions from old**
  Composition, shifting, stretching.
- **Power functions and proportionality**
- **Polynomials**
- **Periodic functions**
  Sine, cosine, amplitude, period.

REVIEW PROBLEMS FOR CHAPTER ONE

1. The time T, in minutes, that it takes Dan to run x kilometers is a function $T = f(x)$. Explain the meaning of the statement $f(5) = 23$ in terms of running.

2. Describe what Figure 1.103 tells you about an assembly line whose productivity is represented as a function of the number of workers on the line.

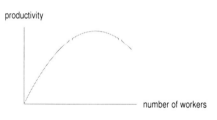

Figure 1.103

3. It warmed up throughout the morning, and then suddenly got much cooler around noon, when a storm came through. After the storm, it warmed up before cooling off at sunset. Sketch temperature as a function of time.

4. A gas tank 6 meters underground springs a leak. Gas seeps out and contaminates the soil around it. Graph the amount of contamination as a function of the depth (in meters) below ground.

5. The yield, Y, of an apple orchard (in bushels) as a function of the amount, a, of fertilizer (in pounds) used on the orchard is shown in Figure 1.104.

   (a) Describe the effect of the amount of fertilizer on the yield of the orchard.
   (b) What is the vertical intercept? Explain what it means in terms of apples and fertilizer.
   (c) What is the horizontal intercept? Explain what it means in terms of apples and fertilizer.
   (d) What is the range of this function for $0 \le a \le 80$?
   (e) Is the function increasing or decreasing at $a = 60$?

   (f) Is the graph concave up or down near $a = 40$?

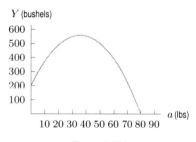

Figure 1.104

6. Let $y = f(x) = 3x - 5$.

   (a) What is $f(1)$?
   (b) Find the value of y when x is 5.
   (c) Find the value of x when y is 4.
   (d) Find the average rate of change of f between $x = 2$ and $x = 4$.

7. You drive at a constant speed from Chicago to Detroit, a distance of 275 miles. About 120 miles from Chicago you pass through Kalamazoo, Michigan. Sketch a graph of your distance from Kalamazoo as a function of time.

Find the equation of the line passing through the points in Problems 8–11.

8. $(0, -1)$ and $(2, 3)$ 9. $(-1, 3)$ and $(2, 2)$

10. $(0, 2)$ and $(2, 2)$ 11. $(-1, 3)$ and $(-1, 4)$

12. Match the graphs in Figure 1.105 with the following equations. (Note that the x and y scales may be unequal.)

(a) $y = x - 5$

(b) $-3x + 4 = y$

(c) $5 = y$

(d) $y = -4x - 5$

(e) $y = x + 6$

(f) $y = x/2$

Figure 1.105

13. Find a linear function that generates the values in Table 1.41.

Table 1.41

x	5.2	5.3	5.4	5.5	5.6
y	27.8	29.2	30.6	32.0	33.4

14. A controversial 1992 Danish study[78] reported that men's average sperm count has decreased from 113 million per milliliter in 1940 to 66 million per milliliter in 1990.

(a) Express the average sperm count, S, as a linear function of the number of years, t, since 1940.

(b) A man's fertility is affected if his sperm count drops below about 20 million per milliliter. If the linear model found in part (a) is accurate, in what year will the average male sperm count fall below this level?

15. Residents of the town of Maple Grove who are connected to the municipal water supply are billed a fixed amount monthly plus a charge for each cubic foot of water used. A household using 1000 cubic feet was billed $40, while one using 1600 cubic feet was billed $55.

(a) What is the charge per cubic foot?

(b) Write an equation for the total cost of a resident's water as a function of cubic feet of water used.

(c) How many cubic feet of water used would lead to a bill of $100?

In Problems 16–21, find the average velocity for the position function $s(t)$, in mm, over the interval $1 \leq t \leq 3$, where t is in seconds.

16. $s(t) = 12t - t^2$

17. $s(t) = \ln(t)$

18.
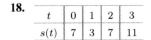

t	0	1	2	3
$s(t)$	7	3	7	11

19.
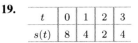

t	0	1	2	3
$s(t)$	8	4	2	4

20.
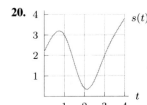

21.
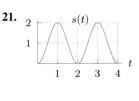

22. The graphs in Figure 1.106 represent the temperature, H, of four loaves of bread each put into an oven at time $t = 0$.

(a) Which curve corresponds to the bread that was put into the hottest oven?

(b) Which curve corresponds to the bread that had the lowest temperature at the time that it was put into the oven?

(c) Which two curves correspond to loaves of bread that were at the same temperature when they were put into the oven?

(d) Write a sentence describing any differences between the curves shown in (II) and (III). In terms of bread, what might cause this difference?

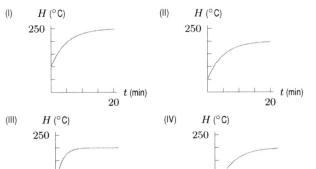

Figure 1.106

23. Sketch reasonable graphs for the following. Pay particular attention to the concavity of the graphs.

(a) The total revenue generated by a car rental business, plotted against the amount spent on advertising.

(b) The temperature of a cup of hot coffee standing in a room, plotted as a function of time.

[78]"Investigating the Next Silent Spring," *US News and World Report*, pp. 50–52 (March 11, 1996).

24. Each of the functions g, h, k in Table 1.42 is increasing, but each increases in a different way. Which of the graphs in Figure 1.107 best fits each function?

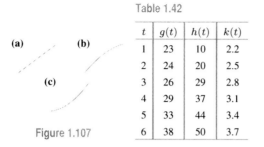

Figure 1.107

Table 1.42

t	$g(t)$	$h(t)$	$k(t)$
1	23	10	2.2
2	24	20	2.5
3	26	29	2.8
4	29	37	3.1
5	33	44	3.4
6	38	50	3.7

25. When a new product is advertised, more and more people try it. However, the rate at which new people try it slows as time goes on.

  (a) Graph the total number of people who have tried such a product against time.

  (b) What do you know about the concavity of the graph?

26. Figure 1.108 shows the age-adjusted death rates from different types of cancer among US males.[79]

  (a) Discuss how the death rate has changed for the different types of cancers.

  (b) For which type of cancer has the average rate of change between 1930 and 1967 been the largest? Estimate the average rate of change for this cancer type. Interpret your answer.

  (c) For which type of cancer has the average rate of change between 1930 and 1967 been the most negative? Estimate the average rate of change for this cancer type. Interpret your answer.

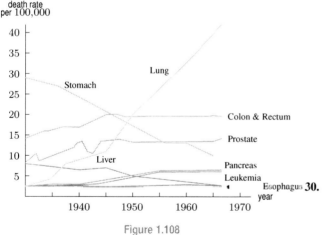

Figure 1.108

27. The volume of water in a pond over a period of 20 weeks is shown in Figure 1.109.

  (a) Is the average rate of change of volume positive or negative over the following intervals?

  (i) $t = 0$ and $t = 5$ (ii) $t = 0$ and $t = 10$
  (iii) $t = 0$ and $t = 15$ (iv) $t = 0$ and $t = 20$

  (b) During which of the following time intervals was the average rate of change larger?

  (i) $0 \le t \le 5$ or $0 \le t \le 10$

  (ii) $0 \le t \le 10$ or $0 \le t \le 20$

  (c) Estimate the average rate of change between $t = 0$ and $t = 10$. Interpret your answer in terms of water.

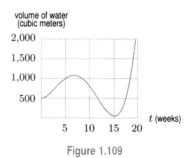

Figure 1.109

28. Find the average rate of change between $x = 0$ and $x = 10$ of each of the following functions: $y = x$, $y = x^2$, $y = x^3$, and $y = x^4$. Which has the largest average rate of change? Graph the four functions, and draw lines whose slopes represent these average rates of change.

29. **(a)** What are the fixed costs and the marginal cost for the cost function in Figure 1.110?

  (b) Explain what $C(100) = 2500$ tells you about costs.

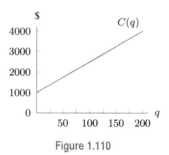

Figure 1.110

30. The Quick-Food company provides a college meal-service plan. Quick-Food has fixed costs of $350,000 per term and variable costs of $400 per student. Quick-Food charges $800 per student per term. How many students must sign up with the Quick-Food plan in order for the company to make a profit?

[79] Abraham M. Lilienfeld, *Foundations of Epidemiology*, p. 67 (New York; Oxford University Press, 1976).

31. For tax purposes, you may have to report the value of your assets, such as cars or refrigerators. The value you report drops with time. "Straight-line depreciation" assumes that the value is a linear function of time. If a \$950 refrigerator depreciates completely in seven years, find a formula for its value as a function of time.

32. One of the graphs in Figure 1.111 is a supply curve, and the other is a demand curve. Which is which? Explain how you made your decision using what you know about the effect of price on supply and demand.

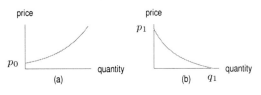

Figure 1.111

33. Figure 1.112 shows supply and demand curves.

  (a) What is the equilibrium price for this product? At this price, what quantity is produced?

  (b) Choose a price above the equilibrium price—for example, $p = 300$. At this price, how many items are suppliers willing to produce? How many items do consumers want to buy? Use your answers to these questions to explain why, if prices are above the equilibrium price, the market tends to push prices lower (toward the equilibrium).

  (c) Now choose a price below the equilibrium price—for example, $p = 200$. At this price, how many items are suppliers willing to produce? How many items do consumers want to buy? Use your answers to these questions to explain why, if prices are below the equilibrium price, the market tends to push prices higher (toward the equilibrium).

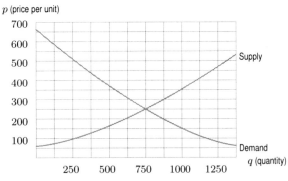

Figure 1.112

Find possible formulas for the graphs in Problems 34–39.

34.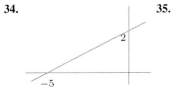

35.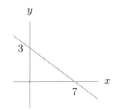

36.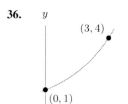

37.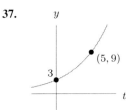

38.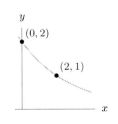

39.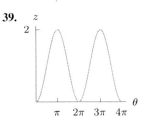

40. The worldwide carbon dioxide emission[80], C, from consumption of fossil fuels was 22.0 billion tons in 1995 and 28.2 billion tons in 2005. Find a formula for the emission C in t years after 1995 if:

  (a) C is a linear function of t. What is the annual rate of increase in carbon dioxide emission?

  (b) C is an exponential function of t. What is the annual percent rate of increase in carbon dioxide emission?

41. Table 1.43 gives values for three functions. Which functions could be linear? Which could be exponential? Which are neither? For those which could be linear or exponential, give a possible formula for the function.

Table 1.43

x	$f(x)$	$g(x)$	$h(x)$
0	25	30.8	15,000
1	20	27.6	9,000
2	14	24.4	5,400
3	7	21.2	3,240

For Problems 42–45, solve for x using logs.

42. $3^x = 11$

43. $20 = 50(1.04)^x$

44. $e^{5x} = 100$

45. $25e^{3x} = 10$

---

[80]*Statistical Abstracts of the United States 2009*, Table 1304.

46. Write the exponential functions $P = e^{0.08t}$ and $Q = e^{-0.3t}$ in the form $P = a^t$ and $Q = b^t$.

47. **(a)** What is the continuous percent growth rate for the function $P = 10e^{0.15t}$?
 (b) Write this function in the form $P = P_0a^t$.
 (c) What is the annual (not continuous) percent growth rate for this function?
 (d) Graph $P = 10e^{0.15t}$ and your answer to part (b) on the same axes. Explain what you see.

48. You need \$10,000 in your account 3 years from now and the interest rate is 8% per year, compounded continuously. How much should you deposit now?

49. If Q_0 is the quantity of radioactive carbon-14 in an organism at the time of death, the quantity, Q, remaining t years later is given by

$$Q = Q_0e^{-0.000121t}.$$

 (a) A skull uncovered at an archeological dig has 15% of the original amount of carbon-14 present. Estimate its age.
 (b) Calculate the half-life of carbon-14.

50. A radioactive substance has a half-life of 8 years. If 200 grams are present initially, how much remains at the end of 12 years? How long until only 10% of the original amount remains?

51. The size of an exponentially growing bacteria colony doubles in 5 hours. How long will it take for the number of bacteria to triple?

52. When the Olympic Games were held outside Mexico City in 1968, there was much discussion about the effect the high altitude (7340 feet) would have on the athletes. Assuming air pressure decays exponentially by 0.4% every 100 feet, by what percentage is air pressure reduced by moving from sea level to Mexico City?

53. You have the option of renewing the service contract on your three-year old dishwasher. The new service contract is for three years at a price of \$200. The interest rate is 7.25% per year, compounded annually, and you estimate that the costs of repairs if you do not buy the service contract will be \$50 at the end of the first year, \$100 at the end of the second year, and \$150 at the end of the third year. Should you buy the service contract? Explain.

54. If $h(x) = x^3 + 1$ and $g(x) = \sqrt{x}$, find
 (a) $g(h(x))$ **(b)** $h(g(x))$
 (c) $h(h(x))$ **(d)** $g(x) + 1$
 (e) $g(x + 1)$

55. Let $f(x) = 2x + 3$ and $g(x) = \ln x$. Find formulas for each of the following functions.
 (a) $g(f(x))$ **(b)** $f(g(x))$ **(c)** $f(f(x))$

In Problems 56–59, use Figure 1.113 to graph the functions.

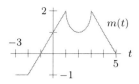

Figure 1.113

56. $n(t) = m(t) + 2$

57. $p(t) = m(t - 1)$

58. $k(t) = m(t + 1.5)$

59. $w(t) = m(t - 0.5) - 2.5$

In Problems 60–62, use Figure 1.114 to graph the function.

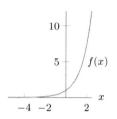

Figure 1.114

60. $5f(x)$ **61.** $f(x + 5)$ **62.** $f(x) + 5$

63. A plan is adopted to reduce the pollution in a lake to the legal limit. The quantity Q of pollutants in the lake after t weeks of clean-up is modeled by the function $Q = f(t)$ where $f(t) = A + Be^{Ct}$.
 (a) What are the signs of A, B and C?
 (b) What is the initial quantity of pollution in the lake?
 (c) What is the legal limit of pollution in the lake?

In Problems 64–65, use shifts of a power function to find a possible formula for the graph.

64.   **65.**

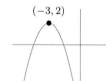
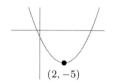

66. The following table gives values for a function $p = f(t)$. Could p be proportional to t?

t	0	10	20	30	40	50
p	0	25	60	100	140	200

67. Zipf's Law, developed by George Zipf in 1949, states that in a given country, the population of a city is inversely proportional to the city's rank by size in the country.[81] Assuming Zipf's Law:

   (a) Write a formula for the population, P, of a city as a function of its rank, R.

   (b) If the constant of proportionality k is 300,000, what is the approximate population of the largest city (rank 1)? The second largest city (rank 2)? The third largest city?

   (c) Answer the questions of part (b) if $k = 6$ million.

   (d) Interpret the meaning of the constant of proportionality k in this context.

Find the period and amplitude in Problems 68–70.

68. $y = 7 \sin(3t)$

69. $z = 3 \cos(u/4) + 5$

70. $r = 0.1 \sin(\pi t) + 2$

71. Figure 1.115 shows the number of reported[82] cases of mumps by month, in the US, for 1972–73.

   (a) Find the period and amplitude of this function, and interpret each in terms of mumps.

   (b) Predict the number of cases of mumps 30 months and 45 months after January 1, 1972.

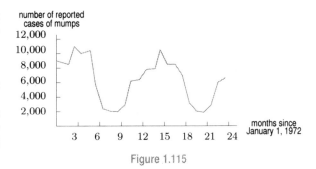

Figure 1.115

72. A population of animals varies periodically between a low of 700 on January 1 and a high of 900 on July 1. Graph the population against time.

For Problems 73–74, find a possible formula for each graph.

73.

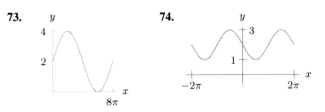

74.

CHECK YOUR UNDERSTANDING

In Problems 1–105, indicate whether the statement is true or false.

1. The domain is the set of outputs of a function.

2. If $V = f(a)$, where V is the value of a car (in thousands of dollars) and a is the car's age (in years), then $f(10)$ is the age of a car valued at $10,000.

3. If $f(x) = x^2$, then the point $(2, 4)$ is on the graph of $f(x)$.

4. The set of numbers between 3 and 4, including 3 and 4, is written $(3, 4)$.

5. The function $D = f(r)$ given by $D = -3r + 10$ has vertical intercept 10.

6. A function is always given by a formula.

7. If $f(x) = x^2 + 2x + 1$ then $f(3) = 16$.

8. The graph of a function can have more than one horizontal intercept.

9. The graph of a function can have more than one vertical intercept.

10. The vertical intercept on the graph of $C = f(q)$, where C is the cost to produce q items, represents the cost to produce no items.

11. The slope of the graph of a linear function $f(p)$ is $(f(p_2) - f(p_1))/(p_2 - p_1)$.

12. The graph of the linear function with formula $m(x) = 3x + 2$ has slope 2.

13. The slope of the graph of a linear function can be zero.

14. An equation of a line with slope -1 that passes through the point $(2, 5)$ is $y = -x + 7$.

15. The function whose values are shown in the following table could be linear:

s	2	4	6	8
$h(s)$	-1	-5	-9	-13

[81] Strogatz, S., "Math and the City", The New York Times, May 20, 2009.

[82] Center for Disease Control, 1974, *Reported Morbidity and Mortality in the United States 1973*, Vol. 22, No. 53. Prior to the licensing of the vaccine in 1967, 100,000–200,000 cases of mumps were reported annually. Since 1995, fewer than 1000 cases are reported annually. Source: CDC.

16. The graphs of two different linear functions must intersect in a point.

17. A line with positive slope must also have a positive y-intercept.

18. The linear function $f(x)$ with $f(4) = 2$ and $f(9) = 3$ has slope 5.

19. If number of acres of harvested land is a linear function of elevation in meters, the units of the slope are meters per acre.

20. If $(2,5)$ is a point on the graph of a line with slope 3, then $(3,8)$ is also a point on the line.

21. The function $D = f(r)$ given by $D = -3r + 10$ is an increasing function of r.

22. The average rate of change of a linear function $f(t)$ between $t = a$ and $t = b$ is $(f(b) - f(a))/(b - a)$.

23. If $C(n)$ is the total cost, in dollars, to feed n students in the campus cafeteria, then the average rate of change of $C(n)$ has units of students per dollar.

24. If $s(z) = z^2$, then the average rate of change between $z = -1$ and $z = 2$ is positive.

25. A function $f(x)$ can be both increasing and concave down over the interval $0 \le x \le 1$.

26. The function $Q(r)$ given in the following table appears to be concave up:

r	1	2	3	4
$Q(r)$	10	15	17	18

27. The speed and velocity of a moving object are the same.

28. The average rate of change of a function is the slope of the line between two points on the graph of the function.

29. A particle whose position s, in feet, at time t, in seconds, is given by $s(t) = 3t + 2$ has average velocity 3 feet per second.

30. If the graph of a function is concave up, the function must be increasing.

31. The relative change in a quantity is the change divided by the size of the quantity before the change.

32. If production is measured in tons, the units of the percent change of production is also measured in tons.

33. If cost is $1000 for a quantity of 500 and percent change is 15% when the quantity increases by 100 units, then the cost to produce 600 units is $1150.

34. Relative change is always positive.

35. If P in grams is a function of t in hours, then the relative rate of change is measured in grams per hour.

36. Profit is the sum of cost and revenue.

37. Revenue from selling a product is the selling price times the quantity sold.

38. The graph of the cost function $C(q)$ always passes through the origin.

39. The cost function $C(q)$ is a decreasing function of quantity q.

40. If the selling price is constant, the revenue function $R(q)$ is an increasing function of quantity q.

41. Demand is always greater than supply.

42. At equilibrium price p^* and quantity q^*, the supply and demand curves intersect.

43. The units of marginal cost are the same as the units of marginal revenue.

44. The marginal profit is the marginal revenue minus the marginal cost.

45. The imposition of a sales tax won't change the equilibrium price and quantity.

46. The function $Q(t) = 5 \cdot 3^t$ is exponential.

47. An exponential function has a constant percent growth or decay rate.

48. Exponential functions are always increasing.

49. The function $P(t) = 10 \cdot (1.03)^t$ has a 30% growth rate.

50. The function $Q(x) = 35(1/3)^x$ has $Q(2)/Q(1) = 1/3$.

51. The function $R(s) = 16 \cdot 5^s$ has a vertical intercept of 16.

52. The number e satisfies $2 < e < 3$.

53. The function $f(t) = 5^t$ grows more quickly than the function $g(t) = e^t$.

54. If $P = 25(1.15)^t$ gives the size of a population in year t, then the population is growing by 15% per year.

55. The function $Q(r)$ given in the following table could be exponential:

r	0	0.1	0.2	0.3
$Q(r)$	5	6.2	7.4	8.6

56. The value of $\ln(0)$ is 1.

57. The value of $\ln(1)$ is 0.

58. For all $a > 0$ and $b > 0$, we have $\ln(a + b) = \ln(a) \cdot \ln(b)$.

59. The value of $\ln(e^2)$ is 2.

60. The function $f(x) = \ln(x)$ is an increasing function of x.

61. If $5^t = 36$, then $t \ln 5 = \ln 36$.

62. The function $P = 10e^{5t}$ is an exponential growth function.

63. When $b > 0$, we have $e^{\ln b} = \ln(e^b)$.

64. When $B > 0$, we have $\ln(2B) = 2\ln(B)$.

65. When $A > 0$ and $B > 0$, we have $\ln(A^B) = B\ln(A)$.

66. The doubling time of $P(t) = 3e^{5t}$ and $Q(t) = 6e^{5t}$ is the same.

67. If the half-life of a quantity is 6 years, then 100 mg of the substance will decay to 25 mg in 9 years.

68. The half-life of $P(t) = 2e^{-0.6t}$ is twice the half-life of $Q(t) = e^{-0.6t}$.

69. An amount of $1000 invested in a bank account at an interest rate of 3% compounded annually has a balance after t years of $1000e^{0.03t}$.

70. The doubling time of a continuous rate of 6% is more than the doubling time of a continuous rate of 3%.

71. To find the doubling time of $P(t) = 5e^{2t}$, we can solve $10 = 5e^{2t}$ for t.

72. Present value is always less than future value if interest rates are greater than zero.

73. Assuming continuous growth rate of 3%, the future value five years from now of a $1000 payment today is $1000e^{0.15}$.

74. The future value in five years of a $1000 payment made today is less than its future value in ten years (assuming an annual interest rate of 2%).

75. The present value of a payment of $1000 five years from now is less than the present value of a $1000 payment made ten years from now (assuming an annual interest rate of 2%).

76. The graph of $f(x + 5)$ is the same as the graph of $f(x) + 5$.

77. If the graph of $f(x)$ is always increasing, then so is the graph of $f(x + k)$.

78. If the graph of $g(t)$ is concave up, then the graph of $-2g(t)$ is concave down.

79. If the graph of $f(x)$ crosses the x-axis at $x = 1$, then so does the graph of $5f(x)$.

80. If the graph of $f(x)$ crosses the x-axis at $x = 1$, then the graph of $f(x + 1)$ crosses the x-axis at $x = 0$.

81. If $g(s) = s^2$ then $g(3 + h) = 9 + h^2$.

82. If $f(t) = t^2$ and $g(t) = t + 1$ then $f(g(t)) = g(f(t))$.

83. If $f(x) = x^3 - 5$ and $g(x) = \ln x$ then $f(g(x)) = (\ln x)^3 - 5$.

84. The function $h(x) = (3x^2 + 2)^3$ can be written $g(u(x))$ where $g(t) = t^3$ and $u(x) = 3x^2 + 2$.

85. If $f(x) = x^2 - 1$ then $f(x + h) - f(x) = h^2$.

86. If A is proportional to B, then $A = kB$ for some nonzero constant k.

87. If A is inversely proportional to B, then $A = -kB$ for some nonzero constant k.

88. The function $f(x) = 3x^{10}$ is a power function.

89. The function $h(s) = 3 \cdot 10^s$ is a power function.

90. The function $h(x) = 3/\sqrt{x}$ can be written as a power function in the form $h(x) = 3x^{-2}$.

91. The function $g(x) = 3/(2x^2)$ can be written as a power function in the form $g(x) = 6x^{-2}$.

92. The function $f(x) = (3\sqrt{x})/2$ can be written as the power function $f(x) = 1.5x^{1/2}$.

93. If $w = 10.25r^3$, then w is proportional to the cube of r.

94. If $S = 25/\sqrt[3]{t}$, then S is inversely proportional to the cube root of t.

95. If p is proportional to q, then the ratio p/q is constant.

96. The amplitude of $f(x) = 3\sin x$ is $3/2$.

97. The period of $g(x) = \cos x + 2$ is 2π.

98. The value of $\sin(3t)/\sin(5t)$ is $3/5$.

99. The graph of $y = \cos x$ is a horizontal shift of the graph of $y = \sin x$.

100. The period of $y = 3\cos(5t) + 7$ is 5.

101. The period of $y = \sin(2t)$ is twice the period of $y = \sin(t)$.

102. The functions $f(t) = 5\sin t$ and $g(t) = 8 + 5\sin t$ have the same amplitude.

103. The graphs of $y = (\sin x)^2$ and $y = \sin(x^2)$ are the same.

104. For all x, we have $0 \leq \sin(x) \leq 1$.

105. For all x, we have $\sin^2 x + \cos^2 x = 1$.

PROJECTS FOR CHAPTER ONE

1. Compound Interest

The newspaper article below is from *The New York Times*, May 27, 1990. Fill in the three blanks. (For the first blank, assume that daily compounding is essentially the same as continuous compounding. For the last blank, assume the interest has been compounded yearly, and give your answer in dollars. Ignore the occurrence of leap years.)

213 Years After Loan, Uncle Sam Is Dunned

By LISA BELKIN

Special to The New York Times

SAN ANTONIO, May 26 — More than 200 years ago, a wealthy Pennsylvania merchant named Jacob DeHaven lent $450,000 to the Continental Congress to rescue the troops at Valley Forge. That loan was apparently never repaid.

So Mr. DeHaven's descendants are taking the United States Government to court to collect what they believe they are owed.

The total: ____ in today's dollars if the interest is compounded daily at 6 percent, the going rate at the time. If compounded yearly, the bill is only ____.

Family Is Flexible

The descendants say that they are willing to be flexible about the amount of a settlement and that they might even accept a heartfelt thank you or perhaps a DeHaven statue. But they also note that interest is accumulating at ____ a second.

2. Population Center of the US

Since the opening up of the West, the US population has moved westward. To observe this, we look at the "population center" of the US, which is the point at which the country would balance if it were a flat plate with no weight, and every person had equal weight. In 1790 the population center was east of Baltimore, Maryland. It has been moving westward ever since, and in 2000 it was in Edgar Springs, Missouri. During the second half of the 20th century, the population center has moved about 50 miles west every 10 years.

(a) Let us measure position westward from Edgar Springs along the line running through Baltimore. For the years since 2000, express the approximate position of the population center as a function of time in years from 2000.

(b) The distance from Baltimore to Edgar Springs is a bit over 1000 miles. Could the population center have been moving at roughly the same rate for the last two centuries?

(c) Could the function in part (a) continue to apply for the next four centuries? Why or why not? [Hint: You may want to look at a map. Note that distances are in air miles and are not driving distances.]

Chapter Two

RATE OF CHANGE: THE DERIVATIVE

Contents

2.1 INSTANTANEOUS RATE OF CHANGE

Chapter 1 introduced the average rate of change of a function over an interval. In this section, we consider the rate of change of a function at a point. We saw in Chapter 1 that when an object is moving along a straight line, the average rate of change of position with respect to time is the average velocity. If position is expressed as $y = f(t)$, where t is time, then

$$\text{Average rate of change in position between } t = a \text{ and } t = b = \frac{\Delta y}{\Delta t} = \frac{f(b) - f(a)}{b - a}.$$

If you drive 200 miles in 4 hours, your average velocity is $200/4 = 50$ miles per hour. Of course, this does not mean that you travel at exactly 50 mph the entire trip. Your velocity at a given instant during the trip is shown on your speedometer, and this is the quantity that we investigate now.

Instantaneous Velocity

We throw a grapefruit straight upward into the air. Table 2.1 gives its height, y, at time t. What is the velocity of the grapefruit at exactly $t = 1$? We use average velocities to estimate this quantity.

Table 2.1   *Height of the grapefruit above the ground*

t (sec)	0	1	2	3	4	5	6
$y = s(t)$ (feet)	6	90	142	162	150	106	30

The average velocity on the interval $0 \le t \le 1$ is 84 ft/sec and the average velocity on the interval $1 \le t \le 2$ is 52 ft/sec. Notice that the average velocity before $t = 1$ is larger than the average velocity after $t = 1$ since the grapefruit is slowing down. We expect the velocity *at $t = 1$* to be between these two average velocities. How can we find the velocity at *exactly $t = 1$*? We look at what happens near $t = 1$ in more detail. Suppose that we find the average velocities on either side of $t = 1$ over smaller and smaller intervals, as in Figure 2.1. Then, for example,

$$\text{Average velocity between } t = 1 \text{ and } t = 1.01 = \frac{\Delta y}{\Delta t} = \frac{s(1.01) - s(1)}{1.01 - 1} = \frac{90.678 - 90}{0.01} = 67.8 \text{ ft/sec.}$$

We expect the instantaneous velocity at $t = 1$ to be between the average velocities on either side of $t = 1$. In Figure 2.1, the values of the average velocity before $t = 1$ and the average velocity after $t = 1$ get closer together as the size of the interval shrinks. For the smallest intervals in Figure 2.1, both velocities are 68.0 ft/sec (to one decimal place), so we say the velocity at $t = 1$ is 68.0 ft/sec (to one decimal place).

t	0	0.9	0.99	0.999	1	1.001	1.01	1.1	2
$y = s(t)$	6.000	83.040	89.318	89.932	90.000	90.068	90.678	96.640	142.000

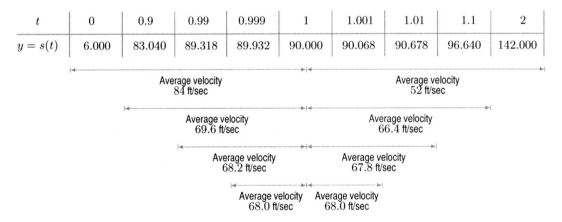

Average velocity
84 ft/sec

Average velocity
52 ft/sec

Average velocity
69.6 ft/sec

Average velocity
66.4 ft/sec

Average velocity
68.2 ft/sec

Average velocity
67.8 ft/sec

Average velocity
68.0 ft/sec

Average velocity
68.0 ft/sec

Figure 2.1: Average velocities over intervals on either side of $t = 1$ showing successively smaller intervals

Of course, if we showed more decimal places, the average velocities before and after $t = 1$ would no longer agree. To calculate the velocity at $t = 1$ to more decimal places of accuracy, we take smaller and smaller intervals on either side of $t = 1$ until the average velocities agree to the number of decimal places we want. In this way, we can estimate the velocity at $t = 1$ to any accuracy.

Defining Instantaneous Velocity Using the Idea of a Limit

When we take smaller intervals near $t = 1$, it turns out that the average velocities for the grapefruit are always just above or just below 68 ft/sec. It seems natural, then, to define velocity at the instant $t = 1$ to be 68 ft/sec. This is called the *instantaneous velocity* at this point. Its definition depends on our being convinced that smaller and smaller intervals provide average velocities that come arbitrarily close to 68. This process is referred to as *taking the limit*.

> The **instantaneous velocity** of an object at time t is defined to be the limit of the average velocity of the object over shorter and shorter time intervals containing t.

Notice that the instantaneous velocity seems to be exactly 68, but what if it were 68.000001? How can we be sure that we have taken small enough intervals? Showing that the limit is exactly 68 requires more precise knowledge of how the velocities were calculated and of the limiting process; see the Focus on Theory section.

Instantaneous Rate of Change

We can define the *instantaneous rate of change* of any function $y = f(t)$ at a point $t = a$. We mimic what we did for velocity and look at the average rate of change over smaller and smaller intervals.

> The **instantaneous rate of change** of f at a, also called the **rate of change** of f at a, is defined to be the limit of the average rates of change of f over shorter and shorter intervals around a.

Since the average rate of change is a difference quotient of the form $\Delta y / \Delta t$, the instantaneous rate of change is a limit of difference quotients. In practice, we often approximate a rate of change by one of these difference quotients.

Example 1   The quantity (in mg) of a drug in the blood at time t (in minutes) is given by $Q = 25(0.8)^t$. Estimate the rate of change of the quantity at $t = 3$ and interpret your answer.

Solution   We estimate the rate of change at $t = 3$ by computing the average rate of change over intervals near $t = 3$. We can make our estimate as accurate as we like by choosing our intervals small enough. Let's look at the average rate of change over the interval $3 \leq t \leq 3.01$:

$$\text{Average rate of change} = \frac{\Delta Q}{\Delta t} = \frac{25(0.8)^{3.01} - 25(0.8)^3}{3.01 - 3.00} = \frac{12.7715 - 12.80}{3.01 - 3.00} = -2.85.$$

A reasonable estimate for the rate of change of the quantity at $t = 3$ is -2.85. Since Q is in mg and t in minutes, the units of $\Delta Q / \Delta t$ are mg/minute. Since the rate of change is negative, the quantity of the drug is decreasing. After 3 minutes, the quantity of the drug in the body is decreasing at 2.85 mg/minute.

In Example 1, we estimated the rate of change using an interval to the right of the point ($t = 3$ to $t = 3.01$). We could use an interval to the left of the point, or we could average the rates of change to the left and the right. In this text, we usually use an interval to the right of the point.

The Derivative at a Point

The instantaneous rate of change of a function f at a point a is so important that it is given its own name, the *derivative of f at a*, denoted $f'(a)$ (read "f-prime of a"). If we want to emphasize that $f'(a)$ is the rate of change of $f(x)$ as the variable x increases, we call $f'(a)$ the derivative of f *with respect to x at $x = a$*. Notice that the derivative is just a new name for the rate of change of a function.

> The **derivative of f at a**, written $f'(a)$, is defined to be the instantaneous rate of change of f at the point a.

A definition of the derivative using a formula is given in the Focus on Theory section.

Example 2   Estimate $f'(2)$ if $f(x) = x^3$.

Solution   Since $f'(2)$ is the derivative, or rate of change, of $f(x) = x^3$ at 2, we look at the average rate of change over intervals near 2. Using the interval $2 \le x \le 2.001$, we see that

$$\text{Average rate of change on } 2 \le x \le 2.001 = \frac{(2.001)^3 - 2^3}{2.001 - 2} = \frac{8.0120 - 8}{0.001} = 12.0.$$

The rate of change of $f(x)$ at $x = 2$ appears to be approximately 12, so we estimate $f'(2) = 12$.

Visualizing the Derivative: Slope of the Graph and Slope of the Tangent Line

Figure 2.2 shows the average rate of change of a function represented by the slope of the secant line joining points A and B. The derivative is found by taking the average rate of change over smaller and smaller intervals. In Figure 2.3, as point B moves toward point A, the secant line becomes the tangent line at point A. Thus, the derivative is represented by the slope of the tangent line to the graph at the point.

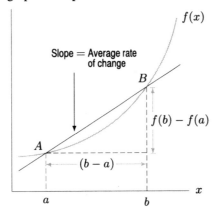

Figure 2.2: Visualizing the average rate of change of f between a and b

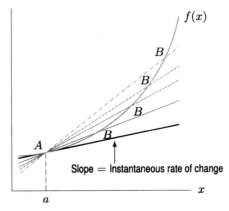

Figure 2.3: Visualizing the instantaneous rate of change of f at a

Alternatively, take the graph of a function around a point and "zoom in" to get a close-up view. (See Figure 2.4.) The more we zoom in, the more the graph appears to be straight. We call the slope of this line the *slope of the graph* at the point; it also represents the derivative.

> The derivative of a function at the point A is equal to
>
> - The slope of the graph of the function at A.
> - The slope of the line tangent to the curve at A.

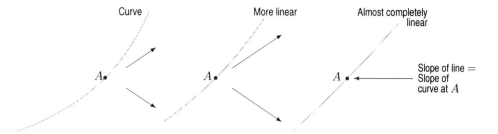

Figure 2.4: Finding the slope of a curve at a point by "zooming in"

The slope interpretation is often useful in gaining rough information about the derivative, as the following examples show.

Example 3   Use a graph of $f(x) = x^2$ to determine whether each of the following quantities is positive, negative, or zero: (a) $f'(1)$ (b) $f'(-1)$ (c) $f'(2)$ (d) $f'(0)$

Solution   Figure 2.5 shows tangent line segments to the graph of $f(x) = x^2$ at the points $x = 1$, $x = -1$, $x = 2$, and $x = 0$. Since the derivative is the slope of the tangent line at the point, we have:

(a) $f'(1)$ is positive.
(b) $f'(-1)$ is negative.
(c) $f'(2)$ is positive (and larger than $f'(1)$).
(d) $f'(0) = 0$ since the graph has a horizontal tangent at $x = 0$.

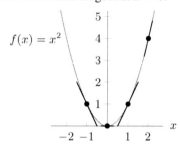

Figure 2.5: Tangent lines showing sign of derivative of $f(x) = x^2$

Example 4   Estimate the derivative of $f(x) = 2^x$ at $x = 0$ graphically and numerically.

Solution   Graphically: If we draw a tangent line at $x = 0$ to the exponential curve in Figure 2.6, we see that it has a positive slope between 0.5 and 1.

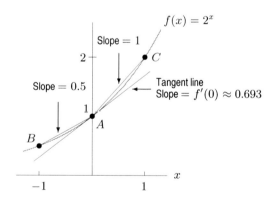

Figure 2.6: Graph of $f(x) = 2^x$ showing the derivative at $x = 0$

Numerically: To estimate the derivative at $x = 0$, we compute the average rate of change on an interval around 0.

$$\begin{array}{c} \text{Average rate of change} \\ \text{on } 0 \leq x \leq 0.0001 \end{array} = \frac{2^{0.0001} - 2^0}{0.0001 - 0} = \frac{1.000069317 - 1}{0.0001} = 0.69317.$$

Since using smaller intervals gives approximately the same values, it appears that the derivative is approximately 0.69317; that is, $f'(0) \approx 0.693$.

Example 5   The graph of a function $y = f(x)$ is shown in Figure 2.7. Indicate whether each of the following quantities is positive or negative, and illustrate your answers graphically.

(a) $f'(1)$ (b) $\dfrac{f(3) - f(1)}{3 - 1}$ (c) $f(4) - f(2)$

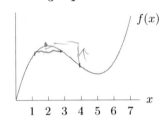

Figure 2.7

Solution   (a) Since $f'(1)$ is the slope of the graph at $x = 1$, we see in Figure 2.8 that $f'(1)$ is positive.

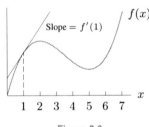

Figure 2.8

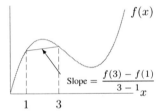

Figure 2.9

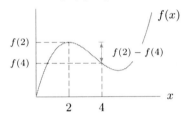

Figure 2.10

(b) The difference quotient $(f(3) - f(1))/(3 - 1)$ is the slope of the secant line between $x = 1$ and $x = 3$. We see from Figure 2.9 that this slope is positive.

(c) Since $f(4)$ is the value of the function at $x = 4$ and $f(2)$ is the value of the function at $x = 2$, the expression $f(4) - f(2)$ is the change in the function between $x = 2$ and $x = 4$. Since $f(4)$ lies below $f(2)$, this change is negative. See Figure 2.10.

Estimating the Derivative of a Function Given Numerically

If we are given a table of values for a function, we can estimate values of its derivative. To do this, we have to assume that the points in the table are close enough together that the function does not change wildly between them.

Example 6   The total acreage of farms in the US[1] has decreased since 1980. See Table 2.2.

Table 2.2   *Total farm land in million acres*

Year	1980	1985	1990	1995	2000
Farm land (million acres)	1039	1012	987	963	945

(a) What was the average rate of change in farm land between 1980 and 2000?

(b) Estimate $f'(1995)$ and interpret your answer in terms of farm land.

[1] *Statistical Abstracts of the United States 2004–2005*, Table 796.

Solution   (a) Between 1980 and 2000,

$$\text{Average rate of change} = \frac{945 - 1039}{2000 - 1980} = \frac{-94}{20} = -4.7 \text{ million acres per year.}$$

Between 1980 and 2000, the amount of farm land was decreasing at an average rate of 4.7 million acres per year.

(b) We use the interval from 1995 to 2000 to estimate the instantaneous rate of change at 1995:

$$f'(1995) = \begin{array}{c} \text{Rate of change} \\ \text{in 1995} \end{array} \approx \frac{945 - 963}{2000 - 1995} = \frac{-18}{5} = -3.6 \text{ million acres per year.}$$

In 1995, the amount of farm land was decreasing at a rate of approximately 3.6 million acres per year.

Problems for Section 2.1

1. Use the graph in Figure 2.7 to decide if each of the following quantities is positive, negative or approximately zero. Illustrate your answers graphically.

   (a) The average rate of change of $f(x)$ between $x = 3$ and $x = 7$.
   (b) The instantaneous rate of change of $f(x)$ at $x = 3$.

2. Figure 2.11 shows $N = f(t)$, the number of farms in the US[2] between 1930 and 2000 as a function of year, t.

   (a) Is $f'(1950)$ positive or negative? What does this tell you about the number of farms?
   (b) Which is more negative: $f'(1960)$ or $f'(1980)$? Explain.

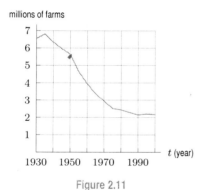

millions of farms

Figure 2.11

3. In a time of t seconds, a particle moves a distance of s meters from its starting point, where $s = 3t^2$.

   (a) Find the average velocity between $t = 1$ and $t = 1 + h$ if:
       (i) $h = 0.1$, (ii) $h = 0.01$, (iii) $h = 0.001$.
   (b) Use your answers to part (a) to estimate the instantaneous velocity of the particle at time $t = 1$.

4. Find the average velocity over the interval $0 \le t \le 0.8$, and estimate the velocity at $t = 0.2$ of a car whose position, s, is given by the following table.

t (sec)	0	0.2	0.4	0.6	0.8	1.0
s (ft)	0	0.5	1.8	3.8	6.5	9.6

5. The distance (in feet) of an object from a point is given by $s(t) = t^2$, where time t is in seconds.

   (a) What is the average velocity of the object between $t = 3$ and $t = 5$?
   (b) By using smaller and smaller intervals around 3, estimate the instantaneous velocity at time $t = 3$.

6. Figure 2.12 shows the cost, $y = f(x)$, of manufacturing x kilograms of a chemical.

   (a) Is the average rate of change of the cost greater between $x = 0$ and $x = 3$, or between $x = 3$ and $x = 5$? Explain your answer graphically.
   (b) Is the instantaneous rate of change of the cost of producing x kilograms greater at $x = 1$ or at $x = 4$? Explain your answer graphically.
   (c) What are the units of these rates of change?

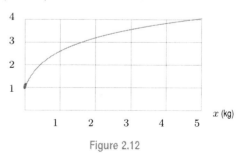

y (thousand $)

Figure 2.12

---

[2] www.nass.usda.gov:81/ipedb/farmnum.htm, accessed April 11, 2005.

7. The size, S, of a tumor (in cubic millimeters) is given by $S = 2^t$, where t is the number of months since the tumor was discovered. Give units with your answers.

    (a) What is the total change in the size of the tumor during the first six months?

    (b) What is the average rate of change in the size of the tumor during the first six months?

    (c) Estimate the rate at which the tumor is growing at $t = 6$. (Use smaller and smaller intervals.)

8. **(a)** Using Table 2.3, find the average rate of change in the world's population between 1975 and 2005. Give units.

    (b) If $P = f(t)$ with t in years, estimate $f'(2005)$ and give units.

Table 2.3   *World population, in billions of people*

Year	1975	1980	1985	1990	1995	2000	2005
Population	4.08	4.45	4.84	5.27	5.68	6.07	6.45

9. Let $f(x) = 5^x$. Use a small interval to estimate $f'(2)$. Now improve your accuracy by estimating $f'(2)$ again, using an even smaller interval.

10. **(a)** Let $g(t) = (0.8)^t$. Use a graph to determine whether $g'(2)$ is positive, negative, or zero.

    (b) Use a small interval to estimate $g'(2)$.

11. **(a)** The function f is given in Figure 2.13. At which of the labeled points is $f'(x)$ positive? Negative? Zero?

    (b) At which labeled point is f' largest? At which labeled point is f' most negative?

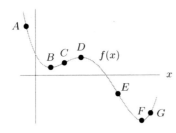

Figure 2.13

12. Estimate $f'(2)$ for $f(x) = 3^x$. Explain your reasoning.

13. Figure 2.14 shows the graph of f. Match the derivatives in the table with the points a, b, c, d, e.

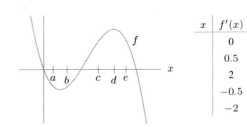

x	$f'(x)$
	0
	0.5
	2
	−0.5
	−2

Figure 2.14

14. The following table gives the percent of the US population living in urban areas as a function of year.[3]

Year	1800	1830	1860	1890	1920
Percent	6.0	9.0	19.8	35.1	51.2
Year	1950	1980	1990	2000	
Percent	64.0	73.7	75.2	79.0	

    (a) Find the average rate of change of the percent of the population living in urban areas between 1890 and 1990.

    (b) Estimate the rate at which this percent is increasing at the year 1990.

    (c) Estimate the rate of change of this function for the year 1830 and explain what it is telling you.

    (d) Is this function increasing or decreasing?

15. **(a)** Graph $f(x) = x^2$ and $g(x) = x^2 + 3$ on the same axes. What can you say about the slopes of the tangent lines to the two graphs at the point $x = 0$? $x = 1$? $x = 2$? $x = a$, where a is any value?

    (b) Explain why adding a constant to any function will not change the value of the derivative at any point.

16. Table 2.4 gives $P = f(t)$, the percent of households in the US with cable television t years since 1990.[4]

    (a) Does $f'(6)$ appear to be positive or negative? What does this tell you about the percent of households with cable television?

    (b) Estimate $f'(2)$. Estimate $f'(10)$. Explain what each is telling you, in terms of cable television.

Table 2.4

t (years since 1990)	0	2	4	6	8	10	12
P (% with cable)	59.0	61.5	63.4	66.7	67.4	67.8	68.9

17. Estimate $P'(0)$ if $P(t) = 200(1.05)^t$. Explain how you obtained your answer.

[3] *Statistical Abstracts of the US*, 1985, US Department of Commerce, Bureau of the Census, p. 22, and *World Almanac and Book of Facts 2005*, p. 624 (New York).
[4] *The World Almanac and Book of Facts 2005*, p. 310 (New York).

18. The function in Figure 2.15 has $f(4) = 25$ and $f'(4) = 1.5$. Find the coordinates of the points A, B, C.

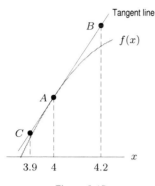

Figure 2.15

19. Use Figure 2.16 to fill in the blanks in the following statements about the function g at point B.

   (a) $g(\underline{}) = \underline{}$ (b) $g'(\underline{}) = \underline{}$

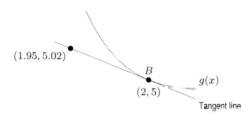

Figure 2.16

20. Show how to represent the following on Figure 2.17.

   (a) $f(4)$ (b) $f(4) - f(2)$
   (c) $\dfrac{f(5) - f(2)}{5 - 2}$ (d) $f'(3)$

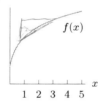

Figure 2.17

21. For each of the following pairs of numbers, use Figure 2.17 to decide which is larger. Explain your answer.

   (a) $f(3)$ or $f(4)$?
   (b) $f(3) - f(2)$ or $f(2) - f(1)$?

   (c) $\dfrac{f(2) - f(1)}{2 - 1}$ or $\dfrac{f(3) - f(1)}{3 - 1}$?
   (d) $f'(1)$ or $f'(4)$?

22. Estimate the instantaneous rate of change of the function $f(x) = x \ln x$ at $x = 1$ and at $x = 2$. What do these values suggest about the concavity of the graph between 1 and 2?

23. On October 17, 2006, in an article called "US Population Reaches 300 Million," the BBC reported that the US gains 1 person every 11 seconds. If $f(t)$ is the US population in millions t years after October 17, 2006, find $f(0)$ and $f'(0)$.

24. The following table shows the number of hours worked in a week, $f(t)$, hourly earnings, $g(t)$, in dollars, and weekly earnings, $h(t)$, in dollars, of production workers as functions of t, the year.[5]

   (a) Indicate whether each of the following derivatives is positive, negative, or zero: $f'(t), g'(t), h'(t)$. Interpret each answer in terms of hours or earnings.
   (b) Estimate each of the following derivatives, and interpret your answers:

   (i) $f'(1970)$ and $f'(1995)$

   (ii) $g'(1970)$ and $g'(1995)$

   (iii) $h'(1970)$ and $h'(1995)$

t	1970	1975	1980	1985	1990	1995	2000
$f(t)$	37.0	36.0	35.2	34.9	34.3	34.3	34.3
$g(t)$	3.40	4.73	6.84	8.73	10.09	11.64	14.00
$h(t)$	125.80	170.28	240.77	304.68	349.29	399.53	480.41

2.2 THE DERIVATIVE FUNCTION

In Section 2.1 we looked at the derivative of a function at a point. In general, the derivative takes on different values at different points and is itself a function. Recall that the derivative is the slope of the tangent line to the graph at the point.

Finding the Derivative of a Function Given Graphically

Example 1   Estimate the derivative of the function $f(x)$ graphed in Figure 2.18 at $x = -2, -1, 0, 1, 2, 3, 4, 5$.

---

[5]*The World Almanac and Book of Facts 2005,* p. 151 (New York). Production workers includes nonsupervisory workers in mining, manufacturing, construction, transportation, public utilities, wholesale and retail trade, finance, insurance, real estate, and services.

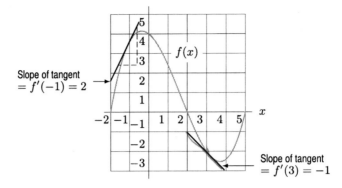

Figure 2.18: Estimating the derivative graphically as the slope of a tangent line

Solution   From the graph, we estimate the derivative at any point by placing a straightedge so that it forms the tangent line at that point, and then using the grid to estimate the slope of the tangent line. For example, the tangent at $x = -1$ is drawn in Figure 2.18, and has a slope of about 2, so $f'(-1) \approx 2$. Notice that the slope at $x = -2$ is positive and fairly large; the slope at $x = -1$ is positive but smaller. At $x = 0$, the slope is negative, by $x = 1$ it has become more negative, and so on. Some estimates of the derivative, to the nearest integer, are listed in Table 2.5. You should check these values yourself. Is the derivative positive where you expect? Negative?

Table 2.5   *Estimated values of derivative of function in Figure 2.18*

x	-2	-1	0	1	2	3	4	5
Derivative at x	6	2	-1	-2	-2	-1	1	4

The important point to notice is that for every x-value, there is a corresponding value of the derivative. The derivative, therefore, is a function of x.

> For a function f, we define the **derivative function**, f', by
>
> $$f'(x) = \text{Instantaneous rate of change of } f \text{ at } x.$$

Example 2   Plot the values of the derivative function calculated in Example 1. Compare the graphs of f' and f.

Solution   Graphs of f and f' are in Figures 2.19 and 2.20, respectively. Notice that f' is positive (its graph is above the x-axis) where f is increasing, and f' is negative (its graph is below the x-axis) where f is decreasing. The value of $f'(x)$ is 0 where f has a maximum or minimum value (at approximately $x = -0.4$ and $x = 3.7$).

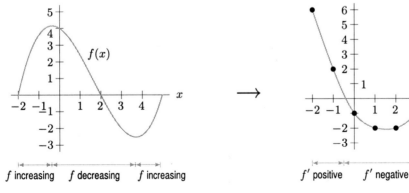

Figure 2.19: The function f

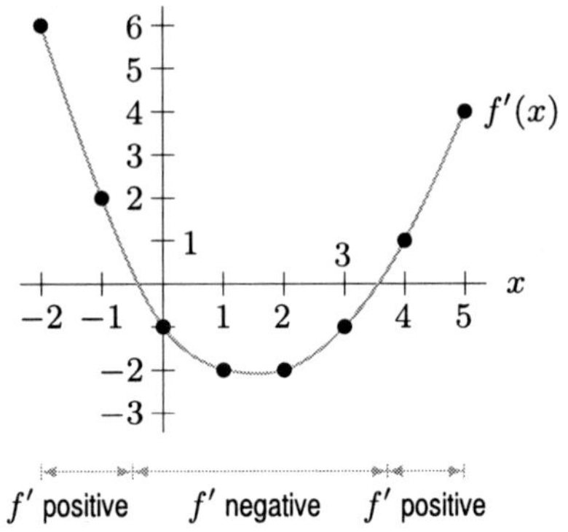

Figure 2.20: Estimates of the derivative, f'

Example 3   The graph of f is in Figure 2.21. Which of the graphs (a)–(c) is a graph of the derivative, f'?

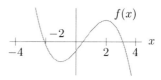

Figure 2.21

(a) 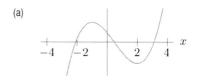   (b) 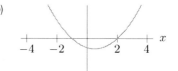   (c)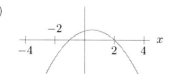

Solution   Since the graph of $f(x)$ is horizontal at $x = -1$ and $x = 2$, the derivative is zero there. Therefore, the graph of $f'(x)$ has x-intercepts at $x = -1$ and $x = 2$.

The function f is decreasing for $x < -1$, increasing for $-1 < x < 2$, and decreasing for $x > 2$. The derivative is positive (its graph is above the x-axis) where f is increasing, and the derivative is negative (its graph is below the x-axis) where f is decreasing. The correct graph is (c).

What Does the Derivative Tell Us Graphically?

Where the derivative, f', of a function is positive, the tangent to the graph of f is sloping up; where f' is negative, the tangent is sloping down. If $f' = 0$ everywhere, then the tangent is horizontal everywhere and so f is constant. The sign of the derivative f' tells us whether the function f is increasing or decreasing.

> If $f' > 0$ on an interval, then f is *increasing* over that interval.
> If $f' < 0$ on an interval, then f is *decreasing* over that interval.
> If $f' = 0$ on an interval, then f is *constant* over that interval.

The magnitude of the derivative gives us the magnitude of the rate of change of f. If f' is large in magnitude, then the graph of f is steep (up if f' is positive or down if f' is negative); if f' is small in magnitude, the graph of f is gently sloping.

Estimating the Derivative of a Function Given Numerically

If we are given a table of function values instead of a graph of the function, we can estimate values of the derivative.

Example 4   Table 2.6 gives values of $c(t)$, the concentration (mg/cc) of a drug in the bloodstream at time t (min). Construct a table of estimated values for $c'(t)$, the rate of change of $c(t)$ with respect to t.

Table 2.6   *Concentration of a drug as a function of time*

t (min)	0	0.1	0.2	0.3	0.4	0.5	0.6	0.7	0.8	0.9	1.0
$c(t)$ (mg/cc)	0.84	0.89	0.94	0.98	1.00	1.00	0.97	0.90	0.79	0.63	0.41

Solution   To estimate the derivative of c using the values in the table, we assume that the data points are close enough together that the concentration does not change wildly between them. From the table, we see that the concentration is increasing between $t = 0$ and $t = 0.4$, so we expect a positive derivative there. From $t = 0.5$ to $t = 1.0$, the concentration starts to decrease, and the rate of decrease gets larger and larger, so we would expect the derivative to be negative and of greater and greater magnitude.

We estimate the derivative for each value of t using a difference quotient. For example,

$$c'(0) \approx \frac{c(0.1) - c(0)}{0.1 - 0} = \frac{0.89 - 0.84}{0.1} = 0.5 \text{ (mg/cc) per minute.}$$

Similarly, we get the estimates

$$c'(0.1) \approx \frac{c(0.2) - c(0.1)}{0.2 - 0.1} = \frac{0.94 - 0.89}{0.1} = 0.5$$

$$c'(0.2) \approx \frac{c(0.3) - c(0.2)}{0.3 - 0.2} = \frac{0.98 - 0.94}{0.1} = 0.4$$

and so on. These values are tabulated in Table 2.7. Notice that the derivative has small positive values up until $t = 0.4$, and then it gets more and more negative, as we expected.

Table 2.7 *Derivative of concentration*

t	0	0.1	0.2	0.3	0.4	0.5	0.6	0.7	0.8	0.9
$c'(t)$	0.5	0.5	0.4	0.2	0.0	−0.3	−0.7	−1.1	−1.6	−2.2

Improving Numerical Estimates for the Derivative

In the previous example, our estimate for the derivative of $c(t)$ at $t = 0.2$ used the point to the right. We found the average rate of change between $t = 0.2$ and $t = 0.3$. However, we could equally well have gone to the left and used the rate of change between $t = 0.1$ and $t = 0.2$ to approximate the derivative at 0.2. For a more accurate result, we could average these slopes, getting the approximation

$$c'(0.2) \approx \frac{1}{2} \left(\begin{array}{c} \text{Slope to left} \\ \text{of 0.2} \end{array} + \begin{array}{c} \text{Slope to right} \\ \text{of 0.2} \end{array} \right) = \frac{0.5 + 0.4}{2} = 0.45.$$

Each of these methods of approximating the derivative gives a reasonable answer. We will usually estimate the derivative by going to the right.

Finding the Derivative of a Function Given by a Formula

If we are given a formula for a function f, can we come up with a formula for f'? Using the definition of the derivative, we often can. Indeed, much of the power of calculus depends on our ability to find formulas for the derivatives of all the familiar functions. This is explained in detail in Chapter 3. In the next example, we see how to guess a formula for the derivative.

Example 5   Guess a formula for the derivative of $f(x) = x^2$.

Solution   We use difference quotients to estimate the values of $f'(1)$, $f'(2)$, and $f'(3)$. Then we look for a pattern in these values which we use to guess a formula for $f'(x)$.

Near $x = 1$, we have

$$f'(1) \approx \frac{1.001^2 - 1^2}{0.001} = \frac{1.002 - 1}{0.001} = \frac{0.002}{0.001} = 2.$$

Similarly,

$$f'(2) \approx \frac{2.001^2 - 2^2}{0.001} = \frac{4.004 - 4}{0.001} = \frac{0.004}{0.001} = 4$$

$$f'(3) \approx \frac{3.001^2 - 3^2}{0.001} = \frac{9.006 - 9}{0.001} = \frac{0.006}{0.001} = 6.$$

Knowing the value of f' at specific points cannot tell us the formula for f', but it can be suggestive: knowing $f'(1) \approx 2$, $f'(2) \approx 4$, $f'(3) \approx 6$ suggests that $f'(x) = 2x$. In Chapter 3, we show that this is indeed the case.

Problems for Section 2.2

1. The graph of $f(x)$ is given in Figure 2.22. Draw tangent lines to the graph at $x = -2$, $x = -1$, $x = 0$, and $x = 2$. Estimate $f'(-2)$, $f'(-1)$, $f'(0)$, and $f'(2)$.

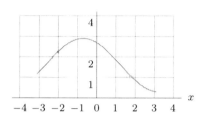

Figure 2.22

For Problems 2–7, graph the derivative of the given functions.

2.
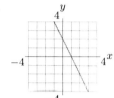

3.
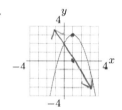

4.
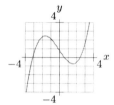

5.
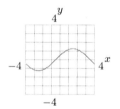

6.
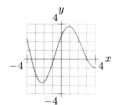

7.
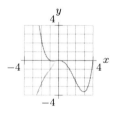

8. Given the numerical values shown, find approximate values for the derivative of $f(x)$ at each of the x-values given. Where is the rate of change of $f(x)$ positive? Where is it negative? Where does the rate of change of $f(x)$ seem to be greatest?

x	0	1	2	3	4	5	6	7	8
$f(x)$	18	13	10	9	9	11	15	21	30

9. Find approximate values for $f'(x)$ at each of the x-values given in the following table.

x	0	5	10	15	20
$f(x)$	100	70	55	46	40

10. In the graph of f in Figure 2.23, at which of the labeled x-values is

(a) $f(x)$ greatest?   **(b)** $f(x)$ least?

(c) $f'(x)$ greatest?   **(d)** $f'(x)$ least?

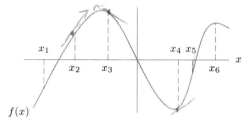

Figure 2.23

Sketch the graphs of the derivatives of the functions shown in Problems 11–18. Be sure your sketches are consistent with the important features of the graphs of the original functions.

11.
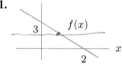

12.
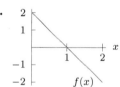

13.
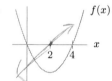

14.
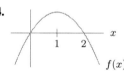

15.

16.

17.

18.

19. A city grew in population throughout the 1980s. The population was at its largest in 1990, and then shrank throughout the 1990s. Let $P = f(t)$ represent the population of the city t years since 1980. Sketch graphs of $f(t)$ and $f'(t)$, labeling the units on the axes.

20. Values of x and $g(x)$ are given in the table. For what value of x does $g'(x)$ appear to be closest to 3?

x	2.7	3.2	3.7	4.2	4.7	5.2	5.7	6.2
$g(x)$	3.4	4.4	5.0	5.4	6.0	7.4	9.0	11.0

Match the functions in Problems 21–24 with one of the derivatives in Figure 2.24.

(I)
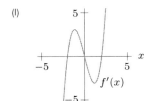

(II)
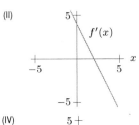

(III)
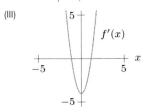

(IV)
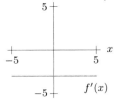

(V)
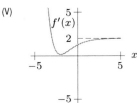

(VI)
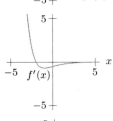

(VII)
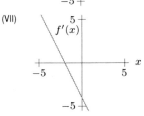

(VIII)
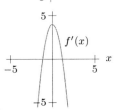

Figure 2.24

21.
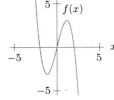

22.
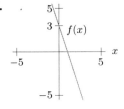

23.
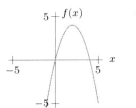

24.
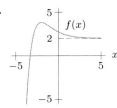

25. Draw a possible graph of $y = f(x)$ given the following information about its derivative.

- $f'(x) > 0$ for $x < -1$
- $f'(x) < 0$ for $x > -1$
- $f'(x) = 0$ at $x = -1$

26. Draw the graph of a continuous function $y = f(x)$ that satisfies the following three conditions:

- $f'(x) > 0$ for $1 < x < 3$
- $f'(x) < 0$ for $x < 1$ and $x > 3$
- $f'(x) = 0$ at $x = 1$ and $x = 3$

27. A vehicle moving along a straight road has distance $f(t)$ from its starting point at time t. Which of the graphs in Figure 2.25 could be $f'(t)$ for the following scenarios? (Assume the scales on the vertical axes are all the same.)

(a) A bus on a popular route, with no traffic
(b) A car with no traffic and all green lights
(c) A car in heavy traffic conditions

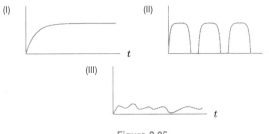

Figure 2.25

28. (a) Let $f(x) = \ln x$. Use small intervals to estimate $f'(1)$, $f'(2)$, $f'(3)$, $f'(4)$, and $f'(5)$.
(b) Use your answers to part (a) to guess a formula for the derivative of $f(x) = \ln x$.

29. Suppose $f(x) = \frac{1}{3}x^3$. Estimate $f'(2)$, $f'(3)$, and $f'(4)$. What do you notice? Can you guess a formula for $f'(x)$?

30. Match each of the following descriptions of a function with one of the following graphs.

(a) $f'(1) > 0$ and f' is always decreasing
(b) $f'(1) > 0$ and f' is always increasing
(c) $f'(1) < 0$ and f' is always decreasing
(d) $f'(1) < 0$ and f' is always increasing

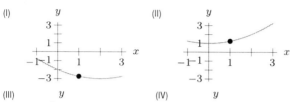
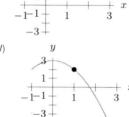

31. A child inflates a balloon, admires it for a while and then lets the air out at a constant rate. If $V(t)$ gives the volume of the balloon at time t, then Figure 2.26 shows $V'(t)$ as a function of t. At what time does the child:

(a) Begin to inflate the balloon?
(b) Finish inflating the balloon?
(c) Begin to let the air out?
(d) What would the graph of $V'(t)$ look like if the child had alternated between pinching and releasing the open end of the balloon, instead of letting the air out at a constant rate?

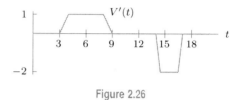

Figure 2.26

2.3 INTERPRETATIONS OF THE DERIVATIVE

We have seen the derivative interpreted as a slope and as a rate of change. In this section, we see other interpretations. The purpose of these examples is not to make a catalog of interpretations but to illustrate the process of obtaining them. There is another notation for the derivative that is often helpful.

An Alternative Notation for the Derivative

So far we have used the notation f' to stand for the derivative of the function f. An alternative notation for derivatives was introduced by the German mathematician Gottfried Wilhelm Leibniz (1646–1716) when calculus was first being developed. We know that $f'(x)$ is approximated by the average rate of change over a small interval. If $y = f(x)$, then the average rate of change is given by $\Delta y/\Delta x$. For small Δx, we have

$$f'(x) \approx \frac{\Delta y}{\Delta x}.$$

Leibniz's notation for the derivative, dy/dx, is meant to remind us of this. If $y = f(x)$, then we write

$$f'(x) = \frac{dy}{dx}.$$

Leibniz's notation is quite suggestive, especially if we think of the letter d in dy/dx as standing for "small difference in . . . ." The notation dy/dx reminds us that the derivative is a limit of ratios of the form

$$\frac{\text{Difference in } y\text{-values}}{\text{Difference in } x\text{-values}}.$$

The notation dy/dx is useful for determining the units for the derivative: the units for dy/dx are the units for y divided by (or "per") the units for x.

The separate entities dy and dx officially have no independent meaning: they are part of one notation. In fact, a good formal way to view the notation dy/dx is to think of d/dx as a single symbol meaning "the derivative with respect to x of . . .". Thus, dy/dx could be viewed as

$$\frac{d}{dx}(y), \quad \text{meaning "the derivative with respect to } x \text{ of } y."$$

On the other hand, many scientists and mathematicians really do think of dy and dx as separate entities representing "infinitesimally" small differences in y and x, even though it is difficult to say exactly how small "infinitesimal" is. It may not be formally correct, but it is very helpful intuitively to think of dy/dx as a very small change in y divided by a very small change in x.

For example, recall that if $s = f(t)$ is the position of a moving object at time t, then $v = f'(t)$ is the velocity of the object at time t. Writing

$$v = \frac{ds}{dt}$$

reminds us that v is a velocity since the notation suggests a distance, ds, over a time, dt, and we know that distance over time is velocity. Similarly, we recognize

$$\frac{dy}{dx} = f'(x)$$

as the slope of the graph of $y = f(x)$ by remembering that slope is vertical rise, dy, over horizontal run, dx.

The disadvantage of the Leibniz notation is that it is awkward to specify the value at which a derivative is evaluated. To specify $f'(2)$, for example, we have to write

$$\left.\frac{dy}{dx}\right|_{x=2}.$$

Using Units to Interpret the Derivative

Suppose a body moves along a straight line. If $s = f(t)$ gives the position in meters of the body from a fixed point on the line as a function of time, t, in seconds, then knowing that

$$\frac{ds}{dt} = f'(2) = 10 \text{ meters/sec}$$

tells us that when $t = 2$ sec, the body is moving at a velocity of 10 meters/sec. If the body continues to move at this velocity for a whole second (from $t = 2$ to $t = 3$), it would move an additional 10 meters.

In other words, if Δs represents the change in position during a time interval Δt, and if the body continues to move at this velocity, we have $\Delta s = 10\Delta t$, so

$$\Delta s = f'(2)\,\Delta t = 10\,\Delta t.$$

If the velocity is varying, this relationship is no longer exact. For small values of Δt, we have the *Tangent Line Approximation*

$$\Delta s \approx f'(t)\Delta t.$$

(See also Local Linear Approximation in this section.) Notice that, for a given positive Δt, a large derivative gives large change in s; a small derivative gives small change in s. In general:

- The units of the derivative of a function are the units of the dependent variable divided by the units of the independent variable. In other words, the units of dA/dB are the units of A divided by the units of B.

- If the derivative of a function is not changing rapidly near a point, then the derivative is approximately equal to the change in the function when the independent variable increases by 1 unit.

The following examples illustrate how useful units can be in suggesting interpretations of the derivative.

Example 1   The cost C (in dollars) of building a house A square feet in area is given by the function $C = f(A)$. What are the units and the practical interpretation of the function $f'(A)$?

Solution   In the Leibniz notation,

$$f'(A) = \frac{dC}{dA}.$$

This is a cost divided by an area, so it is measured in dollars per square foot. You can think of dC as the extra cost of building an extra dA square feet of house. So if you are planning to build a house with area A square feet, $f'(A)$ is approximately the cost per square foot of the *extra* area involved in building a slightly larger house, and is called the *marginal cost*.

Example 2   The cost of extracting T tons of ore from a copper mine is $C = f(T)$ dollars. What does it mean to say that $f'(2000) = 100$?

Solution   In the Leibniz notation,

$$f'(2000) = \frac{dC}{dT}\bigg|_{T=2000}.$$

Since C is measured in dollars and T is measured in tons, dC/dT is measured in dollars per ton. You can think of dC as the extra cost of extracting an extra dT tons of ore. So the statement

$$\frac{dC}{dT}\bigg|_{T=2000} = 100$$

says that when 2000 tons of ore have already been extracted from the mine, the cost of extracting the next ton is approximately $100. Another way of saying this is that it costs about $100 to extract the 2001st ton.

Example 3   If $q = f(p)$ gives the number of thousands of tons of zinc produced when the price is p dollars per ton, then what are the units and the meaning of

$$\frac{dq}{dp}\bigg|_{p=900} - 0.2?$$

Solution   The units of dq/dp are the units of q over the units of p, or thousands of tons per dollar. You can think of dq as the extra zinc produced when the price increases by dp. The statement

$$\frac{dq}{dp}\bigg|_{p=900} = f'(900) = 0.2 \text{ thousand tons per dollar}$$

tells us that the instantaneous rate of change of q with respect to p is 0.2 when $p = 900$. This means that when the price is $900, the quantity produced increases by about 0.2 thousand tons, or 200 tons for a one-dollar increase in price.

Example 4   The time, L (in hours), that a drug stays in a person's system is a function of the quantity administered, q, in mg, so $L = f(q)$.

(a) Interpret the statement $f(10) = 6$. Give units for the numbers 10 and 6.
(b) Write the derivative of the function $L = f(q)$ in Leibniz notation. If $f'(10) = 0.5$, what are the units of the 0.5?
(c) Interpret the statement $f'(10) = 0.5$ in terms of dose and duration.

Solution   (a) We know that $f(q) = L$. In the statement $f(10) = 6$, we have $q = 10$ and $L = 6$, so the units are 10 mg and 6 hours. The statement $f(10) = 6$ tells us that a dose of 10 mg lasts 6 hours.
(b) Since $L = f(q)$, we see that L depends on q. The derivative of this function is dL/dq. Since L is in hours and q is in mg, the units of the derivative are hours per mg. In the statement $f'(10) = 0.5$, the 0.5 is the derivative and the units are hours per mg.
(c) The statement $f'(10) = 0.5$ tells us that, at a dose of 10 mg, the instantaneous rate of change of duration is 0.5 hour per mg. In other words, if we increase the dose by 1 mg, the drug stays in the body approximately 30 minutes longer.

In the previous example, notice that $f'(10) = 0.5$ tells us that a 1 mg increase in dose leads to about a 0.5-hour increase in duration. If, on the other hand, we had had $f'(10) = 20$, we would have known that a 1-mg increase in dose leads to about a 20-hour increase in duration. Thus the derivative is the multiplier relating changes in dose to changes in duration. The magnitude of the derivative tells us how sensitive the time is to changes in dose.

We define the derivative of velocity, dv/dt, as *acceleration*.

Example 5   If the velocity of a body at time t seconds is measured in meters/sec, what are the units of the acceleration?

Solution   Since acceleration, dv/dt, is the derivative of velocity, the units of acceleration are units of velocity divided by units of time, or (meters/sec)/sec, written meters/sec^2.

Using the Derivative to Estimate Values of a Function

Since the derivative tells us how fast the value of a function is changing, we can use the derivative at a point to estimate values of the function at nearby points.

Example 6   Fertilizers can improve agricultural production. A Cornell University study[6] on maize (corn) production in Kenya found that the average value, $y = f(x)$, in Kenyan shillings of the yearly maize production from an average plot of land is a function of the quantity, x, of fertilizer used in kilograms. (The shilling is the Kenyan unit of currency.)

(a) Interpret the statements $f(5) = 11,500$ and $f'(5) = 350$.
(b) Use the statements in part (a) to estimate $f(6)$ and $f(10)$.
(c) The value of the derivative, f', is increasing for x near 5. Which estimate in part (b) is more reliable?

Solution   (a) The statement $f(5) = 11,500$ tells us that $y = 11,500$ when $x = 5$. This means that if 5 kg of fertilizer are applied, maize worth 11,500 Kenyan shillings is produced. Since the derivative is dy/dx, the statement $f'(5) = 350$ tells us that

$$\frac{dy}{dx} = 350 \quad \text{when } x = 5.$$

This means that if the amount of fertilizer used is 5 kg and increases by 1 kg, then maize production increases by about 350 Kenyan shillings.

(b) We want to estimate $f(6)$, that is the production when 6 kg of fertilizer are used. If instead of 5 kg of fertilizer, one more kilogram is used, giving 6 kg altogether, we expect production, in Kenyan shillings, to increase from 11,500 by about 350. Thus,

$$f(6) \approx 11,500 + 350 = 11,850.$$

Similarly, if 5 kg more fertilizer is used, so 10 kg are used altogether, we expect production to increase by about $5 \cdot 350 = 1750$ Kenyan shillings, so production is approximately

$$f(10) \approx 11,500 + 1750 = 13,250.$$

(c) To estimate $f(6)$, we assume that production increases at rate of 350 Kenyan shillings per kilogram between $x = 5$ and $x = 6$ kg. To estimate $f(10)$, we assume that production continues to increase at the same rate all the way from $x = 5$ to $x = 10$ kg. Since the derivative is increasing for x near 5, the estimate of $f(6)$ is more reliable.

[6]"State-conditional fertilizer yield response on western Kenyan farms", by P. Marenya, C. Barrett, Social Science Research Network, abstract = 1141937.

In Example 6, representing the change in y by Δy and the change in x by Δx, we used the result introduced earlier in this section:

Local Linear Approximation

If $y = f(x)$ and Δx is near 0, then $\Delta y \approx f'(x)\Delta x$. Then for x near a and $\Delta x = x - a$,

$$f(x) \approx f(a) + f'(a)\Delta x.$$

This is called the Tangent Line Approximation.

Relative Rate of Change

In Section 1.5, we saw that an exponential function has a constant percent rate of change. Now we link this idea to derivatives. Analogous to the relative change, we look at the rate of change as a fraction of the original quantity.

The **relative rate of change** of $y = f(t)$ at $t = a$ is defined to be

$$\text{Relative rate of change of } y \text{ at } a = \frac{dy/dt}{y} = \frac{f'(a)}{f(a)}.$$

We see in Section 3.3 that an exponential function has a constant relative rate of change. If the independent variable is time, the relative rate is often given as a percent change per unit time.

Example 7   Annual world soybean production, $W = f(t)$, in million tons, is a function of t years since the start of 2000.

(a) Interpret the statements $f(8) = 253$ and $f'(8) = 17$ in terms of soybean production.
(b) Calculate the relative rate of change of W at $t = 8$; interpret it in terms of soybean production.

Solution   (a) The statement $f(8) = 253$ tells us that 253 million tons of soybeans were produced in the year 2008. The statement $f'(8) = 17$ tells us that in 2008 annual soybean production was increasing at a rate of 17 million tons per year.
(b) We have

$$\text{Relative rate of change of soybean production} = \frac{f'(8)}{f(8)} = \frac{17}{253} = 0.067.$$

In 2008, annual soybean production was increasing at a rate of 6.7% per year.

Example 8   Solar photovoltaic (PV) cells are the world's fastest growing energy source.[7] Annual production of PV cells, S, in megawatts, is approximated by $S = 277e^{0.368t}$, where t is in years since 2000. Estimate the relative rate of change of PV cell production in 2010 using

(a) $\Delta t = 1$ (b) $\Delta t = 0.1$ (c) $\Delta t = 0.01$

Solution   Let $S = f(t)$. The relative rate of change of f at $t = 10$ is $f'(10)/f(10)$. We estimate $f'(10)$ using a difference quotient.

(a) Estimating the relative rate of change using $\Delta t = 1$ at $t = 10$, we have

$$\frac{dS/dt}{S} = \frac{f'(10)}{f(10)} \approx \frac{1}{f(10)} \frac{f(11) - f(10)}{1} = 0.445 = 44.5\% \text{ per year}$$

---

[7]*Vital Signs 2007-2008*, The Worldwatch Institute, W.W. Norton & Company, 2007, p. 38.

(b) With $\Delta t = 0.1$ and $t = 10$, we have

$$\frac{dS/dt}{S} = \frac{f'(10)}{f(10)} \approx \frac{1}{f(10)} \frac{f(10.1) - f(10)}{0.1} = 0.375 = 37.5\% \text{ per year}$$

(c) With $\Delta t = 0.01$ and $t = 10$, we have

$$\frac{dS/dt}{S} = \frac{f'(10)}{f(10)} \approx \frac{1}{f(10)} \frac{f(10.01) - f(10)}{0.01} = 0.369 = 36.9\% \text{ per year}$$

The relative rate of change is approximately 36.9% per year. From Section 1.6, we know that the exponential function $S = 277e^{0.368t}$ has a continuous rate of change for all t of 36.8% per year, which is the exact relative rate of change of this function.

Example 9   In April 2009, the US Bureau of Economic Analysis announced that the US gross domestic product (GDP) was decreasing at an annual rate of 6.1%. The GDP of the US at that time was 13.84 trillion dollars. Calculate the annual rate of change of the US GDP in April 2009.

Solution   The Bureau of Economic Analysis is reporting the relative rate of change. In April 2009, the relative rate of change of GDP was -0.061 per year. To find the rate of change, we use:

$$\text{Relative rate of change in April 2009} = \frac{\text{Rate of change in April 2009}}{\text{GDP in April 2009}}$$

$$-0.061 = \frac{\text{Rate of change}}{13.84}$$

$$\text{Rate of change in April 2009} = -0.061 \cdot 13.84 = -0.84424 \text{ trillion dollars per year.}$$

The GDP of the US was decreasing at a rate of 844.24 billion dollars per year in April 2009.

Problems for Section 2.3

In Problems 1–4, write the Leibniz notation for the derivative of the given function and include units.

1. The cost, C, of a steak, in dollars, is a function of the weight, W, of the steak, in pounds.

2. The distance to the ground, D, in feet, of a skydiver is a function of the time t in minutes since the skydiver jumped out of the airplane.

3. An employee's pay, P, in dollars, for a week is a function of the number of hours worked, H.

4. The number, N, of gallons of gas left in a gas tank is a function of the distance, D, in miles, the car has been driven.

5. The cost, $C = f(w)$, in dollars of buying a chemical is a function of the weight bought, w, in pounds.

(a) In the statement $f(12) = 5$, what are the units of the 12? What are the units of the 5? Explain what this is saying about the cost of buying the chemical.

(b) Do you expect the derivative f' to be positive or negative? Why?

(c) In the statement $f'(12) = 0.4$, what are the units of the 12? What are the units of the 0.4? Explain what this is saying about the cost of buying the chemical.

6. The time for a chemical reaction, T (in minutes), is a function of the amount of catalyst present, a (in milliliters), so $T = f(a)$.

(a) If $f(5) = 18$, what are the units of 5? What are the units of 18? What does this statement tell us about the reaction?

(b) If $f'(5) = -3$, what are the units of 5? What are the units of -3? What does this statement tell us?

7. An economist is interested in how the price of a certain item affects its sales. At a price of \$$p$, a quantity, q, of the item is sold. If $q = f(p)$, explain the meaning of each of the following statements:

(a) $f(150) = 2000$ (b) $f'(150) = -25$

8. Figure 2.27 shows the length, L, in cm, of a sturgeon (a type of fish) as a function of the time, t, in years.[8] Estimate $f'(10)$. Give units and interpret your answer.

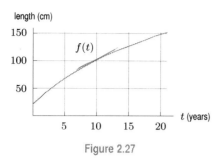

length (cm)

Figure 2.27

9. The temperature, T, in degrees Fahrenheit, of a cold yam placed in a hot oven is given by $T = f(t)$, where t is the time in minutes since the yam was put in the oven.

(a) What is the sign of $f'(t)$? Why?

(b) What are the units of $f'(20)$? What is the practical meaning of the statement $f'(20) = 2$?

10. On May 9, 2007, CBS Evening News had a 4.3 point rating. (Ratings measure the number of viewers.) News executives estimated that a 0.1 drop in the ratings for the CBS Evening News corresponds to a $5.5 million drop in revenue.[9] Express this information as a derivative. Specify the function, the variables, the units, and the point at which the derivative is evaluated.

11. When you breathe, a muscle (called the diaphragm) reduces the pressure around your lungs and they expand to fill with air. The table shows the volume of a lung as a function of the reduction in pressure from the diaphragm. Pulmonologists (lung doctors) define the *compliance* of the lung as the derivative of this function.[10]

(a) What are the units of compliance?

(b) Estimate the maximum compliance of the lung.

(c) Explain why the compliance gets small when the lung is nearly full (around 1 liter).

Pressure reduction (cm of water)	Volume (liters)
0	0.20
5	0.29
10	0.49
15	0.70
20	0.86
25	0.95
30	1.00

12. Meteorologists define the temperature lapse rate to be $-dT/dz$ where T is the air temperature in Celsius at altitude z kilometers above the ground.

(a) What are the units of the lapse rate?

(b) What is the practical meaning of a lapse rate of 6.5?

13. Investing $1000 at an annual interest rate of $r\%$, compounded continuously, for 10 years gives you a balance of B, where $B = g(r)$. Give a financial interpretation of the statements:

(a) $g(5) \approx 1649$.

(b) $g'(5) \approx 165$. What are the units of $g'(5)$?

14. Let $f(x)$ be the elevation in feet of the Mississippi River x miles from its source. What are the units of $f'(x)$? What can you say about the sign of $f'(x)$?

15. The average weight, W, in pounds, of an adult is a function, $W = f(c)$, of the average number of Calories per day, c, consumed.

(a) Interpret the statements $f(1800) = 155$ and $f'(2000) = 0$ in terms of diet and weight.

(b) What are the units of $f'(c) = dW/dc$?

16. The cost, C (in dollars), to produce g gallons of a chemical can be expressed as $C = f(g)$. Using units, explain the meaning of the following statements in terms of the chemical:

(a) $f(200) = 1300$ (b) $f'(200) = 6$

17. The weight, W, in lbs, of a child is a function of its age, a, in years, so $W = f(a)$.

(a) Do you expect $f'(a)$ to be positive or negative? Why?

(b) What does $f(8) = 45$ tell you? Give units for the numbers 8 and 45.

(c) What are the units of $f'(a)$? Explain what $f'(a)$ tells you in terms of age and weight.

(d) What does $f'(8) = 4$ tell you about age and weight?

(e) As a increases, do you expect $f'(a)$ to increase or decrease? Explain.

18. Let G be annual US government purchases, T be annual US tax revenues, and Y be annual US output of all goods and services. All three quantities are given in dollars. Interpret the statements about the two derivatives, called fiscal policy multipliers.

(a) $dY/dG = 0.60$ (b) $dY/dT = -0.26$

19. A recent study reports that men who retired late developed Alzheimer's at a later stage than those who stopped work earlier. Each additional year of employment was associated with about a six-week later age of onset. Express these results as a statement about the derivative of a function. State clearly what function you use, including the units of the dependent and independent variables.

[8] Data from von Bertalanffy, L., *General System Theory*, p. 177 (New York: Braziller, 1968).

[9] OC Register, May 9, 2007; *The New York Times*, May 14, 2007.

[10] Adapted from John B. West, *Respiratory Physiology*, 4th Ed. (New York: Williams and Wilkins, 1990).

20. The thickness, P, in mm, of pelican eggshells depends on the concentration, c, of PCBs in the eggshell, measured in ppm (parts per million); that is, $P = f(c)$.

    (a) The derivative $f'(c)$ is negative. What does this tell you?

    (b) Give units and interpret $f(200) = 0.28$ and $f'(200) = -0.0005$ in terms of PCBs and eggs.

Problems 21–24 concern $g(t)$ in Figure 2.28, which gives the weight of a human fetus as a function of its age.

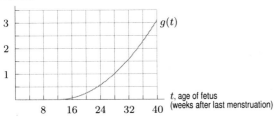

weight (kg)

Figure 2.28

21. (a) What are the units of $g'(24)$?

    (b) What is the biological meaning of $g'(24) = 0.096$?

22. (a) Which is greater, $g'(20)$ or $g'(36)$?

    (b) What does your answer say about fetal growth?

23. Is the instantaneous weight growth rate greater or less than the average rate of change of weight over the 40-week period

    (a) At week 20?   **(b)** At week 36?

24. Estimate **(a)** $g'(20)$ **(b)** $g'(36)$

    (c) The average rate of change of weight for the entire 40-week gestation.

25. Suppose that $f(t)$ is a function with $f(25) = 3.6$ and $f'(25) = -0.2$. Estimate $f(26)$ and $f(30)$.

26. Suppose that $f(x)$ is a function with $f(20) = 345$ and $f'(20) = 6$. Estimate $f(22)$.

27. Annual net sales, in billion of dollars, for the Hershey Company, the largest US producer of chocolate, is a function $S = f(t)$ of time, t, in years since 2000.

    (a) Interpret the statements $f(8) = 5.1$ and $f'(8) = 0.22$ in terms of Hershey sales.[11]

    (b) Estimate $f(12)$ and interpret it in terms of Hershey sales.

28. World meat[12] production, $M = f(t)$, in millions of metric tons, is a function of t, years since 2000.

    (a) Interpret $f(5) = 249$ and $f'(5) = 6.5$ in terms of meat production.

    (b) Estimate $f(10)$ and interpret it in terms of meat production.

29. For some painkillers, the size of the dose, D, given depends on the weight of the patient, W. Thus, $D = f(W)$, where D is in milligrams and W is in pounds.

    (a) Interpret the statements $f(140) = 120$ and $f'(140) = 3$ in terms of this painkiller.

    (b) Use the information in the statements in part (a) to estimate $f(145)$.

30. The quantity, Q mg, of nicotine in the body t minutes after a cigarette is smoked is given by $Q = f(t)$.

    (a) Interpret the statements $f(20) = 0.36$ and $f'(20) = -0.002$ in terms of nicotine. What are the units of the numbers 20, 0.36, and -0.002?

    (b) Use the information given in part (a) to estimate $f(21)$ and $f(30)$. Justify your answers.

31. A mutual fund is currently valued at \$80 per share and its value per share is increasing at a rate of \$0.50 a day. Let $V = f(t)$ be the value of the share t days from now.

    (a) Express the information given about the mutual fund in term of f and f'.

    (b) Assuming that the rate of growth stays constant, estimate and interpret $f(10)$.

32. Figure 2.29 shows how the contraction velocity, $v(x)$, of a muscle changes as the load on it changes.

    (a) Find the slope of the line tangent to the graph of contraction velocity at a load of 2 kg. Give units.

    (b) Using your answer to part (a), estimate the change in the contraction velocity if the load is increased from 2 kg by adding 50 grams.

    (c) Express your answer to part (a) as a derivative of $v(x)$.

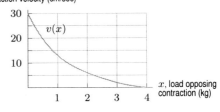

Figure 2.29

---

[11]2008 Annual Report to Stockholders, accessed at www.thehersheycompany.com.

[12]*FAO Statistical Yearbook 2007–2008*, Food and Agriculture Organization of the United Nations, http://www.fao.org/economic/ess/publications-studies/statistical-yearbook/fao-statistical-yearbook-2007-2008.

33. Figure 2.30 shows how the pumping rate of a person's heart changes after bleeding.

   (a) Find the slope of the line tangent to the graph at time 2 hours. Give units.
   (b) Using your answer to part (a), estimate how much the pumping rate increases during the minute beginning at time 2 hours.
   (c) Express your answer to part (a) as a derivative of $g(t)$.

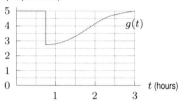

Figure 2.30

34. Suppose $C(r)$ is the total cost of paying off a car loan borrowed at an annual interest rate of $r\%$. What are the units of $C'(r)$? What is the practical meaning of $C'(r)$? What is its sign?

35. A company's revenue from car sales, C (in thousands of dollars), is a function of advertising expenditure, a, in thousands of dollars, so $C = f(a)$.

   (a) What does the company hope is true about the sign of f'?
   (b) What does the statement $f'(100) = 2$ mean in practical terms? How about $f'(100) = 0.5$?
   (c) Suppose the company plans to spend about $100,000 on advertising. If $f'(100) = 2$, should the company spend more or less than $100,000 on advertising? What if $f'(100) = 0.5$?

36. A person with a certain liver disease first exhibits larger and larger concentrations of certain enzymes (called SGOT and SGPT) in the blood. As the disease progresses, the concentration of these enzymes drops, first to the predisease level and eventually to zero (when almost all of the liver cells have died). Monitoring the levels of these enzymes allows doctors to track the progress of a patient with this disease. If $C = f(t)$ is the concentration of the enzymes in the blood as a function of time,

   (a) Sketch a possible graph of $C = f(t)$.
   (b) Mark on the graph the intervals where $f' > 0$ and where $f' < 0$.
   (c) What does $f'(t)$ represent, in practical terms?

Problems 37–41 refer to Figure 2.31, which shows the depletion of food stores in the human body during starvation.

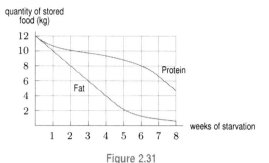

Figure 2.31

37. Which is being consumed at a greater rate, fat or protein, during the

   (a) Third week?   **(b)** Seventh week?

38. The fat storage graph is linear for the first four weeks. What does this tell you about the use of stored fat?

39. Estimate the rate of fat consumption after

   (a) 3 weeks   **(b)** 6 weeks   **(c)** 8 weeks

40. What seems to happen during the sixth week? Why do you think this happens?

41. Figure 2.32 shows the derivatives of the protein and fat storage functions. Which graph is which?

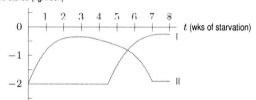

Figure 2.32

42. The area of Brazil's rain forest, $R = f(t)$, in million acres, is a function of the number of years, t, since 2000.

   (a) Interpret $f(9) = 740$ and $f'(9) = -2.7$ in terms of Brazil's rain forests.[13]
   (b) Find and interpret the relative rate of change of $f(t)$ when $t = 9$.

43. The number of active Facebook users hit 175 million at the end of February 2009 and 200 million[14] at the end of April 2009. With t in months since the start of 2009, let $f(t)$ be the number of active users in millions. Estimate $f(4)$ and $f'(4)$ and the relative rate of change of f at $t = 4$. Interpret your answers in terms of Facebook users.

44. The weight, w, in kilograms, of a baby is a function $f(t)$ of her age, t, in months.

   (a) What does $f(2.5) = 5.67$ tell you?
   (b) What does $f'(2.5)/f(2.5) = 0.13$ tell you?

[13]www.rain-tree.com/facts.htm, accessed June 2009.
[14]www.facebook.com/press, accessed June 2009.

45. Estimate the relative rate of change of $f(t) = t^2$ at $t = 4$. Use $\Delta t = 0.01$.

46. The population, P, of a city (in thousands) at time t (in years) is $P = 700e^{0.035t}$. Estimate the relative rate of change of the population at $t = 3$ using

(a) $\Delta t = 1$ (b) $\Delta t = 0.1$ (c) $\Delta t = 0.01$

2.4 THE SECOND DERIVATIVE

glabelsec:2second-derivative

Since the derivative is itself a function, we can calculate its derivative. For a function f, the derivative of its derivative is called the *second derivative*, and written f''. If $y = f(x)$, the second derivative can also be written as $\dfrac{d^2y}{dx^2}$, which means $\dfrac{d}{dx}\left(\dfrac{dy}{dx}\right)$, the derivative of $\dfrac{dy}{dx}$.

What Does the Second Derivative Tell Us?

Recall that the derivative of a function tells us whether the function is increasing or decreasing:

If $f' > 0$ on an interval, then f is increasing over that interval.
If $f' < 0$ on an interval, then f is decreasing over that interval.

Since f'' is the derivative of f', we have

If $f'' > 0$ on an interval, then f' is increasing over that interval.
If $f'' < 0$ on an interval, then f' is decreasing over that interval.

So the question becomes: What does it mean for f' to be increasing or decreasing? The case in which f' is increasing is shown in Figure 2.33, where the graph of f is bending upward, or is *concave up*. In the case when f' is decreasing, shown in Figure 2.34, the graph is bending downward, or is *concave down*.

$f'' > 0$ on an interval means f' is increasing, so the graph of f is concave up there.
$f'' < 0$ on an interval means f' is decreasing, so the graph of f is concave down there.

Figure 2.33: Meaning of f'': The slope increases from negative to positive as you move from left to right, so f'' is positive and f is concave up

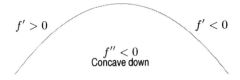

Figure 2.34: Meaning of f'': The slope decreases from positive to negative as you move from left to right, so f'' is negative and f is concave down

---

Example 1   For the functions whose graphs are given in Figure 2.35, decide where their second derivatives are positive and where they are negative.

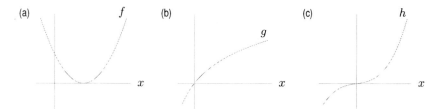

Figure 2.35: What signs do the second derivatives have?

Solution   From the graphs it appears that

(a) $f'' > 0$ everywhere, because the graph of f is concave up everywhere.

(b) $g'' < 0$ everywhere, because the graph is concave down everywhere.

(c) $h'' > 0$ for $x > 0$, because the graph of h is concave up there; $h'' < 0$ for $x < 0$, because the graph of h is concave down there.

Interpretation of the Second Derivative as a Rate of Change

If we think of the derivative as a rate of change, then the second derivative is a rate of change of a rate of change. If the second derivative is positive, the rate of change is increasing; if the second derivative is negative, the rate of change is decreasing.

The second derivative is often a matter of practical concern. In 1985 a newspaper headline reported the Secretary of Defense as saying that Congress and the Senate had cut the defense budget. As his opponents pointed out, however, Congress had merely cut the rate at which the defense budget was increasing.[15] In other words, the derivative of the defense budget was still positive (the budget was increasing), but the second derivative was negative (the budget's rate of increase had slowed).

Example 2   A population, P, growing in a confined environment often follows a *logistic* growth curve, like the graph shown in Figure 2.36. Describe how the rate at which the population is increasing changes over time. What is the sign of the second derivative d^2P/dt^2? What is the practical interpretation of t^* and L?

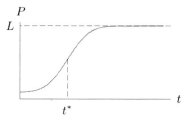

Figure 2.36: Logistic growth curve

Solution   Initially, the population is increasing, and at an increasing rate. So, initially dP/dt is increasing and $d^2P/dt^2 > 0$. At t^*, the rate at which the population is increasing is a maximum; the population is growing fastest then. Beyond t^*, the rate at which the population is growing is decreasing, so $d^2P/dt^2 < 0$. At t^*, the graph changes from concave up to concave down and $d^2P/dt^2 = 0$.

The quantity L represents the limiting value of the population that is approached as t tends to infinity; L is called the *carrying capacity* of the environment and represents the maximum population that the environment can support.

[15]In the *Boston Globe*, March 13, 1985, Representative William Gray (D–Pa.) was reported as saying: "It's confusing to the American people to imply that Congress threatens national security with reductions when you're really talking about a reduction in the increase."

Example 3   Table 2.8 shows the number of abortions per year, A, reported in the US[16] in the year t.

Table 2.8   *Abortions reported in the US (1972–2000)*

Year, t	1972	1975	1980	1985	1990	1995	2000	2005
1000s of abortions reported, A	587	1034	1554	1589	1609	1359	1313	1206

(a) Calculate the average rate of change for the time intervals shown between 1972 and 2005.

(b) What can you say about the sign of d^2A/dt^2 during the period 1972–1995?

Solution   (a) For each time interval we can calculate the average rate of change of the number of abortions per year over this interval. For example, between 1972 and 1975

$$\begin{array}{c} \text{Average rate} \\ \text{of change} \end{array} = \frac{\Delta A}{\Delta t} = \frac{1034 - 587}{1975 - 1972} = \frac{447}{3} = 149.$$

Thus, between 1972 and 1975, there were approximately 149,000 more abortions reported each year. Values of $\Delta A/\Delta t$ are listed in Table 2.9.

Table 2.9   *Rate of change of number of abortions reported*

Time	1972–75	1975–80	1980–85	1985–90	1990–95	1995–2000	2000–05
Average rate of change, $\Delta A/\Delta t$ (1000s/year)	149	104	7	4	−50	−9.2	−21.4

(b) We assume the data lies on a smooth curve. Since the values of $\Delta A/\Delta t$ are decreasing dramatically for 1975–1995, we can be pretty certain that dA/dt also decreases, so d^2A/dt^2 is negative for this period. For 1972–1975, the sign of d^2A/dt^2 is less clear; abortion data from 1968 would help. Figure 2.37 confirms this; the graph appears to be concave down for 1975–1995. The fact that dA/dt is positive during the period 1972–1980 corresponds to the fact that the number of abortions reported increased from 1972 to 1980. The fact that dA/dt is negative during the period 1990–2005 corresponds to the fact that the number of abortions reported decreased from 1990 to 2005. The fact that d^2A/dt^2 is negative for 1975–1995 reflects the fact that the rate of increase slowed over this period.

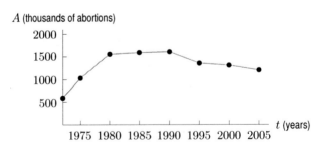

Figure 2.37: How the number of reported abortions in the US is changing with time

---

[16]*Statistical Abstracts of the United States 2004–2009*, Table 99.

Problems for Section 2.4

1. For the function graphed in Figure 2.38, are the following nonzero quantities positive or negative?

   (a) $f(2)$ **(b)** $f'(2)$ **(c)** $f''(2)$

   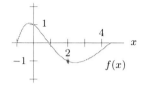

   Figure 2.38

2. At one of the labeled points on the graph in Figure 2.39 both dy/dx and d^2y/dx^2 are positive. Which is it?

   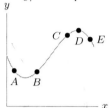

   Figure 2.39

For Problems 3–8, give the signs of the first and second derivatives for the following functions. Each derivative is either positive everywhere, zero everywhere, or negative everywhere.

3.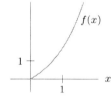

4.

5.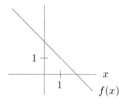

6.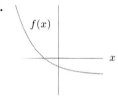

7.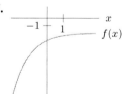

8.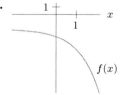

9. Graph the functions described in parts (a)–(d).

   (a) First and second derivatives everywhere positive.
   (b) Second derivative everywhere negative; first derivative everywhere positive.
   (c) Second derivative everywhere positive; first derivative everywhere negative.
   (d) First and second derivatives everywhere negative.

In Problems 10–11, use the values given for each function.

 (a) Does the derivative of the function appear to be positive or negative over the given interval? Explain.

 (b) Does the second derivative of the function appear to be positive or negative over the given interval? Explain.

10.

t	100	110	120	130	140
$w(t)$	10.7	6.3	4.2	3.5	3.3

11.

t	0	1	2	3	4	5
$s(t)$	12	14	17	20	31	55

12. Sketch the graph of a function whose first derivative is everywhere negative and whose second derivative is positive for some x-values and negative for other x-values.

13. IBM-Peru uses second derivatives to assess the relative success of various advertising campaigns. They assume that all campaigns produce some increase in sales. If a graph of sales against time shows a positive second derivative during a new advertising campaign, what does this suggest to IBM management? Why? What does a negative second derivative suggest?

14. Values of $f(t)$ are given in the following table.

   (a) Does this function appear to have a positive or negative first derivative? Second derivative? Explain.
   (b) Estimate $f'(2)$ and $f'(8)$.

t	0	2	4	6	8	10
$f(t)$	150	145	137	122	98	56

15. The table gives the number of passenger cars, $C = f(t)$, in millions,[17] in the US in the year t.

   (a) Do $f'(t)$ and $f''(t)$ appear to be positive or negative during the period 1940–1980?
   (b) Estimate $f'(1975)$. Using units, interpret your answer in terms of passenger cars.

t	1940	1950	1960	1970	1980	1990	2000
C	27.5	40.3	61.7	89.2	121.6	133.7	133.6

---

[17] www.bts.gov/publications/national_transportation_statistics/html/table_01_11.html. Accessed June 22, 2008.

In Problems 16–17, use the graph given for each function.

(a) Estimate the intervals on which the derivative is positive and the intervals on which the derivative is negative.

(b) Estimate the intervals on which the second derivative is positive and the intervals on which the second derivative is negative.

16.

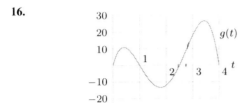

17.

$(-\infty, -\frac{1}{2})$ INC.

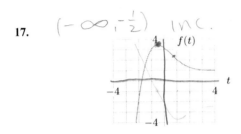

18. Sketch a graph of a continuous function f with the following properties:

- $f'(x) > 0$ for all x
- $f''(x) < 0$ for $x < 2$ and $f''(x) > 0$ for $x > 2$.

19. At exactly two of the labeled points in Figure 2.40, the derivative f' is 0; the second derivative f'' is not zero at any of the labeled points. On a copy of the table, give the signs of f, f', f'' at each marked point.

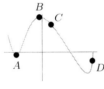

Point	f	f'	f''
A			
B			
C			
D			

Figure 2.40

20. For three minutes the temperature of a feverish person has had positive first derivative and negative second derivative. Which of the following is correct?

(a) The temperature rose in the last minute more than it rose in the minute before.

(b) The temperature rose in the last minute, but less than it rose in the minute before.

(c) The temperature fell in the last minute but less than it fell in the minute before.

(d) The temperature rose two minutes ago but fell in the last minute.

21. Yesterday's temperature at t hours past midnight was $f(t)$ °C. At noon the temperature was 20°C. The first derivative, $f'(t)$, decreased all morning, reaching a low of 2°C/hour at noon, then increased for the rest of the day. Which one of the following must be correct?

(a) The temperature fell in the morning and rose in the afternoon.

(b) At 1 pm the temperature was 18°C.

(c) At 1 pm the temperature was 22°C.

(d) The temperature was lower at noon than at any other time.

(e) The temperature rose all day.

22. Sketch the graph of a function f such that $f(2) = 5$, $f'(2) = 1/2$, and $f''(2) > 0$.

23. A function f has $f(5) = 20$, $f'(5) = 2$, and $f''(x) < 0$, for $x \geq 5$. Which of the following are possible values for $f(7)$ and which are impossible?

(a) 26   (b) 24   (c) 22

24. An industry is being charged by the Environmental Protection Agency (EPA) with dumping unacceptable levels of toxic pollutants in a lake. Over a period of several months, an engineering firm makes daily measurements of the rate at which pollutants are being discharged into the lake. The engineers produce a graph similar to either Figure 2.41(a) or Figure 2.41(b). For each case, give an idea of what argument the EPA might make in court against the industry and of the industry's defense.

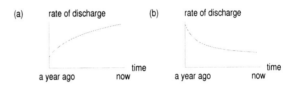

Figure 2.41

25. "Winning the war on poverty" has been described cynically as slowing the rate at which people are slipping below the poverty line. Assuming that this is happening:

(a) Graph the total number of people in poverty against time.

(b) If N is the number of people below the poverty line at time t, what are the signs of dN/dt and d^2N/dt^2? Explain.

26. Let $P(t)$ represent the price of a share of stock of a corporation at time t. What does each of the following statements tell us about the signs of the first and second derivatives of $P(t)$?

(a) "The price of the stock is rising faster and faster."

(b) "The price of the stock is close to bottoming out."

27. In economics, *total utility* refers to the total satisfaction from consuming some commodity. According to the economist Samuelson:[18]

> As you consume more of the same good, the total (psychological) utility increases. However, ...with successive new units of the good, your total utility will grow at a slower and slower rate because of a fundamental tendency for your psychological ability to appreciate more of the good to become less keen.

(a) Sketch the total utility as a function of the number of units consumed.

(b) In terms of derivatives, what is Samuelson saying?

28. Each of the graphs in Figure 2.42 shows the position of a particle moving along the x-axis as a function of time, $0 \le t \le 5$. The vertical scales of the graphs are the same. During this time interval, which particle has

(a) Constant velocity?

(b) The greatest initial velocity?

(c) The greatest average velocity?

(d) Zero average velocity?

(e) Zero acceleration?

(f) Positive acceleration throughout?

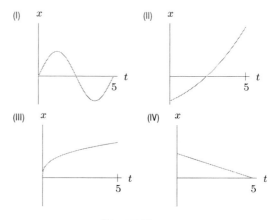

Figure 2.42

2.5 MARGINAL COST AND REVENUE

Management decisions within a particular firm or industry usually depend on the costs and revenues involved. In this section we look at the cost and revenue functions.

Graphs of Cost and Revenue Functions

The graph of a cost function may be linear, as in Figure 2.43, or it may have the shape shown in Figure 2.44. The intercept on the C-axis represents the fixed costs, which are incurred even if nothing is produced. (This includes, for instance, the cost of the machinery needed to begin production.) In Figure 2.44, the cost function increases quickly at first and then more slowly because producing larger quantities of a good is usually more efficient than producing smaller quantities—this is called *economy of scale*. At still higher production levels, the cost function increases faster again as resources become scarce; sharp increases may occur when new factories have to be built. Thus, the graph of a cost function, C, may start out concave down and become concave up later on.

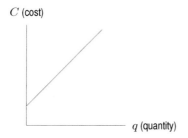

Figure 2.43: A linear cost function

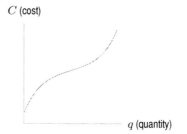

Figure 2.44: A nonlinear cost function

The revenue function is $R = pq$, where p is price and q is quantity. If the price, p, is a constant, the graph of R against q is a straight line through the origin with slope equal to the price. (See

[18]From Paul A. Samuelson, *Economics*, 11th edition (New York: McGraw-Hill, 1981).

Figure 2.45.) In practice, for large values of q, the market may become glutted, causing the price to drop and giving R the shape in Figure 2.46.

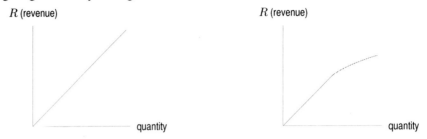

Figure 2.45: Revenue: Constant price   Figure 2.46: Revenue: Decreasing price

Example 1   If cost, C, and revenue, R, are given by the graph in Figure 2.47, for what production quantities does the firm make a profit?

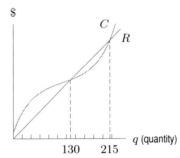

Figure 2.47: Costs and revenues for Example 1

Solution   The firm makes a profit whenever revenues are greater than costs, that is, when $R > C$. The graph of R is above the graph of C approximately when $130 < q < 215$. Production between 130 units and 215 units will generate a profit.

Marginal Analysis

Many economic decisions are based on an analysis of the costs and revenues "at the margin." Let's look at this idea through an example.

Suppose you are running an airline and you are trying to decide whether to offer an additional flight. How should you decide? We'll assume that the decision is to be made purely on financial grounds: if the flight will make money for the company, it should be added. Obviously you need to consider the costs and revenues involved. Since the choice is between adding this flight and leaving things the way they are, the crucial question is whether the *additional costs* incurred are greater or smaller than the *additional revenues* generated by the flight. These additional costs and revenues are called *marginal costs* and *marginal revenues*.

Suppose $C(q)$ is the function giving the cost of running q flights. If the airline had originally planned to run 100 flights, its costs would be $C(100)$. With the additional flight, its costs would be $C(101)$. Therefore,

$$\text{Additional cost "at the margin"} = C(101) - C(100).$$

Now

$$C(101) - C(100) = \frac{C(101) - C(100)}{101 - 100},$$

and this quantity is the average rate of change of cost between 100 and 101 flights. In Figure 2.48 the average rate of change is the slope of the secant line. If the graph of the cost function is not curving too fast near the point, the slope of the secant line is close to the slope of the tangent line

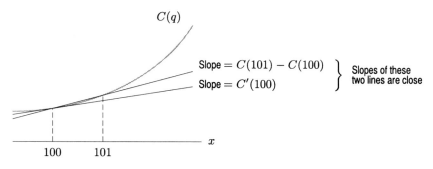

Figure 2.48: Marginal cost: Slope of one of these lines

there. Therefore, the average rate of change is close to the instantaneous rate of change. Since these rates of change are not very different, many economists choose to define marginal cost, MC, as the instantaneous rate of change of cost with respect to quantity:

$$\text{Marginal cost} = MC = C'(q) \qquad \text{so} \qquad \text{Marginal cost} \approx C(q+1) - C(q).$$

Marginal cost is represented by the slope of the cost curve.

Similarly if the revenue generated by q flights is $R(q)$ and the number of flights increases from 100 to 101, then

$$\text{Additional revenue "at the margin"} = R(101) - R(100).$$

Now $R(101) - R(100)$ is the average rate of change of revenue between 100 and 101 flights. As before, the average rate of change is approximately equal to the instantaneous rate of change, so economists often define

$$\text{Marginal revenue} = MR = R'(q) \qquad \text{so} \qquad \text{Marginal revenue} \approx R(q+1) - R(q).$$

Example 2 If $C(q)$ and $R(q)$ for the airline are given in Figure 2.49, should the company add the 101st flight?

Solution The marginal revenue is the slope of the revenue curve at $q = 100$. The marginal cost is the slope of the graph of C at $q = 100$. Figure 2.49 suggests that the slope at point A is smaller than the slope at B, so $MC < MR$ for $q = 100$. This means that the airline will make more in extra revenue than it will spend in extra costs if it runs another flight, so it should go ahead and run the 101st flight.

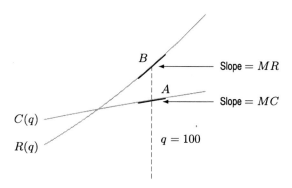

Figure 2.49: Cost and revenue for Example 2

117

Example 3   The graph of a cost function is given in Figure 2.50. Does it cost more to produce the 500th item or the 2000th? Does it cost more to produce the 3000th item or the 4000th? At approximately what production level is marginal cost smallest? What is the total cost at this production level?

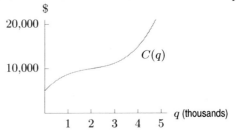

Figure 2.50: Estimating marginal cost: Where is marginal cost smallest?

Solution   The cost to produce an additional item is the marginal cost, which is represented by the slope of the cost curve. Since the slope of the cost function in Figure 2.50 is greater at $q = 0.5$ (when the quantity produced is 0.5 thousand, or 500) than at $q = 2$, it costs more to produce the 500th item than the 2000th item. Since the slope is greater at $q = 4$ than $q = 3$, it costs more to produce the 4000th item than the 3000th item.

The slope of the cost function is close to zero at $q = 2$, and is positive everywhere else, so the slope is smallest at $q = 2$. The marginal cost is smallest at a production level of 2000 units. Since $C(2) \approx 10,000$, the total cost to produce 2000 units is about $10,000.

Example 4   If the revenue and cost functions, R and C, are given by the graphs in Figure 2.51, sketch graphs of the marginal revenue and marginal cost functions, MR and MC.

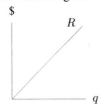   

Figure 2.51: Total revenue and total cost for Example 4

Solution   The revenue graph is a line through the origin, with equation

$$R = pq$$

where p represents the constant price, so the slope is p and

$$MR = R'(q) = p.$$

The total cost is increasing, so the marginal cost is always positive. For small q values, the graph of the cost function is concave down, so the marginal cost is decreasing. For larger q, say $q > 100$, the graph of the cost function is concave up and the marginal cost is increasing. Thus, the marginal cost has a minimum at about $q = 100$. (See Figure 2.52.)

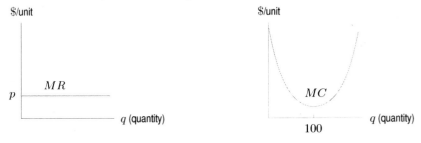

Figure 2.52: Marginal revenue and costs for Example 4

Problems for Section 2.5

1. The function $C(q)$ gives the cost in dollars to produce q barrels of olive oil.

   (a) What are the units of marginal cost?

   (b) What is the practical meaning of the statement $MC = 3$ for $q = 100$?

2. It costs \$4800 to produce 1295 items and it costs \$4830 to produce 1305 items. What is the approximate marginal cost at a production level of 1300 items?

3. In Figure 2.53, is marginal cost greater at $q = 5$ or at $q = 30$? At $q = 20$ or at $q = 40$? Explain.

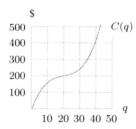

Figure 2.53

4. In Figure 2.54, estimate the marginal cost when the level of production is 10,000 units and interpret it.

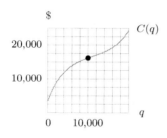

Figure 2.54

5. In Figure 2.55, estimate the marginal revenue when the level of production is 600 units and interpret it.

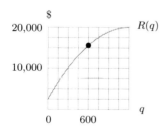

Figure 2.55

6. For q units of a product, a manufacturer's cost is $C(q)$ dollars and revenue is $R(q)$ dollars, with $C(500) = 7200$, $R(500) = 9400$, $MC(500) = 15$, and $MR(500) = 20$.

   (a) What is the profit or loss at $q = 500$?

   (b) If production is increased from 500 to 501 units, by approximately how much does profit change?

7. The cost of recycling q tons of paper is given in the following table. Estimate the marginal cost at $q = 2000$. Give units and interpret your answer in terms of cost. At approximately what production level does marginal cost appear smallest?

q (tons)	1000	1500	2000	2500	3000	3500
$C(q)$ (dollars)	2500	3200	3640	3825	3900	4400

8. Figure 2.56 shows part of the graph of cost and revenue for a car manufacturer. Which is greater, marginal cost or marginal revenue, at

   (a) q_1?   **(b)** q_2?

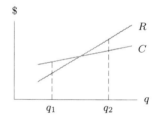

Figure 2.56

9. Let $C(q)$ represent the total cost of producing q items. Suppose $C(15) = 2300$ and $C'(15) = 108$. Estimate the total cost of producing: **(a)** 16 items **(b)** 14 items.

10. To produce 1000 items, the total cost is \$5000 and the marginal cost is \$25 per item. Estimate the costs of producing 1001 items, 999 items, and 1100 items.

11. Let $C(q)$ represent the cost and $R(q)$ represent the revenue, in dollars, of producing q items

   (a) If $C(50) = 4300$ and $C'(50) = 24$, estimate $C(52)$.

   (b) If $C'(50) = 24$ and $R'(50) = 35$, approximately how much profit is earned by the 51st item?

   (c) If $C'(100) = 38$ and $R'(100) = 35$, should the company produce the 101st item? Why or why not?

12. Cost and revenue functions for a charter bus company are shown in Figure 2.57. Should the company add a 50th bus? How about a 90th? Explain your answers using marginal revenue and marginal cost.

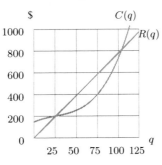

Figure 2.57

13. A company's cost of producing q liters of a chemical is $C(q)$ dollars; this quantity can be sold for $R(q)$ dollars. Suppose $C(2000) = 5930$ and $R(2000) = 7780$.

(a) What is the profit at a production level of 2000?

(b) If $MC(2000) = 2.1$ and $MR(2000) = 2.5$, what is the approximate change in profit if q is increased from 2000 to 2001? Should the company increase or decrease production from $q = 2000$?

(c) If $MC(2000) = 4.77$ and $MR(2000) = 4.32$, should the company increase or decrease production from $q = 2000$?

14. An industrial production process costs $C(q)$ million dollars to produce q million units; these units then sell for $R(q)$ million dollars. If $C(2.1) = 5.1$, $R(2.1) = 6.9$, $MC(2.1) = 0.6$, and $MR(2.1) = 0.7$, calculate

(a) The profit earned by producing 2.1 million units

(b) The approximate change in revenue if production increases from 2.1 to 2.14 million units.

(c) The approximate change in revenue if production decreases from 2.1 to 2.05 million units.

(d) The approximate change in profit in parts (b) and (c).

15. Let $C(q)$ be the total cost of producing a quantity q of a certain product. See Figure 2.58.

(a) What is the meaning of $C(0)$?

(b) Describe in words how the marginal cost changes as the quantity produced increases.

(c) Explain the concavity of the graph (in terms of economics).

(d) Explain the economic significance (in terms of marginal cost) of the point at which the concavity changes.

(e) Do you expect the graph of $C(q)$ to look like this for all types of products?

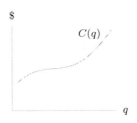

Figure 2.58

CHAPTER SUMMARY

- **Rate of change**
  Average, instantaneous

- **Estimating derivatives**
  Estimate derivatives from a graph, table of values, or formula

- **Interpretation of derivatives**
  Rate of change, slope, using units, instantaneous velocity

- **Relative rate of change**
  Calculation and interpretation

- **Marginality**
  Marginal cost and marginal revenue

- **Second derivative**
  Concavity

- **Derivatives and graphs**
  Understand relation between sign of f' and whether f is increasing or decreasing. Sketch graph of f' from graph of f. Marginal analysis

REVIEW PROBLEMS FOR CHAPTER TWO

1. In a time of t seconds, a particle moves a distance of s meters from its starting point, where $s = 4t^2 + 3$.

   (a) Find the average velocity between $t = 1$ and $t = 1 + h$ if:

   (i) $h = 0.1$,   (ii) $h = 0.01$, (iii) $h = 0.001$.

   (b) Use your answers to part (a) to estimate the instantaneous velocity of the particle at time $t = 1$.

2. Match the points labeled on the curve in Figure 2.59 with the given slopes.

Slope	Point
-3	
-1	
0	
$1/2$	
1	
2	

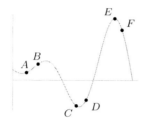

Figure 2.59

3. (a) Use a graph of $f(x) = 2 - x^3$ to decide whether $f'(1)$ is positive or negative. Give reasons.
   (b) Use a small interval to estimate $f'(1)$.

4. For the function $f(x) = 3^x$, estimate $f'(1)$. From the graph of $f(x)$, would you expect your estimate to be greater than or less than the true value of $f'(1)$?

5. For the function shown in Figure 2.60, at what labeled points is the slope of the graph positive? Negative? At which labeled point does the graph have the greatest (i.e., most positive) slope? The least slope (i.e., negative and with the largest magnitude)?

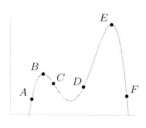

Figure 2.60

6. Use Figure 2.61 to fill in the blanks in the following statements about the function f at point A.
   (a) $f(\underline{}) = \underline{}$ (b) $f'(\underline{}) = \underline{}$

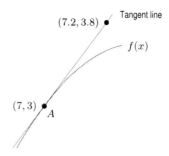

Figure 2.61

7. In a time of t seconds, a particle moves a distance of s meters from its starting point, where $s = \sin(2t)$.

   (a) Find the average velocity between $t = 1$ and $t = 1 + h$ if:

   (i) $h = 0.1$,   (ii) $h = 0.01$, (iii) $h = 0.001$.

   (b) Use your answers to part (a) to estimate the instantaneous velocity of the particle at time $t = 1$.

For Problems 8–13, sketch the graph of $f'(x)$.

8.

9.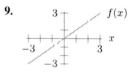

10.

11.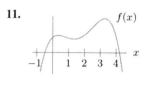

12.

13.

14. (a) Estimate $f'(2)$ using the values of f in the table.
    (b) For what values of x does $f'(x)$ appear to be positive? Negative?

x	0	2	4	6	8	10	12
$f(x)$	10	18	24	21	20	18	15

15. Figure 2.62 is the graph of f', the derivative of a function f. On what interval(s) is the function f

(a) Increasing?   (b) Decreasing?

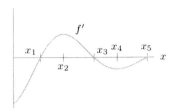

Figure 2.62: Graph of f', not f

16. The average weight W of an oak tree in kilograms that is x meters tall is given by the function $W = f(x)$. What are the units of measurement of $f'(x)$?

17. The percent, P, of US households with a personal computer is a function of the number of years, t, since 1982 (when the percent was essentially zero), so $P = f(t)$. Interpret the statements $f(20) = 57$ and $f'(20) = 3$.

18. A yam has just been taken out of the oven and is cooling off before being eaten. The temperature, T, of the yam (measured in degrees Fahrenheit) is a function of how long it has been out of the oven, t (measured in minutes). Thus, we have $T = f(t)$.

(a) Is $f'(t)$ positive or negative? Why?
(b) What are the units for $f'(t)$?

19. The quantity sold, q, of a certain product is a function of the price, p, so $q = f(p)$. Interpret each of the following statements in terms of demand for the product:

(a) $f(15) = 200$ (b) $f'(15) = -25$.

20. Figure 2.63 shows world solar energy output, in megawatts, as a function of years since 1990.[19] Estimate $f'(6)$. Give units and interpret your answer.

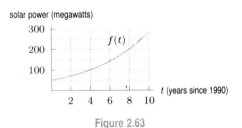

Figure 2.63

21. For a function $f(x)$, we know that $f(20) = 68$ and $f'(20) = -3$. Estimate $f(21)$, $f(19)$ and $f(25)$.

22. Table 2.10 shows world gold production,[20] $G = f(t)$, as a function of year, t.

(a) Does $f'(t)$ appear to be positive or negative? What does this mean in terms of gold production?
(b) In which time interval does $f'(t)$ appear to be greatest?
(c) Estimate $f'(2002)$. Give units and interpret your answer in terms of gold production.
(d) Use the estimated value of $f'(2002)$ to estimate $f(2003)$ and $f(2010)$, and interpret your answers.

Table 2.10   *World gold production*

t (year)	1990	1993	1996	1999	2002
G (mn troy ounces)	70.2	73.3	73.6	82.6	82.9

23. The wind speed W in meters per second at a distance x kilometers from the center of a hurricane is given by the function $W = h(x)$. What does the fact that $h'(15) > 0$ tell you about the hurricane?

24. You drop a rock from a high tower. After it falls x meters its speed S in meters per second is $S = h(x)$. What is the meaning of $h'(20) = 0.5$?

25. If t is the number of years since 2003, the population, P, of China, in billions, can be approximated by the function

$$P = f(t) = 1.291(1.006)^t.$$

Estimate $f(6)$ and $f'(6)$, giving units. What do these two numbers tell you about the population of China?

26. After investing \$1000 at an annual interest rate of 7% compounded continuously for t years, your balance is \$$B$, where $B = f(t)$. What are the units of dB/dt? What is the financial interpretation of dB/dt?

27. Let $f(t)$ be the depth, in centimeters, of water in a tank at time t, in minutes.

(a) What does the sign of $f'(t)$ tell us?
(b) Explain the meaning of $f'(30) = 20$. Include units.
(c) Using the information in part (b), at time $t = 30$ minutes, find the rate of change of depth, in meters, with respect to time in hours. Give units.

28. A climber on Mount Everest is 6000 meters from the start of a trail and at elevation 8000 meters above sea level. At x meters from the start, the elevation of the trail is $h(x)$ meters above sea level. If $h'(x) = 0.5$ for x near 6000, what is the approximate elevation another 3 meters along the trail?

29. Let $f(v)$ be the gas consumption (in liters/km) of a car going at velocity v (in km/hr). In other words, $f(v)$ tells you how many liters of gas the car uses to go one kilometer at velocity v. Explain what the following statements tell you about gas consumption:

$$f(80) = 0.05 \quad \text{and} \quad f'(80) = 0.0005.$$

[19]The Worldwatch Institute, *Vital Signs* 2001, p. 47 (New York: W.W. Norton, 2001).
[20]*The World Almanac and Book of Facts 2005*, p. 135 (New York).

30. To study traffic flow, a city installs a device which records $C(t)$, the total number of cars that have passed by t hours after 4:00 am. The graph of $C(t)$ is in Figure 2.64.

(a) When is the traffic flow the greatest?
(b) Estimate $C'(2)$.
(c) What does $C'(2)$ mean in practical terms?

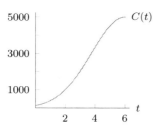

Figure 2.64

31. The temperature, H, in degrees Celsius, of a cup of coffee placed on the kitchen counter is given by $H = f(t)$, where t is in minutes since the coffee was put on the counter.

(a) Is $f'(t)$ positive or negative? Give a reason for your answer.
(b) What are the units of $f'(20)$? What is its practical meaning in terms of the temperature of the coffee?

32. Suppose that $f(x)$ is a function with $f(100) = 35$ and $f'(100) = 3$. Estimate $f(102)$.

33. Suppose $P(t)$ is the monthly payment, in dollars, on a mortgage which will take t years to pay off. What are the units of $P'(t)$? What is the practical meaning of $P'(t)$? What is its sign?

34. The table[21] shows $f(t)$, total sales of music compact discs (CDs), in millions, and $g(t)$, total sales of music cassettes, in millions, as a function of year t.

(a) Estimate $f'(2002)$ and $g'(2002)$. Give units with your answers and interpret each answer in terms of sales of CDs or cassettes.
(b) Use $f'(2002)$ to estimate $f(2003)$ and $f(2010)$. Interpret your answers in terms of sales of CDs.

Year, t	1994	1996	1998	2000	2002
CD sales, $f(t)$	662.1	778.9	847.0	942.5	803.3
Cassette sales, $g(t)$	345.4	225.3	158.5	76.0	31.1

35. For each part below, sketch a graph of a function that satisfies the given conditions. There are many possible correct answers.

(a) $f' > 0$ and $f'' > 0$.
(b) $f' > 0$ and $f'' < 0$.
(c) $f' < 0$ and $f'' > 0$.
(d) $f' < 0$ and $f'' < 0$.

36. For each function, do the signs of the first and second derivatives of the function appear to be positive or negative over the given interval?

(a)

x	1.0	1.1	1.2	1.3	1.4	1.5
f(x)	10.1	11.2	13.7	16.0	21.2	27.7

(b)

x	1.0	1.1	1.2	1.3	1.4	1.5
f(x)	10.1	9.9	8.1	6.0	3.9	0.1

(c)

x	1.0	1.1	1.2	1.3	1.4	1.5
f(x)	1000	1010	1015	1018	1020	1021

37. The length L of the day in minutes (sunrise to sunset) x kilometers north of the equator on June 21 is given by $L = f(x)$. What are the units of

(a) $f'(3000)$?   (b) $f''(3000)$?

38. Sketch the graph of a function whose first and second derivatives are everywhere positive.

39. At which of the marked x-values in Figure 2.65 can the following statements be true?

(a) $f(x) < 0$
(b) $f'(x) < 0$
(c) $f(x)$ is decreasing
(d) $f'(x)$ is decreasing
(e) Slope of $f(x)$ is positive
(f) Slope of $f(x)$ is increasing

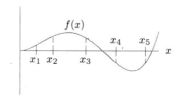

Figure 2.65

40. Sketch the graph of the height of a particle against time if velocity is positive and acceleration is negative.

41. Figure 2.66 gives the position, $f(t)$, of a particle at time t. At which of the marked values of t can the following statements be true?

(a) The position is positive
(b) The velocity is positive
(c) The acceleration is positive
(d) The position is decreasing
(e) The velocity is decreasing

[21] *The World Almanac and Book of Facts 2005*, p. 309 (New York).

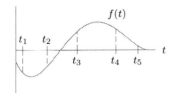

Figure 2.66

42. A high school principal is concerned about the drop in the percentage of students who graduate from her school, shown in the following table.

Year entered school, t	1992	1995	1998	2001	2004
Percent graduating, P	62.4	54.1	48.0	43.5	41.8

(a) Calculate the average rate of change of P for each of the three-year intervals between 1992 and 2004.
(b) Does d^2P/dt^2 appear to be positive or negative between 1992 and 2004?
(c) Explain why the values of P and dP/dt are troublesome to the principal.
(d) Explain why the sign of d^2P/dt^2 and the magnitude of dP/dt in the year 2001 may give the principal some cause for optimism.

43. Students were asked to evaluate $f'(4)$ from the following table which shows values of the function f:

x	1	2	3	4	5	6
$f(x)$	4.2	4.1	4.2	4.5	5.0	5.7

- Student A estimated the derivative as $f'(4) \approx \dfrac{f(5) - f(4)}{5 - 4} = 0.5$.

- Student B estimated the derivative as $f'(4) \approx \dfrac{f(4) - f(3)}{4 - 3} = 0.3$.
- Student C suggested that they should split the difference and estimate the average of these two results, that is, $f'(4) \approx \frac{1}{2}(0.5 + 0.3) = 0.4$.

(a) Sketch the graph of f, and indicate how the three estimates are represented on the graph.
(b) Explain which answer is likely to be best.

44. Figure 2.67 shows the rate at which energy, $f(v)$, is consumed by a bird flying at speed v meters/sec.

(a) What rate of energy consumption is needed by the bird to keep aloft, without moving forward?
(b) What does the shape of the graph tell you about how birds fly?
(c) Sketch $f'(v)$.

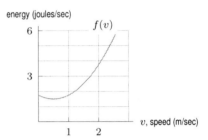

Figure 2.67

CHECK YOUR UNDERSTANDING

In Problems 1–55, indicate whether the statement is true or false.

1. The derivative of f at a is the instantaneous rate of change of f at the point a.

2. If $f(x) = x^2$ then $f'(1)$ is positive.

3. If $g(t) = 2^t$ then $g'(1) \approx (2^{1.001} - 2^1)/(1.001 - 1)$.

4. The slope of the graph of $H(x) = \sqrt{x}$ is negative at $x = 2$.

5. A function f cannot have both $f(a) = 0$ and $f'(a) = 0$.

6. The average rate of change of f between 0 and 1 is always less than the slope of the graph of f at 0.

7. The function r in the following table appears to have a negative derivative at $t = 0.5$:

t	0.3	0.5	0.7	0.9
$r(t)$	1.2	6.5	9.7	13.4

8. The derivative of f at 3 is given by the difference quotient $(f(3) - f(0))/(3 - 0)$.

9. If $R(w) = w^3$ then $R'(-2)$ is negative.

10. A function can have more than one point where the derivative is 0.

11. If $f' > 0$ on an interval, then f is increasing on that interval.

12. The f is negative on an interval, then the derivative of f is decreasing on that interval.

13. If f is always increasing, then f' is always increasing.

14. If $f(2) > g(2)$ then $f'(2) > g'(2)$.

15. The derivative of the constant function $q(x) = 5$ is $q'(x) = 0$.

16. The derivative of a function that is always decreasing is also always decreasing.

17. The function $g(t) = \ln(t)$ has $g'(1) < 0$.

18. If f is a function with $f(3) = 0$ then $f'(3) = 0$.

19. If f' is zero at all points in an interval, then the graph of f is a horizontal line on that interval.

20. If f' is negative at all points in an interval, then f is decreasing on that interval.

21. If the cost C (in dollars) of feeding x students in the dining center is given by $C = f(x)$, then the units of dC/dx are dollars per student.

22. If $y = f(x)$ then dy/dx and $f'(x)$ mean the same thing.

23. If the cost C (in dollars) of extracting T tons of ore is given by $C = f(T)$ then $f'(2000)$ is the approximate cost to extract the 2001st ton.

24. If $f(10) = 20$ and $f'(10) = 3$, then 23 is a reasonable approximation of $f(11)$.

25. The units of dA/dB are the units of A divided by the units of B.

26. If $A = f(B)$ then the units of $f'(B)$ are the units of B divided by the units of A.

27. If $f'(100) = 2$ and the derivative of f is not changing rapidly, then $f(101) - f(100) \approx 2$.

28. If t (in minutes) is the time that a dose of D milligrams of morphine is effective as a painkiller, so that $t = f(D)$, then the units of $f'(D)$ are minutes per milligram.

29. If $h = f(a)$ gives height h (in inches) of a child aged a years, then dh/da is positive when $0 < a < 10$.

30. If $W = f(R)$ and $f'(R)$ is negative, then we expect W to decrease as R increases.

31. If $f'' > 0$ on an interval, then f is increasing on that interval.

32. If $f'' < 0$ on an interval, then f is concave down on that interval.

33. If $f'' > 0$ on an interval, then $f' > 0$ on that interval.

34. If f' is decreasing on an interval, then f is concave down on that interval.

35. If a car is slowing down, then the second derivative of its distance as a function of time is negative.

36. There is a function with $f > 0$ and $f' > 0$ and $f'' > 0$ everywhere.

37. There is a function with $f > 0$ and $f' < 0$ and $f'' > 0$ everywhere.

38. There is a function with $f'' = 0$ everywhere.

39. The function $f(x) = e^x$ has $f'' > 0$ everywhere.

40. The function $f(x) = \ln(x)$ has $f'' > 0$ for $x > 0$.

41. Marginal cost is the instantaneous rate of change of cost with respect to quantity.

42. Marginal revenue is the derivative of the revenue function with respect to quantity.

43. If marginal revenue is greater than marginal cost when quantity is 1000, then increasing the quantity to 1001 will increase the profit.

44. Marginal cost is always greater than or equal to zero.

45. If the revenue function is linear with slope 5, then marginal revenue is always zero.

46. The units of marginal cost are quantity per dollar.

47. The units of marginal revenue are the same as the units of marginal cost.

48. Marginal cost is never equal to marginal revenue.

49. If the graphs of the cost and revenue functions $C(q)$ and $R(q)$ cross at q^*, then marginal revenue is equal to marginal cost at q^*

50. If the graphs of the marginal cost and marginal revenue functions $C'(q)$ and $R'(q)$ cross at q^*, then marginal revenue is equal to marginal cost at q^*.

51. If P is a function of t, then the relative rate of change of P is $(1/P)(dP/dt)$.

52. If $f(25) = 10$ gallons and $f'(25) = 2$ gallons per minute, then at $t = 25$, the function f is changing at a rate of 20% per minute.

53. If a quantity is 100 and has relative rate of change -3% per minute, we expect the quantity to be about 97 one minute later.

54. If a population of animals changes from 500 to 400 in one month, then the relative rate of change is about -100 animals per month.

55. Linear functions have constant rate of change and exponential functions have constant relative rate of change.

PROJECTS FOR CHAPTER TWO

1. Estimating the Temperature of a Yam

Suppose you put a yam in a hot oven, maintained at a constant temperature of $200°$C. As the yam picks up heat from the oven, its temperature rises.[22]

---

[22]From Peter D. Taylor, *Calculus: The Analysis of Functions* (Toronto: Wall & Emerson, Inc., 1992).

(a) Draw a possible graph of the temperature T of the yam against time t (minutes) since it is put into the oven. Explain any interesting features of the graph, and in particular explain its concavity.

(b) Suppose that, at $t = 30$, the temperature T of the yam is $120°$ and increasing at the (instantaneous) rate of $2°$/min. Using this information, plus what you know about the shape of the T graph, estimate the temperature at time $t = 40$.

(c) Suppose in addition you are told that at $t = 60$, the temperature of the yam is $165°$. Can you improve your estimate of the temperature at $t = 40$?

(d) Assuming all the data given so far, estimate the time at which the temperature of the yam is $150°$.

2. Temperature and Illumination

Alone in your dim, unheated room, you light a single candle rather than curse the darkness. Depressed with the situation, you walk directly away from the candle, sighing. The temperature (in degrees Fahrenheit) and illumination (in % of one candle power) decrease as your distance (in feet) from the candle increases. In fact, you have tables showing this information.

Distance (feet)	Temperature (°F)	Distance (feet)	Illumination (%)
0	55	0	100
1	54.5	1	85
2	53.5	2	75
3	52	3	67
4	50	4	60
5	47	5	56
6	43.5	6	53

You are cold when the temperature is below $40°$. You are in the dark when the illumination is at most 50% of one candle power.

(a) Two graphs are shown in Figures 2.68 and 2.69. One is temperature as a function of distance and one is illumination as a function of distance. Which is which? Explain.

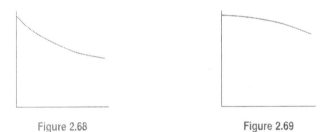

Figure 2.68   Figure 2.69

(b) What is the average rate at which the temperature is changing when the illumination drops from 75% to 56%?

(c) You can still read your watch when the illumination is about 65%. Can you still read your watch at 3.5 feet? Explain.

(d) Suppose you know that at 6 feet the instantaneous rate of change of the temperature is $-4.5°$F/ft and the instantaneous rate of change of illumination is -3% candle power/ft. Estimate the temperature and the illumination at 7 feet.

(e) Are you in the dark before you are cold, or vice versa?

FOCUS ON THEORY

LIMITS, CONTINUITY, AND THE DEFINITION OF THE DERIVATIVE

The velocity at a single instant in time is surprisingly difficult to define precisely. Consider the statement "At the instant it crossed the finish line, the horse was traveling at 42 mph." How can such a claim be substantiated? A photograph taken at that instant will show the horse motionless—it is no help at all. There is some paradox in trying to quantify the property of motion at a particular instant in time, since by focusing on a single instant we stop the motion!

A similar difficulty arises whenever we attempt to measure the rate of change of anything—for example, oil leaking out of a damaged tanker. The statement "One hour after the ship's hull ruptured, oil was leaking at a rate of 200 barrels per second" seems not to make sense. We could argue that at any given instant *no* oil is leaking.

Problems of motion were of central concern to Zeno and other philosophers as early as the fifth century BC. The approach that we took, made famous by Newton's calculus, is to stop looking for a simple notion of speed at an instant, and instead to look at speed over small intervals containing the instant. This method sidesteps the philosophical problems mentioned earlier but brings new ones of its own.

Definition of the Derivative Using Average Rates

In Section 2.1, we defined the derivative as the instantaneous rate of change of a function. We can estimate a derivative by computing average rates of change over smaller and smaller intervals. We use this idea to give a symbolic definition of the derivative. Letting h represent the size of the interval, we have

$$\text{Average rate of change between } x \text{ and } x + h = \frac{f(x+h) - f(x)}{(x+h) - x} = \frac{f(x+h) - f(x)}{h}.$$

To find the derivative, or instantaneous rate of change at the point x, we use smaller and smaller intervals. To find the derivative exactly, we take the limit as h, the size of the interval, shrinks to zero, so we say

$$\text{Derivative } = \text{Limit, as } h \text{ approaches zero, of } \frac{f(x+h) - f(x)}{h}.$$

Finally, instead of writing the phrase "limit, as h approaches 0," we use the notation $\lim_{h \to 0}$. This leads to the following symbolic definition:

For any function f, we define the **derivative function**, f', by

$$f'(x) = \lim_{h \to 0} \frac{f(x+h) - f(x)}{h},$$

provided the limit exists. The function f is said to be **differentiable** at any point x at which the derivative function is defined.

Notice that we have replaced the original difficulty of computing velocity at a point by an argument that the average rates of change approach a number as the time intervals shrink in size. In a sense, we have traded one hard question for another, since we don't yet have any idea how to be certain what number the average velocities are approaching.

The Idea of a Limit

We used a limit to define the derivative. Now we look a bit more at the idea of the limit of a function at the point c. Provided the limit exists:

> We write $\lim\limits_{x \to c} f(x)$ to represent the number approached by $f(x)$ as x approaches c.

Example 1   Investigate $\lim\limits_{x \to 2} x^2$.

Solution   Notice that we can make x^2 as close to 4 as we like by taking x sufficiently close to 2. (Look at the values of 1.9^2, 1.99^2, 1.999^2, and 2.1^2, 2.01^2, 2.001^2 in Table 2.11; they seem to be approaching 4.) We write

$$\lim_{x \to 2} x^2 = 4,$$

which is read "the limit, as x approaches 2, of x^2 is 4." Notice that the limit does not ask what happens *at* $x = 2$, so it is not sufficient to substitute 2 to find the answer. The limit describes behavior of a function *near* a point, not *at* the point.

Table 2.11   *Values of x^2 near $x = 2$*

x	1.9	1.99	1.999	2.001	2.01	2.1
x^2	3.61	3.96	3.996	4.004	4.04	4.41

Example 2   Use a graph to estimate $\lim\limits_{x \to 0} \dfrac{2^x - 1}{x}$.

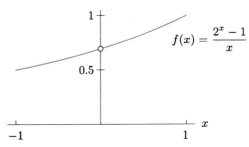

Figure 2.70: Find the limit as $x \to 0$ of $\dfrac{2^x - 1}{x}$

Solution   Notice that the expression $\dfrac{2^x - 1}{x}$ is undefined at $x = 0$. To find out what happens to this expression as x approaches 0, look at a graph of $f(x) = \dfrac{2^x - 1}{x}$. Figure 2.70 shows that as x approaches 0 from either side, the value of $\dfrac{2^x - 1}{x}$ appears to approach 0.7. If we zoom in on the graph near $x = 0$, we can estimate the limit with greater accuracy, giving

$$\lim_{x \to 0} \frac{2^x - 1}{x} \approx 0.693.$$

Example 3   Estimate $\displaystyle\lim_{h \to 0} \frac{(3+h)^2 - 9}{h}$ numerically.

Solution   The limit is the value approached by this expression as h approaches 0. The values in Table 2.12 seem to be approaching 6 as $h \to 0$. So it is a reasonable guess that

$$\lim_{h \to 0} \frac{(3+h)^2 - 9}{h} = 6.$$

However, we cannot be sure that the limit is *exactly* 6 by looking at the table. To calculate the limit exactly requires algebra.

Table 2.12   *Values of* $\left((3+h)^2 - 9 \right) / h$

h	-0.1	-0.01	-0.001	0.001	0.01	0.1
$\left((3+h)^2 - 9 \right) / h$	5.9	5.99	5.999	6.001	6.01	6.1

Example 4   Use algebra to find $\displaystyle\lim_{h \to 0} \frac{(3+h)^2 - 9}{h}$.

Solution   Expanding the numerator gives

$$\frac{(3+h)^2 - 9}{h} = \frac{9 + 6h + h^2 - 9}{h} = \frac{6h + h^2}{h}.$$

Since taking the limit as $h \to 0$ means looking at values of h near, but not equal, to 0, we can cancel a common factor of h, giving

$$\lim_{h \to 0} \frac{(3+h)^2 - 9}{h} = \lim_{h \to 0} \frac{6h + h^2}{h} = \lim_{h \to 0} (6 + h).$$

As h approaches 0, the values of $(6 + h)$ approach 6, so

$$\lim_{h \to 0} \frac{(3+h)^2 - 9}{h} = \lim_{h \to 0} (6 + h) = 6.$$

Continuity

Roughly speaking, a function is said to be *continuous* on an interval if its graph has no breaks, jumps, or holes in that interval. A continuous function has a graph that can be drawn without lifting the pencil from the paper.

Example: The function $f(x) = 3x^2 - x^2 + 2x + 1$ is continuous on any interval. (See Figure 2.71.)

Example: The function $f(x) = 1/x$ is not defined at $x = 0$. It is continuous on any interval not containing the origin. (See Figure 2.72.)

Example: Suppose $p(x)$ is the price of mailing a first-class letter weighing x ounces. It costs 34¢ for one ounce or less, 57¢ between the first and second ounces, and so on. So the graph (in Figure 2.73) is a series of steps. This function is not continuous on intervals such as $(0, 2)$ because the graph jumps at $x = 1$.

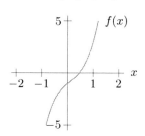

Figure 2.71: The graph of $f(x) = 3x^3 - x^2 + 2x - 1$

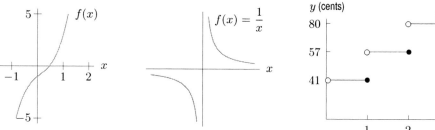

Figure 2.72: Graph of $f(x) = 1/x$: Not defined at 0

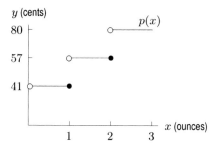

Figure 2.73: Cost of mailing a letter

What Does Continuity Mean Numerically?

Continuity is important in practical work because it means that small errors in the independent variable lead to small errors in the value of the function.

Example: Suppose that $f(x) = x^2$ and that we want to compute $f(\pi)$. Knowing f is continuous tells us that taking $x = 3.14$ should give a good approximation to $f(\pi)$, and that we can get a better approximation to $f(\pi)$ by using more decimals of π.

Example: If $p(x)$ is the cost of mailing a letter weighing x ounces, then $p(0.99) = p(1) = 34¢$, whereas $p(1.01) = 57¢$, because as soon as we get over 1 ounce, the price jumps up to 57¢. So a small difference in the weight of a letter can lead to a significant difference in its mailing cost. Hence p is not continuous at $x = 1$.

Definition of Continuity

We now define continuity using limits. The idea of continuity rules out breaks, jumps, or holes by demanding that the behavior of a function *near* a point be consistent with its behavior *at* the point:

> The function f is **continuous** at $x = c$ if f is defined at $x = c$ and
>
> $$\lim_{x \to c} f(x) = f(c).$$
>
> The function is **continuous on an interval** (a, b) if it is continuous at every point in the interval.

Which Functions Are Continuous?

Requiring a function to be continuous on an interval is not asking very much, as any function whose graph is an unbroken curve over the interval is continuous. For example, exponential functions, polynomials, and sine and cosine are continuous on every interval. Functions created by adding, multiplying, or composing continuous functions are also continuous.

Using the Definition to Calculate Derivatives

By estimating the derivative of the function $f(x) = x^2$ at several points, we guessed in Example 5 of Section 2.2 that the derivative of x^2 is $f'(x) = 2x$. In order to show that this formula is correct, we have to use the symbolic definition of the derivative given earlier in this section.

In evaluating the expression

$$\lim_{h \to 0} \frac{f(x+h) - f(x)}{h},$$

we simplify the difference quotient first, and then take the limit as h approaches zero.

Example 5   Show that the derivative of $f(x) = x^2$ is $f'(x) = 2x$.

Solution   Using the definition of the derivative with $f(x) = x^2$, we have

$$f'(x) = \lim_{h \to 0} \frac{f(x+h) - f(x)}{h} = \lim_{h \to 0} \frac{(x+h)^2 - x^2}{h}$$

$$= \lim_{h \to 0} \frac{x^2 + 2xh + h^2 - x^2}{h} = \lim_{h \to 0} \frac{2xh + h^2}{h}$$

$$= \lim_{h \to 0} \frac{h(2x + h)}{h}.$$

To take the limit, look at what happens when h is close to 0, but do not let $h = 0$. Since $h \neq 0$, we cancel the common factor of h, giving

$$f'(x) = \lim_{h \to 0} \frac{h(2x + h)}{h} = \lim_{h \to 0} (2x + h) = 2x,$$

because as h gets close to zero, $2x + h$ gets close to $2x$. So

$$f'(x) = \frac{d}{dx}(x^2) = 2x.$$

Example 6   Show that if $f(x) = 3x - 2$, then $f'(x) = 3$.

Solution   Since the slope of the linear function $f(x) = 3x - 2$ is 3 and the derivative is the slope, we see that $f'(x) = 3$. We can also use the definition to get this result:

$$f'(x) = \lim_{h \to 0} \frac{f(x + h) - f(x)}{h} = \lim_{h \to 0} \frac{(3(x + h) - 2) - (3x - 2)}{h}$$
$$= \lim_{h \to 0} \frac{3x + 3h - 2 - 3x + 2}{h} = \lim_{h \to 0} \frac{3h}{h}.$$

To find the limit, look at what happens when h is close to, but not equal to, 0. Simplifying, we get

$$f'(x) = \lim_{h \to 0} \frac{3h}{h} = \lim_{h \to 0} 3 = 3.$$

Problems on Limits and the Definition of the Derivative

1. On Figure 2.74, mark lengths that represent the quantities in parts (a) – (e). (Pick any h, with $h > 0$.)

(a) $a + h$ (b) h (c) $f(a)$
(d) $f(a + h)$ (e) $f(a+h) - f(a)$

(f) Using your answers to parts (a)–(e), show how the quantity $\dfrac{f(a + h) - f(a)}{h}$ can be represented as the slope of a line on the graph.

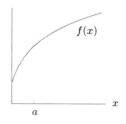

Figure 2.74

2. On Figure 2.75, mark lengths that represent the quantities in parts (a)–(e). (Pick any h, with $h > 0$.)

(a) $a + h$ (b) h (c) $f(a)$
(d) $f(a + h)$ (e) $f(a+h) - f(a)$

(f) Using your answers to parts (a)–(e), represent the quantity $\dfrac{f(a + h) - f(a)}{h}$ as the slope of a line on the graph.

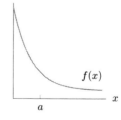

Figure 2.75

131

Use a graph to estimate the limits in Problems 3–4.

3. $\lim\limits_{x \to 0} \dfrac{\sin x}{x}$ (with x in radians)

4. $\lim\limits_{x \to 0} \dfrac{5^x - 1}{x}$

Estimate the limits in Problems 5–8 by substituting smaller and smaller values of h. For trigonometric functions, use radians. Give answers to one decimal place.

5. $\lim\limits_{h \to 0} \dfrac{(3+h)^3 - 27}{h}$ **6.** $\lim\limits_{h \to 0} \dfrac{7^h - 1}{h}$

7. $\lim\limits_{h \to 0} \dfrac{e^{1+h} - e}{h}$ **8.** $\lim\limits_{h \to 0} \dfrac{\cos h - 1}{h}$

In Problems 9–12, does the function $f(x)$ appear to be continuous on the interval $0 \le x \le 2$? If not, what about on the interval $0 \le x \le 0.5$?

9.
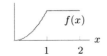

10.
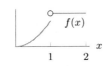

11.
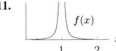

12.

Are the functions in Problems 13–18 continuous on the given intervals?

13. $f(x) = x + 2$ on $-3 \le x \le 3$

14. $f(x) = 2^x$ on $0 \le x \le 10$

15. $f(x) = x^2 + 2$ on $0 \le x \le 5$

16. $f(x) = \dfrac{1}{x - 1}$ on $2 \le x \le 3$

17. $f(x) = \dfrac{1}{x - 1}$ on $0 \le x \le 2$

18. $f(x) = \dfrac{1}{x^2 + 1}$ on $0 \le x \le 2$

Which of the functions described in Problems 19–23 are continuous?

19. The number of people in a village as a function of time.

20. The weight of a baby as a function of time during the second month of the baby's life.

21. The number of pairs of pants as a function of the number of yards of cloth from which they are made. Each pair requires 3 yards.

22. The distance traveled by a car in stop-and-go traffic as a function of time.

23. You start in North Carolina and go westward on Interstate 40 toward California. Consider the function giving the local time of day as a function of your distance from your starting point.

Use the definition of the derivative to show how the formulas in Problems 24–33 are obtained.

24. If $f(x) = 5x$, then $f'(x) = 5$.

25. If $f(x) = 3x - 2$, then $f'(x) = 3$.

26. If $f(x) = x^2 + 4$, then $f'(x) = 2x$.

27. If $f(x) = 3x^2$, then $f'(x) = 6x$.

28. If $f(x) = -2x^3$, then $f'(x) = -6x^2$.

29. If $f(x) = x - x^2$, then $f'(x) = 1 - 2x$.

30. If $f(x) = 1 - x^3$, then $f'(x) = -3x^2$.

31. If $f(x) = 5x^2 + 1$, then $f'(x) = 10x$.

32. If $f(x) = 2x^2 + x$, then $f'(x) = 4x + 1$.

33. If $f(x) = 1/x$, then $f'(x) = -1/x^2$.

Chapter Three

SHORTCUTS TO DIFFERENTIATION

Contents

3.1 DERIVATIVE FORMULAS FOR POWERS AND POLYNOMIALS

The derivative of a function at a point represents a slope and a rate of change. In Chapter 2, we learned how to estimate values of the derivative of a function given by a graph or by a table. Now, we learn how to find a formula for the derivative of a function given by a formula.

Derivative of a Constant Function

The graph of a constant function $f(x) = k$ is a horizontal line, with a slope of 0 everywhere. Therefore, its derivative is 0 everywhere. (See Figure 3.1.)

$$\boxed{\text{If } f(x) = k, \text{then } f'(x) = 0.}$$

For example, $\dfrac{d}{dx}(5) = 0$.

Figure 3.1: A constant function

Derivative of a Linear Function

We already know that the slope of a line is constant. This tells us that the derivative of a linear function is constant.

$$\boxed{\text{If } f(x) = b + mx, \text{then } f'(x) = \text{Slope} = m.}$$

For example, $\dfrac{d}{dx}\left(5 - \dfrac{3}{2}x\right) = -\dfrac{3}{2}$.

Derivative of a Constant Times a Function

Figure 3.2 shows the graph of $y = f(x)$ and of three multiples: $y = 3f(x)$, $y = \frac{1}{2}f(x)$, and $y = -2f(x)$. What is the relationship between the derivatives of these functions? In other words, for a particular x-value, how are the slopes of these graphs related?

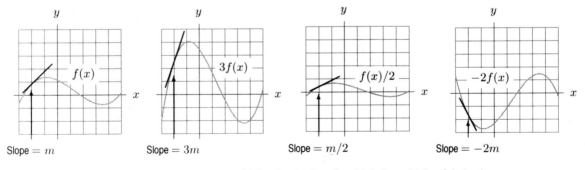

Figure 3.2: A function and its multiples: Derivative of multiple is multiple of derivative

Multiplying by a constant stretches or shrinks the graph (and reflects it about the x-axis if the constant is negative). This changes the slope of the curve at each point. If the graph has been stretched, the "rises" have all been increased by the same factor, whereas the "runs" remain the same. Thus, the slopes are all steeper by the same factor. If the graph has been shrunk, the slopes are all smaller by the same factor. If the graph has been reflected about the x-axis, the slopes will all have their signs reversed. Thus, if a function is multiplied by a constant, c, so is its derivative:

Derivative of a Constant Multiple

If c is a constant,

$$\frac{d}{dx}[cf(x)] = cf'(x).$$

Derivatives of Sums and Differences

Values of two functions, $f(x)$ and $g(x)$, and their sum $f(x) + g(x)$, are listed in Table 3.1.

Table 3.1   *Sum of functions*

x	$f(x)$	$g(x)$	$f(x) + g(x)$
0	100	0	100
1	110	0.2	110.2
2	130	0.4	130.4
3	160	0.6	160.6
4	200	0.8	200.8

We see that adding the increments of $f(x)$ and the increments of $g(x)$ gives the increments of $f(x) + g(x)$. For example, as x increases from 0 to 1, $f(x)$ increases by 10 and $g(x)$ increases by 0.2, while $f(x) + g(x)$ increases by $110.2 - 100 = 10.2$. Similarly, as x increases from 3 to 4, $f(x)$ increases by 40 and $g(x)$ by 0.2, while $f(x) + g(x)$ increases by $200.8 - 160.6 = 40.2$.

From this example, we see that the rate at which $f(x) + g(x)$ is increasing is the sum of the rates at which $f(x)$ and $g(x)$ are increasing. Similar reasoning applies to the difference, $f(x) - g(x)$. In terms of derivatives:

Derivative of Sum and Difference

$$\frac{d}{dx}[f(x) + g(x)] = f'(x) + g'(x) \qquad \text{and} \qquad \frac{d}{dx}[f(x) - g(x)] = f'(x) - g'(x).$$

Powers of x

We start by looking at $f(x) = x^2$ and $g(x) = x^3$. We show in the Focus on Theory section at the end of this chapter that

$$f'(x) = \frac{d}{dx}\left(x^2\right) = 2x \quad \text{and} \quad g'(x) = \frac{d}{dx}\left(x^3\right) = 3x^2.$$

The graphs of $f(x) = x^2$ and $g(x) = x^3$ and their derivatives are shown in Figures 3.3 and 3.4. Notice $f'(x) = 2x$ has the behavior we expect. It is negative for $x < 0$ (when f is decreasing), zero for $x = 0$, and positive for $x > 0$ (when f is increasing). Similarly, $g'(x) = 3x^2$ is zero when $x = 0$, but positive everywhere else, as g is increasing everywhere else. These examples are special cases of the power rule.

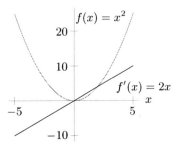

Figure 3.3: Graphs of $f(x) = x^2$ and its derivative $f'(x) = 2x$

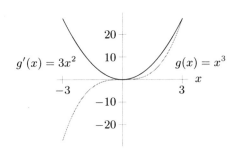

Figure 3.4: Graphs of $g(x) = x^3$ and its derivative $g'(x) = 3x^2$

The Power Rule

For any constant real number n,

$$\frac{d}{dx}(x^n) = nx^{n-1}.$$

Example 1 Find the derivative of (a) $h(x) = x^8$ (b) $P(t) = t^7$.

Solution   (a) $h'(x) = 8x^7$.   (b) $P'(t) = 7t^6$.

Derivatives of Polynomials

Using the derivatives of powers, constant multiples, and sums, we can differentiate any polynomial.

Example 2 Differentiate:

(a) $A(t) = 3t^5$

(b) $r(p) = p^5 + p^3$

(c) $f(x) = 5x^2 - 7x^3$

(d) $g(t) = \dfrac{t^2}{4} + 3$

Solution   (a) Using the constant multiple rule: $A'(t) = \dfrac{d}{dt}(3t^5) = 3\dfrac{d}{dt}(t^5) = 3 \cdot 5t^4 = 15t^4$.

(b) Using the sum rule: $r'(p) = \dfrac{d}{dp}(p^5 + p^3) = \dfrac{d}{dp}(p^5) + \dfrac{d}{dp}(p^3) = 5p^4 + 3p^2$.

(c) Using both rules together:

$$\begin{aligned}
f'(x) &= \frac{d}{dx}(5x^2 - 7x^3) = \frac{d}{dx}(5x^2) - \frac{d}{dx}(7x^3) \quad \text{\small Derivative of difference} \\
&= 5\frac{d}{dx}(x^2) - 7\frac{d}{dx}(x^3) \quad \text{\small Derivative of multiple} \\
&= 5(2x) - 7(3x^2) = 10x - 21x^2.
\end{aligned}$$

(d) Using both rules:

$$g'(t) = \frac{d}{dt}\left(\frac{t^2}{4} + 3\right) = \frac{1}{4}\frac{d}{dt}(t^2) + \frac{d}{dt}(3)$$

$$= \frac{1}{4}(2t) + 0 = \frac{t}{2}. \quad \text{Since the derivative of a constant, } \frac{d}{dt}(3), \text{ is zero.}$$

We can use the power rule to differentiate negative and fractional powers.

Example 3   Use the power rule to differentiate (a) $\dfrac{1}{x^3}$ (b) $\sqrt{x}$ (c) $2t^{4.5}$.

Solution   (a) For $n = -3$: $\dfrac{d}{dx}\left(\dfrac{1}{x^3}\right) = \dfrac{d}{dx}(x^{-3}) = -3x^{-3-1} = -3x^{-4} = -\dfrac{3}{x^4}.$

   (b) For $n = 1/2$: $\dfrac{d}{dx}(\sqrt{x}) = \dfrac{d}{dx}\left(x^{1/2}\right) = \dfrac{1}{2}x^{(1/2)-1} = \dfrac{1}{2}x^{-1/2} = \dfrac{1}{2\sqrt{x}}.$

   (c) For $n = 4.5$: $\dfrac{d}{dt}\left(2t^{4.5}\right) = 2\left(4.5t^{4.5-1}\right) = 9t^{3.5}.$

Using the Derivative Formulas

Since the slope of the tangent line to a curve is given by the derivative, we use differentiation to find the equation of the tangent line.

Example 4   Find an equation for the tangent line at $x = 1$ to the graph of

$$y = x^3 + 2x^2 - 5x + 7.$$

Sketch the graph of the curve and its tangent line on the same axes.

Solution   Differentiating gives

$$\frac{dy}{dx} = 3x^2 + 2(2x) - 5(1) + 0 = 3x^2 + 4x - 5,$$

so the slope of the tangent line at $x = 1$ is

$$m = \left.\frac{dy}{dx}\right|_{x=1} = 3(1)^2 + 4(1) - 5 = 2.$$

When $x = 1$, we have $y = 1^3 + 2(1^2) - 5(1) + 7 = 5$, so the point $(1, 5)$ lies on the tangent line. Using the formula $y - y_0 = m(x - x_0)$ gives

$$y - 5 = 2(x - 1)$$
$$y = 3 + 2x.$$

The equation of the tangent line is $y = 3 + 2x$. See Figure 3.5.

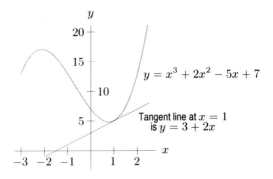

Figure 3.5: Find the equation for this tangent line

Example 5   Find and interpret the second derivatives of (a) $f(x) = x^2$ (b) $g(x) = x^3$.

Solution   (a) Differentiating $f(x) = x^2$ gives $f'(x) = 2x$, so $f''(x) = \dfrac{d}{dx}(2x) = 2$. Since f'' is always positive, the graph of f is concave up, as expected for a parabola opening upward. (See Figure 3.6.)

(b) Differentiating $g(x) = x^3$ gives $g'(x) = 3x^2$, so $g''(x) = \dfrac{d}{dx}(3x^2) = 3\dfrac{d}{dx}(x^2) = 3 \cdot 2x = 6x$. This is positive for $x > 0$ and negative for $x < 0$, which means that the graph of $g(x) = x^3$ is concave up for $x > 0$ and concave down for $x < 0$. (See Figure 3.7.)

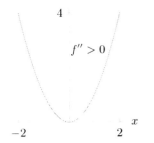

Figure 3.6: Graph of $f(x) = x^2$ with $f''(x) = 2$ Figure 3.7: Graph of $g(x) = x^3$ with $g''(x) = 6x$

Example 6   The revenue (in dollars) from producing q units of a product is given by
$$R(q) = 1000q - 3q^2.$$
Find $R(125)$ and $R'(125)$. Give units and interpret your answers.

Solution   We have
$$R(125) = 100 \cdot 125 - 3 \cdot 125^2 = 78{,}125 \text{ dollars.}$$
Since $R'(q) = 1000 - 6q$, we have
$$R'(125) = 1000 - 6 \cdot 125 = 250 \text{ dollars per unit.}$$

If 125 units are produced, the revenue is 78,125 dollars. If one additional unit is produced, revenue increases by about $250.

Example 7   Figure 3.8 shows the graph of a cubic polynomial. Both graphically and algebraically, describe the behavior of the derivative of this cubic.

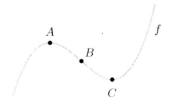

Figure 3.8: The cubic of Example 7

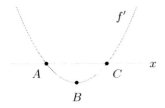

Figure 3.9: Derivative of the cubic of Example 7

Solution   Graphical approach: Suppose we move along the curve from left to right. To the left of A, the slope is positive; it starts very positive and decreases until the curve reaches A, where the slope is 0. Between A and C the slope is negative. Between A and B the slope is decreasing (getting more negative); it is most negative at B. Between B and C the slope is negative but increasing; at C the slope is zero. From C to the right, the slope is positive and increasing. The graph of the derivative function is shown in Figure 3.9.

Algebraic approach: f is a cubic that goes to $+\infty$ as $x \to +\infty$, so

$$f(x) = ax^3 + bx^2 + cx + d$$

with $a > 0$. Hence,

$$f'(x) = 3ax^2 + 2bx + c,$$

whose graph is a parabola opening upward, as in Figure 3.9.

Problems for Section 3.1

For Problems 1–36, find the derivative. Assume a, b, c, k are constants.

1. $y = 5$

2. $y = 3x$

3. $y = x^{12}$

4. $y = x^{-12}$

5. $y = x^{4/3}$

6. $y = 8t^3$

7. $y = 3t^4 - 2t^2$

8. $y = 5x + 13$

9. $f(x) = \dfrac{1}{x^4}$

10. $f(q) = q^3 + 10$

11. $y = x^2 + 5x + 9$

12. $y = 6x^3 + 4x^2 - 2x$

13. $y = 3x^2 + 7x - 9$

14. $y = 8t^3 - 4t^2 + 12t - 3$

15. $y = 4.2q^2 - 0.5q + 11.27$

16. $y = -3x^4 - 4x^3 - 6x + 2$

17. $g(t) = \dfrac{1}{t^5}$

18. $f(z) = -\dfrac{1}{z^{6.1}}$

19. $y = \dfrac{1}{r^{7/2}}$

20. $y = \sqrt{x}$

21. $h(\theta) = \dfrac{1}{\sqrt[3]{\theta}}$

22. $f(x) = \sqrt{\dfrac{1}{x^3}}$

23. $y = 3t^5 - 5\sqrt{t} + \dfrac{7}{t}$

24. $y = z^2 + \dfrac{1}{2z}$

25. $y = 3t^2 + \dfrac{12}{\sqrt{t}} - \dfrac{1}{t^2}$

26. $h(t) = \dfrac{3}{t} + \dfrac{4}{t^2}$

27. $y = \sqrt{x}(x + 1)$

28. $h(\theta) = \theta(\theta^{-1/2} - \theta^{-2})$

29. $f(x) = kx^2$

30. $y = ax^2 + bx + c$

31. $Q = aP^2 + bP^3$

32. $v = at^2 + \dfrac{b}{t^2}$

33. $P = a + b\sqrt{t}$

34. $V = \frac{4}{3}\pi r^2 b$

35. $w = 3ab^2q$

36. $h(x) = \dfrac{ax + b}{c}$

37. (a) Use a graph of $P(q) = 6q - q^2$ to determine whether each of the following derivatives is positive, negative, or zero: $P'(1)$, $P'(3)$, $P'(4)$. Explain.
(b) Find $P'(q)$ and the three derivatives in part (a).

38. Let $f(x) = x^3 - 4x^2 + 7x - 11$. Find $f'(0)$, $f'(2)$, $f'(-1)$.

39. Let $f(t) = t^2 - 4t + 5$.
(a) Find $f'(t)$.
(b) Find $f'(1)$ and $f'(2)$.
(c) Use a graph of $f(t)$ to check that your answers to part (b) are reasonable. Explain.

40. Find the rate of change of a population of size $P(t) = t^3 + 4t + 1$ at time $t = 2$.

41. The height of a sand dune (in centimeters) is represented by $f(t) = 700 - 3t^2$, where t is measured in years since 2005. Find $f(5)$ and $f'(5)$. Using units, explain what each means in terms of the sand dune.

42. Zebra mussels are freshwater shellfish that first appeared in the St. Lawrence River in the early 1980s and have spread throughout the Great Lakes. Suppose that t months after they appeared in a small bay, the number of zebra mussels is given by $Z(t) = 300t^2$. How many zebra mussels are in the bay after four months? At what rate is the population growing at that time? Give units.

43. The quantity, Q, in tons, of material at a municipal waste site is a function of the number of years since 2000, with

$$Q = f(t) = 3t^2 + 100.$$

Find $f(10)$, $f'(10)$, and the relative rate of change f'/f at $t = 10$. Interpret your answers in terms of waste.

44. The number, N, of acres of harvested land in a region is given by

$$N = f(t) = 120\sqrt{t},$$

where t is the number of years since farming began in the region. Find $f(9)$, $f'(9)$, and the relative rate of change f'/f at $t = 9$. Interpret your answers in terms of harvested land.

45. If $f(t) = 2t^3 - 4t^2 + 3t - 1$, find $f'(t)$ and $f''(t)$.

46. If $f(t) = t^4 - 3t^2 + 5t$, find $f'(t)$ and $f''(t)$.

47. The time, T, in seconds for one complete oscillation of a pendulum is given by $T = f(L) = 1.111\sqrt{L}$, where L is the length of the pendulum in feet. Find the following quantities, with units, and interpret in terms of the pendulum.

   (a) $f(100)$ **(b)** $f'(100)$.

48. Kleiber's Law states that the daily calorie requirement, $C(w)$, of a mammal is proportional to the mammal's body weight w raised to the 0.75 power.[1] If body weight is measured in pounds, the constant of proportionality is approximately 42.

   (a) Give formulas for $C(w)$ and $C'(w)$.

   (b) Find and interpret

      (i) $C(10)$ and $C'(10)$

      (ii) $C(100)$ and $C'(100)$

      (iii) $C(1000)$ and $C'(1000)$

49. Find the equation of the line tangent to the graph of f at $(1, 1)$, where f is given by $f(x) = 2x^3 - 2x^2 + 1$.

50. **(a)** Find the equation of the tangent line to $f(x) = x^3$ at the point where $x = 2$.

   (b) Graph the tangent line and the function on the same axes. If the tangent line is used to estimate values of the function, will the estimates be overestimates or underestimates?

51. Find the equation of the line tangent to the graph of $f(t) = 6t - t^2$ at $t = 4$. Sketch the graph of $f(t)$ and the tangent line on the same axes.

52. If you are outdoors, the wind may make it feel a lot colder than the thermometer reads. You feel the windchill temperature, which, if the air temperature is $20°$F, is given in $°$F by $W(v) = 48.17 - 27.2v^{0.16}$, where v is the wind velocity in mph for $5 \leq v \leq 60$.[2]

   (a) If the air temperature is $20°$F, and the wind is blowing at 40 mph, what is the windchill temperature, to the nearest degree?

   (b) Find $W'(40)$, and explain what this means in terms of windchill.

53. **(a)** Use the formula for the area of a circle of radius r, $A = \pi r^2$, to find dA/dr.

   (b) The result from part (a) should look familiar. What does dA/dr represent geometrically?

   (c) Use the difference quotient to explain the observation you made in part (b).

54. Suppose W is proportional to r^3. The derivative dW/dr is proportional to what power of r?

55. The cost to produce q items is $C(q) = 1000 + 2q^2$ dollars. Find the marginal cost of producing the 25th item. Interpret your answer in terms of costs.

56. The demand curve for a product is given by $q = 300 - 3p$, where p is the price of the product and q is the quantity that consumers buy at this price.

   (a) Write the revenue as a function, $R(p)$, of price.

   (b) Find $R'(10)$ and interpret your answer in terms of revenue.

   (c) For what prices is $R'(p)$ positive? For what prices is it negative?

57. The yield, Y, of an apple orchard (measured in bushels of apples per acre) is a function of the amount x of fertilizer in pounds used per acre. Suppose

$$Y = f(x) = 320 + 140x - 10x^2.$$

   (a) What is the yield if 5 pounds of fertilizer is used per acre?

   (b) Find $f'(5)$. Give units with your answer and interpret it in terms of apples and fertilizer.

   (c) Given your answer to part (b), should more or less fertilizer be used? Explain.

58. The demand for a product is given, for $p, q \geq 0$, by

$$p = f(q) = 50 - 0.03q^2.$$

   (a) Find the p- and q-intercepts for this function and interpret them in terms of demand for this product.

   (b) Find $f(20)$ and give units with your answer. Explain what it tells you in terms of demand.

   (c) Find $f'(20)$ and give units with your answer. Explain what it tells you in terms of demand.

59. The cost (in dollars) of producing q items is given by $C(q) = 0.08q^3 + 75q + 1000$.

   (a) Find the marginal cost function.

   (b) Find $C(50)$ and $C'(50)$. Give units with your answers and explain what each is telling you about costs of production.

60. A ball is dropped from the top of the Empire State Building. The height, y, of the ball above the ground (in feet) is given as a function of time, t, (in seconds) by

$$y = 1250 - 16t^2.$$

   (a) Find the velocity of the ball at time t. What is the sign of the velocity? Why is this to be expected?

   (b) When does the ball hit the ground, and how fast is it going at that time? Give your answer in feet per second and in miles per hour (1 ft/sec = 15/22 mph).

61. Let $f(x) = x^3 - 6x^2 - 15x + 20$. Find $f'(x)$ and all values of x for which $f'(x) = 0$. Explain the relationship between these values of x and the graph of $f(x)$.

62. Show that for any power function $f(x) = x^n$, we have $f'(1) = n$.

63. If the demand curve is a line, we can write $p = b + mq$, where p is the price of the product, q is the quantity sold at that price, and b and m are constants.

   (a) Write the revenue as a function of quantity sold.

   (b) Find the marginal revenue function.

---

[1] Strogatz, S., "Math and the City", *The New York Times*, May 20, 2009.

[2] www.weather.gov/om/windchill/index.shtml Accessed on 4/21/09.

3.2 EXPONENTIAL AND LOGARITHMIC FUNCTIONS

The Exponential Function

What do we expect the graph of the derivative of the exponential function $f(x) = a^x$ to look like? The graph of an exponential function with $a > 1$ is shown in Figure 3.10. The function increases slowly for $x < 0$ and more rapidly for $x > 0$, so the values of f' are small for $x < 0$ and larger for $x > 0$. Since the function is increasing for all values of x, the graph of the derivative must lie above the x-axis. In fact, the graph of f' resembles the graph of f itself. We will see how this observation holds for $f(x) = 2^x$ and $g(x) = 3^x$.

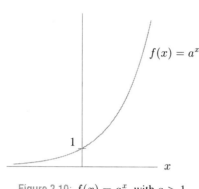

Figure 3.10: $f(x) = a^x$, with $a > 1$

The Derivatives of 2^x and 3^x

In Section 2.1, we estimated the derivative of $f(x) = 2^x$ at $x = 0$:

$$f'(0) \approx 0.693.$$

By estimating the derivative at other values of x, we obtain the graph in Figure 3.11. Since the graph of f' looks like the graph of f stretched vertically, we assume that f' is a multiple of f. Since $f'(0) \approx 0.693 = 0.693 \cdot 1 = 0.693 f(0)$, the multiplier is approximately 0.693, which suggests that

$$\frac{d}{dx}(2^x) = f'(x) \approx (0.693)2^x.$$

Similarly, in Figure 3.12, the derivative of $g(x) = 3^x$ is a multiple of g, with multiplier $g'(0) \approx 1.0986$. So

$$\frac{d}{dx}(3^x) = g'(x) \approx (1.0986)3^x.$$

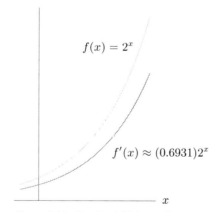

Figure 3.11: Graph of $f(x) = 2^x$ and its derivative

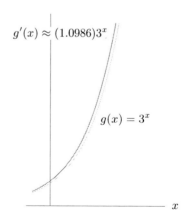

Figure 3.12: Graph of $g(x) = 3^x$ and its derivative

The Derivative of a^x and the Number e

The calculation of the derivative of $f(x) = a^x$, for $a > 0$, is similar to that of 2^x and 3^x. The derivative is again proportional to the original function. When $a = 2$, the constant of proportionality (0.6931) is less than 1, and the derivative is smaller than the original function. When $a = 3$, the constant of proportionality (1.0986) is more than 1, and the derivative is greater than the original function. Is there an in-between case, when derivative and function are exactly equal? In other words:

$$\text{Is there a value of } a \text{ that makes } \frac{d}{dx}(a^x) = a^x?$$

The answer is yes: the value is $a \approx 2.718\ldots$, the number e introduced in Chapter 1. This means that the function e^x is its own derivative:

$$\frac{d}{dx}(e^x) = e^x.$$

It turns out that the constants involved in the derivatives of 2^x and 3^x are natural logarithms. In fact, since $0.6931 \approx \ln 2$ and $1.0986 \approx \ln 3$, we (correctly) guess that

$$\frac{d}{dx}(2^x) = (\ln 2)2^x \quad \text{and} \quad \frac{d}{dx}(3^x) = (\ln 3)3^x.$$

In the Focus on Theory section at the end of this chapter, we show that, in general:

The Exponential Rule

For any positive constant a,

$$\frac{d}{dx}(a^x) = (\ln a)a^x.$$

Since $\ln a$ is a constant, the derivative of a^x is proportional to a^x. Many quantities have rates of change that are proportional to themselves; for example, the simplest model of population growth has this property. The fact that the constant of proportionality is 1 when $a = e$ makes e a particularly useful base for exponential functions.

Example 1 Differentiate $2 \cdot 3^x + 5e^x$.

Solution We have $\dfrac{d}{dx}(2 \cdot 3^x + 5e^x) = 2\dfrac{d}{dx}(3^x) + 5\dfrac{d}{dx}(e^x) = 2\ln 3 \cdot 3^x + 5e^x.$

The Derivative of e^{kt}

Since functions of the form e^{kt} where k is a constant are often useful, we calculate the derivative of e^{kt}. Since $e^{kt} = (e^k)^t$, the derivative is $(\ln(e^k))(e^k)^t = (k\ln e)(e^{kt}) = ke^{kt}$. Thus, if k is a constant,

$$\frac{d}{dt}(e^{kt}) = ke^{kt}.$$

Example 2 Find the derivative of $P = 5 + 3x^2 - 7e^{-0.2x}$.

Solution The derivative is

$$\frac{dP}{dx} = 0 + 3(2x) - 7(-0.2e^{-0.2x}) = 6x + 1.4e^{-0.2x}.$$

The Derivative of $\ln x$

What does the graph of the derivative of the logarithmic function $f(x) = \ln x$ look like? Figure 3.13 shows that $\ln x$ is increasing, so its derivative is positive. The graph of $f(x) = \ln x$ is concave down, so the derivative is decreasing. Furthermore, the slope of $f(x) = \ln x$ is very large near $x = 0$ and very small for large x, so the derivative tends to $+\infty$ for x near 0 and tends to 0 for very large x. See Figure 3.14. It turns out that

$$\frac{d}{dx}(\ln x) = \frac{1}{x}.$$

We give an algebraic justification for this rule in the Focus on Theory section.

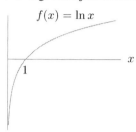

Figure 3.13: Graph of $f(x) = \ln x$

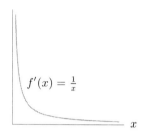

Figure 3.14: Graph of the derivative of $f(x) = \ln x$

Example 3   Differentiate $y = 5 \ln t + 7e^t - 4t^2 + 12$.

Solution   We have

$$\frac{d}{dt}(5 \ln t + 7e^t - 4t^2 + 12) = 5\frac{d}{dt}(\ln t) + 7\frac{d}{dt}(e^t) - 4\frac{d}{dt}(t^2) + \frac{d}{dt}(12)$$

$$= 5\left(\frac{1}{t}\right) + 7(e^t) - 4(2t) + 0$$

$$= \frac{5}{t} + 7e^t - 8t.$$

Using the Derivative Formulas

Example 4   In Chapter 1, we saw that the population of Nevada, P, in millions, can be approximated by

$$P = 2.020(1.036)^t,$$

where t is years since the start of 2000. At what rate was the population growing at the beginning of 2009? Give units with your answer.

Solution   The instantaneous rate of growth is the derivative, so we want dP/dt when $t = 9$. We have:

$$\frac{dP}{dt} = \frac{d}{dt}(2.020(1.036)^t) = 2.020(\ln 1.036)(1.036)^t = 0.0714(1.036)^t.$$

Substituting $t = 9$ gives

$$0.0714(1.036)^9 = 0.0982.$$

The population of Nevada was growing at a rate of about 0.0982 million, or 98,200, people per year at the start of 2009.

Example 5   Find the equation of the tangent line to the graph of $f(x) = \ln x$ at the point where $x = 2$. Draw a graph with $f(x)$ and the tangent line on the same axes.

Solution   Since $f'(x) = 1/x$, the slope of the tangent line at $x = 2$ is $f'(2) = 1/2 = 0.5$. When $x = 2$, $y = \ln 2 = 0.693$, so a point on the tangent line is $(2, 0.693)$. Substituting into the equation for a line, we have:

$$y - 0.693 = 0.5(x - 2)$$
$$y = -0.307 + 0.5x.$$

The equation of the tangent line is $y = -0.307 + 0.5x$. See Figure 3.15.

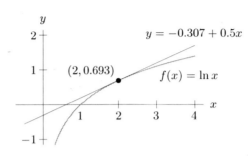

Figure 3.15: Graph of $f(x) = \ln x$ and a tangent line

Example 6   Suppose $1000 is deposited into a bank account that pays 8% annual interest, compounded continuously.

(a) Find a formula $f(t)$ for the balance t years after the initial deposit.
(b) Find $f(10)$ and $f'(10)$ and explain what your answers mean in terms of money.

Solution   (a) The balance is $f(t) = 1000e^{0.08t}$.
(b) Substituting $t = 10$ gives

$$f(10) = 1000e^{(0.08)(10)} = 2225.54.$$

This means that the balance is $2225.54 after 10 years.
To find $f'(10)$, we compute $f'(t) = 1000(0.08e^{0.08t}) = 80e^{0.08t}$. Therefore,

$$f'(10) = 80e^{(0.08)(10)} = 178.04.$$

This means that after 10 years, the balance is growing at the rate of about $178 per year.

Problems for Section 3.2

Differentiate the functions in Problems 1–28. Assume that A, B, and C are constants.

1. $f(x) = 2e^x + x^2$

2. $P = 3t^3 + 2e^t$

3. $y = 5t^2 + 4e^t$

4. $f(x) = x^3 + 3^x$

5. $y = 2^x + \dfrac{2}{x^3}$

6. $y = 5 \cdot 5^t + 6 \cdot 6^t$

7. $f(x) = 2^x + 2 \cdot 3^x$

8. $y = 4 \cdot 10^x - x^3$

9. $y = 3x - 2 \cdot 4^x$

10. $y = 5 \cdot 2^x - 5x + 4$

11. $f(t) = e^{3t}$

12. $y = e^{0.7t}$

13. $y = e^{-4t}$

14. $P = e^{-0.2t}$

15. $P = 50e^{-0.6t}$

16. $P = 200e^{0.12t}$

17. $P(t) = 3000(1.02)^t$

18. $P(t) = 12.41(0.94)^t$

19. $P(t) = Ce^t$.

20. $y = B + Ae^t$

21. $f(x) = Ae^x - Bx^2 + C$

22. $y = 10^x + \dfrac{10}{x}$

23. $R = 3 \ln q$

24. $D = 10 - \ln p$

25. $y = t^2 + 5 \ln t$

26. $R(q) = q^2 - 2 \ln q$

27. $y = x^2 + 4x - 3 \ln x$

28. $f(t) = Ae^t + B \ln t$

29. For $f(t) = 4 - 2e^t$, find $f'(-1)$, $f'(0)$, and $f'(1)$. Graph $f(t)$, and draw tangent lines at $t = -1$, $t = 0$, and $t = 1$. Do the slopes of the lines match the derivatives you found?

30. Find the equation of the tangent line to the graph of $y = 3^x$ at $x = 1$. Check your work by sketching a graph of the function and the tangent line on the same axes.

31. Find the equation of the tangent line to $y = e^{-2t}$ at $t = 0$. Check by sketching the graphs of $y = e^{-2t}$ and the tangent line on the same axes.

32. Find the equation of the tangent line to $f(x) = 10e^{-0.2x}$ at $x = 4$.

33. A fish population is approximated by $P(t) = 10e^{0.6t}$, where t is in months. Calculate and use units to explain what each of the following tells us about the population:

(a) $P(12)$ (b) $P'(12)$

34. The world's population is about $f(t) = 6.8e^{0.012t}$ billion,[3] where t is time in years since 2009. Find $f(0)$, $f'(0)$, $f(10)$, and $f'(10)$. Using units, interpret your answers in terms of population.

35. The demand curve for a product is given by

$$q = f(p) = 10{,}000e^{-0.25p},$$

where q is the quantity sold and p is the price of the product, in dollars. Find $f(2)$ and $f'(2)$. Explain in economic terms what information each of these answers gives you.

36. Worldwide production of solar power, in megawatts, can be modeled by $f(t) = 1040(1.3)^t$, where t is years[4] since 2000. Find $f(0)$, $f'(0)$, $f(15)$, and $f'(15)$. Give units and interpret your answers in terms of solar power.

37. A new DVD is available for sale in a store one week after its release. The cumulative revenue, $\$R$, from sales of the DVD in this store in week t after its release is

$$R = f(t) = 350 \ln t \quad \text{with} \quad t > 1.$$

Find $f(5)$, $f'(5)$, and the relative rate of change f'/f at $t = 5$. Interpret your answers in terms of revenue.

38. In 2009, the population of Hungary[5] was approximated by

$$P = 9.906(0.997)^t,$$

where P is in millions and t is in years since 2009. Assume the trend continues.

(a) What does this model predict for the population of Hungary in the year 2020?
(b) How fast (in people/year) does this model predict Hungary's population will be decreasing in 2020?

39. With t in years since January 1, 2010, the population P of Slim Chance is predicted by

$$P = 35{,}000(0.98)^t.$$

At what rate will the population be changing on January 1, 2023?

40. Some antique furniture increased very rapidly in price over the past decade. For example, the price of a particular rocking chair is well approximated by

$$V = 75(1.35)^t,$$

where V is in dollars and t is in years since 2000. Find the rate, in dollars per year, at which the price is increasing at time t.

41. Find the value of c in Figure 3.16, where the line l tangent to the graph of $y = 2^x$ at $(0, 1)$ intersects the x-axis.

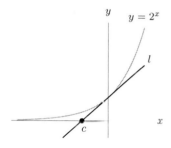

Figure 3.16

42. At a time t hours after it was administered, the concentration of a drug in the body is $f(t) = 27e^{-0.14t}$ ng/ml. What is the concentration 4 hours after it was administered? At what rate is the concentration changing at that time?

43. The cost of producing a quantity, q, of a product is given by
$$C(q) = 1000 + 30e^{0.05q} \quad \text{dollars.}$$

Find the cost and the marginal cost when $q = 50$. Interpret these answers in economic terms.

44. With time, t, in minutes, the temperature, H, in degrees Celsius, of a bottle of water put in the refrigerator at $t = 0$ is given by

$$H = 4 + 16e^{-0.02t}.$$

How fast is the water cooling initially? After 10 minutes? Give units.

---

[3] www.census/gov/ipc/www/popclockworld.html and www.cia.gov/library/publications/the-world-factbook/print/xx.html.
[4] https://www.solarbuzz.com/FastFactsIndustry.htm, accessed 4/14/09.
[5] https://www.cia.gov/library/publications/the-world-factbook/print/hu.html, accessed 4/14/09.

45. Carbon-14 is a radioactive isotope used to date objects. If A_0 represents the initial amount of carbon-14 in the object, then the quantity remaining at time t, in years, is

$$A(t) = A_0 e^{-0.000121t}.$$

    (a) A tree, originally containing 185 micrograms of carbon-14, is now 500 years old. At what rate is the carbon-14 decaying?

    (b) In 1988, scientists found that the Shroud of Turin, which was reputed to be the burial cloth of Jesus, contained 91% of the amount of carbon-14 in freshly made cloth of the same material.[6] According to this data, how old was the Shroud of Turin in 1988?

46. For the cost function $C = 1000 + 300 \ln q$ (in dollars), find the cost and the marginal cost at a production level of 500. Interpret your answers in economic terms.

47. In 2009, the population, P, of India was 1.166 billion and growing at 1.5% annually.[7]

    (a) Give a formula for P in terms of time, t, measured in years since 2009.

    (b) Find $\dfrac{dP}{dt}$, $\left.\dfrac{dP}{dt}\right|_{t=0}$, and $\left.\dfrac{dP}{dt}\right|_{t=25}$. What do each of these represent in practical terms?

48. In 2009, the population of Mexico was 111 million and growing 1.13% annually, while the population of the US was 307 million and growing 0.975% annually.[8] If we measure growth rates in people/year, which population was growing faster in 2009?

49. **(a)** Find the equation of the tangent line to $y = \ln x$ at $x = 1$.

    (b) Use it to calculate approximate values for $\ln(1.1)$ and $\ln(2)$.

    (c) Using a graph, explain whether the approximate values are smaller or larger than the true values. Would the same result have held if you had used the tangent line to estimate $\ln(0.9)$ and $\ln(0.5)$? Why?

50. Find the quadratic polynomial $g(x) = ax^2 + bx + c$ which best fits the function $f(x) = e^x$ at $x = 0$, in the sense that

$$g(0) = f(0), \text{ and } g'(0) = f'(0), \text{ and } g''(0) = f''(0).$$

Using a computer or calculator, sketch graphs of f and g on the same axes. What do you notice?

3.3 THE CHAIN RULE

We now see how to differentiate composite functions such as $f(t) = \ln(3t)$ and $g(x) = e^{-x^2}$.

The Derivative of a Composition of Functions

Suppose $y = f(z)$ with $z = g(t)$ for some inside function g and outside function f, where f and g are differentiable. A small change in t, called Δt, generates a small change in z, called Δz. In turn, Δz generates a small change in y, called Δy. Provided Δt and Δz are not zero, we can say

$$\frac{\Delta y}{\Delta t} = \frac{\Delta y}{\Delta z} \cdot \frac{\Delta z}{\Delta t}.$$

Since the derivative $\dfrac{dy}{dt}$ is the limit of the quotient $\dfrac{\Delta y}{\Delta t}$ as Δt gets smaller and smaller, this suggests

---

The Chain Rule

If $y = f(z)$ and $z = g(t)$ are differentiable, then the derivative of $y = f(g(t))$ is given by

$$\frac{dy}{dt} = \frac{dy}{dz} \cdot \frac{dz}{dt}.$$

In words, the derivative of a composite function is the derivative of the outside function times the derivative of the inside function:

$$\frac{d}{dt}(f(g(t))) = f'(g(t)) \cdot g'(t).$$

---

[6] The New York Times, October 18, 1988.

[7] https://www.cia.gov/library/publications/the-world-factbook/print/in.html, accessed 4/14/09.

[8] https://www.cia.gov/library/publications/the-world-factbook/print/ms.html and https://www.cia.gov/library/publications/the-world-factbook/print/us.html, accessed 4/14/09.

The following example shows us how to interpret the chain rule in practical terms.

Example 1
The amount of gas, G, in gallons, consumed by a car depends on the distance traveled, s, in miles, and s depends on the time, t, in hours. If 0.05 gallons of gas is consumed for each mile traveled, and the car is traveling at 30 miles/hr, how fast is gas being consumed? Give units.

Solution
We expect the rate of gas consumption to be in gallons/hr. We are told that

$$\text{Rate gas is consumed with respect to distance} = \frac{dG}{ds} = 0.05 \text{ gallons/mile}$$

$$\text{Rate distance is increasing with respect to time} = \frac{ds}{dt} = 30 \text{ miles/hr.}$$

We want to calculate the rate at which gas is being consumed with respect to time, or dG/dt. We think of G as a function of s, and s as a function of t. By the chain rule we know that

$$\frac{dG}{dt} = \frac{dG}{ds} \cdot \frac{ds}{dt} = \left(0.05 \frac{\text{gallons}}{\text{mile}}\right) \cdot \left(30 \frac{\text{miles}}{\text{hour}}\right) = 1.5 \text{ gallons/hour.}$$

Thus, gas is being consumed at a rate of 1.5 gallons/hour.

The Chain Rule for Functions Given by Formulas

In order to use the chain rule to differentiate a composite function, we first rewrite the function using a new variable z to represent the inside function:

$$y = (t+1)^4 \quad \text{is the same as} \quad y = z^4 \quad \text{where} \quad z = t+1.$$

Example 2
Use a new variable z for the inside function to express each of the following as a composite function:
(a) $y = \ln(3t)$ (b) $P = e^{-0.03t}$ (c) $w = 5(2r+3)^2$.

Solution
(a) The inside function is $3t$, so we have $y = \ln z$ with $z = 3t$.
(b) The inside function is $-0.03t$, so we have $P = e^z$ with $z = -0.03t$.
(c) The inside function is $2r+3$, so we have $w = 5z^2$ with $z = 2r+3$.

Example 3
Find the derivative of the following functions: (a) $y = (4t^2+1)^7$ (b) $P = e^{3t}$.

Solution
(a) Here $z = 4t^2+1$ is the inside function; $y = z^7$ is the outside function. Since $dy/dz = 7z^6$ and $dz/dt = 8t$, we have

$$\frac{dy}{dt} = \frac{dy}{dz} \cdot \frac{dz}{dt} = (7z^6)(8t) = 7(4t^2+1)^6(8t) = 56t(4t^2+1)^6.$$

(b) Let $z = 3t$ and $P = e^z$. Then $dP/dz = e^z$ and $dz/dt = 3$, so

$$\frac{dP}{dt} = \frac{dP}{dz} \cdot \frac{dz}{dt} = e^z \cdot 3 = e^{3t} \cdot 3 = 3e^{3t}.$$

Notice that the derivative formula for e^{kt} introduced in Section 3.2 is just a special case of the chain rule.

The derivative rules give us

$$\frac{d}{dt}(t^n) = nt^{n-1} \qquad \frac{d}{dt}(e^t) = e^t \qquad \frac{d}{dt}(\ln t) = \frac{1}{t}.$$

Using the chain rule in addition, we have the following results.

If z is a differentiable function of t, then

$$\frac{d}{dt}(z^n) = nz^{n-1}\frac{dz}{dt}, \qquad \frac{d}{dt}(e^z) = e^z\frac{dz}{dt}, \qquad \frac{d}{dt}(\ln z) = \frac{1}{z}\frac{dz}{dt}.$$

Example 4   Differentiate (a) $(3t^3 - t)^5$ (b) $\ln(q^2 + 1)$ (c) e^{-x^2}.

Solution   (a) Let $z = 3t^3 - t$, giving

$$\frac{d}{dt}(3t^3 - t)^5 = \frac{d}{dt}(z^5) = 5z^4\frac{dz}{dt} = 5(3t^3 - t)^4(9t^2 - 1).$$

$5(3t^2 - 4$

(b) We have $z = q^2 + 1$, so

$$\frac{d}{dq}(\ln(q^2 + 1)) = \frac{d}{dq}(\ln z) = \frac{1}{z}\frac{dz}{dq} = \frac{1}{q^2 + 1}(2q).$$

(c) Since $z = -x^2$, the derivative is

$$\frac{d}{dx}(e^{-x^2}) = \frac{d}{dx}(e^z) = e^z\frac{dz}{dx} = e^{-x^2}(-2x) = -2xe^{-x^2}.$$

As we see in the following example, it is often faster to use the chain rule without introducing the new variable, z.

Example 5   Differentiate
(a) $(x^2 + 4)^3$ (b) $5\ln(2t^2 + 3)$ (c) $\sqrt{1 + 2e^{5t}}$

Solution   (a) We have

$$\frac{d}{dx}((x^2 + 4)^3) = 3(x^2 + 4)^2 \cdot \frac{d}{dx}(x^2 + 4)$$
$$= 3(x^2 + 4)^2 \cdot 2x$$
$$= 6x(x^2 + 4)^2.$$

(b) We have

$$\frac{d}{dt}(5\ln(2t^2 + 3)) = 5 \cdot \frac{1}{2t^2 + 3} \cdot \frac{d}{dt}(2t^2 + 3)$$
$$= 5 \cdot \frac{1}{2t^2 + 3} \cdot 4t$$
$$= \frac{20t}{2t^2 + 3}.$$

(c) Here we use the chain rule twice, giving

$$\frac{d}{dt}((1 + 2e^{5t})^{1/2}) = \frac{1}{2}(1 + 2e^{5t})^{-1/2} \cdot \frac{d}{dt}(1 + 2e^{5t})$$
$$= \frac{1}{2}(1 + 2e^{5t})^{-1/2} \cdot 2e^{5t} \cdot \frac{d}{dt}(5t)$$
$$= \frac{1}{2}(1 + 2e^{5t})^{-1/2} \cdot 2e^{5t} \cdot 5$$
$$= \frac{5e^{5t}}{\sqrt{1 + 2e^{5t}}}.$$

Example 6   Let $h(x) = f(g(x))$ and $k(x) = g(f(x))$. Use Figure 3.17 to estimate: (a) $h'(1)$ (b) $k'(2)$

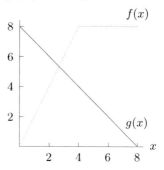

Figure 3.17: Graphs of f and g for Example 6

Solution   (a) The chain rule tells us that $h'(x) = f'(g(x)) \cdot g'(x)$, so

$$h'(1) = f'(g(1)) \cdot g'(1)$$
$$= f'(7) \cdot g'(1)$$
$$= 0 \cdot (-1)$$
$$= 0.$$

We use the slopes of the lines in Figure 3.17 to find the derivatives $f'(7) = 0$ and $g'(1) = -1$.

(b) The chain rule tells us that $k'(x) = g'(f(x)) \cdot f'(x)$, so

$$k'(2) = g'(f(2)) \cdot f'(2)$$
$$= g'(4) \cdot f'(2)$$
$$= (-1) \cdot 2$$
$$= -2.$$

We use slopes to compute the derivatives $g'(4) = -1$ and $f'(2) = 2$.

Relative Rates and Logarithms

In Section 2.3 we defined the relative rate of change of a function $z = f(t)$ to be

$$\text{Relative rate of change} \; = \frac{f'(t)}{f(t)} = \frac{1}{z}\frac{dz}{dt}.$$

Since

$$\frac{d}{dt}(\ln z) = \frac{1}{z}\frac{dz}{dt},$$

we have the following result:

For any positive function $f(t)$,

$$\begin{matrix}\text{Relative rate of change} \\ \text{of } f(t)\end{matrix} \quad = \frac{d}{dt}(\ln f(t)).$$

Just as linear functions have constant rates of change, in the following example we see that exponential functions have constant relative rates of change.

Example 7   Find the relative rate of change of the exponential function $z = P_0 e^{kt}$.

Solution   Since

$$\ln z = \ln(P_0 e^{kt}) = \ln P_0 + \ln(e^{kt}) = \ln P_0 + kt,$$

we have

$$\frac{d}{dt}(\ln z) = k.$$

The relative rate of change of the exponential function $P_0 e^{kt}$ is the constant k.

Example 8   The surface area S of a mammal, in cm^2, is a function of the body mass, M, of the mammal, in kilograms, and is given by $S = 1095 \cdot M^{2/3}$. Find the relative rate of change of S with respect to M and evaluate for a human with body mass 70 kilograms. Interpret your answer.

Solution   We have

$$\ln S = \ln(1095 \cdot M^{2/3}) = \ln(1095) + \ln(M^{2/3}) = \ln(1095) + \frac{2}{3}\ln M.$$

Thus,

$$\begin{aligned}
\text{Relative rate of change} &= \frac{d}{dM}(\ln S) \\
&= \frac{d}{dM}\left(\ln(1095) + \frac{2}{3}\ln M\right) \\
&= \frac{2}{3} \cdot \frac{1}{M}.
\end{aligned}$$

For a human with body mass $M = 70$ kilograms, we have

$$\text{Relative rate} = \frac{2}{3}\frac{1}{70} = 0.0095 = 0.95\%.$$

The surface area of a human with body mass 70 kilograms increases by about 0.95% if body mass increases by 1 kilogram.

Problems for Section 3.3

Find the derivative of the functions in Problems 1–28.

1. $(4x^2 + 1)^7$

2. $f(x) = (x + 1)^{99}$

3. $R = (q^2 + 1)^4$

4. $w = (t^2 + 1)^{100}$

5. $w = (t^3 + 1)^{100}$

6. $w = (5r - 6)^3$

7. $y = \sqrt{s^3 + 1}$

8. $y = 12 - 3x^2 + 2e^{3x}$

9. $C = 12(3q^2 - 5)^3$

10. $f(x) = 6e^{5x} + e^{-x^2}$

11. $y = 5e^{5t+1}$

12. $w = e^{-3t^2}$

13. $w = e^{\sqrt{s}}$

14. $y = \ln(5t + 1)$

15. $f(x) = \ln(1 - x)$

16. $f(t) = \ln(t^2 + 1)$

17. $f(x) = \ln(1 - e^{-x})$

18. $f(x) = \ln(e^x + 1)$

19. $f(t) = 5\ln(5t + 1)$

20. $g(t) = \ln(4t + 9)$

21. $y = 5 + \ln(3t + 2)$

22. $Q = 100(t^2 + 5)^{0.5}$

23. $y = 5x + \ln(x + 2)$

24. $y = (5 + e^x)^2$

25. $P = (1 + \ln x)^{0.5}$

26. $\sqrt{e^x + 1}$

27. $f(x) = \sqrt{1 - x^2}$

28. $f(\theta) = (e^\theta + e^{-\theta})^{-1}$

In Problems 29–34, find the relative rate of change, $f'(t)/f(t)$, of the function $f(t)$.

29. $f(t) = 10t + 5$

30. $f(t) = 15t + 12$

31. $f(t) = 8e^{5t}$

32. $f(t) = 30e^{-7t}$

33. $f(t) = 6t^2$

34. $f(t) = 35t^{-4}$

In Problems 35–38, find the relative rate of change of $f(t)$ using the formula $\frac{d}{dt}\ln f(t)$.

35. $f(t) = 5e^{1.5t}$

36. $f(t) = 6.8e^{-0.5t}$

37. $f(t) = 3t^2$

38. $f(t) = 4.5t^{-4}$

39. Find the equation of the tangent line to $f(x) = (x-1)^3$ at the point where $x = 2$.

40. A firm estimates that the total revenue, R, received from the sale of q goods is given by

$$R = \ln(1 + 1000q^2).$$

Calculate the marginal revenue when $q = 10$.

41. The distance, s, of a moving body from a fixed point is given as a function of time by $s = 20e^{t/2}$. Find the velocity, v, of the body as a function of t.

42. If you invest P dollars in a bank account at an annual interest rate of $r\%$, then after t years you will have B dollars, where

$$B = P\left(1 + \frac{r}{100}\right)^t.$$

(a) Find dB/dt, assuming P and r are constant. In terms of money, what does dB/dt represent?

(b) Find dB/dr, assuming P and t are constant. In terms of money, what does dB/dr represent?

43. The distance traveled, D in feet, is a function of time, t, in seconds, with $D = f(t) = \sqrt{t^3 + 1}$. Find $f(10)$, $f'(10)$, and the relative rate of change f'/f at $t = 10$. Interpret your answers in terms of distance traveled.

In Problems 44–47, use Figure 3.18 to evaluate the derivative.

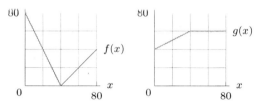

Figure 3.18

44. $\frac{d}{dx}f(g(x))\big|_{x=30}$

45. $\frac{d}{dx}f(g(x))\big|_{x=70}$

46. $\frac{d}{dx}g(f(x))\big|_{x=30}$

47. $\frac{d}{dx}g(f(x))\big|_{x=70}$

For Problems 48–51, let $h(x) = f(g(x))$ and $k(x) = g(f(x))$. Use Figure 3.19 to estimate the derivatives.

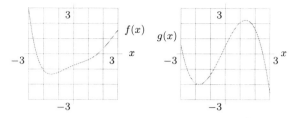

Figure 3.19

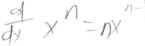

48. $h'(1)$

49. $k'(1)$

50. $h'(2)$

51. $k'(2)$

52. Some economists suggest that an extra year of education increases a person's wages, on average, by about 14%. Assume you could make \$10 per hour with your current level of education and that inflation increases wages at a continuous rate of 3.5% per year.

(a) How much would you make per hour with four additional years of education?

(b) What is the difference between your wages in 20 years' time with and without the additional four years of education?

(c) Is the difference you found in part (b) increasing with time? If so, at what rate? (Assume the number of additional years of education stays fixed at four.)

53. Show that if the graphs of $f(t)$ and $h(t) = Ae^{kt}$ are tangent at $t = a$, then k is the relative rate of change of f at $t = a$.

3.4   THE PRODUCT AND QUOTIENT RULES

This section shows how to find the derivatives of products and quotients of functions.

The Product Rule

Suppose we know the derivatives of $f(x)$ and $g(x)$ and want to calculate the derivative of the product, $f(x)g(x)$. We start by looking at an example. Let $f(x) = x$ and $g(x) = x^2$. Then

$$f(x)g(x) = x \cdot x^2 = x^3,$$

so the derivative of the product is $3x^2$. Notice that the derivative of the product is *not* equal to the product of the derivatives, since $f'(x) = 1$ and $g'(x) = 2x$, so $f'(x)g'(x) = (1)(2x) = 2x$. In general, we have the following rule, which is justified in the Focus on Theory section at the end of this chapter.

The Product Rule

If $u = f(x)$ and $v = g(x)$ are differentiable functions, then

$$(fg)' = f'g + fg'.$$

The product rule can also be written

$$\frac{d(uv)}{dx} = \frac{du}{dx} \cdot v + u \cdot \frac{dv}{dx}.$$

In words:

The derivative of a product is the derivative of the first times the second, plus the first times the derivative of the second.

We check that this rule gives the correct answers for $f(x) = x$ and $g(x) = x^2$. The derivative of $f(x)g(x)$ is

$$f'(x)g(x) + f(x)g'(x) = 1(x^2) + x(2x) = x^2 + 2x^2 = 3x^2.$$

This is the answer we expect for the derivative of $f(x)g(x) = x \cdot x^2 = x^3$.

Example 1   Differentiate (a) $x^2 e^{2x}$ (b) $t^3 \ln(t + 1)$ (c) $(3x^2 + 5x)e^x$.

Solution   (a) Using the product rule, we have

$$\frac{d}{dx}(x^2 e^{2x}) = \frac{d}{dx}(x^2) \cdot e^{2x} + x^2 \frac{d}{dx}(e^{2x})$$
$$= (2x)e^{2x} + x^2(2e^{2x})$$
$$= 2xe^{2x} + 2x^2 e^{2x}.$$

(b) Differentiating using the product rule gives

$$\frac{d}{dt}(t^3 \ln(t + 1)) = \frac{d}{dt}(t^3) \cdot \ln(t + 1) + t^3 \frac{d}{dt}(\ln(t + 1))$$
$$= (3t^2) \ln(t + 1) + t^3 \left(\frac{1}{t + 1} \right)$$
$$= 3t^2 \ln(t + 1) + \frac{t^3}{t + 1}.$$

(c) The product rule gives

$$\frac{d}{dx}((3x^2 + 5x)e^x) = \left(\frac{d}{dx}(3x^2 + 5x) \right) e^x + (3x^2 + 5x)\frac{d}{dx}(e^x)$$
$$= (6x + 5)e^x + (3x^2 + 5x)e^x$$
$$= (3x^2 + 11x + 5)e^x.$$

Example 2   Find the derivative of $C = \dfrac{e^{2t}}{t}$.

Solution   We write $C = e^{2t}t^{-1}$ and use the product rule:

$$\frac{d}{dt}(e^{2t}t^{-1}) = \frac{d}{dt}(e^{2t}) \cdot t^{-1} + e^{2t}\frac{d}{dt}(t^{-1})$$
$$= (2e^{2t}) \cdot t^{-1} + e^{2t}(-1)t^{-2}$$
$$= \frac{2e^{2t}}{t} - \frac{e^{2t}}{t^2}.$$

Example 3   A demand curve for a product has the equation $p = 80e^{-0.003q}$, where p is price and q is quantity.

(a) Find the revenue as a function of quantity sold.
(b) Find the marginal revenue function.

Solution   (a) Since Revenue = Price × Quantity, we have $R = pq = (80e^{-0.003q})q = 80qe^{-0.003q}$.

(b) The marginal revenue function is the derivative of revenue with respect to quantity. The product rule gives

$$\text{Marginal Revenue} = \frac{d}{dq}(80qe^{-0.003q})$$

$$= \left(\frac{d}{dq}(80q)\right)e^{-0.003q} + 80q\left(\frac{d}{dq}(e^{-0.003q})\right)$$

$$= (80)e^{-0.003q} + 80q(-0.003e^{-0.003q})$$

$$= (80 - 0.24q)e^{-0.003q}.$$

The Quotient Rule

Suppose we want to differentiate a function of the form $Q(x) = f(x)/g(x)$. (Of course, we have to avoid points where $g(x) = 0$.) We want a formula for Q' in terms of f' and g'. We have the following rule, which is justified in the Focus on Theory section at the end of this chapter.

The Quotient Rule

If $u = f(x)$ and $v = g(x)$ are differentiable functions, then

$$\left(\frac{f}{g}\right)' = \frac{f'g - fg'}{g^2},$$

or equivalently,

$$\frac{d}{dx}\left(\frac{u}{v}\right) = \frac{\dfrac{du}{dx}\cdot v - u \cdot \dfrac{dv}{dx}}{v^2}.$$

In words:
  The derivative of a quotient is the derivative of the numerator times the denominator minus the numerator times the derivative of the denominator, all over the denominator squared.

Example 4   Differentiate (a) $\dfrac{5x^2}{x^3 + 1}$ (b) $\dfrac{1}{1 + e^x}$ (c) $\dfrac{e^x}{x^2}$.

Solution   (a) Using the quotient rule

$$\frac{d}{dx}\left(\frac{5x^2}{x^3 + 1}\right) = \frac{\left(\dfrac{d}{dx}(5x^2)\right)(x^3 + 1) - 5x^2\dfrac{d}{dx}(x^3 + 1)}{(x^3 + 1)^2} = \frac{10x(x^3 + 1) - 5x^2(3x^2)}{(x^3 + 1)^2}$$

$$= \frac{-5x^4 + 10x}{(x^3 + 1)^2}.$$

$10x^4 + 10x \quad - 15x$

$- 5x + 10x$

(b) Differentiating using the quotient rule yields

$$\frac{d}{dx}\left(\frac{1}{1+e^x}\right) = \frac{\left(\frac{d}{dx}(1)\right)(1+e^x) - 1\frac{d}{dx}(1+e^x)}{(1+e^x)^2} = \frac{0(1+e^x) - 1(0+e^x)}{(1+e^x)^2}$$
$$= \frac{-e^x}{(1+e^x)^2}.$$

(c) The quotient rule gives

$$\frac{d}{dx}\left(\frac{e^x}{x^2}\right) = \frac{\left(\frac{d}{dx}(e^x)\right)x^2 - e^x\left(\frac{d}{dx}(x^2)\right)}{(x^2)^2} = \frac{e^x x^2 - e^x(2x)}{x^4}$$
$$= e^x\left(\frac{x^2 - 2x}{x^4}\right) = e^x\left(\frac{x-2}{x^3}\right).$$

Problems for Section 3.4

1. If $f(x) = (2x+1)(3x-2)$, find $f'(x)$ two ways: by using the product rule and by multiplying out. Do you get the same result?

2. If $f(x) = x^2(x^3 + 5)$, find $f'(x)$ two ways: by using the product rule and by multiplying out before taking the derivative. Do you get the same result? Should you?

For Problems 3–33, find the derivative. Assume that a, b, c, and k are constants.

3. $f(x) = xe^x$

4. $f(t) = te^{-2t}$

5. $y = 5xe^{x^2}$

6. $y = t^2(3t+1)^3$

7. $y = x\ln x$

8. $y = (t^2+3)e^t$

9. $z = (3t+1)(5t+2)$

10. $y = (t^3 - 7t^2 + 1)e^t$

11. $P = t^2\ln t$

12. $R = 3qe^{-q}$

13. $f(t) = \frac{5}{t} + \frac{6}{t^2}$

14. $f(x) = \frac{x^2+3}{x}$

15. $y = te^{-t^2}$

16. $f(z) = \sqrt{z}e^{-z}$

17. $g(p) = p\ln(2p+1)$

18. $f(t) = te^{5-2t}$

19. $f(w) = (5w^2 + 3)e^{w^2}$

20. $y = x \cdot 2^x$

21. $w = (t^3 + 5t)(t^2 - 7t + 2)$

22. $z = (te^{3t} + e^{5t})^9$

23. $f(x) = \frac{x}{e^x}$

24. $w = \frac{3z}{1+2z}$

25. $z = \frac{1-t}{1+t}$

26. $y = \frac{e^x}{1+e^x}$

27. $w = \frac{3y+y^2}{5+y}$

28. $y = \frac{1+z}{\ln z}$

29. $f(x) = \frac{ax+b}{cx+k}$

30. $f(x) = (ax^2 + b)^3$

31. $f(x) = axe^{-bx}$

32. $f(t) = ae^{bt}$

33. $g(\alpha) = e^{\alpha e^{-2\alpha}}$

34. If $f(x) = (3x+8)(2x-5)$, find $f'(x)$ and $f''(x)$.

35. Find the equation of the tangent line to the graph of $f(x) = x^2e^{-x}$ at $x = 0$. Check by graphing this function and the tangent line on the same axes.

36. Find the equation of the tangent line to the graph of $f(x) = \frac{2x-5}{x+1}$ at the point at which $x = 0$.

37. The quantity of a drug, Q mg, present in the body t hours after an injection of the drug is given is

$$Q = f(t) = 100te^{-0.5t}.$$

Find $f(1)$, $f'(1)$, $f(5)$, and $f'(5)$. Give units and interpret the answers.

38. A drug concentration curve is given by $C = f(t) = 20te^{-0.04t}$, with C in mg/ml and t in minutes.

(a) Graph C against t. Is $f'(15)$ positive or negative? Is $f'(45)$ positive or negative? Explain.

(b) Find $f(30)$ and $f'(30)$ analytically. Interpret them in terms of the concentration of the drug in the body.

39. For positive constants c and k, the *Monod growth curve* describes the growth of a population, P, as a function of the available quantity of a resource, r:

$$P = \frac{cr}{k+r}.$$

Find dP/dr and interpret it in terms of the growth of the population.

40. If p is price in dollars and q is quantity, demand for a product is given by

$$q = 5000e^{-0.08p}.$$

   (a) What quantity is sold at a price of \$10?
   (b) Find the derivative of demand with respect to price when the price is \$10 and interpret your answer in terms of demand.

41. The demand for a product is given in Problem 40. Find the revenue and the derivative of revenue with respect to price at a price of \$10. Interpret your answers in economic terms.

42. The quantity demanded of a certain product, q, is given in terms of p, the price, by

$$q = 1000e^{-0.02p}$$

   (a) Write revenue, R, as a function of price.
   (b) Find the rate of change of revenue with respect to price.
   (c) Find the revenue and rate of change of revenue with respect to price when the price is \$10. Interpret your answers in economic terms.

43. If $\dfrac{d}{dt}(tf(t)) = 1 + f(t)$, what is $f'(t)$?

44. The quantity, q, of a certain skateboard sold depends on the selling price, p, in dollars, so we write $q = f(p)$. You are given that $f(140) = 15{,}000$ and $f'(140) = -100$.

   (a) What do $f(140) = 15{,}000$ and $f'(140) = -100$ tell you about the sales of skateboards?
   (b) The total revenue, R, earned by the sale of skateboards is given by $R = pq$. Find $\left.\dfrac{dR}{dp}\right|_{p=140}$.
   (c) What is the sign of $\left.\dfrac{dR}{dp}\right|_{p=140}$? If the skateboards are currently selling for \$140, what happens to revenue if the price is increased to \$141?

45. Show that the relative rate of change of a product fg is the sum of the relative rates of change of f and g.

46. Show that the relative rate of change of a quotient f/g is the difference between the relative rates of change of f and g.

47. If $h = f^n$, show that

$$\frac{(f^n)'}{f^n} = n\frac{f'}{f}.$$

3.5 DERIVATIVES OF PERIODIC FUNCTIONS

Since the sine and cosine functions are periodic, their derivatives must be periodic also. (Why?) Let's look at the graph of $f(x) = \sin x$ in Figure 3.20 and estimate the derivative function graphically.

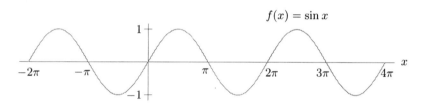

$$f(x) = \sin x$$

Figure 3.20: The sine function

First we might ask ourselves where the derivative is zero. (At $x = \pm\pi/2, \pm3\pi/2, \pm5\pi/2$, etc.) Then ask where the derivative is positive and where it is negative. (Positive for $-\pi/2 < x < \pi/2$; negative for $\pi/2 < x < 3\pi/2$, etc.) Since the largest positive slopes are at $x = 0, 2\pi$, and so on, and the largest negative slopes are at $x = \pi, 3\pi$, and so on, we get something like the graph in Figure 3.21.

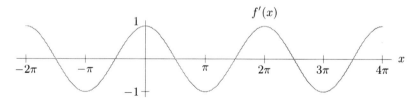

$$f'(x)$$

Figure 3.21: Derivative of $f(x) = \sin x$

The graph of the derivative in Figure 3.21 looks suspiciously like the graph of the cosine function. This might lead us to conjecture, quite correctly, that the derivative of the sine is the cosine.

Of course, we cannot be sure, just from the graphs, that the derivative of the sine really is the cosine. However, this is in fact true.

One thing we can do is to check that the derivative function in Figure 3.21 has amplitude 1 (as it must if it is the cosine). That means we have to convince ourselves that the derivative of $f(x) = \sin x$ is 1 when $x = 0$. The next example suggests that this is true when x is in radians.

Example 1   Using a calculator, estimate the derivative of $f(x) = \sin x$ at $x = 0$. Make sure your calculator is set in radians.

Solution   We use the average rate of change of $\sin x$ on the small interval $0 \leq x \leq 0.01$ to compute

$$f'(0) \approx \frac{\sin(0.01) - \sin(0)}{0.01 - 0} = \frac{0.0099998 - 0}{0.01} = 0.99998 \approx 1.0.$$

The derivative of $f(x) = \sin x$ at $x = 0$ is approximately 1.0.

Warning: It is important to notice that in the previous example x was in *radians*; any conclusions we have drawn about the derivative of $\sin x$ are valid *only* when x is in radians.

Example 2   Starting with the graph of the cosine function, sketch a graph of its derivative.

Solution   The graph of $g(x) = \cos x$ is in Figure 3.22(a). Its derivative is 0 at $x = 0, \pm\pi, \pm 2\pi$, and so on; it is positive for $-\pi < x < 0$, $\pi < x < 2\pi$, and so on, and it is negative for $0 < x < \pi$, $2\pi < x < 3\pi$, and so on. The derivative is in Figure 3.22(b).

(a)

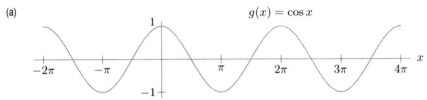

(b)

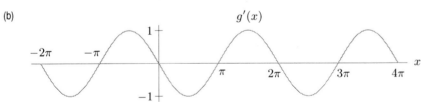

Figure 3.22: $g(x) = \cos x$ and its derivative, $g'(x)$

As we did with the sine, we'll use the graphs to make a conjecture. The derivative of the cosine in Figure 3.22(b) looks exactly like the graph of sine, except reflected about the x-axis. It turns out that the derivative of $\cos x$ is $-\sin x$.

For x in radians,

$$\frac{d}{dx}(\sin x) = \cos x \quad \text{and} \quad \frac{d}{dx}(\cos x) = -\sin x.$$

Example 3   Differentiate (a) $5 \sin t - 8 \cos t$ (b) $5 - 3 \sin x + x^3$

Solution   (a) Differentiating gives

$$\frac{d}{dt}(5 \sin t - 8 \cos t) = 5\frac{d}{dt}(\sin t) - 8\frac{d}{dt}(\cos t) - 5(\cos t) \quad 8(\quad \sin t) - 5\cos t + 8 \sin t$$

(b) We have

$$\frac{d}{dx}(5 - 3 \sin x + x^3) = \frac{d}{dx}(5) - 3\frac{d}{dx}(\sin x) + \frac{d}{dx}(x^3) = 0 - 3(\cos x) + 3x^2 = -3 \cos x + 3x^2.$$

The chain rule tells us how to differentiate composite functions involving the sine and cosine. Suppose $y = \sin(3t)$, so $y = \sin z$ and $z = 3t$, so

$$\frac{dy}{dt} = \frac{dy}{dz} \cdot \frac{dz}{dt} = \cos z\frac{dz}{dt} = \cos(3t) \cdot 3 = 3\cos(3t).$$

In general,

If z is a differentiable function of t, then

$$\frac{d}{dt}(\sin z) = \cos z\frac{dz}{dt} \quad \text{and} \quad \frac{d}{dt}(\cos z) = -\sin z\frac{dz}{dt}$$

In many applications, $z = kt$ for some constant k. Then we have:

If k is a constant, then

$$\frac{d}{dt}(\sin kt) = k \cos kt \quad \text{and} \quad \frac{d}{dt}(\cos kt) = -k \sin kt$$

Example 4   Differentiate:
(a) $\sin(t^2)$ (b) $5 \cos(2t)$ (c) $t \sin t$.

Solution   (a) We have $y = \sin z$ with $z = t^2$, so

$$\frac{d}{dt}(\sin(t^2)) = \frac{d}{dt}(\sin z) = \cos z\frac{dz}{dt} = \cos(t^2) \cdot 2t = 2t \cos(t^2).$$

(b) We have $y = 5 \cos z$ with $z = 2t$, so

$$\frac{d}{dt}(5 \cos(2t)) = 5\frac{d}{dt}(\cos(2t)) = 5(-2 \sin(2t)) = -10 \sin(2t).$$

(c) We use the product rule:

$$\frac{d}{dt}(t \sin t) = \frac{d}{dt}(t) \cdot \sin t + t\frac{d}{dt}(\sin t) = 1 \cdot \sin t + t(\cos t) = \sin t + t \cos t.$$

Problems for Section 3.5

Differentiate the functions in Problems 1–20. Assume that A and B are constants.

1. $y = 5 \sin x$

2. $P = 3 + \cos t$

3. $y = t^2 + 5 \cos t$

4. $y = B + A \sin t$

5. $R(q) = q^2 - 2 \cos q$

6. $y = 5 \sin x - 5x + 4$

7. $f(x) = \sin(3x)$

8. $R = \sin(5t)$

9. $W = 4 \cos(t^2)$

10. $y = 2 \cos(5t)$

11. $y = \sin(x^2)$

12. $y = A \sin(Bt)$

13. $z = \cos(4\theta)$

14. $y = 6 \sin(2t) + \cos(4t)$

15. $f(x) = x^2 \cos x$

16. $f(x) = 2x \sin(3x)$

17. $f(\theta) = \theta^3 \cos \theta$

18. $z = \dfrac{e^{t^2} + t}{\sin(2t)}$

19. $f(t) = \dfrac{t^2}{\cos t}$

20. $f(\theta) = \dfrac{\sin \theta}{\theta}$

21. Find the equation of the tangent line to the graph of $y = \sin x$ at $x = \pi$. Graph the function and the tangent line on the same axes.

22. If t is the number of months since June, the number of bird species, N, found in an Ohio forest oscillates approximately according to the formula
$$N = f(t) = 19 + 9 \cos\left(\frac{\pi}{6}t\right).$$
   (a) Graph $f(t)$ for $0 \le t \le 24$ and describe what it shows. Use the graph to decide whether $f'(1)$ and $f'(10)$ are positive or negative.
   (b) Find $f'(t)$.
   (c) Find and interpret $f(1)$, $f'(1)$, $f(10)$, and $f'(10)$.

23. Is the graph of $y = \sin(x^4)$ increasing or decreasing when $x = 10$? Is it concave up or concave down?

24. Find equations of the tangent lines to the graph of $f(x) = \sin x$ at $x = 0$ and at $x = \pi/3$. Use each tangent line to approximate $\sin(\pi/6)$. Would you expect these results to be equally accurate, since they are taken equally far away from $x = \pi/6$ but on opposite sides? If the accuracy is different, can you account for the difference?

25. A company's monthly sales, $S(t)$, are seasonal and given as a function of time, t, in months, by
$$S(t) = 2000 + 600 \sin\left(\frac{\pi}{6}t\right).$$
   (a) Graph $S(t)$ for $t = 0$ to $t = 12$. What is the maximum monthly sales? What is the minimum monthly sales? If $t = 0$ is January 1, when during the year are sales highest?
   (b) Find $S(2)$ and $S'(2)$. Interpret in terms of sales.

26. A boat at anchor is bobbing up and down in the sea. The vertical distance, y, in feet, between the sea floor and the boat is given as a function of time, t, in minutes, by
$$y = 15 + \sin(2\pi t).$$
   (a) Find the vertical velocity, v, of the boat at time t.
   (b) Make rough sketches of y and v against t.

27. The depth of the water, y, in meters, in the Bay of Fundy, Canada, is given as a function of time, t, in hours after midnight, by the function
$$y = 10 + 7.5 \cos(0.507t).$$
How quickly is the tide rising or falling (in meters/hour) at each of the following times?
   (a) 6:00 am   **(b)** 9:00 am
   (c) Noon   **(d)** 6:00 pm

28. The average adult takes about 12 breaths per minute. As a patient inhales, the volume of air in the lung increases. As the patient exhales, the volume of air in the lung decreases. For t in seconds since start of the breathing cycle, the volume of air inhaled or exhaled since $t = 0$ is given [9], in hundreds of cubic centimeters, by
$$A(t) = -2 \cos\left(\frac{2\pi}{5}t\right) + 2.$$
   (a) How long is one breathing cycle?
   (b) Find $A'(1)$ and explain what it means.

29. On July 7, 2009, there was a full moon in the Eastern time zone.[10] If t is the number of days since July 7th, the percent of moon illuminated can be represented by
$$H(t) = 50 + 50 \cos\left(\frac{\pi}{15}t\right).$$
   (a) Find $H'(t)$. Explain what this tells about the moon.
   (b) For $0 \le t \le 30$, when is $H'(t) = 0$? What does this tell us about the moon?
   (c) For $0 \le t \le 30$, when is $H'(t)$ negative? When is it positive? Explain what positive and negative values of $H'(t)$ tell us about the moon.

30. Paris, France, has a latitude of approximately $49°$ N. If t is the number of days since the start of 2009, the number of hours of daylight in Paris can be approximated by
$$D(t) = 4 \cos\left(\frac{2\pi}{365}(t - 172)\right) + 12.$$
   (a) Find $D(40)$ and $D'(40)$. Explain what this tells about daylight in Paris.
   (b) Find $D(172)$ and $D'(172)$. Explain what this tells about daylight in Paris.

---

[9] Based upon information obtained from Dr. Gadi Avshalomov on August 14, 2008.
[10] www.aa.usno.navy.mil/feq/docs/moon_phases.php, accessed on April 14, 2009.

CHAPTER SUMMARY

- **Derivatives of elementary functions**
  Powers, polynomials, exponential functions, logarithms, periodic functions
- **Derivatives of sums, differences, constant multiples**
- **Chain rule**

- **Product and quotient rules**
- **Relative rates of change**
  Using logarithms
- **Tangent line approximation**
- **Interpreting derivatives found using formulas**

REVIEW PROBLEMS FOR CHAPTER THREE

Find the derivatives for the functions in Problems 1–40. Assume k is a constant.

1. $f(t) = 6t^4$

2. $f(x) = x^3 - 3x^2 + 5x$

3. $P(t) = e^{2t}$

4. $W = r^3 + 5r - 12$

5. $C = e^{0.08q}$

6. $y = 5e^{-0.2t}$

7. $y = xe^{3x}$

8. $s(t) = (t^2 + 4)(5t - 1)$

9. $g(t) = e^{(1+3t)^2}$

10. $f(x) = x^2 + 3\ln x$

11. $Q(t) = 5t + 3e^{1.2t}$

12. $g(z) = (z^2 + 5)^3$

13. $f(x) = 6(5x - 1)^3$

14. $f(z) = \ln(z^2 + 1)$

15. $h(x) = (1 + e^x)^{10}$

16. $q = 100e^{-0.05p}$

17. $y = x^2 \ln x$

18. $s(t) = t^2 + 2\ln t$

19. $P = 4t^2 + 7\sin t$

20. $R(t) = (\sin t)^5$

21. $h(t) = \ln\left(e^{-t} - t\right)$

22. $f(x) = \sin(2x)$

23. $g(x) = \dfrac{25x^2}{e^x}$

24. $h(t) = \dfrac{t + 4}{t - 4}$

25. $y = x^2 \cos x$

26. $h(x) = \ln(1 + e^x)$

27. $h(w) = e^{\ln w + 1}$

28. $h(x) = \ln(x^3 + x)$

29. $q(x) = \dfrac{1 + e^x}{1 - e^{-x}}$

30. $q(x) = \dfrac{x}{1 + x}$

31. $f(x) = xe^x$

32. $h(x) = \sin(e^x)$

33. $h(x) = \cos(x^3)$

34. $z = \dfrac{3t + 1}{5t + 2}$

35. $z = \dfrac{t^2 + 5t + 2}{t + 3}$

36. $h(p) = \dfrac{1 + p^2}{3 + 2p^2}$

37. $y = \dfrac{3^x}{3} + \dfrac{33}{\sqrt{x}}$

38. $f(t) = \sin\sqrt{e^t + 1}$

39. $g(y) = e^{2e^{(y^3)}}$

40. $g(t) = \dfrac{\ln(kt) + t}{\ln(kt) - t}$

41. Let $f(x) = x^2 + 1$. Compute the derivatives $f'(0)$, $f'(1)$, $f'(2)$, and $f'(-1)$. Check your answers graphically.

42. Let $f(x) = x^2 + 3x - 5$. Find $f'(0)$, $f'(3)$, $f'(-2)$.

43. Find the equation of the line tangent to the graph of $f(x) = 2x^3 - 5x^2 + 3x - 5$ at $x = 1$.

In Problems 44–47, find the relative rate of change $f'(t)/f(t)$ of the function at the given value of t. Assume t is in years and give your answer as a percent.

44. $f(t) = 3t + 2$; $t = 5$

45. $f(t) = 2e^{0.3t}$; $t = 7$

46. $f(t) = 2t^3 + 10$; $t = 4$

47. $f(t) = \ln(t^2 + 1)$; $t = 2$

48. With a yearly inflation rate of 5%, prices are given by

$$P = P_0(1.05)^t,$$

where P_0 is the price in dollars when $t = 0$ and t is time in years. Suppose $P_0 = 1$. How fast (in cents/year) are prices rising when $t = 10$?

49. According to the US Census, the world population P, in billions, was approximately [11]

$$P = 6.8e^{0.012t}$$

where t is in years since January 1, 2009. At what rate was the world's population increasing on that date? Give your answer in millions of people per year.

50. Let $f(x) = x^2 - 4x + 8$. For what x-values is $f'(x) = 0$?

51. A football player kicks a ball at an angle of $30°$ from the ground with an initial velocity of 64 feet per second. Its height t seconds later is $h(t) = 32t - 16t^2$.

  (a) Graph the height of the ball as a function of the time.
  (b) Find $v(t)$, the velocity of the ball at time t.
  (c) Find $v(1)$, and explain what is happening to the ball at this time. What is the height of the ball at this time?

---

[11] www.census.gov/ipc/www/popclockworld.html and
www.cia.gov/library/publications/the-world-factbook/print/xx.html.

52. The graph of $y = x^3 - 9x^2 - 16x + 1$ has a slope of 5 at two points. Find the coordinates of the points.

53. (a) Find the slope of the graph of $f(x) = 1 - e^x$ at the point where it crosses the x-axis.
(b) Find the equation of the tangent line to the curve at this point.

54. Find the equation of the tangent line to the graph of $P(t) = t \ln t$ at $t = 2$. Graph the function $P(t)$ and the tangent line $Q(t)$ on the same axes.

55. The balance, $\$B$, in a bank account t years after a deposit of \$5000 is given by $B = 5000e^{0.08t}$. At what rate is the balance in the account changing at $t = 5$ years? Use units to interpret your answer in financial terms.

56. If a cup of coffee is left on a counter top, it cools off slowly. The temperature in degrees Fahrenheit t minutes after it was left on the counter is given by

$$C(t) = 74 + 103e^{-0.033t}.$$

(a) What was the temperature of the coffee when it was left on the counter?
(b) If the coffee was left on the counter for a long time, what is the lowest temperature the coffee would reach? What does that temperature represent?
(c) Find $C(5)$ and $C'(5)$. Explain what this tells about the temperature of the coffee.
(d) Without calculation, decide if the magnitude of $C'(50)$ is greater or less than the magnitude of $C'(5)$? Why?

57. With length, l, in meters, the period T, in seconds, of a pendulum is given by

$$T = 2\pi \sqrt{\frac{l}{9.8}}.$$

(a) How fast does the period increase as l increases?
(b) Does this rate of change increase or decrease as l increases?

58. One gram of radioactive carbon-14 decays according to the formula

$$Q = e^{-0.000121t},$$

where Q is the number of grams of carbon-14 remaining after t years.

(a) Find the rate at which carbon-14 is decaying (in grams/year).
(b) Sketch the rate you found in part (a) against time.

59. The temperature, H, in degrees Fahrenheit (°F), of a can of soda that is put into a refrigerator to cool is given as a function of time, t, in hours, by

$$H = 40 + 30e^{-2t}.$$

(a) Find the rate at which the temperature of the soda is changing (in °F/hour).

(b) What is the sign of dH/dt? Explain.
(c) When, for $t \geq 0$, is the magnitude of dH/dt largest? In terms of the can of soda, why is this?

60. Explain for which values of a the function a^x is increasing and for which values it is decreasing. Use the fact that, for $a > 0$,

$$\frac{d}{dx}(a^x) = (\ln a)a^x.$$

61. The value of an automobile purchased in 2009 can be approximated by the function $V(t) = 25(0.85)^t$, where t is the time, in years, from the date of purchase, and $V(t)$ is the value, in thousands of dollars.

(a) Evaluate and interpret $V(4)$.
(b) Find an expression for $V'(t)$, including units.
(c) Evaluate and interpret $V'(4)$.
(d) Use $V(t)$, $V'(t)$, and any other considerations you think are relevant to write a paragraph in support of or in opposition to the following statement: "From a monetary point of view, it is best to keep this vehicle as long as possible."

62. The temperature Y in degrees Fahrenheit of a yam in a hot oven t minutes after it is placed there is given by

$$Y(t) = 350(1 - 0.7e^{-0.008t}).$$

(a) What was the temperature of the yam when it was placed in the oven?
(b) What is the temperature of the oven?
(c) When does the yam reach 175° F?
(d) Estimate the rate at which the temperature of the yam is increasing when $t = 20$.

63. Kepler's third law of planetary motion states that, $P^2 = kd^3$, where P represents the time, in earth days, it takes a planet to orbit the sun once, d is the planet's average distance, in miles, from the sun, and k is a constant.[12]

(a) If $k = 1.65864 \cdot 10^{-19}$, write P as a function of d.
(b) Mercury is approximately 36 million miles from the sun. How long does it take Mercury to orbit the sun?
(c) Find $P'(d)$. What does the sign of the derivative tell us about the time it takes a planet to orbit the sun? Why does this make sense from a practical viewpoint?

64. Imagine you are zooming in on the graph of each of the following functions near the origin:

$$y = x \qquad y = \sqrt{x} \qquad y = x^2$$
$$y = x^3 + \tfrac{1}{2}x^2 \qquad y = x^3 \qquad y = \ln(x+1)$$
$$y = \tfrac{1}{2}\ln(x^2+1) \qquad y = \sqrt{2x - x^2}$$

Which of them look the same? Group together those functions which become indistinguishable near the origin, and give the equations of the lines they look like.

---

[12]www.exploratorium.edu/ronh/age/index.html, accessed on 5/3/09.

65. Given a power function of the form $f(x) = ax^n$, with $f'(2) = 3$ and $f'(4) = 24$, find n and a.

66. Given $r(2) = 4$, $s(2) = 1$, $s(4) = 2$, $r'(2) = -1$, $s'(2) = 3$, and $s'(4) = 3$, compute the following derivatives, or state what additional information you would need to be able to compute the derivative.

(a) $H'(2)$ if $H(x) = r(x) + s(x)$
(b) $H'(2)$ if $H(x) = 5s(x)$
(c) $H'(2)$ if $H(x) = r(x) \cdot s(x)$
(d) $H'(2)$ if $H(x) = \sqrt{r(x)}$

67. Given $F(2) = 1$, $F'(2) = 5$, $F(4) = 3$, $F'(4) = 7$ and $G(4) = 2$, $G'(4) = 6$, $G(3) = 4$, $G'(3) = 8$, find:

(a) $H(4)$ if $H(x) = F(G(x))$
(b) $H'(4)$ if $H(x) = F(G(x))$
(c) $H(4)$ if $H(x) = G(F(x))$
(d) $H'(4)$ if $H(x) = G(F(x))$
(e) $H'(4)$ if $H(x) = F(x)/G(x)$

68. A dose, D, of a drug causes a temperature change, T, in a patient. For C a positive constant, T is given by

$$T = \left(\frac{C}{2} - \frac{D}{3}\right)D^3.$$

(a) What is the rate of change of temperature change with respect to dose?
(b) For what doses does the temperature change increase as the dose increases?

69. A yam is put in a hot oven, maintained at a constant temperature $200°C$. At time $t = 30$ minutes, the temperature T of the yam is $120°$ and is increasing at an (instantaneous) rate of $2°/min$. Newton's law of cooling (or, in our case, warming) implies that the temperature at time t is given by

$$T(t) = 200 - ae^{-bt}.$$

Find a and b.

For Problems 70–75, let $h(x) = f(x) \cdot g(x)$, and $k(x) = f(x)/g(x)$, and $l(x) = g(x)/f(x)$. Use Figure 3.23 to estimate the derivatives.

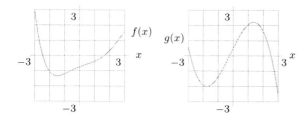

Figure 3.23

70. $h'(1)$
72. $h'(2)$
74. $l'(1)$

71. $k'(1)$
73. $k'(2)$
75. $l'(2)$

76. On what intervals is the function $f(x) = x^4 - 4x^3$ both decreasing and concave up?

77. Given $p(x) = x^n - x$, find the intervals over which p is a decreasing function when:

(a) $n = 2$ (b) $n = \frac{1}{2}$ (c) $n = -1$

78. Using the equation of the tangent line to the graph of e^x at $x = 0$, show that

$$e^x \geq 1 + x$$

for all values of x. A sketch may be helpful.

79. In Section 1.10 the depth, y, in feet, of water in Portland, Maine is given in terms of t, the number of hours since midnight, by

$$y = 4.9 + 4.4\cos\left(\frac{\pi}{6}t\right).$$

(a) Find dy/dt. What does dy/dt represent, in terms of water level?
(b) For $0 \leq t \leq 24$, when is dy/dt zero? (Figure 1.98 may be helpful.) Explain what it means (in terms of water level) for dy/dt to be zero.

80. Using a graph to help you, find the equations of all lines through the origin tangent to the parabola

$$y = x^2 - 2x + 4.$$

Sketch the lines on the graph.

81. A museum has decided to sell one of its paintings and to invest the proceeds. If the picture is sold between the years 2000 and 2020 and the money from the sale is invested in a bank account earning 5% interest per year compounded annually, then $B(t)$, the balance in the year 2020, depends on the year, t, in which the painting is sold and the sale price $P(t)$. If t is measured from the year 2000 so that $0 < t < 20$ then

$$B(t) = P(t)(1.05)^{20-t}.$$

(a) Explain why $B(t)$ is given by this formula.
(b) Show that the formula for $B(t)$ is equivalent to

$$B(t) = (1.05)^{20}\frac{P(t)}{(1.05)^t}.$$

(c) Find $B'(10)$, given that $P(10) = 150{,}000$ and $P'(10) = 5000$.

82. Find the mean and variance of the normal distribution of statistics using parts (a) and (b) with $m(t) = e^{\mu t + \sigma^2 t^2/2}$.

(a) Mean $= m'(0)$
(b) Variance $= m''(0) - (m'(0))^2$

83. Given a number $a > 1$, the equation

$$a^x = 1 + x$$

has the solution $x = 0$. Are there any other solutions? How does your answer depend on the value of a? [Hint: Graph the functions on both sides of the equation.]

84. Suppose that f is a positive function such that $f(t_0) = P_0$ and $f(t_1) = P_1$. If $h(t) = Ae^{kt}$ is the exponential function taking the same values as f at $t = t_0$ and $t = t_1$, show that $k = (\ln P_1 - \ln P_0)/(t_1 - t_0)$.

85. Figure 3.24 shows the number of gallons, G, of gasoline used on a trip of M miles.

   (a) The function f is linear on each of the intervals $0 < M < 70$ and $70 < M < 100$. What is the slope of these lines? What are the units of these slopes?
   (b) What is gas consumption (in miles per gallon) during the first 70 miles of this trip? During the next 30 miles?
   (c) Figure 3.25 shows distance traveled, M (in miles), as a function of time t, in hours since the start of the trip. Describe this trip in words. Give a possible explanation for what happens one hour into the trip. What do your answers to part (b) tell you about the trip?
   (d) If we let $G = k(t) = f(h(t))$, estimate $k(0.5)$ and interpret your answer in terms of the trip.
   (e) Find $k'(0.5)$ and $k'(1.5)$. Give units and interpret your answers.

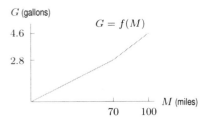

Figure 3.24

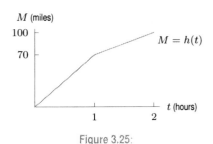

Figure 3.25:

86. On October 17, 2006, the US population was 300 million and growing exponentially. If the population was increasing at a rate of 2.9 million a year on that date, find a formula for the population as a function of time, t, in years since that date.

87. The speed of sound in dry air is

$$f(T) = 331.3\sqrt{1 + \frac{T}{273.15}} \text{ meters/second}$$

where T is the temperature in ° Celsius. Find a linear function that approximates the speed of sound for temperatures near $0°C$.

CHECK YOUR UNDERSTANDING

In Problems 1–50, indicate whether the statement is true or false.

1. If $f(x) = 5x^2 + 1$ then $f'(-1) = -10$.

2. The two functions $f(x) = 3x^5$ and $g(x) = 3x^5 + 7$ have the same derivative.

3. The derivative of $h(t) = (3t^2 + 1)(2t)$ is $h'(t) = (6t)(2) = 12t$.

4. If $k(s) = \sqrt{s^3}$ then $k'(s) = \sqrt{3s^2}$.

5. The equation of the line tangent to $f(x) = x^5 + 5$ at $x = 1$ is $y = 9x - 3$.

6. The derivative of $f(r) = 1/r^5$ is $f'(r) = 1/5r^4$.

7. The function $g(w) = w^3 - 3w$ has exactly two places where $g'(w) = 0$.

8. If $f(x) = 3x^3 - x^2 + 2x$ then f is decreasing at $x = 1$.

9. If $f(x) = 3x^3 - x^2 + 2x$ then the graph of f is concave up at $x = 1$.

10. If $g(t) = t^\pi$ then $g'(1) = \pi$.

11. If $f(x) = e^x$ then $f'(x) = xe^{x-1}$.

12. If $g(s) = 5\ln(s)$ then $g'(2) = 5/2$.

13. The function $f(x) = 3e^x + x$ has tangent line at $x = 0$ with equation $y = 4x + 3$.

14. The function $h(x) = 2\ln x - x^2$ is decreasing at $x = 1$.

15. The graph of $h(x) = 2\ln x - x^2$ is concave down at $x = 1$.

16. If $f(x) = \ln 2$ then $f'(x) = 1/2$.

17. If $f(x) = e^{2x}$ then $f'(x) = e^{2x}$.

18. If $w(q) = 10^q$ then $w'(q) = \log(10) \cdot 10^q$.

19. If $k(p) = 5 \cdot e^p$ then $k'(p) = \ln 5 \cdot e^p$.

20. If $f(x) = 3e^{5x}$ then $f'(x) = 15e^{5x}$.

21. The derivative of $f(t) = e^{t^2}$ is $f'(t) = 2e^{2t}$.

22. If $y = (x + x^2)^5$ then $dy/dx = 5(1 + 2x)^4$.

23. If $y = \ln(x^2 + 4)$ then $dy/dx = (2x)/(x^2 + 4)$.

24. If $y = \sqrt{1 - 2t}$ then $y' = 1/(2\sqrt{1 - 2t})$.

25. If $y = e^{-x}$ then $dy/dx = -e^{-x}$.

26. The chain rule says $d/dt(f(g(t))) = f(g'(t)) \cdot g(t)$.

27. If $g(x) = \ln(x^2 + 3x)$ then $g'(x) = (1/x)(2x + 3)$.

28. The function $f(x) = e^{1-x}$ is decreasing at $x = 1$.

29. The graph of $f(x) = e^{1-x}$ is concave up at $x = 1$.

30. If $B = 30(1 + 2r)^5$ then $dB/dr = 150(1 + 2r)^4$.

31. If $y = e^x \ln(x)$ then $y' = e^x/x$.

32. If $y = (x^2 + 1)2^x$ then $y' = 2x(\ln 2)2^x$.

33. If $y = x/e^x$ then $y' = (e^x - xe^x)/e^{2x} = (1 - x)/e^x$.

34. If $z = 2/(1 + t^2)$ then $z' = -4t/(1 + t^2)^2$.

35. If $s = w^2 e^w$ then $s' = 2we^w$.

36. If $P = q\ln(q^2 + 1)$ then $P' = q/(q^2 + 1) + \ln(q^2 + 1)$.

37. If $y = e^{x \ln x}$ then $y' = e^{x \ln x}(1 + \ln x)$.

38. Using the product rule to differentiate $x^2 \cdot x^2$ gives the same result as differentiating x^4 directly.

39. The derivative of the product of two functions is the product of their derivatives.

40. The derivative of the quotient of two functions is the quotient of their derivatives.

41. The derivative of $\sin t + \cos t$ is $\cos t - \sin t$.

42. The second derivative of $f(t) = \sin t$ is $f''(t) = \sin t$.

43. The second derivative of $g(t) = \cos t$ is $g''(t) = -\cos t$.

44. If $y = \sin 2t$ then $y' = \cos 2t$.

45. If $y = \cos t^2$ then $y' = -\sin 2t$.

46. If $z = (\sin 2t)\cos 3t$ then $z' = -6(\cos 2t)\sin 3t$.

47. If $y = \sin(\cos t)$ then $y' = \cos(\cos t) + \sin(-\sin t)$.

48. If $P = 1/\sin q$ then $dP/dq = 1/\cos q$.

49. If $Q = \cos(\pi - t)$ then $dQ/dt = \sin(\pi - t)$.

50. If $Q = \sin(\pi t + 1)$ then $dQ/dt = \pi\cos(\pi t + 1)$.

PROJECTS FOR CHAPTER THREE

1. Coroner's Rule of Thumb

Coroners estimate time of death using the rule of thumb that a body cools about 2°F during the first hour after death and about 1°F for each additional hour. Assuming an air temperature of 68°F and a living body temperature of 98.6°F, the temperature $T(t)$ in °F of a body at a time t hours since death is given by

$$T(t) = 68 + 30.6e^{-kt}.$$

(a) For what value of k will the body cool by 2°F in the first hour?

(b) Using the value of k found in part (a), after how many hours will the temperature of the body be decreasing at a rate of 1°F per hour?

(c) Using the value of k found in part (a), show that, 24 hours after death, the coroner's rule of thumb gives approximately the same temperature as the formula.

2. Air Pressure and Altitude

Air pressure at sea level is 30 inches of mercury. At an altitude of h feet above sea level, the air pressure, P, in inches of mercury, is given by

$$P = 30e^{-3.23 \times 10^{-5} h}$$

(a) Sketch a graph of P against h.

(b) Find the equation of the tangent line at $h = 0$.

(c) A rule of thumb used by travelers is that air pressure drops about 1 inch for every 1000-foot increase in height above sea level. Write a formula for the air pressure given by this rule of thumb.

(d) What is the relation between your answers to parts (b) and (c)? Explain why the rule of thumb works.

(e) Are the predictions made by the rule of thumb too large or too small? Why?

3. **Relative Growth Rates: Population, GDP, and GDP per capita**

(a) Let Y be the world's annual production (GDP). The world GDP per capita is given by Y/P where P is world population. Figure 3.26 shows relative growth rate of GDP and GDP per capita for 1952–2000.[13]

(i) Explain why the vertical distance between the two curves gives the relative rate of growth of the world population.

(ii) Estimate the relative rate of population growth in 1970 and 2000.

(b) In 2006 the relative rate of change of the GDP in developing countries was 4.5% per year, and the relative rate of change of the population was 1.2%. What was the relative rate of change of the per capita GDP in developing countries?

(c) In 2006 the relative rate of change of the world's total production (GDP) was 3.8% per year, and the relative rate of change of the world's per capita production was 2.6% per year. What was the relative rate of change of the world population?

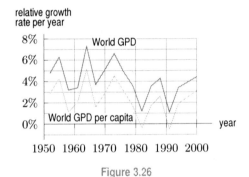

Figure 3.26

---

[13] Angus Maddison, *The World Economy: Historical Statistics*, OECD, 2003.

FOCUS ON THEORY

ESTABLISHING THE DERIVATIVE FORMULAS

The graph of $f(x) = x^2$ suggests that the derivative of x^2 is $f'(x) = 2x$. However, as we saw in the Focus on Theory section in Chapter 2, to be sure that this formula is correct, we have to use the definition:

$$f'(x) = \lim_{h \to 0} \frac{f(x+h) - f(x)}{h}.$$

As in Chapter 2, we simplify the difference quotient and then take the limit as h approaches zero.

Example 1   Confirm that the derivative of $g(x) = x^3$ is $g'(x) = 3x^2$.

Solution   Using the definition, we calculate $g'(x)$:

$$g'(x) = \lim_{h \to 0} \frac{g(x+h) - g(x)}{h} = \lim_{h \to 0} \frac{(x+h)^3 - x^3}{h}$$

$$\text{Multiplying out} \longrightarrow = \lim_{h \to 0} \frac{x^3 + 3x^2 h + 3xh^2 + h^3 - x^3}{h}$$

$$= \lim_{h \to 0} \frac{3x^2 h + 3xh^2 + h^3}{h}$$

$$\text{Simplifying} \longrightarrow = \lim_{h \to 0} (3x^2 + 3xh + h^2) = 3x^2.$$

$$\nwarrow \text{Looking at what happens as } h \to 0$$

So $g'(x) = \dfrac{d}{dx}(x^3) = 3x^2$.

Example 2   Give an informal justification that the derivative of $f(x) = e^x$ is $f'(x) = e^x$.

Solution   Using $f(x) = e^x$, we have

$$f'(x) = \lim_{h \to 0} \frac{f(x+h) - f(x)}{h} = \lim_{h \to 0} \frac{e^{x+h} - e^x}{h}$$

$$= \lim_{h \to 0} \frac{e^x e^h - e^x}{h} = \lim_{h \to 0} e^x \left(\frac{e^h - 1}{h} \right).$$

What is the limit of $\dfrac{e^h - 1}{h}$ as $h \to 0$? The graph of $\dfrac{e^h - 1}{h}$ in Figure 3.27 suggests that $\dfrac{e^h - 1}{h}$ approaches 1 as $h \to 0$. In fact, it can be proved that the limit equals 1, so

$$f'(x) = \lim_{h \to 0} e^x \left(\frac{e^h - 1}{h} \right) = e^x \cdot 1 = e^x.$$

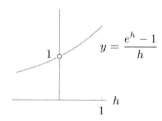

Figure 3.27: What is $\lim\limits_{h \to 0} \frac{e^h - 1}{h}$?

Example 3   Show that if $f(x) = 2x^2 + 1$, then $f'(x) = 4x$.

Solution   We use the definition of the derivative with $f(x) = 2x^2 + 1$:

$$f'(x) = \lim_{h \to 0} \frac{f(x + h) - f(x)}{h} = \lim_{h \to 0} \frac{(2(x + h)^2 + 1) - (2x^2 + 1)}{h}$$

$$= \lim_{h \to 0} \frac{2(x^2 + 2xh + h^2) + 1 - 2x^2 - 1}{h} = \lim_{h \to 0} \frac{2x^2 + 4xh + 2h^2 + 1 - 2x^2 - 1}{h}$$

$$= \lim_{h \to 0} \frac{4xh + 2h^2}{h} = \lim_{h \to 0} \frac{h(4x + 2h)}{h}$$

To find the limit, look at what happens when h is close to 0, but $h \neq 0$. Simplifying, we have

$$f'(x) = \lim_{h \to 0} \frac{h(4x + 2h)}{h} = \lim_{h \to 0} (4x + 2h) = 4x$$

because as h gets close to 0, we know that $4x + 2h$ gets close to $4x$.

Using the Chain Rule to Establish Derivative Formulas

We use the chain rule to justify the formulas for derivatives of $\ln x$ and of a^x.

Derivative of $\ln x$

We'll differentiate an identity that involves $\ln x$. In Section 1.6, we have $e^{\ln x} = x$. Differentiating gives

$$\frac{d}{dx}(e^{\ln x}) = \frac{d}{dx}(x) = 1.$$

On the left side, since e^x is the outside function and $\ln x$ is the inside function, the chain rule gives

$$\frac{d}{dx}(e^{\ln x}) = e^{\ln x} \cdot \frac{d}{dx}(\ln x).$$

Thus, as we said in Sections 3.2,

$$\frac{d}{dx}(\ln x) = \frac{1}{e^{\ln x}} = \frac{1}{x}.$$

Derivative of a^x

Graphical arguments suggest that the derivative of a^x is proportional to a^x. Now we show that the constant of proportionality is $\ln a$. For $a > 0$, we use the identity from Section 1.6:

$$\ln(a^x) = x \ln a.$$

On the left side, using $\frac{d}{dx}(\ln x) = \frac{1}{x}$ and the chain rule gives

$$\frac{d}{dx}(\ln a^x) = \frac{1}{a^x} \cdot \frac{d}{dx}(a^x).$$

Since $\ln a$ is a constant, differentiating the right side gives

$$\frac{d}{dx}(x \ln a) = \ln a.$$

Since the two sides are equal, we have

$$\frac{1}{a^x} \frac{d}{dx}(a^x) = \ln a.$$

Solving for $\frac{d}{dx}(a^x)$ gives the result of Section 3.2. For $a > 0$,

$$\frac{d}{dx}(a^x) = (\ln a)a^x.$$

The Product Rule

Suppose we want to calculate the derivative of the product of differentiable functions, $f(x)g(x)$, using the definition of the derivative. Notice that in the second step below, we are adding and subtracting the same quantity: $f(x)g(x + h)$.

$$\frac{d[f(x)g(x)]}{dx} = \lim_{h \to 0} \frac{f(x + h)g(x + h) - f(x)g(x)}{h}$$

$$= \lim_{h \to 0} \frac{f(x + h)g(x + h) - f(x)g(x + h) + f(x)g(x + h) - f(x)g(x)}{h}$$

$$= \lim_{h \to 0} \left[\frac{f(x + h) - f(x)}{h} \cdot g(x + h) + f(x) \cdot \frac{g(x + h) - g(x)}{h} \right]$$

Taking the limit as $h \to 0$ gives the product rule:

$$(f(x)g(x))' = f'(x) \cdot g(x) + f(x) \cdot g'(x).$$

The Quotient Rule

Let $Q(x) = f(x)/g(x)$ be the quotient of differentiable functions. Assuming that $Q(x)$ is differentiable, we can use the product rule on $f(x) = Q(x)g(x)$:

$$f'(x) = Q'(x)g(x) + Q(x)g'(x).$$

Substituting for $Q(x)$ gives

$$f'(x) = Q'(x)g(x) + \frac{f(x)}{g(x)}g'(x).$$

Solving for $Q'(x)$ gives

$$Q'(x) = \frac{f'(x) - \frac{f(x)}{g(x)}g'(x)}{g(x)}.$$

Multiplying the top and bottom by $g(x)$ to simplify gives the quotient rule:

$$\left(\frac{f(x)}{g(x)} \right)' = \frac{f'(x)g(x) - f(x)g'(x)}{(g(x))^2}.$$

Problems on Establishing the Derivative Formulas

For Problems 1–7, use the definition of the derivative to obtain the following results.

1. If $f(x) = 2x + 1$, then $f'(x) = 2$.
2. If $f(x) = 5x^2$, then $f'(x) = 10x$.
3. If $f(x) = 2x^2 + 3$, then $f'(x) = 4x$.
4. If $f(x) = x^2 + x$, then $f'(x) = 2x + 1$.
5. If $f(x) = 4x^2 + 1$, then $f'(x) = 8x$.
6. If $f(x) = x^4$, then $f'(x) = 4x^3$. [Hint: $(x + h)^4 = x^4 + 4x^3h + 6x^2h^2 + 4xh^3 + h^4$.]
7. If $f(x) = x^5$, then $f'(x) = 5x^4$. [Hint: $(x + h)^5 = x^5 + 5x^4h + 10x^3h^2 + 10x^2h^3 + 5xh^4 + h^5$.]

8. (a) Use a graph of $g(h) = \dfrac{2^h - 1}{h}$ to explain why we believe that $\lim_{h \to 0} \dfrac{2^h - 1}{h} \approx 0.6931$.

(b) Use the definition of the derivative and the result from part (a) to explain why, if $f(x) = 2^x$, we believe that $f'(x) \approx (0.6931)2^x$.

9. Use the definition of the derivative to show that if $f(x) = C$, where C is a constant, then $f'(x) = 0$.

10. Use the definition of the derivative to show that if $f(x) = b + mx$, for constants m and b, then $f'(x) = m$.

11. Use the definition of the derivative to show that if $f(x) = k \cdot u(x)$, where k is a constant and $u(x)$ is a function, then $f'(x) = k \cdot u'(x)$.

12. Use the definition of the derivative to show that if $f(x) = u(x) + v(x)$, for functions $u(x)$ and $v(x)$, then $f'(x) = u'(x) + v'(x)$.

FOCUS ON PRACTICE

Find derivatives for the functions in Problems 1–63. Assume a, b, c, and k are constants.

1. $f(t) = t^2 + t^4$

2. $g(x) = 5x^4$

3. $y = 5x^3 + 7x^2 - 3x + 1$

4. $s(t) = 6t^{-2} + 3t^3 - 4t^{1/2}$

5. $f(x) = \dfrac{1}{x^2} + 5\sqrt{x} - 7$

6. $P(t) = 100e^{0.05t}$

7. $f(x) = 5e^{2x} - 2 \cdot 3^x$

8. $P(t) = 1{,}000(1.07)^t$

9. $D(p) = e^{p^2} + 5p^2$

10. $y = t^2 e^{5t}$

11. $y = x^2 \sqrt{x^2 + 1}$

12. $f(x) = \ln\left(x^2 + 1\right)$

13. $s(t) = 8\ln(2t + 1)$

14. $g(w) = w^2 \ln(w)$

15. $f(x) = 2^x + x^2 + 1$

16. $P(t) = \sqrt{t^2 + 4}$

17. $C(q) = (2q + 1)^3$

18. $g(x) = 5x(x + 3)^2$

19. $P(t) = be^{kt}$

20. $f(x) = ax^2 + bx + c$

21. $y = x^2 \ln(2x + 1)$

22. $f(t) = \left(e^t + 4\right)^3$

23. $f(x) = 5\sin(2x)$

24. $W(r) = r^2 \cos r$

25. $g(t) = 3\sin(5t) + 4$

26. $y = e^{3t}\sin(2t)$

27. $y = 2e^x + 3\sin x + 5$

28. $f(t) = 3t^2 - 4t + 1$

29. $y = 17x + 24x^{1/2}$

30. $g(x) = -\frac{1}{2}(x^5 + 2x - 9)$

31. $f(x) = 5x^4 + \dfrac{1}{x^2}$

32. $y = \dfrac{e^{2x}}{x^2 + 1}$

33. $f(x) = \dfrac{x^2 + 3x + 2}{x + 1}$

34. $y = \left(\dfrac{x^2 + 2}{3}\right)^2$

35. $g(x) = \sin(2 - 3x)$

36. $f(z) = \dfrac{z^2 + 1}{3z}$

37. $q(r) = \dfrac{3r}{5r + 2}$

38. $y = x\ln x - x + 2$

39. $j(x) = \ln(e^{ax} + b)$

40. $g(t) = \dfrac{t - 4}{t + 4}$

41. $h(w) = (w^4 - 2w)^5$

42. $h(w) = w^3 \ln(10w)$

43. $f(x) = \ln(\sin x + \cos x)$

44. $w(r) = \sqrt{r^4 + 1}$

45. $h(w) = -2w^{-3} + 3\sqrt{w}$

46. $h(x) = \sqrt{\dfrac{x^2 + 9}{x + 3}}$

47. $v(t) = t^2 e^{-ct}$

48. $f(x) = \dfrac{x}{1 + \ln x}$

49. $g(\theta) = e^{\sin\theta}$

50. $p(t) = e^{4t+2}$

51. $j(x) = \dfrac{x^3}{a} + \dfrac{a}{b}x^2 - cx$

52. $f(z) = \dfrac{z^2 + 1}{\sqrt{z}}$

53. $h(r) = \dfrac{r^2}{2r + 1}$

54. $g(x) = 2x - \dfrac{1}{\sqrt[3]{x}} + 3^x - e$

55. $f(t) = 2te^t - \dfrac{1}{\sqrt{t}}$

56. $w = \dfrac{5 - 3z}{5 + 3z}$

57. $f(x) = \dfrac{x^3}{9}(3\ln x - 1)$

58. $g(x) = \dfrac{x^2 + \sqrt{x} + 1}{x^{3/2}}$

59. $y = \left(x^2 + 5\right)^3 \left(3x^3 - 2\right)^2$

60. $f(x) = \dfrac{a^2 - x^2}{a^2 + x^2}$

61. $w(r) = \dfrac{ar^2}{b + r^3}$

62. $H(t) = (at^2 + b)e^{-ct}$

63. $g(w) = \dfrac{5}{(a^2 - w^2)^2}$

Chapter Four

USING THE DERIVATIVE

Contents

4.1 LOCAL MAXIMA AND MINIMA

What Derivatives Tell Us About a Function and Its Graph

When we graph a function on a computer or calculator, we often see only part of the picture. Information given by the first and second derivatives can help identify regions with interesting behavior.

Example 1 Use a computer or calculator to sketch a useful graph of the function
$$f(x) = x^3 - 9x^2 - 48x + 52.$$

Solution Since f is a cubic polynomial, we expect a graph that is roughly S-shaped. Graphing this function with $-10 \le x \le 10$, $-10 \le y \le 10$ gives the two nearly vertical lines in Figure 4.1. We know that there is more going on than this, but how do we know where to look?

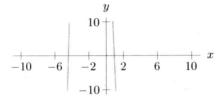

Figure 4.1: Unhelpful graph of $f(x) = x^3 - 9x^2 - 48x + 52$

We use the derivative to determine where the function is increasing and where it is decreasing. The derivative of f is
$$f'(x) = 3x^2 - 18x - 48.$$
To find where $f' > 0$ or $f' < 0$, we first find where $f' = 0$, that is, where $3x^2 - 18x - 48 = 0$. Factoring gives $3(x - 8)(x + 2) = 0$, so $x = -2$ or $x = 8$. Since $f' = 0$ *only* at $x = -2$ and $x = 8$, and since f' is continuous, f' cannot change sign on any of the three intervals $x < -2$, or $-2 < x < 8$, or $8 < x$. How can we tell the sign of f' on each of these intervals? The easiest way is to pick a point and substitute into f'. For example, since $f'(-3) = 33 > 0$, we know f' is positive for $x < -2$, so f is increasing for $x < -2$. Similarly, since $f'(0) = -48$ and $f'(10) = 72$, we know that f decreases between $x = -2$ and $x = 8$ and increases for $x > 8$. Summarizing:

	$x = -2$		$x = 8$	
f increasing $\nearrow$		f decreasing $\searrow$		f increasing $\nearrow$
$f' > 0$	$f' = 0$	$f' < 0$	$f' = 0$	$f' > 0$

We find that $f(-2) = 104$ and $f(8) = -396$. Hence, on the interval $-2 < x < 8$ the function decreases from a high of 104 to a low of -396. (Now we see why not much showed up in our first calculator graph.) One more point on the graph is easy to get: the y-intercept, $f(0) = 52$. With just these three points we can get a much more helpful graph. By setting the plotting window to $-10 \le x \le 20$ and $-400 \le y \le 400$, we get Figure 4.2, which gives much more insight into the behavior of $f(x)$ than the graph in Figure 4.1.

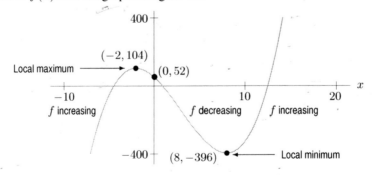

Figure 4.2: Useful graph of $f(x) = x^3 - 9x^2 - 48x + 52$. Notice that the scales on the x-and y-axes are different

Local Maxima and Minima

We are often interested in points such as those marked local maximum and local minimum in Figure 4.2. We have the following definition:

> Suppose p is a point in the domain of f:
> - f has a **local minimum** at p if $f(p)$ is less than or equal to the values of f for points near p.
> - f has a **local maximum** at p if $f(p)$ is greater than or equal to the values of f for points near p.

We use the adjective "local" because we are describing only what happens near p.

How Do We Detect a Local Maximum or Minimum?

In the preceding example, the points $x = -2$ and $x = 8$, where $f'(x) = 0$, played a key role in leading us to local maxima and minima. We give a name to such points:

> For any function f, a point p in the domain of f where $f'(p) = 0$ or $f'(p)$ is undefined is called a **critical point** of the function. In addition, the point $(p, f(p))$ on the graph of f is also called a critical point. A **critical value** of f is the value, $f(p)$, of the function at a critical point, p.

Notice that "critical point of f" can refer either to points in the domain of f or to points on the graph of f. You will know which meaning is intended from the context.

Geometrically, at a critical point where $f'(p) = 0$, the line tangent to the graph of f at p is horizontal. At a critical point where $f'(p)$ is undefined, there is no horizontal tangent to the graph—there is either a vertical tangent or no tangent at all. (For example, $x = 0$ is a critical point for the absolute value function $f(x) = |x|$.) However, most of the functions we will work with will be differentiable everywhere, and therefore most of our critical points will be of the $f'(p) = 0$ variety.

The critical points divide the domain of f into intervals on which the sign of the derivative remains the same, either positive or negative. Therefore, if f is defined on the interval between two successive critical points, its graph cannot change direction on that interval; it is either going up or it is going down. We have the following result:

> If a function, continuous on an interval (its domain), has a local maximum or minimum at p, then p is a critical point or an endpoint of the interval.

A function may have any number of critical points or none at all. (See Figures 4.3–4.5.)

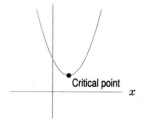

Figure 4.3: A quadratic: One critical point

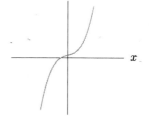

Figure 4.4: $f(x) = x^3 + x + 1$: No critical points

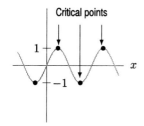

Figure 4.5: Many critical points

Testing For Local Maxima and Minima

If f' has different signs on either side of a critical point p with $f'(p) = 0$, then the graph changes direction at p and looks like one of those in Figure 4.6. We have the following criteria:

First Derivative Test for Local Maxima and Minima

Suppose p is a critical point of a continuous function f. Then, as we go from left to right:

- If f changes from decreasing to increasing at p, then f has a local minimum at p.
- If f changes from increasing to decreasing at p, then f has a local maximum at p.

Alternatively, the concavity of the graph of f gives another way of distinguishing between local maxima and minima:

Second Derivative Test for Local Maxima and Minima

Suppose p is a critical point of a continuous function f, and $f'(p) = 0$.

- If f is concave up at p, then f has a local minimum at p.
- If f is concave down at p, then f has a local maximum at p.

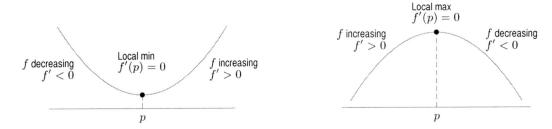

Figure 4.6: Changes in direction at a critical point, p: Local maxima and minima

---

Example 2   Use the second derivative test to confirm that $f(x) = x^3 - 9x^2 - 48x + 52$ has a local maximum at $x = -2$ and a local minimum at $x = 8$.

Solution   In Example 1, we calculated $f'(x) = 3x^2 - 18x - 48 = 3(x - 8)(x + 2)$, so $f'(8) = f'(-2) = 0$. Differentiating again gives $f''(x) = 6x - 18$. Since $f''(8) = 6 \cdot 8 - 18 = 30$ and $f''(-2) = 6(-2) - 18 = -30$, the second derivative test confirms that $x = 8$ is a local minimum and $x = -2$ is a local maximum.

Example 3   (a) Graph a function f with the following properties:
- $f(x)$ has critical points at $x = 2$ and $x = 5$;
- $f'(x)$ is positive to the left of 2 and positive to the right of 5;
- $f'(x)$ is negative between 2 and 5.

(b) Identify the critical points as local maxima, local minima, or neither.

Solution   (a) We know that $f(x)$ is increasing when $f'(x)$ is positive, and $f(x)$ is decreasing when $f'(x)$ is negative. The function is increasing to the left of 2 and increasing to the right of 5, and it is decreasing between 2 and 5. A possible sketch is given in Figure 4.7.

(b) We see that the function has a local maximum at $x = 2$ and a local minimum at $x = 5$.

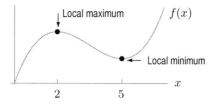

Figure 4.7: A function with critical points at $x = 2$ and $x = 5$

Warning!

Not every critical point of a function is a local maximum or minimum. For instance, consider $f(x) = x^3$, graphed in Figure 4.8. The derivative is $f'(x) = 3x^2$ so $x = 0$ is a critical point. But $f'(x) = 3x^2$ is positive on both sides of $x = 0$, so f increases on both sides of $x = 0$. There is neither a local maximum nor a local minimum for $f(x)$ at $x = 0$.

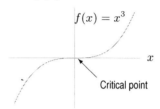

Figure 4.8: A critical point which is neither a local maximum nor minimum.

Example 4   The value of an investment at time t is given by $S(t)$. The rate of change, $S'(t)$, of the value of the investment is shown in Figure 4.9.

(a) What are the critical points of the function $S(t)$?

(b) Identify each critical point as a local maximum, a local minimum, or neither.

(c) Explain the financial significance of each of the critical points.

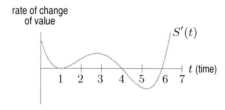

Figure 4.9: Graph of $S'(t)$, the rate of change of the value of the investment

Solution   (a) The critical points of S occur at times t when $S'(t) = 0$. We see in Figure 4.9 that $S'(t) = 0$ at $t = 1, 4,$ and 6, so the critical points occur at $t = 1, 4,$ and 6.

(b) In Figure 4.9, we see that $S'(t)$ is positive to the left of 1 and between 1 and 4, that $S'(t)$ is negative between 4 and 6, and that $S'(t)$ is positive to the right of 6. Therefore $S(t)$ is increasing to the left of 1 and between 1 and 4 (with a slope of zero at 1), decreasing between 4 and 6, and increasing again to the right of 6. A possible sketch of $S(t)$ is given in Figure 4.10. We see that S has neither a local maximum nor a local minimum at the critical point $t = 1$, but that it has a local maximum at $t = 4$ and a local minimum at $t = 6$.

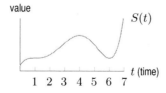

Figure 4.10: Possible graph of the function representing the value of the investment at time t

173

(c) At time $t = 1$ the investment momentarily stopped increasing in value, though it started increasing again immediately afterward. At $t = 4$, the value peaked and began to decline. At $t = 6$, it started increasing again.

Example 5   Find the critical point of the function $f(x) = x^2 + bx + c$. What is its graphical significance?

Solution   Since $f'(x) = 2x + b$, the critical point x satisfies the equation $2x + b = 0$. Thus, the critical point is at $x = -b/2$. The graph of f is a parabola and the critical point is its vertex. See Figure 4.11.

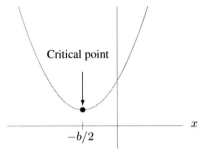

Figure 4.11: Critical point of the parabola $f(x) = x^2 + bx + c$. (Sketched with $b, c > 0$)

Problems for Section 4.1

In Problems 1–4, indicate all critical points of the function f. How many critical points are there? Identify each critical point as a local maximum, a local minimum, or neither.

1.

2.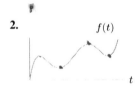

3.

4.

5. (a) Graph a function with two local minima and one local maximum.
   (b) Graph a function with two critical points. One of these critical points should be a local minimum, and the other should be neither a local maximum nor a local minimum.

6. During an illness a person ran a fever. His temperature rose steadily for eighteen hours, then went steadily down for twenty hours. When was there a critical point for his temperature as a function of time?

7. Graph two continuous functions f and g, each of which has exactly five critical points, the points A–E in Figure 4.12, and which satisfy the following conditions:
   (a) $f(x) \to \infty$ as $x \to -\infty$ and
       $f(x) \to \infty$ as $x \to \infty$
   (b) $g(x) \to -\infty$ as $x \to -\infty$ and
       $g(x) \to 0$ as $x \to \infty$

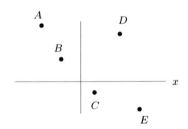

Figure 4.12

Problems 8–9 show the graph of a derivative function f'. Indicate on a sketch the x-values that are critical points of the function f itself. Identify each critical point as a local maximum, a local minimum, or neither.

8.   9.

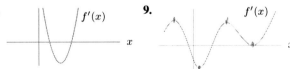

Using a calculator or computer, graph the functions in Problems 10–15. Describe in words the interesting features of the graph, including the location of the critical points and where the function is monotonic (that is, increasing or decreasing). Then use the derivative and algebra to explain the shape of the graph.

10. $f(x) = x^3 - 6x + 1$ **11.** $f(x) = x^3 + 6x + 1$

12. $f(x) = 3x^5 - 5x^3$ **13.** $f(x) = e^x - 10x$

14. $f(x) = x \ln x, \quad x > 0$ **15.** $f(x) = x + 2 \sin x$

16. Figure 4.13 is a graph of f'. For what values of x does f have a local maximum? A local minimum?

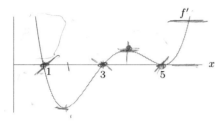

Figure 4.13: Graph of f' (not f)

17. On the graph of f' in Figure 4.14, indicate the x-values that are critical points of the function f itself. Are they local maxima, local minima, or neither?

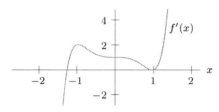

Figure 4.14: Graph of f' (not f)

18. The derivative of $f(t)$ is given by $f'(t) = t^3 - 6t^2 + 8t$ for $0 \le t \le 5$. Graph $f'(t)$, and describe how the function $f(t)$ changes over the interval $t = 0$ to $t = 5$. When is $f(t)$ increasing and when is it decreasing? Where does $f(t)$ have a local maximum and where does it have a local minimum?

19. If U and V are positive constants, find all critical points of
$$F(t) = Ue^t + Ve^{-t}.$$

20. Consumer demand for a certain product is changing over time, and the rate of change of this demand, $f'(t)$, in units/week, is given, in week t, in the following table.

t	0	1	2	3	4	5	6	7	8	9	10
$f'(t)$	12	10	4	-2	-3	-1	3	7	11	15	10

(a) When is the demand for this product increasing? When is it decreasing?
(b) Approximately when is demand at a local maximum? A local minimum?

21. Suppose f has a continuous derivative whose values are given in the following table.

(a) Estimate the x-coordinates of critical points of f for $0 \le x \le 10$.
(b) For each critical point, indicate if it is a local maximum of f, local minimum, or neither.

x	0	1	2	3	4	5	6	7	8	9	10
$f'(x)$	5	2	1	-2	-5	-3	-1	2	3	1	-1

22. The function $f(x) = x^4 - 4x^3 + 8x$ has a critical point at $x = 1$. Use the second derivative test to identify it as a local maximum or local minimum.

23. Find and classify the critical points of $f(x) = x^3(1-x)^4$ as local maxima and minima.

In Problems 24–26, investigate the one-parameter family of functions. Assume that a is positive.

(a) Graph $f(x)$ using three different values for a.
(b) Using your graph in part (a), describe the critical points of f and how they appear to move as a increases.
(c) Find a formula for the x-coordinates of the critical point(s) of f in terms of a.

24. $f(x) = (x - a)^2$
25. $f(x) = x^3 - ax$
26. $f(x) = x^2 e^{-ax}$

In Problems 27–28, find constants a and b so that the minimum for the parabola $f(x) = x^2 + ax + b$ is at the given point. [Hint: Begin by finding the critical point in terms of a.]

27. $(3, 5)$ **28.** $(-2, -3)$

29. Sketch several members of the family $y = x^3 - ax^2$ on the same axes. Discuss the effect of the parameter a on the graph. Find all critical points for this function.

30. For what values of a and b does $f(x) = a(x - b \ln x)$ have a local minimum at the point $(2, 5)$? Figure 4.15 shows a graph of $f(x)$ with $a = 1$ and $b = 1$.

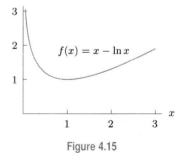

Figure 4.15

31. Find the value of a so that the function $f(x) = xe^{ax}$ has a critical point at $x = 3$.

32. (a) If b is a positive constant and $x > 0$, find all critical points of $f(x) = x - b \ln x$.
   (b) Use the second derivative test to determine whether the function has a local maximum or local minimum at each critical point.

33. (a) For a a positive constant, find all critical points of $f(x) = x - a\sqrt{x}$.
   (b) What value of a gives a critical point at $x = 5$? Does $f(x)$ have a local maximum or a local minimum at this critical point?

34. Let $g(x) = x - ke^x$, where k is any constant. For what value(s) of k does the function g have a critical point?

35. If a and b are nonzero constants, find the domain and all critical points of

$$f(x) = \frac{ax^2}{x - b}.$$

36. Assume f has a derivative everywhere and has just one critical point, at $x = 3$. In parts (a)–(d), you are given additional conditions. In each case decide whether $x = 3$ is a local maximum, a local minimum, or neither. Explain your reasoning. Sketch possible graphs for all four cases.
   (a) $f'(1) = 3$ and $f'(5) = -1$
   (b) $f(x) \to \infty$ as $x \to \infty$ and as $x \to -\infty$
   (c) $f(1) = 1$, $f(2) = 2$, $f(4) = 4$, $f(5) = 5$
   (d) $f'(2) = -1$, $f(3) = 1$, $f(x) \to 3$ as $x \to \infty$

37. (a) On a computer or calculator, graph $f(\theta) = \theta - \sin \theta$. Can you tell whether the function has any zeros in the interval $0 \leq \theta \leq 1$?
   (b) Find f'. What does the sign of f' tell you about the zeros of f in the interval $0 \leq \theta \leq 1$?

4.2 INFLECTION POINTS

Concavity and Inflection Points

A study of the points on the graph of a function where the slope changes sign led us to critical points. Now we will study the points on the graph where the concavity changes, either from concave up to concave down, or from concave down to concave up.

> A point at which the graph of a function f changes concavity is called an **inflection point** of f.

The words "inflection point of f" can refer either to a point in the domain of f or to a point on the graph of f. The context of the problem will tell you which is meant.

How Do You Locate an Inflection Point?

Since the concavity of the graph of f changes at an inflection point, the sign of f'' changes there: it is positive on one side of the inflection point and negative on the other. Thus, at the inflection point, f'' is zero or undefined. (See Figure 4.16.)

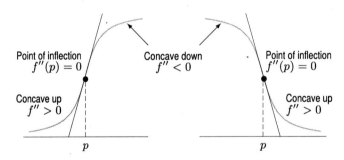

Figure 4.16: Change in concavity (from positive to negative or vice versa) at point p

---

Example 1   Find the inflection points of $f(x) = x^3 - 9x^2 - 48x + 52$.

Solution   In Figure 4.17, part of the graph of f is concave up and part is concave down, so the function must have an inflection point. However, it is difficult to locate the inflection point accurately by examining the graph. To find the inflection point exactly, calculate where the second derivative is zero.[1] Since
$$f'(x) = 3x^2 - 18x - 48,$$

---
[1] For a polynomial, the second derivative cannot be undefined.

$$f''(x) = 6x - 18 \qquad \text{so} \qquad f''(x) = 0 \quad \text{when} \quad x = 3.$$

The graph of $f(x)$ changes concavity at $x = 3$, so $x = 3$ is an inflection point.

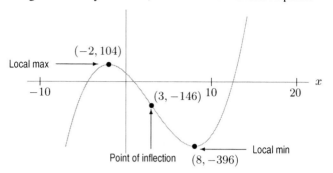

Figure 4.17: Graph of $f(x) = x^3 - 9x^2 - 48x + 52$ showing the inflection point at $x = 3$

Example 2   Graph a function f with the following properties: f has a critical point at $x = 4$ and an inflection point at $x = 8$; the value of f' is negative to the left of 4 and positive to the right of 4; the value of f'' is positive to the left of 8 and negative to the right of 8.

Solution   Since f' is negative to the left of 4 and positive to the right of 4, the value of $f(x)$ is decreasing to the left of 4 and increasing to the right of 4. The values of f'' tell us that the graph of $f(x)$ is concave up to the left of 8 and concave down to the right of 8. A possible sketch is given in Figure 4.18.

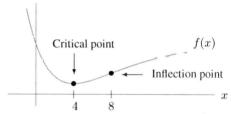

Figure 4.18: A function with a critical point at $x = 4$ and an inflection point at $x = 8$

Example 3   Figure 4.19 shows a population growing toward a limiting population, L. There is an inflection point on the graph at the point where the population reaches $L/2$. What is the significance of the inflection point to the population?

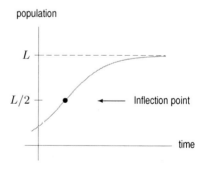

Figure 4.19: Inflection point on graph of a population growing toward a limiting population, L

Solution   At times before the inflection point, the population is increasing faster every year. At times after the inflection point the population is increasing slower every year. At the inflection point, the population is growing fastest.

Example 4   (a) How many critical points and how many inflection points does the function $f(x) = xe^{-x}$ have?
(b) Use derivatives to find the critical points and inflection points exactly.

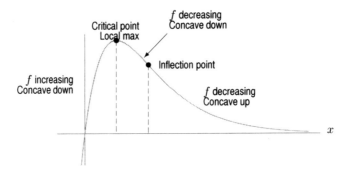

Figure 4.20: Graph of $f(x) = xe^{-x}$

Solution   (a) Figure 4.20 shows the graph of $f(x) = xe^{-x}$. It appears to have one critical point, which is a local maximum. Are there any inflection points? Since the graph of the function is concave down at the critical point and concave up for large x, the graph of the function changes concavity, so there must be an inflection point to the right of the critical point.

(b) To find the critical point, find the point where the first derivative of f is zero or undefined. The product rule gives

$$f'(x) = x(-e^{-x}) + (1)(e^{-x}) = (1 - x)e^{-x}.$$

We have $f'(x) = 0$ when $x = 1$, so the critical point is at $x = 1$. To find the inflection point, we find where the second derivative of f changes sign. Using the product rule on the first derivative, we have

$$f''(x) = (1 - x)(-e^{-x}) + (-1)(e^{-x}) = (x - 2)e^{-x}.$$

We have $f''(x) = 0$ when $x = 2$. Since $f''(x) > 0$ for $x > 2$ and $f''(x) < 0$ for $x < 2$, the concavity changes sign at $x = 2$. So the inflection point is at $x = 2$.

Warning!

Not every point x where $f''(x) = 0$ (or f'' is undefined) is an inflection point (just as not every point where $f' = 0$ is a local maximum or minimum). For instance, $f(x) = x^4$ has $f''(x) = 12x^2$ so $f''(0) = 0$, but $f'' > 0$ when $x > 0$ and when $x < 0$, so the graph of f is concave up on both sides of $x = 0$. There is *no* change in concavity at $x = 0$. (See Figure 4.21.)

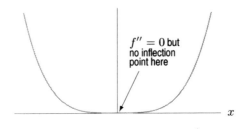

Figure 4.21: Graph of $f(x) = x^4$

Example 5   Suppose that water is being poured into the vase in Figure 4.22 at a constant rate measured in liters per minute. Graph $y = f(t)$, the depth of the water against time, t. Explain the concavity, and indicate the inflection points.

Solution   Notice that the volume of water in the vase increases at a constant rate.

At first the water level, y, rises quite slowly because the base of the vase is wide, and so it takes a lot of water to make the depth increase. However, as the vase narrows, the rate at which the water level rises increases. This means that initially y is increasing at an increasing rate, and the graph is concave up. The water level is rising fastest, so the rate of change of the depth y is at a maximum, when the water reaches the middle of the vase, where the diameter is smallest; this is an inflection point. (See Figure 4.23.) After that, the rate at which the water level changes starts to decrease, and so the graph is concave down.

Figure 4.22: A vase

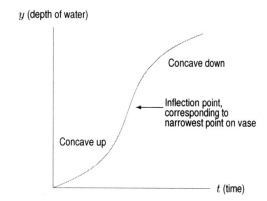

Figure 4.23: Graph of depth of water in the vase, y, against time, t

---

Example 6   What is the concavity of the graph of $f(x) = ax^2 + bx + c$?

Solution   We have $f'(x) = 2ax + b$ and $f''(x) = 2a$. The second derivative of f has the same sign as a. If $a > 0$, the graph is concave up everywhere, an upward-opening parabola. If $a < 0$, the graph is concave down everywhere, a downward-opening parabola. (See Figure 4.24.)

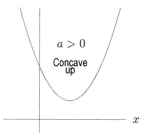

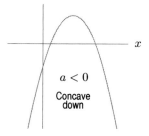

Figure 4.24: Concavity of $f(x) = ax^2 + bx + c$

Problems for Section 4.2

In Problems 1–4, indicate the approximate locations of all inflection points. How many inflection points are there?

1.

2.

3.

4.

5. **(a)** Graph a polynomial with two local maxima and two local minima.
   (b) What is the least number of inflection points this function must have? Label the inflection points.

6. Graph a function with only one critical point (at $x = 5$) and one inflection point (at $x = 10$). Label the critical point and the inflection point on your graph.

7. Graph a function which has a critical point and an inflection point at the same place.

8. During a flood, the water level in a river first rose faster and faster, then rose more and more slowly until it reached its highest point, then went back down to its pre-flood level. Consider water depth as a function of time.

   (a) Is the time of highest water level a critical point or an inflection point of this function?
   (b) Is the time when the water first began to rise more slowly a critical point or an inflection point?

9. When I got up in the morning I put on only a light jacket because, although the temperature was dropping, it seemed that the temperature would not go much lower. But I was wrong. Around noon a northerly wind blew up and the temperature began to drop faster and faster. The worst was around 6 pm when, fortunately, the temperature started going back up.

   (a) When was there a critical point in the graph of temperature as a function of time?
   (b) When was there an inflection point in the graph of temperature as a function of time?

10. For $f(x) = x^3 - 18x^2 - 10x + 6$, find the inflection point algebraically. Graph the function with a calculator or computer and confirm your answer.

In each of Problems 11–20, use the first derivative to find all critical points and use the second derivative to find all inflection points. Use a graph to identify each critical point as a local maximum, a local minimum, or neither.

11. $f(x) = x^2 - 5x + 3$

12. $f(x) = x^3 - 3x + 10$

13. $f(x) = 2x^3 + 3x^2 - 36x + 5$

14. $f(x) = \dfrac{x^3}{6} + \dfrac{x^2}{4} - x + 2$

15. $f(x) = x^4 - 2x^2$

16. $f(x) = 3x^4 - 4x^3 + 6$

17. $f(x) = x^4 - 8x^2 + 5$

18. $f(x) = x^4 - 4x^3 + 10$

19. $f(x) = x^5 - 5x^4 + 35$

20. $f(x) = 3x^5 - 5x^3$

21. Find the inflection points of $f(x) = x^4 + x^3 - 3x^2 + 2$.

22. **(a)** Find all critical points and all inflection points of the function $f(x) = x^4 - 2ax^2 + b$. Assume a and b are positive constants.
    (b) Find values of the parameters a and b if f has a critical point at the point $(2, 5)$.
    (c) If there is a critical point at $(2, 5)$, where are the inflection points?

For Problems 23–26, sketch a possible graph of $y = f(x)$, using the given information about the derivatives $y' = f'(x)$ and $y'' = f''(x)$. Assume that the function is defined and continuous for all real x.

23.

24.

25.

26.

27. In 1774, Captain James Cook left 10 rabbits on a small Pacific island. The rabbit population is approximated by

$$P(t) = \frac{2000}{1 + e^{5.3 - 0.4t}}$$

with t measured in years since 1774. Using a calculator or computer:

   (a) Graph P. Does the population level off?

(b) Estimate when the rabbit population grew most rapidly. How large was the population at that time?

(c) Find the inflection point on the graph and explain its significance for the rabbit population.

(d) What natural causes could lead to the shape of the graph of P?

28. (a) Water is flowing at a constant rate (i.e., constant volume per unit time) into a cylindrical container standing vertically. Sketch a graph showing the depth of water against time.

(b) Water is flowing at a constant rate into a cone-shaped container standing on its point. Sketch a graph showing the depth of the water against time.

29. If water is flowing at a constant rate (i.e., constant volume per unit time) into the Grecian urn in Figure 4.25, sketch a graph of the depth of the water against time. Mark on the graph the time at which the water reaches the widest point of the urn.

Figure 4.25

30. If water is flowing at a constant rate (i.e., constant volume per unit time) into the vase in Figure 4.26, sketch a graph of the depth of the water against time. Mark on the graph the time at which the water reaches the corner of the vase.

Figure 4.26

31. Water flows at a constant rate into the left side of the W-shaped container in Figure 4.27. Sketch a graph of the height, H, of the water in the left side of the container as a function of time, t. The container starts empty.

Figure 4.27

32. The vase in Figure 4.28 is filled with water at a constant rate (i.e., constant volume per unit time).

(a) Graph $y = f(t)$, the depth of the water, against time, t. Show on your graph the points at which the concavity changes.

(b) At what depth is $y = f(t)$ growing most quickly? Most slowly? Estimate the ratio between the growth rates at these two depths.

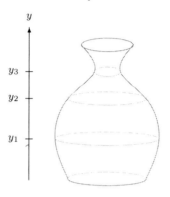

Figure 4.28

Find formulas for the functions described in Problems 33–34.

33. A cubic polynomial, $ax^3 + bx^2 + cx + d$, with a critical point at $x = 2$, an inflection point at $(1, 4)$, and a leading coefficient of 1.

34. A function of the form $y = bxe^{-ax}$ with a local maximum at $(3, 6)$.

35. Indicate on Figure 4.29 approximately where the inflection points of $f(x)$ are if the graph shows

(a) The function $f(x)$ **(b)** The derivative $f'(x)$

(c) The second derivative $f''(x)$

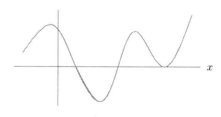

Figure 4.29

36. Assume that the polynomial f has exactly two local maxima and one local minimum, and that these are the only critical points of f.

(a) Sketch a possible graph of f.

(b) What is the largest number of zeros f could have?

(c) What is the least number of zeros f could have?

(d) What is the least number of inflection points f could have?

(e) What is the smallest degree f could have?

(f) Find a possible formula for $f(x)$.

4.3   GLOBAL MAXIMA AND MINIMA

Global Maxima and Minima

The techniques for finding maximum and minimum values make up the field called *optimization*. Local maxima and minima occur where a function takes larger or smaller values than at nearby points. However, we are often interested in where a function is larger or smaller than at all other points. For example, a firm trying to maximize its profit may do so by minimizing its costs. We make the following definition:

> For any function f:
> - f has a **global minimum** at p if $f(p)$ is less than or equal to all values of f.
> - f has a **global maximum** at p if $f(p)$ is greater than or equal to all values of f.

How Do We Find Global Maxima and Minima?

If f is a continuous function defined on an interval $a \leq x \leq b$ (including its endpoints), Figure 4.30 illustrates that the global maximum or minimum of f occurs either at a local maximum or a local minimum, respectively, or at one of the endpoints, $x = a$ or $x = b$, of the interval.

> **To find the global maximum and minimum of a continuous function on an interval including endpoints:** Compare values of the function at all the critical points in the interval and at the endpoints.

What if the continuous function is defined on an interval $a < x < b$ (excluding its endpoints), or on the entire real line which has no endpoints? The function graphed in Figure 4.31 has no global maximum because the function has no largest value. The global minimum of this function coincides with one of the local minima and is marked. A function defined on the entire real line or on an interval excluding endpoints may or may not have a global maximum or a global minimum.

> **To find the global maximum and minimum of a continuous function on an interval excluding endpoints or on the entire real line:** Find the values of the function at all the critical points and sketch a graph.

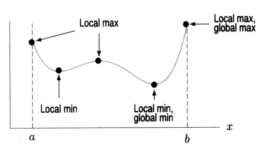

Figure 4.30: Global maximum and minimum on an interval domain, $a \leq x \leq b$

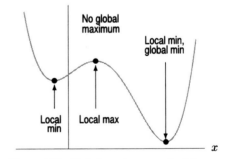

Figure 4.31: Global maximum and minimum on the entire real line

Example 1   Find the global maximum and minimum of $f(x) = x^3 - 9x^2 - 48x + 52$ on the interval $-5 \leq x \leq 14$.

Solution   We have calculated the critical points of this function previously using

$$f'(x) = 3x^2 - 18x - 48 = 3(x + 2)(x - 8),$$

so $x = -2$ and $x = 8$ are critical points. Since the global maxima and minima occur at a critical point or at an endpoint of the interval, we evaluate f at these four points:

$$f(-5) = -58, \qquad f(-2) = 104, \qquad f(8) = -396, \qquad f(14) = 360.$$

Comparing these four values, we see that the global maximum is 360 and occurs at $x = 14$, and that the global minimum is -396 and occurs at $x = 8$. See Figure 4.32.

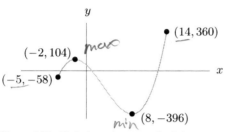

Figure 4.32: Global maximum and minimum on the interval $-5 \leq x \leq 14$

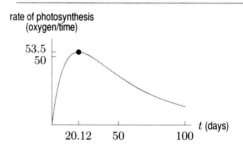

Figure 4.33: Maximum rate of photosynthesis

Example 2   For time, $t \geq 0$, in days, the rate at which photosynthesis takes place in the leaf of a plant, represented by the rate at which oxygen is produced, is approximated by[2]

$$p(t) = 100(e^{-0.02t} - e^{-0.1t}).$$

When is photosynthesis occurring fastest? What is that rate?

Solution   To find the global maximum value of $p(t)$, we first find critical points. We differentiate, set equal to zero, and solve for t:

$$p'(t) = 100(-0.02e^{-0.02t} + 0.1e^{-0.1t}) = 0$$
$$-0.02e^{-0.02t} = -0.1e^{-0.1t}$$
$$\frac{e^{-0.02t}}{e^{-0.1t}} = \frac{0.1}{0.02}$$
$$e^{-0.02t+0.1t} = 5$$
$$e^{0.08t} = 5$$
$$0.08t = \ln 5$$
$$t = \frac{\ln 5}{0.08} = 20.12 \text{ days.}$$

Differentiating again gives

$$p''(t) = 100(0.0004e^{-0.02t} - 0.01e^{-0.1t})$$

and substituting $t = 20.12$ gives $p''(20.12) = -0.107$, so $t = 20.12$ is a local maximum. However, there is only one critical point, so this local maximum is the global maximum. See Figure 4.33.

When $t = 20.12$ days, the rate, in units of oxygen per unit time, is

$$p(20.12) = 100\left(e^{-0.02(20.12)} - e^{-0.1(20.12)}\right) = 53.50.$$

[2]Examples adapted from Rodney Gentry, *Introduction to Calculus for the Biological and Health Sciences* (Reading: Addison-Wesley, 1978).

A Graphical Example: Minimizing Gas Consumption

Next we look at an example in which a function is given graphically and the optimum values are read from a graph. You already know how to estimate the optimum values of $f(x)$ from a graph of $f(x)$—read off the highest and lowest values. In this example, we see how to estimate the optimum value of the quantity $f(x)/x$ from a graph of $f(x)$ against x.

The question we investigate is how to set driving speeds to maximize fuel efficiency.[3] We assume that gas consumption, g (in gallons/hour), as a function of velocity, v (in mph) is as shown in Figure 4.34. We want to minimize the gas consumption per *mile*, not the gas consumption per hour. Let $G = g/v$ represent the average gas consumption per mile. (The units of G are gallons/mile.)

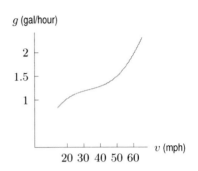

Figure 4.34: Gas consumption versus velocity

Example 3   Using Figure 4.34, estimate the velocity which minimizes $G = g/v$.

Solution   We want to find the minimum value of $G = g/v$ when g and v are related by the graph in Figure 4.34. We could use Figure 4.34 to sketch a graph of G against v and estimate a critical point. But there is an easier way. Figure 4.35 shows that g/v is the slope of the line from the origin to the point P. Where on the curve should P be to make the slope a minimum? From the possible positions of the line shown in Figure 4.35, we see that the slope of the line is both a local and global minimum when the line is tangent to the curve. From Figure 4.36, we can see that the velocity at this point is about 50 mph. Thus to minimize gas consumption per mile, we should drive about 50 mph.

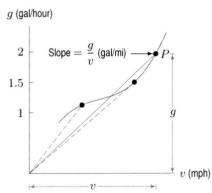

Figure 4.35: Graphical representation of gas consumption per mile, $G = g/v$

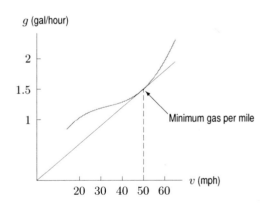

Figure 4.36: Velocity for maximum fuel efficiency

---

[3] Adapted from Peter D. Taylor, *Calculus: The Analysis of Functions* (Toronto: Wall & Emerson, 1992).

Problems for Section 4.3

For Problems 1–2, indicate all critical points on the given graphs. Which correspond to local minima, local maxima, global maxima, global minima, or none of these? (Note that the graphs are on closed intervals.)

1.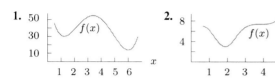

2.

3. For each interval, use Figure 4.37 to choose the statement that gives the location of the global maximum and global minimum of f on the interval.

(a) $4 \leq x \leq 12$ (b) $11 \leq x \leq 16$

(c) $4 \leq x \leq 9$ (d) $8 \leq x \leq 18$

(I) Maximum at right endpoint, minimum at left endpoint.

(II) Maximum at right endpoint, minimum at critical point.

(III) Maximum at left endpoint, minimum at right endpoint.

(IV) Maximum at left endpoint, minimum at critical point.

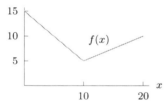

Figure 4.37

In Problems 4–7, graph a function with the given properties.

4. Has local minimum and global minimum at $x = 3$ but no local or global maximum.

5. Has local minimum at $x = 3$, local maximum at $x = 8$, but no global maximum or minimum.

6. Has local and global minimum at $x = 3$, local and global maximum at $x = 8$.

7. Has no local or global maxima or minima.

8. True or false? Give an explanation for your answer. The global maximum of $f(x) = x^2$ on every closed interval is at one of the endpoints of the interval.

In Problems 9–12, sketch the graph of a function on the interval $0 \leq x \leq 10$ with the given properties.

9. Has local minimum at $x = 3$, local maximum at $x = 8$, but global maximum and global minimum at the endpoints of the interval.

10. Has local and global maximum at $x = 3$, local and global minimum at $x = 10$.

11. Has local and global minimum at $x = 3$, local and global maximum at $x = 8$.

12. Has global maximum at $x = 0$, global minimum at $x = 10$, and no other local maxima or minima.

13. A grapefruit is tossed straight up with an initial velocity of 50 ft/sec. The grapefruit is 5 feet above the ground when it is released. Its height at time t is given by

$$y = -16t^2 + 50t + 5.$$

How high does it go before returning to the ground?

14. Find the value of x that maximizes $y = 12 + 18x - 5x^2$ and the corresponding value of y, by

(a) Estimating the values from a graph of y.

(b) Finding the values using calculus.

15. Plot the graph of $f(x) = x^3 - e^x$ using a graphing calculator or computer to find all local and global maxima and minima for: **(a)** $-1 \leq x \leq 4$ **(b)** $-3 \leq x \leq 2$

16. Figure 4.38 shows the rate at which photosynthesis is taking place in a leaf.

(a) At what time, approximately, is photosynthesis proceeding fastest for $t \geq 0$?

(b) If the leaf grows at a rate proportional to the rate of photosynthesis, for what part of the interval $0 \leq t \leq 200$ is the leaf growing? When is it growing fastest?

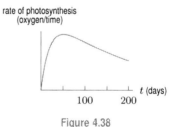

Figure 4.38

17. For some positive constant C, a patient's temperature change, T, due to a dose, D, of a drug is given by

$$T - \left(\frac{C}{2} - \frac{D}{3} \right) D^2.$$

(a) What dosage maximizes the temperature change?

(b) The sensitivity of the body to the drug is defined as dT/dD. What dosage maximizes sensitivity?

For the functions in Problems 18–22, do the following:

(a) Find f' and f''.

(b) Find the critical points of f.

(c) Find any inflection points of f.

(d) Evaluate f at its critical points and at the endpoints of the given interval. Identify local and global maxima and minima of f in the interval.

(e) Graph f.

18. $f(x) = x^3 - 3x^2 \quad (-1 \le x \le 3)$

19. $f(x) = 2x^3 - 9x^2 + 12x + 1 \, (-0.5 \le x \le 3)$

20. $f(x) = x^3 - 3x^2 - 9x + 15 \quad (-5 \le x \le 4)$

21. $f(x) = x + \sin x \quad (0 \le x \le 2\pi)$

22. $f(x) = e^{-x} \sin x \quad (0 \le x \le 2\pi)$

In Problems 23–28, find the exact global maximum and minimum values of the function. The domain is all real numbers unless otherwise specified.

23. $g(x) = 4x - x^2 - 5$

24. $f(x) = x + 1/x$ for $x > 0$

25. $g(t) = te^{-t}$ for $t > 0$

26. $f(x) = x - \ln x$ for $x > 0$

27. $f(t) = \dfrac{t}{1 + t^2}$

28. $f(t) = (\sin^2 t + 2) \cos t$

29. Find the value(s) of x that give critical points of $y = ax^2 + bx + c$, where a, b, c are constants. Under what conditions on a, b, c is the critical value a maximum? A minimum?

30. What value of w minimizes S if $S - 5pw = 3qw^2 - 6pq$ and p and q are positive constants?

31. Figure 4.39 gives the derivative of $g(x)$ on $-2 \le x \le 2$.

(a) Write a few sentences describing the behavior of $g(x)$ on this interval.

(b) Does the graph of $g(x)$ have any inflection points? If so, give the approximate x-coordinates of their locations. Explain your reasoning.

(c) What are the global maxima and minima of g on $[-2, 2]$?

(d) If $g(-2) = 5$, what do you know about $g(0)$ and $g(2)$? Explain.

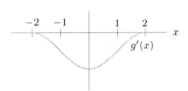

Figure 4.39

32. The energy expended by a bird per day, E, depends on the time spent foraging for food per day, F hours. Foraging for a shorter time requires better territory, which then requires more energy for its defense.[4] Find the foraging time that minimizes energy expenditure if

$$E = 0.25F + \frac{1.7}{F^2}.$$

33. If you have 100 feet of fencing and want to enclose a rectangular area up against a long, straight wall, what is the largest area you can enclose?

34. A closed box has a fixed surface area A and a square base with side x.

(a) Find a formula for its volume, V, as a function of x.

(b) Sketch a graph of V against x.

(c) Find the maximum value of V.

35. On the west coast of Canada, crows eat whelks (a shell-fish). To open the whelks, the crows drop them from the air onto a rock. If the shell does not smash the first time, the whelk is dropped again.[5] The average number of drops, n, needed when the whelk is dropped from a height of x meters is approximated by

$$n(x) = 1 + \frac{27}{x^2}.$$

(a) Give the total vertical distance the crow travels upward to open a whelk as a function of drop height, x.

(b) Crows are observed to drop whelks from the height that minimizes the total vertical upward distance traveled per whelk. What is this height?

36. During a flu outbreak in a school of 763 children, the number of infected children, I, was expressed in terms of the number of susceptible (but still healthy) children, S, by the expression[6]

$$I = 192 \ln\left(\frac{S}{762}\right) - S + 763.$$

What is the maximum possible number of infected children?

37. An apple tree produces, on average, 400 kg of fruit each season. However, if more than 200 trees are planted per km^2, crowding reduces the yield by 1 kg for each tree over 200.

(a) Express the total yield from one square kilometer as a function of the number of trees on it. Graph this function.

(b) How many trees should a farmer plant on each square kilometer to maximize yield?

[4] Adapted from Graham Pyke, reported by J. R. Krebs and N. B. Davis in *An Introduction to Behavioural Ecology* (Oxford: Blackwell, 1987).

[5] Adapted from Reto Zach, reported by J. R. Krebs and N. B. Davis in *An Introduction to Behavioural Ecology* (Oxford: Blackwell, 1987).

[6] Data from Communicable Disease Surveillance Centre (UK), reported in "Influenza in a Boarding School", *British Medical Journal*, March 4, 1978.

38. The number of offspring in a population may not be a linear function of the number of adults. The Ricker curve, used to model fish populations, claims that $y = axe^{-bx}$, where x is the number of adults, y is the number of offspring, and a and b are positive constants.

(a) Find and classify all critical points of the Ricker curve.

(b) Is there a global maximum? What does this imply about populations?

39. The oxygen supply, S, in the blood depends on the hematocrit, H, the percentage of red blood cells in the blood:

$$S = aHe^{-bH} \quad \text{for positive constants } a, b.$$

(a) What value of H maximizes the oxygen supply? What is the maximum oxygen supply?

(b) How does increasing the value of the constants a and b change the maximum value of S?

40. The quantity of a drug in the bloodstream t hours after a tablet is swallowed is given, in mg, by

$$q(t) = 20(e^{-t} - e^{-2t}).$$

(a) How much of the drug is in the bloodstream at time $t = 0$?

(b) When is the maximum quantity of drug in the bloodstream? What is that maximum?

(c) In the long run, what happens to the quantity?

41. When birds lay eggs, they do so in clutches of several at a time. When the eggs hatch, each clutch gives rise to a brood of baby birds. We want to determine the clutch size which maximizes the number of birds surviving to adulthood per brood. If the clutch is small, there are few baby birds in the brood; if the clutch is large, there are so many baby birds to feed that most die of starvation. The number of surviving birds per brood as a function of clutch size is shown by the benefit curve in Figure 4.40.[7]

(a) Estimate the clutch size which maximizes the number of survivors per brood.

(b) Suppose also that there is a biological cost to having a larger clutch: the female survival rate is reduced by large clutches. This cost is represented by the dotted line in Figure 4.40. If we take cost into account by assuming that the optimal clutch size in fact maximizes the vertical distance between the curves, what is the new optimal clutch size?

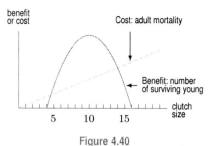

Figure 4.40

42. Let $f(v)$ be the amount of energy consumed by a flying bird, measured in joules per second (a joule is a unit of energy), as a function of its speed v (in meters/sec). Let $a(v)$ be the amount of energy consumed by the same bird, measured in joules per meter.

(a) Suggest a reason (in terms of the way birds fly) for the shape of the graph of $f(v)$ in Figure 4.41.

(b) What is the relationship between $f(v)$ and $a(v)$?

(c) Where is $a(v)$ a minimum?

(d) Should the bird try to minimize $f(v)$ or $a(v)$ when it is flying? Why?

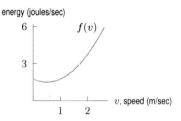

Figure 4.41

43. As an epidemic spreads through a population, the number of infected people, I, is expressed as a function of the number of susceptible people, S, by

$$I = k \ln\left(\frac{S}{S_0}\right) - S + S_0 + I_0, \quad \text{for } k, S_0, I_0 > 0.$$

(a) Find the maximum number of infected people.

(b) The constant k is a characteristic of the particular disease; the constants S_0 and I_0 are the values of S and I when the disease starts. Which of the following affects the maximum possible value of I? Explain.

- The particular disease, but not how it starts.
- How the disease starts, but not the particular disease.
- Both the particular disease and how it starts.

44. The hypotenuse of a right triangle has one end at the origin and one end on the curve $y = x^2 e^{-3x}$, with $x \geq 0$. One of the other two sides is on the x-axis, the other side is parallel to the y-axis. Find the maximum area of such a triangle. At what x-value does it occur?

---

[7]Data from C. M. Perrins and D. Lack, reported by J. R. Krebs and N. B. Davies in *An Introduction to Behavioural Ecology* (Oxford: Blackwell, 1987).

45. A person's blood pressure, p, in millimeters of mercury (mm Hg) is given, for t in seconds, by

$$p = 100 + 20\sin(2.5\pi t).$$

(a) What are the maximum and minimum values of blood pressure?

(b) What is the interval between successive maxima?

(c) Show your answers on a graph of blood pressure against time.

46. A chemical reaction converts substance A to substance Y; the presence of Y catalyzes the reaction. At the start of the reaction, the quantity of A present is a grams. At time t seconds later, the quantity of Y present is y grams. The rate of the reaction, in grams/sec, is given by

$$\text{Rate} = ky(a - y), \quad k \text{ is a positive constant.}$$

(a) For what values of y is the rate nonnegative? Graph the rate against y.

(b) For what values of y is the rate a maximum?

47. In a chemical reaction, substance A combines with substance B to form substance Y. At the start of the reaction, the quantity of A present is a grams, and the quantity of B present is b grams. At time t seconds after the start of the reaction, the quantity of Y present is y grams. Assume $a < b$ and $y \leq a$. For certain types of reactions, the rate of the reaction, in grams/sec, is given by

$$\text{Rate} = k(a - y)(b - y), \quad k \text{ is a positive constant.}$$

(a) For what values of y is the rate nonnegative? Graph the rate against y.

(b) Use your graph to find the value of y at which the rate of the reaction is fastest.

4.4   PROFIT, COST, AND REVENUE

Maximizing Profit

A fundamental issue for a producer of goods is how to maximize profit. For a quantity, q, the profit $\pi(q)$ is the difference between the revenue, $R(q)$, and the cost, $C(q)$, of supplying that quantity. Thus, $\pi(q) = R(q) - C(q)$. The marginal cost, $MC = C'$, is the derivative of C; marginal revenue is $MR = R'$.

Now we look at how to maximize total profit, given functions for revenue and cost. The next example suggests a criterion for identifying the optimal production level.

Example 1   Estimate the maximum profit if the revenue and cost are given by the curves R and C, respectively, in Figure 4.42.

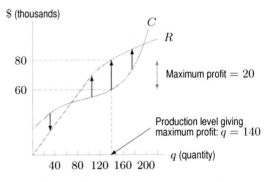

Figure 4.42: Maximum profit at $q = 140$

Solution   Since profit is revenue minus cost, the profit is represented by the vertical distance between the cost and revenue curves, marked by the vertical arrows in Figure 4.42. When revenue is below cost, the company is taking a loss; when revenue is above cost, the company is making a profit. The maximum profit must occur between about $q = 70$ and $q = 200$, which is the interval in which the company is making a profit. Profit is maximized when the vertical distance between the curves is largest (and revenue is above cost). This occurs at approximately $q = 140$.

The profit accrued at $q = 140$ is the vertical distance between the curves, so the maximum profit = \$80,000 − \$60,000 = \$20,000.

Maximum Profit Can Occur Where $MR = MC$

We now analyze the marginal costs and marginal revenues near the optimal point. Zooming in on Figure 4.42 around $q = 140$ gives Figure 4.43.

At a production level q_1 to the left of 140 in Figure 4.43, marginal cost is less than marginal revenue. The company would make more money by producing more units, so production should be increased (toward a production level of 140). At any production level q_2 to the right of 140, marginal cost is greater than marginal revenue. The company would lose money by producing more units and would make more money by producing fewer units. Production should be adjusted down toward 140.

What about the marginal revenue and marginal cost at $q = 140$? Since $MC < MR$ to the left of 140, and $MC > MR$ to the right of 140, we expect $MC = MR$ at 140. In this example, profit is maximized at the point where the slopes of the cost and revenue graphs are equal.

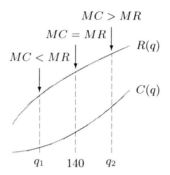

Figure 4.43: Example 1: Maximum profit occurs where $MC = MR$

We can get the same result analytically. Global maxima and minima of a function can only occur at critical points of the function or at the endpoints of the interval. To find critical points of π, look for zeros of the derivative:

$$\pi'(q) = R'(q) - C'(q) = 0.$$

So

$$R'(q) = C'(q),$$

that is, the slopes of the graphs of $R(q)$ and $C(q)$ are equal at q. In economic language,

---

The maximum (or minimum) profit can occur where

$$\text{Marginal profit} = 0,$$

that is, where

$$\text{Marginal revenue} = \text{Marginal cost.}$$

Of course, maximum or minimum profit does not *have* to occur where $MR = MC$; either one could occur at an endpoint. Example 2 shows how to visualize maxima and minima of the profit on a graph of marginal revenue and marginal cost.

Example 2   The total revenue and total cost curves for a product are given in Figure 4.44.

(a) Sketch the marginal revenue and marginal cost, MR and MC, on the same axes. Mark the two quantities where marginal revenue equals marginal cost. What is the significance of these two quantities? At which quantity is profit maximized?

(b) Graph the profit function $\pi(q)$.

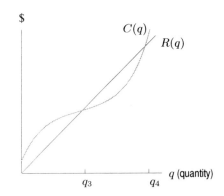

Figure 4.44: Total revenue and total cost

Solution   (a) Since $R(q)$ is a straight line with positive slope, the graph of its derivative, MR, is a horizontal line. (See Figure 4.45.) Since $C(q)$ is always increasing, its derivative, MC, is always positive. As q increases, the cost curve changes from concave down to concave up, so the derivative of the cost function, MC, changes from decreasing to increasing. (See Figure 4.45.) The local minimum on the marginal cost curve corresponds to the inflection point of $C(q)$.

Where is profit maximized? We know that the maximum profit can occur when Marginal revenue = Marginal cost, that is where the curves in Figure 4.45 cross at q_1 and q_2. Do these points give the maximum profit?

We first consider q_1. To the left of q_1, we have $MR < MC$, so $\pi' = MR - MC$ is negative and the profit function is decreasing there. To the right of q_1, we have $MR > MC$, so π' is positive and the profit function is increasing. This behavior, decreasing and then increasing, means that the profit function has a local minimum at q_1. This is certainly not the production level we want.

What happens at q_2? To the left of q_2, we have $MR > MC$, so π' is positive and the profit function is increasing. To the right of q_2, we have $MR < MC$, so π' is negative and the profit function is decreasing. This behavior, increasing and then decreasing, means that the profit function has a local maximum at q_2. The global maximum profit occurs either at the production level q_2 or at an endpoint (the largest and smallest possible production levels). Since the profit is negative at the endpoints (see Figure 4.44), the global maximum occurs at q_2.

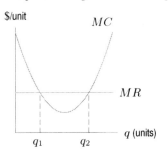

Figure 4.45: Marginal revenue and marginal cost

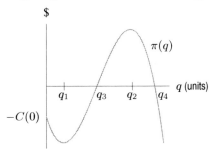

Figure 4.46: Profit function

(b) The graph of the profit function is in Figure 4.46. At the maximum and minimum, the slope of the profit curve is zero:
$$\pi'(q_1) = \pi'(q_2) = 0.$$
Note that since $R(0) = 0$ and $C(0)$ represents the fixed costs of production, we have
$$\pi(0) = R(0) - C(0) = -C(0).$$
Therefore the vertical intercept of the profit function is a negative number, equal in magnitude to the size of the fixed cost.

Example 3   Find the quantity which maximizes profit if the total revenue and total cost (in dollars) are given by

$$R(q) = 5q - 0.003q^2$$
$$C(q) = 300 + 1.1q$$

where q is quantity and $0 \leq q \leq 1000$ units. What production level gives the minimum profit?

Solution   We begin by looking for production levels that give Marginal revenue = Marginal cost. Since

$$MR = R'(q) = 5 - 0.006q$$
$$MC = C'(q) = 1.1,$$

$MR = MC$ leads to

$$5 - 0.006q = 1.1$$
$$q = \frac{3.9}{0.006} = 650 \text{ units.}$$

Does this represent a local maximum or minimum of the profit π? To decide, look to the left and right of 650 units.

When $q = 649$, we have $MR = \$1.106$ per unit, which is greater than $MC = \$1.10$ per unit.

Thus, producing one more unit (the 650th) brings in more revenue than it costs, so profit increases.

When $q = 651$, we have $MR = \$1.094$ per unit, which is less than $MC = \$1.10$ per unit.

It is not profitable to produce the 651st unit. We conclude that $q = 650$ gives a local maximum for the profit function π.

To check whether $q = 650$ gives a global maximum, we compare the profit at the endpoints, $q = 0$ and $q = 1000$, with the profit at $q = 650$.

At $q = 0$, the only cost is \$300 (the fixed costs) and there is no revenue, so $\pi(0) = -\$300$.
At $q = 1000$, we have $R(1000) = \$2000$ and $C(1000) = \$1400$, so $\pi(1000) = \$600$.
At $q = 650$, we have $R(650) = \$1982.50$ and $C(650) = \$1015$, so $\pi(650) = \$967.50$.
Therefore, the maximum profit is obtained at a production level of $q = 650$ units. The minimum profit (a loss) occurs when $q = 0$ and there is no production at all.

Maximizing Revenue

For some companies, costs do not depend on the number of items sold. For example, a city bus company with a fixed schedule has the same costs no matter how many people ride the buses. In such a situation, profit is maximized by maximizing revenue.

Example 4   At a price of \$80 for a half-day trip, a white-water rafting company attracts 300 customers. Every \$5 decrease in price attracts an additional 30 customers.

(a) Find the demand equation.
(b) Express revenue as a function of price.
(c) What price should the company charge per trip to maximize revenue?

Solution   (a) We first find the equation relating price to demand. If price, p, is 80, the number of trips sold, q, is 300. If p is 75, then q is 330, and so on. See Table 4.1. Because demand changes by a constant (30 people) for every \$5 drop in price, q is a linear function of p. Then

$$\text{Slope} = \frac{300 - 330}{80 - 75} = -\frac{30}{5} = -6 \text{ people/dollar,}$$

so the demand equation is $q = -6p + b$. Since $p = 80$ when $q = 300$, we have

$$300 = -6 \cdot 80 + b$$
$$b = 300 + 6 \cdot 80 = 780.$$

The demand equation is $q = -6p + 780$.

(b) Since revenue $R = p \cdot q$, revenue as a function of price is

$$R(p) = p(-6p + 780) = -6p^2 + 780p.$$

(c) Figure 4.47 shows this revenue function has a maximum. To find it, we differentiate:

$$R'(q) = -12p + 780 = 0$$
$$p = \frac{780}{12} = 65.$$

The maximum revenue is achieved when the price is $65.

Table 4.1   *Demand for rafting trips*

Price, p	Number of trips sold, q
80	300
75	330
70	360
65	390
. . .	. . .

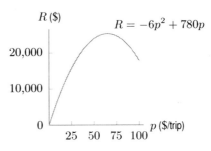

Figure 4.47: Revenue for a rafting company as a function of price

Problems for Section 4.4

1. Figure 4.48 shows cost and revenue. For what production levels is the profit function positive? Negative? Estimate the production at which profit is maximized.

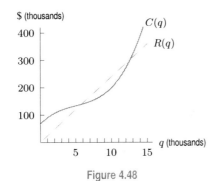

Figure 4.48

2. Using the cost and revenue graphs in Figure 4.49, sketch the following functions. Label the points q_1 and q_2.

(a) Total profit   (b) Marginal cost

(c) Marginal revenue

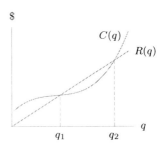

Figure 4.49

3. Table 4.2 shows cost, $C(q)$, and revenue, $R(q)$.

   (a) At approximately what production level, q, is profit maximized? Explain your reasoning.
   (b) What is the price of the product?
   (c) What are the fixed costs?

max

Table 4.2

q	0	500	1000	1500	2000	2500	3000
$R(q)$	0	1500	3000	4500	6000	7500	9000
$C(q)$	3000	3800	4200	4500	4800	5500	7400

$-300 \quad -2300 \quad -1200 \quad 0 \quad 1200 \quad 2000 \quad 1600$

4. A demand function is $p = 400 - 2q$, where q is the quantity of the good sold for price \$$p$.

   (a) Find an expression for the total revenue, R, in terms of q.
   (b) Differentiate R with respect to q to find the marginal revenue, MR, in terms of q. Calculate the marginal revenue when $q = 10$.
   (c) Calculate the change in total revenue when production increases from $q = 10$ to $q = 11$ units. Confirm that a one-unit increase in q gives a reasonable approximation to the exact value of MR obtained in part (b).

5. Let $C(q)$ represent the cost, $R(q)$ the revenue, and $\pi(q)$ the total profit, in dollars, of producing q items.

   (a) If $C'(50) = 75$ and $R'(50) = 84$, approximately how much profit is earned by the 51st item?
   (b) If $C'(90) = 71$ and $R'(90) = 68$, approximately how much profit is earned by the 91st item?
   (c) If $\pi(q)$ is a maximum when $q = 78$, how do you think $C'(78)$ and $R'(78)$ compare? Explain.

6. Figure 4.45 in Section 4.4 shows the points, q_1 and q_2, where marginal revenue equals marginal cost.

   (a) On the graph of the corresponding total cost and total revenue functions in Figure 4.50, label the points q_1 and q_2. Using slopes, explain the significance of these points.
   (b) Explain in terms of profit why one is a local minimum and one is a local maximum.

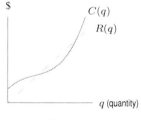

Figure 4.50

7. Table 4.3 shows marginal cost, MC, and marginal revenue, MR.

   (a) Use the marginal cost and marginal revenue at a production of $q = 5000$ to determine whether production should be increased or decreased from 5000.
   (b) Estimate the production level that maximizes profit.

Table 4.3

q	5000	6000	7000	8000	9000	10000
MR	60	58	56	55	54	53
MC	48	52	54	55	58	63

8. Marginal revenue and marginal cost are given in the following table. Estimate the production levels that could maximize profit. Explain.

q	1000	2000	3000	4000	5000	6000
MR	78	76	74	72	70	68
MC	100	80	70	65	75	90

9. A company estimates that the total revenue, R, in dollars, received from the sale of q items is $R = \ln(1 + 1000q^2)$. Calculate and interpret the marginal revenue if $q = 10$.

10. Figure 4.51 shows cost and revenue for a product.

   (a) Estimate the production level that maximizes profit.
   (b) Graph marginal revenue and marginal cost for this product on the same axes. Label on this graph the production level that maximizes profit.

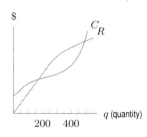

Figure 4.51

11. Figure 4.52 shows graphs of marginal cost and marginal revenue. Estimate the production levels that could maximize profit. Explain your reasoning.

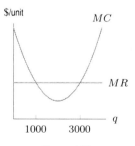

Figure 4.52

12. The marginal cost and marginal revenue of a company are $MC(q) = 0.03q^2 - 1.4q + 34$ and $MR(q) = 30$, where q is the number of items manufactured. To increase profits, should the company increase or decrease production from each of the following levels?

   (a) 25 items   **(b)** 50 items   **(c)** 80 items

13. A manufacturing process has marginal costs given in the table; the item sells for $30 per unit. At how many quantities, q, does the profit appear to be a maximum? In what intervals do these quantities appear to lie?

q	0	10	20	30	40	50	60
MC ($/unit)	34	23	18	19	26	39	58

14. Cost and revenue functions are given in Figure 4.53. Approximately what quantity maximizes profits?

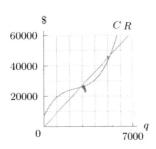

Figure 4.53

15. Cost and revenue functions are given in Figure 4.53.

   (a) At a production level of $q = 3000$, is marginal cost or marginal revenue greater? Explain what this tells you about whether production should be increased or decreased.

   (b) Answer the same questions for $q = 5000$.

16. When production is 2000, marginal revenue is $4 per unit and marginal cost is $3.25 per unit. Do you expect maximum profit to occur at a production level above or below 2000? Explain.

17. Revenue is given by $R(q) = 450q$ and cost is given by $C(q) = 10,000 + 3q^2$. At what quantity is profit maximized? What is the total profit at this production level?

18. The demand equation for a product is $p = 45 - 0.01q$. Write the revenue as a function of q and find the quantity that maximizes revenue. What price corresponds to this quantity? What is the total revenue at this price?

19. Revenue and cost functions for a company are given in Figure 4.54.

   (a) Estimate the marginal cost at $q = 400$.

   (b) Should the company produce the 500th item? Why?

   (c) Estimate the quantity which maximizes profit.

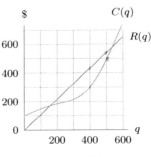

Figure 4.54

20. The following table gives the cost and revenue, in dollars, for different production levels, q.

   (a) At approximately what production level is profit maximized?

   (b) What price is charged per unit for this product?

   (c) What are the fixed costs of production?

q	0	100	200	300	400	500
$R(q)$	0	500	1000	1500	2000	2500
$C(q)$	700	900	1000	1100	1300	1900

21. The demand for tickets to an amusement park is given by $p = 70 - 0.02q$, where p is the price of a ticket in dollars and q is the number of people attending at that price.

   (a) What price generates an attendance of 3000 people? What is the total revenue at that price? What is the total revenue if the price is $20?

   (b) Write the revenue function as a function of attendance, q, at the amusement park.

   (c) What attendance maximizes revenue?

   (d) What price should be charged to maximize revenue?

   (e) What is the maximum revenue? Can we determine the corresponding profit?

22. An ice cream company finds that at a price of $4.00, demand is 4000 units. For every $0.25 decrease in price, demand increases by 200 units. Find the price and quantity sold that maximize revenue.

23. At a price of $8 per ticket, a musical theater group can fill every seat in the theater, which has a capacity of 1500. For every additional dollar charged, the number of people buying tickets decreases by 75. What ticket price maximizes revenue?

24. The demand equation for a quantity q of a product at price p, in dollars, is $p = -5q + 4000$. Companies producing the product report the cost, C, in dollars, to produce a quantity q is $C = 6q + 5$ dollars.

   (a) Express a company's profit, in dollars, as a function of q.

   (b) What production level earns the company the largest profit?

   (c) What is the largest profit possible?

25. **(a)** Production of an item has fixed costs of $10,000 and variable costs of $2 per item. Express the cost, C, of producing q items.
    (b) The relationship between price, p, and quantity, q, demanded is linear. Market research shows that 10,100 items are sold when the price is $5 and 12,872 items are sold when the price is $4.50. Express q as a function of price p.
    (c) Express the profit earned as a function of q.
    (d) How many items should the company produce to maximize profit? (Give your answer to the nearest integer.) What is the profit at that production level?

26. A landscape architect plans to enclose a 3000 square-foot rectangular region in a botanical garden. She will use shrubs costing $45 per foot along three sides and fencing costing $20 per foot along the fourth side. Find the minimum total cost.

27. You run a small furniture business. You sign a deal with a customer to deliver up to 400 chairs, the exact number to be determined by the customer later. The price will be $90 per chair up to 300 chairs, and above 300, the price will be reduced by $0.25 per chair (on the whole order) for every additional chair over 300 ordered. What are the largest and smallest revenues your company can make under this deal?

28. A warehouse selling cement has to decide how often and in what quantities to reorder. It is cheaper, on average, to place large orders, because this reduces the ordering cost per unit. On the other hand, larger orders mean higher storage costs. The warehouse always reorders cement in the same quantity, q. The total weekly cost, C, of ordering and storage is given by

$$C = \frac{a}{q} + bq, \quad \text{where } a, b \text{ are positive constants.}$$

    (a) Which of the terms, a/q and bq, represents the ordering cost and which represents the storage cost?
    (b) What value of q gives the minimum total cost?

29. A business sells an item at a constant rate of r units per month. It reorders in batches of q units, at a cost of $a + bq$ dollars per order. Storage costs are k dollars per item per month, and, on average, $q/2$ items are in storage, waiting to be sold. [Assume r, a, b, k are positive constants.]

    (a) How often does the business reorder?
    (b) What is the average monthly cost of reordering?
    (c) What is the total monthly cost, C of ordering and storage?
    (d) Obtain Wilson's lot size formula, the optimal batch size which minimizes cost.

30. **(a)** A cruise line offers a trip for $2000 per passenger. If at least 100 passengers sign up, the price is reduced for *all* the passengers by $10 for every additional passenger (beyond 100) who goes on the trip. The boat can accommodate 250 passengers. What number of passengers maximizes the cruise line's total revenue? What price does each passenger pay then?
    (b) The cost to the cruise line for n passengers is $80,000 + 400n$. What is the maximum profit that the cruise line can make on one trip? How many passengers must sign up for the maximum to be reached and what price will each pay?

31. A company manufactures only one product. The quantity, q, of this product produced per month depends on the amount of capital, K, invested (i.e., the number of machines the company owns, the size of its building, and so on) and the amount of labor, L, available each month. We assume that q can be expressed as a *Cobb-Douglas production function*:

$$q = cK^\alpha L^\beta$$

where c, α, β are positive constants, with $0 < \alpha < 1$ and $0 < \beta < 1$. In this problem we will see how the Russian government could use a Cobb-Douglas function to estimate how many people a newly privatized industry might employ. A company in such an industry has only a small amount of capital available to it and needs to use all of it, so K is fixed. Suppose L is measured in man-hours per month, and that each man-hour costs the company w rubles (a ruble is the unit of Russian currency). Suppose the company has no other costs besides labor, and that each unit of the good can be sold for a fixed price of p rubles. How many man-hours of labor per month should the company use in order to maximize its profit?

32. A company can produce and sell $f(L)$ tons of a product per month using L hours of labor per month. The wage of the workers is w dollars per hour, and the finished product sells for p dollars per ton.

    (a) The function $f(L)$ is the company's production function. Give the units of $f(L)$. What is the practical significance of $f(1000) = 400$?
    (b) The derivative $f'(L)$ is the company's marginal product of labor. Give the units of $f'(L)$. What is the practical significance of $f'(1000) = 2$?
    (c) The real wage of the workers is the quantity of product that can be bought with one hour's wages. Show that the real wage is w/p tons per hour.
    (d) Show that the monthly profit of the company is

$$\pi(L) = pf(L) - wL.$$

    (e) Show that when operating at maximum profit, the company's marginal product of labor equals the real wage:

$$f'(L) = \frac{w}{p}.$$

4.5 AVERAGE COST

To maximize profit, a company arranges production to equalize marginal cost and marginal revenue. But how do we know if the company makes money? It turns out that whether the maximum profit is positive or negative is determined by the company's average cost of production. Average cost also tells us about the behavior of similar companies in an industry. If average costs are low, more companies will enter the market; if average costs are high, companies will leave the market.

In this section, we see how average cost can be calculated and visualized, and the relationship between average and marginal cost.

What Is Average Cost?

The average cost is the cost per unit of producing a certain quantity; it is the total cost divided by the number of units produced.

> If the cost of producing a quantity q is $C(q)$, then the **average cost**, $a(q)$, of producing a quantity q is given by
> $$a(q) = \frac{C(q)}{q}.$$

Although both are measured in the same units, for example, dollars per item, be careful not to confuse the average cost with the marginal cost (the cost of producing the next item).

Example 1   A salsa company has cost function $C(q) = 0.01q^3 - 0.6q^2 + 13q + 1000$ (in dollars), where q is the number of cases of salsa produced. If 100 cases are produced, find the average cost per case.

Solution   The total cost of producing the 100 cases is given by

$$C(100) = 0.01(100^3) - 0.6(100^2) + 13(100) + 1000 = \$6300.$$

We find the average cost per case by dividing by 100, the number of cases produced.

$$\text{Average cost} = \frac{6300}{100} = 63 \text{ dollars/case}.$$

If 100 cases of salsa are produced, the average cost is $63 per case.

Visualizing Average Cost on the Total Cost Curve

We know that average cost is $a(q) = C(q)/q$. Since we can subtract zero from any number without changing it, we can write

$$a(q) = \frac{C(q)}{q} = \frac{C(q) - 0}{q - 0}.$$

This expression gives the slope of the line joining the points $(0, 0)$ and $(q, C(q))$ on the cost curve. See Figure 4.55.

> $$\frac{\text{Average cost}}{\text{to produce } q \text{ items}} = \frac{C(q)}{q} = \frac{\text{Slope of the line from the origin}}{\text{to point } (q, C(q)) \text{ on cost curve.}}$$

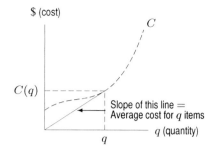

Figure 4.55: Average cost is the slope of the line from the origin to a point on the cost curve

Minimizing Average Cost

We use the graphical representation of average cost to investigate the relationship between average and marginal cost, and to identify the production level which minimizes average cost.

Example 2   A cost function, in dollars, is $C(q) = 1000 + 20q$, where q is the number of units produced. Find and compare the marginal cost to produce the 100th unit and the average cost of producing 100 units. Illustrate your answer on a graph.

Solution   The cost function is linear with fixed costs of $1000 and variable costs of $20 per unit. Thus,

$$\text{Marginal cost} = C'(q) = 20 \text{ dollars per unit.}$$

This means that after 99 units have been produced, it costs an additional $20 to produce the next unit. In contrast,

$$\text{Average cost of producing 100 units} = a(100) = \frac{C(100)}{100} = \frac{3000}{100} = 30 \text{ dollars/unit.}$$

Notice that the average cost includes the fixed costs of $1000 spread over the entire production, whereas marginal cost does not. Thus, the average cost is greater than the marginal cost in this example. See Figure 4.56.

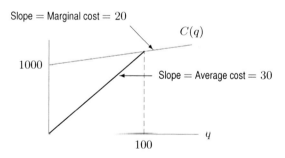

Figure 4.56: Average cost > Marginal cost

197

Example 3   Mark on the cost graph in Figure 4.57 the quantity at which the average cost is minimized.

Solution   In Figure 4.58, the average costs at q_1, q_2, q_3, and q_4 are given by the slopes of the lines from the origin to the curve. These slopes are steep for small q, become less steep as q increases, and then get steeper again. Thus, as q increases, the average cost decreases and then increases, so there is a minimum value. In Figure 4.58 the minimum occurs at the point q_0 where the line from the origin is tangent to the cost curve.

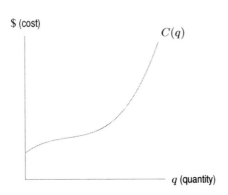

Figure 4.57: A cost function

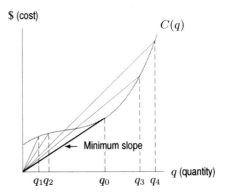

Figure 4.58: Minimum average cost occurs at q_0 where line is tangent to cost curve

In Figure 4.58, notice that average cost is a minimum (at q_0) when average cost equals marginal cost. The next example shows what happens when marginal cost and average cost are not equal.

Example 4   Suppose 100 items are produced at an average cost of $2 per item. Find the average cost of producing 101 items if the marginal cost to produce the 101st item is: (a) $1   (b) $3.

Solution   If 100 items are produced at an average cost of $2 per item, the total cost of producing the items is $100 \cdot \$2 = \$200$.

(a) Since the marginal, or additional, cost to produce the 101st item is $1, the total cost of producing 101 items is $200 + \$1 = \$201$. The average cost to produce these items is $201/101$, or $1.99 per item. The average cost has gone down.

(b) In this case, the marginal cost to produce the 101st item is $3. The total cost to produce 101 items is $203 and the average cost is $203/101$, or $2.01 per item. The average cost has gone up.

Notice that in Example 4 (a), where it costs less than the average to produce an additional item, average cost decreases as production increases. In Example 4 (b), where it costs more than the average to produce an additional item, average cost increases with production. We summarize:

> ### Relationship Between Average Cost and Marginal Cost
> - If marginal cost is less than average cost, then increasing production decreases average cost.
> - If marginal cost is greater than average cost, then increasing production increases average cost.
> - Marginal cost equals average cost at critical points of average cost.

Example 5   Show analytically that critical points of average cost occur when marginal cost equals average cost.

Solution   Since $a(q) = C(q)/q = C(q)q^{-1}$, we use the product rule to find $a'(q)$:

$$a'(q) = C'(q)(q^{-1}) + C(q)(-q^{-2}) = \frac{C'(q)}{q} + \frac{-C(q)}{q^2} = \frac{qC'(q) - C(q)}{q^2}.$$

At critical points we have $a'(q) = 0$, so

$$\frac{qC'(q) - C(q)}{q^2} = 0$$

Therefore, we have

$$qC'(q) - C(q) = 0$$
$$qC'(q) = C(q)$$
$$C'(q) = \frac{C(q)}{q}.$$

In other words, at a critical point:

$$\text{Marginal cost} \;=\; \text{Average cost.}$$

Example 6   A total cost function, in thousands of dollars, is given by $C(q) = q^3 - 6q^2 + 15q$, where q is in thousands and $0 \le q \le 5$.

(a) Graph $C(q)$. Estimate visually the quantity at which average cost is minimized.
(b) Graph the average cost function. Use it to estimate the minimum average cost.
(c) Determine analytically the exact value of q at which average cost is minimized.
(d) Graph the marginal cost function on the same axes as the average cost.
(e) Show that at the minimum average cost, Marginal cost $=$ Average cost. Explain how you can see this result on your graph of average and marginal costs.

Solution   (a) A graph of $C(q)$ is in Figure 4.59. Average cost is minimized at the point where a line from the origin to the point on the curve has minimum slope. This occurs where the line is tangent to the curve, which is at approximately $q = 3$, corresponding to a production of 3000 units.
(b) Since average cost is total cost divided by quantity, we have

$$a(q) = \frac{C(q)}{q} = \frac{q^3 - 6q^2 + 15q}{q} = q^2 - 6q + 15.$$

Figure 4.60 suggests that the minimum average cost occurs at $q = 3$.
(c) Average cost is minimized at a critical point of $a(q) = q^2 - 6q + 15$. Differentiating gives

$$a'(q) = 2q - 6 = 0$$
$$q = 3.$$

The minimum occurs at $q = 3$.
(d) See Figure 4.60. Marginal cost is the derivative of $C(q) = q^3 - 6q^2 + 15q$,

$$MC(q) - 3q^2 - 12q + 15.$$

(e) At $q = 3$, we have

$$\text{Marginal cost} = 3 \cdot 3^2 - 12 \cdot 3 + 15 = 6.$$
$$\text{Average cost} = 3^2 - 6 \cdot 3 + 15 = 6.$$

Thus, marginal and average cost are equal at $q = 3$. This result can be seen in Figure 4.60 since the marginal cost curve cuts the average cost curve at the minimum average cost.

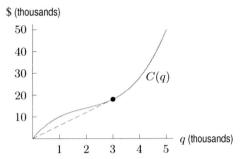

Figure 4.59: Cost function, showing the minimum average cost

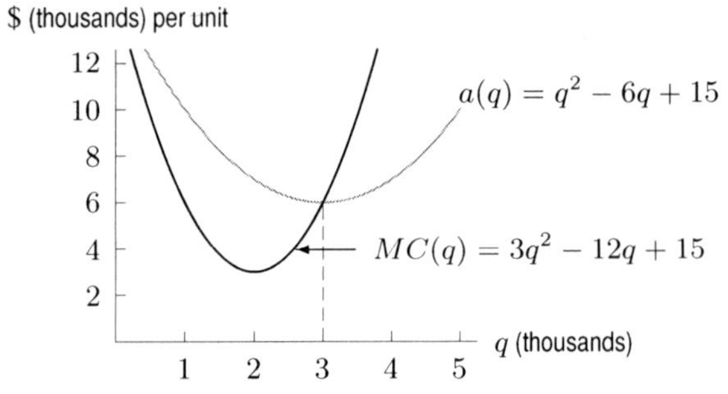

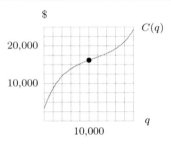

Figure 4.60: Average and marginal cost functions, showing minimum average cost

Problems for Section 4.5

1. For each cost function in Figure 4.61, is there a value of q at which average cost is minimized? If so, approximately where? Explain your answer.

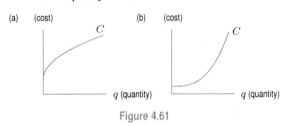

Figure 4.61

2. The graph of a cost function is given in Figure 4.62.

    (a) At $q = 25$, estimate the following quantities and represent your answers graphically.

      (i)   Average cost   (ii)   Marginal cost

    (b) At approximately what value of q is average cost minimized?

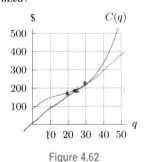

Figure 4.62

3. Figure 4.63 shows cost with $q = 10,000$ marked.

    (a) Find the average cost when the production level is 10,000 units and interpret it.

    (b) Represent your answer to part (a) graphically.

    (c) At approximately what production level is average cost minimized?

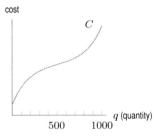

Figure 4.63

4. The cost of producing q items is $C(q) = 2500 + 12q$ dollars.

    (a) What is the marginal cost of producing the 100th item? the 1000th item?

    (b) What is the average cost of producing 100 items? 1000 items?

5. The cost function is $C(q) = 1000 + 20q$. Find the marginal cost to produce the 200th unit and the average cost of producing 200 units.

6. Graph the average cost function corresponding to the total cost function shown in Figure 4.64.

7. The total cost of production, in thousands of dollars, is $C(q) = q^3 - 12q^2 + 60q$, where q is in thousands and $0 \leq q \leq 8$.

    (a) Graph $C(q)$. Estimate visually the quantity at which average cost is minimized.

    (b) Determine analytically the exact value of q at which average cost is minimized.

Figure 4.64

8. You are the manager of a firm that produces slippers that sell for $20 a pair. You are producing 1200 pairs of slippers each month, at an average cost of $2 each. The marginal cost at a production level of 1200 is $3 per pair.

(a) Are you making or losing money?
(b) Will increasing production increase or decrease your average cost? Your profit?
(c) Would you recommend that production be increased or decreased?

9. The average cost per item to produce q items is given by

$$a(q) = 0.01q^2 - 0.6q + 13, \quad \text{for} \quad q > 0.$$

(a) What is the total cost, $C(q)$, of producing q goods?
(b) What is the minimum marginal cost? What is the practical interpretation of this result?
(c) At what production level is the average cost a minimum? What is the lowest average cost?
(d) Compute the marginal cost at $q = 30$. How does this relate to your answer to part (c)? Explain this relationship both analytically and in words.

10. The marginal cost at a production level of 2000 units of an item is $10 per unit and the average cost of producing 2000 units is $15 per unit. If the production level were increased slightly above 2000, would the following quantities increase or decrease, or is it impossible to tell?

(a) Average cost   (b) Profit

11. An agricultural worker in Uganda is planting clover to increase the number of bees making their home in the region. There are 100 bees in the region naturally, and for every acre put under clover, 20 more bees are found in the region.

(a) Draw a graph of the total number, $N(x)$, of bees as a function of x, the number of acres devoted to clover.
(b) Explain, both geometrically and algebraically, the shape of the graph of:

   (i) The marginal rate of increase of the number of bees with acres of clover, $N'(x)$.

   (ii) The average number of bees per acre of clover, $N(x)/x$.

12. A developer has recently purchased a laundromat and an adjacent factory. For years, the laundromat has taken pains to keep the smoke from the factory from soiling the air used by its clothes dryers. Now that the developer owns both the laundromat and the factory, she could install filters in the factory's smokestacks to reduce the emission of smoke, instead of merely protecting the laundromat from it. The cost of filters for the factory and the cost of protecting the laundromat against smoke depend on the number of filters used, as shown in the table.

Number of filters	Total cost of filters	Total cost of protecting laundromat from smoke
0	$0	$127
1	$5	$63
2	$11	$31
3	$18	$15
4	$26	$6
5	$35	$3
6	$45	$0
7	$56	$0

(a) Make a table which shows, for each possible number of filters (0 through 7), the marginal cost of the filter, the average cost of the filters, and the marginal savings in protecting the laundromat from smoke.
(b) Since the developer wishes to minimize the total costs to both her businesses, what should she do? Use the table from part (a) to explain your answer.
(c) What should the developer do if, in addition to the cost of the filters, the filters must be mounted on a rack which costs $100?
(d) What should the developer do if the rack costs $50?

13. Figure 4.65 shows the average cost, $a(q) = b + mq$.

(a) Show that $C'(q) = b + 2mq$.
(b) Graph the marginal cost $C'(q)$.

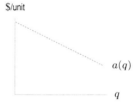

Figure 4.65

14. Show analytically that if marginal cost is less than average cost, then the derivative of average cost with respect to quantity satisfies $a'(q) < 0$.

15. Show analytically that if marginal cost is greater than average cost, then the derivative of average cost with respect to quantity satisfies $a'(q) > 0$.

16. A reasonably realistic model of a firm's costs is given by the *short-run Cobb-Douglas cost curve*

$$C(q) = Kq^{1/a} + F,$$

where a is a positive constant, F is the fixed cost, and K measures the technology available to the firm.

(a) Show that C is concave down if $a > 1$.
(b) Assuming that $a < 1$, find what value of q minimizes the average cost.

4.6 ELASTICITY OF DEMAND

The sensitivity of demand to changes in price varies with the product. For example, a change in the price of light bulbs may not affect the demand for light bulbs much, because people need light bulbs no matter what their price. However, a change in the price of a particular make of car may have a significant effect on the demand for that car, because people can switch to another make.

Elasticity of Demand

We want to find a way to measure this sensitivity of demand to price changes. Our measure should work for products as diverse as light bulbs and cars. The prices of these two items are so different that it makes little sense to talk about absolute changes in price: Changing the price of light bulbs by $1 is a substantial change, whereas changing the price of a car by $1 is not. Instead, we use the percent change in price. How, for example, does a 1% increase in price affect the demand for the product?

Let Δp denote the change in the price p of a product and Δq denote the corresponding change in quantity q demanded. The percent change in price is $\Delta p/p$ and the percent change in quantity demanded is $\Delta q/q$. We assume in this book that Δp and Δq have opposite signs (because increasing the price usually decreases the quantity demanded). Then the effect of a price change on demand is measured by the absolute value of the ratio

$$\left| \frac{\text{Percent change in demand}}{\text{Percent change in price}} \right| = \left| \frac{\Delta q/q}{\Delta p/p} \right| = \left| \frac{\Delta q}{q} \cdot \frac{p}{\Delta p} \right| = \left| \frac{p}{q} \cdot \frac{\Delta q}{\Delta p} \right|$$

For small changes in p, we approximate $\Delta q/\Delta p$ by the derivative dq/dp. We define:

> The **elasticity of demand**[8] for a product, E, is given approximately by
> $$E \approx \left| \frac{\Delta q/q}{\Delta p/p} \right|, \quad \text{or exactly by} \quad E = \left| \frac{p}{q} \cdot \frac{dq}{dp} \right|.$$

Increasing the price of an item by 1% causes a drop of approximately $E\%$ in the quantity of goods demanded. For small changes, Δp, in price,

$$\frac{\Delta q}{q} \approx -E \frac{\Delta p}{p}.$$

If $E > 1$, a 1% increase in price causes demand to drop by more than 1%, and we say that demand is *elastic*. If $0 \leq E < 1$, a 1% increase in price causes demand to drop by less than 1%, and we say that demand is *inelastic*. In general, a larger elasticity causes a larger percent change in demand for a given percent change in price.

Example 1 Raising the price of hotel rooms from $75 to $80 per night reduces weekly sales from 100 rooms to 90 rooms.

(a) Approximate the elasticity of demand for rooms at a price of $75.

(b) Should the owner raise the price?

---

[8] When it is necessary to distinguish it from other elasticities, this quantity is called the elasticity of demand with respect to price, or the price elasticity of demand.

Solution   (a) The percent change in the price is

$$\frac{\Delta p}{p} = \frac{5}{75} = 0.067 = 6.7\%$$

and the percent change in demand is

$$\frac{\Delta q}{q} = \frac{-10}{100} = -0.1 = -10\%.$$

The elasticity of demand is approximated by the ratio

$$E \approx \left| \frac{\Delta q/q}{\Delta p/p} \right| = \frac{0.10}{0.067} = 1.5.$$

The elasticity is greater than 1 because the percent change in the demand is greater than the percent change in the price.

(b) At a price of $75 per room,

$$\text{Revenue} = (100 \text{ rooms})(\$75 \text{ per room}) = \$7500 \text{ per week}.$$

At a price of $80 per room,

$$\text{Revenue} = (90 \text{ rooms})(\$80 \text{ per room}) = \$7200 \text{ per week}.$$

A price increase results in loss of revenue, so the price should not be raised.

Example 2   The demand curve for a product is given by $q = 1000 - 2p^2$, where p is the price. Find the elasticity at $p = 10$ and at $p = 15$. Interpret your answers.

Solution   We first find the derivative $dq/dp = -4p$. At a price of $p = 10$, we have $dq/dp = -4 \cdot 10 = -40$, and the quantity demanded is $q = 1000 - 2 \cdot 10^2 = 800$. At this price, the elasticity is

$$E = \left| \frac{p}{q} \cdot \frac{dq}{dp} \right| = \left| \frac{10}{800}(-40) \right| = 0.5.$$

The demand is inelastic at a price of $p = 10$: a 1% increase in price results in approximately a 0.5% decrease in demand.

At a price of $15, we have $q = 550$ and $dq/dp = -60$. The elasticity is

$$E = \left| \frac{p}{q} \cdot \frac{dq}{dp} \right| = \left| \frac{15}{550}(-60) \right| = 1.64.$$

The demand is elastic: a 1% increase in price results in approximately a 1.64% decrease in demand.

Revenue and Elasticity of Demand

Elasticity enables us to analyze the effect of a price change on revenue. An increase in price usually leads to a fall in demand. However, the revenue may increase or decrease. The revenue $R = pq$ is the product of two quantities, and as one increases, the other decreases. Elasticity measures the relative significance of these two competing changes.

Example 3   Three hundred units of an item are sold when the price of the item is $10. When the price of the item is raised by $1, what is the effect on revenue if the quantity sold drops by

(a)   10 units?   (b)   100 units?

Solution   Since Revenue = Price · Quantity, when the price is $10, we have

$$\text{Revenue} = 10 \cdot 300 = \$3000.$$

(a) At a price of $11, the quantity sold is $300 - 10 = 290$, so

$$\text{Revenue} = 11 \cdot 290 = \$3190.$$

Thus, raising the price has increased revenue.

(b) At a price of $11, the quantity sold is $300 - 100 = 200$, so

$$\text{Revenue} = 11 \cdot 200 = \$2200.$$

Thus, raising the price has decreased revenue.

Elasticity allows us to predict whether revenue increases or decreases with a price increase.

Example 4   The item in Example 3 (a) is wool whose demand equation is $q = 400 - 10p$. The item in Example 3 (b) is houseplants, whose demand equation is $q = 1300 - 100p$. Find the elasticity of wool and houseplants.

Solution   For wool, $q = 400 - 10p$, so $dq/dp = -10$. Thus,

$$E_{\text{Wool}} = \left|\frac{p}{q}\frac{dq}{dp}\right| = \left|\frac{10}{300}(-10)\right| = \frac{1}{3}.$$

For houseplants, $q = 1300 - 100p$, so $dq/dp = -100$. Thus,

$$E_{\text{Houseplants}} = \left|\frac{p}{q}\frac{dq}{dp}\right| = \left|\frac{10}{300}(-100)\right| = \frac{10}{3}.$$

Notice that $E_{\text{Wool}} < 1$ and revenue increases with an increase in price; $E_{\text{Houseplants}} > 1$ and revenue decreases with an increase in price. In the next example we see the relationship between elasticity and maximum revenue.

Example 5   Table 4.4 shows the demand, q, revenue, R, and elasticity, E, for the product in Example 2 at several prices. What price brings in the greatest revenue? What is the elasticity at that price?

Solution   Table 4.4 suggests that maximum revenue is achieved at a price of about $13, and at that price, E is about 1. At prices below $13, we have $E < 1$, so the reduction in demand caused by a price increase is small; thus, raising the price increases revenue. At prices above $13, we have $E > 1$, so the increase in demand caused by a price decrease is relatively large; thus lowering the price increases revenue.

Table 4.4   *Revenue and elasticity at different points*

Price p	10	11	12	13	14	15
Demand q	800	758	712	662	608	550
Revenue R	8000	8338	8544	8606	8512	8250
Elasticity E	0.5	0.64	0.81	1.02	1.29	1.64
	Inelastic	Inelastic	Inelastic	Elastic	Elastic	Elastic

Example 6 shows that revenue does have a local maximum when $E = 1$. We summarize as follows:

Relationship Between Elasticity and Revenue

- If $E < 1$, demand is inelastic and revenue is increased by raising the price.
- If $E > 1$, demand is elastic and revenue is increased by lowering the price.
- $E = 1$ occurs at critical points of the revenue function.

Example 6 Show analytically that critical points of the revenue function occur when $E = 1$.

Solution We think of revenue as a function of price. Using the product rule to differentiate $R = pq$, we have

$$\frac{dR}{dp} = \frac{d}{dp}(pq) = p\frac{dq}{dp} + \frac{dp}{dp}q = p\frac{dq}{dp} + q.$$

At a critical point the derivative dR/dp equals zero, so we have

$$p\frac{dq}{dp} + q = 0$$

$$p\frac{dq}{dp} = -q$$

$$\frac{p}{q}\frac{dq}{dp} = -1$$

$$E = 1.$$

Elasticity of Demand for Different Products

Different products generally have different elasticities. See Table 4.5. If there are close substitutes for a product, or if the product is a luxury rather than a necessity, a change in price generally has a large effect on demand, and the demand for the product is elastic. On the other hand, if there are no close substitutes or if the product is a necessity, changes in price have a relatively small effect on demand, and the demand is inelastic. For example, demand for salt, penicillin, eyeglasses, and lightbulbs is inelastic over the usual range of prices for these products.

Table 4.5 *Elasticity of demand (with respect to price) for selected farm products*[9]

Cabbage	0.25	Oranges	0.62
Potatoes	0.27	Cream	0.69
Wool	0.33	Apples	1.27
Peanuts	0.38	Peaches	1.49
Eggs	0.43	Fresh tomatoes	2.22
Milk	0.49	Lettuce	2.58
Butter	0.62	Fresh peas	2.83

[9]Estimated by the US Department of Agriculture and reported in W. Adams & J. Brock, *The Structure of American Industry*, 10th ed (Englewood Cliffs: Prentice Hall, 2000).

Problems for Section 4.6

1. The elasticity of a good is $E = 0.5$. What is the effect on the quantity demanded of:

   (a) A 3% price increase? (b)   A 3% price decrease?

2. The elasticity of a good is $E = 2$. What is the effect on the quantity demanded of:

   (a) A 3% price increase? (b)   A 3% price decrease?

3. What are the units of elasticity if:

   (a) Price p is in dollars and quantity q is in tons?
   (b) Price p is in yen and quantity q is in liters?
   (c) What can you conclude in general?

4. Dwell time, t, is the time in minutes that shoppers spend in a store. Sales, s, is the number of dollars they spend in the store. The elasticity of sales with respect to dwell time is 1.3. Explain what this means in simple language.

5. What is the elasticity for peaches in Table 4.5? Explain what this number tells you about the effect of price increases on the demand for peaches. Is the demand for peaches elastic or inelastic? Is this what you expect? Explain.

6. What is the elasticity for potatoes in Table 4.5? Explain what this number tells you about the effect of price increases on the demand for potatoes. Is the demand for potatoes elastic or inelastic? Is this what you expect? Explain.

7. There are many brands of laundry detergent. Would you expect the elasticity of demand for any particular brand to be high or low? Explain.

8. Would you expect the demand for high-definition television sets to be elastic or inelastic? Explain.

9. There is only one company offering local telephone service in a town. Would you expect the elasticity of demand for telephone service to be high or low? Explain.

10. The demand for a product is given by $q = 200 - 2p^2$. Find the elasticity of demand when the price is $5. Is the demand inelastic or elastic, or neither?

11. School organizations raise money by selling candy door to door. The table shows p, the price of the candy, and q, the quantity sold at that price.

p	$1.00	$1.25	$1.50	$1.75	$2.00	$2.25	$2.50
q	2765	2440	1980	1660	1175	800	430

   (a) Estimate the elasticity of demand at a price of $1.00. At this price, is the demand elastic or inelastic?
   (b) Estimate the elasticity at each of the prices shown. What do you notice? Give an explanation for why this might be so.

   (c) At approximately what price is elasticity equal to 1?
   (d) Find the total revenue at each of the prices shown. Confirm that the total revenue appears to be maximized at approximately the price where $E = 1$.

12. The demand for a product is given by $p = 90 - 10q$. Find the elasticity of demand when $p = 50$. If this price rises by 2%, calculate the corresponding percentage change in demand.

13. The demand for yams is given by $q = 5000 - 10p^2$, where q is in pounds of yams and p is the price of a pound of yams.

   (a) If the current price of yams is $2 per pound, how many pounds will be sold?
   (b) Is the demand at $2 elastic or inelastic? Is it more accurate to say "People want yams and will buy them no matter what the price" or "Yams are a luxury item and people will stop buying them if the price gets too high"?

14. The demand for yams is given in Problem 13.

   (a) At a price of $2 per pound, what is the total revenue for the yam farmer?
   (b) Write revenue as a function of price, and then find the price that maximizes revenue.
   (c) What quantity is sold at the price you found in part (b), and what is the total revenue?
   (d) Show that $E = 1$ at the price you found in part (b).

15. It has been estimated that the elasticity of demand for slaves in the American South before the civil war was equal to 0.86 (fairly high) in the cities and equal to 0.05 (very low) in the countryside.[10]

   (a) Why might this be?
   (b) Where do you think the staunchest defenders of slavery were from, the cities or the countryside?

16. Find the exact price that maximizes revenue for sales of the product in Example 2.

17. If $E = 2$ for all prices p, how can you maximize revenue?

18. If $E = 0.5$ for all prices p, how can you maximize revenue?

19. (a) If the demand equation is $pq = k$ for a positive constant k, compute the elasticity of demand.
    (b) Explain the answer to part (a) in terms of the revenue function.

20. Show that a demand equation $q = k/p^r$, where r is a positive constant, gives constant elasticity $E = r$.

[10]Donald McCloskey, *The Applied Theory of Price*, p. 134, (New York: Macmillan, 1982).

21. A linear demand function is given in Figure 4.66. Economists compute elasticity of demand E for any quantity q_0 using the formula

$$E = d_1/d_2,$$

where d_1 and d_2 are the vertical distances shown in Figure 4.66.

(a) Explain why this formula works.
(b) Determine the prices, p, at which   (i) $E > 1$ (ii) $E < 1$ (iii) $E = 1$

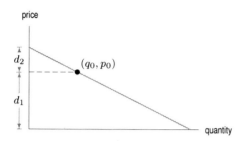

price

d_2

(q_0, p_0)

d_1

quantity

Figure 4.66

22. If p is price and E is the elasticity of demand for a good, show analytically that

$$\text{Marginal revenue} = p(1 - 1/E).$$

23. Suppose cost is proportional to quantity, $C(q) = kq$. Show that a firm earns maximum profit when

$$\frac{\text{Profit}}{\text{Revenue}} = \frac{1}{E}$$

[Hint: Combine the result of Problem 22 with the fact that profit is maximized when $MR = MC$.]

24. Elasticity of cost with respect to quantity is defined as $E_{C,q} = q/C \cdot dC/dq$.

(a) What does this elasticity tell you about sensitivity of cost to quantity produced?
(b) Show that $E_{C,q} = $ Marginal cost/Average cost.

25. If q is the quantity of chicken demanded as a function of the price p of beef, the *cross-price* elasticity of demand for chicken with respect to the price of beef is defined as $E_{\text{cross}} = |p/q \cdot dq/dp|$. What does E_{cross} tell you about the sensitivity of the quantity of chicken bought to changes in the price of beef?

26. The *income* elasticity of demand for a product is defined as $E_{\text{income}} = |I/q \cdot dq/dI|$ where q is the quantity demanded as a function of the income I of the consumer. What does E_{income} tell you about the sensitivity of the quantity of the product purchased to changes in the income of the consumer?

4.7 LOGISTIC GROWTH

In 1923, eighteen koalas were introduced to Kangaroo Island, off the coast of Australia.[11] The koalas thrived on the island and their population grew to about 5000 in 1997. Is it reasonable to expect the population to continue growing exponentially? Since there is only a finite amount of space on the island, the population cannot grow without bound forever. Instead we expect that there is a maximum population that the island can sustain. Population growth with an upper bound can be modeled with a *logistic* or *inhibited growth model*.

Modeling the US Population

Population projections first became important to political philosophers in the late eighteenth century. As concern for scarce resources has grown, so has the interest in accurate population projections. In the US, the population is recorded every ten years by a census. The first such census was in 1790. Table 4.6 contains the census data from 1790 to 2000.

Table 4.6   *US Population,[12] in millions, 1790–2000*

Year	Population	Year	Population	Year	Population	Year	Population
1790	3.9	1850	23.1	1910	92.0	1960	179.3
1800	5.3	1860	31.4	1920	105.7	1970	203.3
1810	7.2	1870	38.6	1930	122.8	1980	226.5
1820	9.6	1880	50.2	1940	131.7	1990	248.7
1830	12.9	1890	62.9	1950	150.7	2000	281.4
1840	17.1	1900	76.0				

[11] *Watertown Daily Times*, April 18, 1997.
[12] *The World Almanac and Book of Facts 2005*, pp. 622–623 (New York).

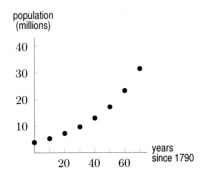

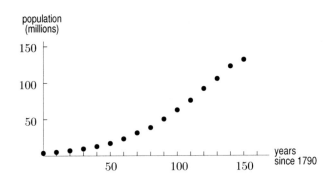

Figure 4.67: US Population, 1790–1860

Figure 4.68: US Population, 1790–1940

Figure 4.67 suggests that the population grew exponentially during the years 1790–1860. However, after 1860 the rate of growth began to decrease. See Figure 4.68.

The Years 1790--1860: An Exponential Model

We begin by modeling the US population for the years 1790–1860 using an exponential function. If t is the number of years since 1790 and P is the population in millions, regression gives the exponential function that fits the data as approximately[13]

$$P = 3.9(1.03)^t.$$

Thus, between 1790 and 1860, the US population was growing at an annual rate of about 3%.

The function $P = 3.9(1.03)^t$ is plotted in Figure 4.69 with the data; it fits the data remarkably well. Of course, since we used the data from throughout the 70-year period, we should expect good agreement throughout that period. What is surprising is that if we had used only the populations in 1790 and 1800 to create our exponential function, the predictions would still be very accurate. It is amazing that a person in 1800 could predict the population 60 years later so accurately, especially when one considers all the wars, recessions, epidemics, additions of new territory, and immigration that took place from 1800 to 1860.

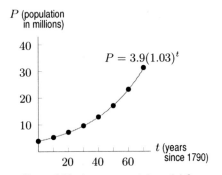

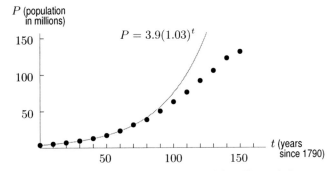

Figure 4.69: An exponential model for
the US population, 1790–1860

Figure 4.70: The exponential model and the US population,
1790-1940. Not a good fit beyond 1860

The Years 1790--1940: A Logistic Model

How well does the exponential function fit the US population beyond 1860? Figure 4.70 shows a graph of the US population from 1790 until 1940 with the exponential function $P = 3.9(1.03)^t$. The exponential function which fit the data so well for the years 1790–1860 does not fit very well beyond 1860. We must look for another way to model this data.

The graph of the function given by the data in Figure 4.68 is concave up for small values of t,

---

[13]See Appendix A: Fitting Formulas to Data. Different algorithms may give different formulas.

but then appears to become concave down and to be leveling off. This kind of growth is modeled with a *logistic function*. If t is in years since 1790, the function

$$P = \frac{187}{1 + 47e^{-0.0318t}},$$

which is graphed in Figure 4.71, fits the data well up to 1940. Such a formula is found by logistic regression on a calculator or computer.[14]

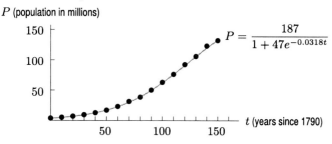

Figure 4.71: A logistic model for US population, 1790–1940

The Logistic Function

A logistic function, such as that used to model the US population, is everywhere increasing. Its graph is concave up at first, then becomes concave down, and levels off at a horizontal asymptote. As we saw in the US population model, a logistic function is approximately exponential for small[15] values of t. A logistic function can be used to model the sales of a new product and the spread of a virus.

> For positive constants L, C, and k, a **logistic function** has the form
>
> $$P = f(t) = \frac{L}{1 + Ce^{-kt}}.$$

The general logistic function has three parameters: L, C, and k. In Example 1, we investigate the effect of two of these parameters on the graph; Problem 5 at the end of the section considers the third.

Example 1   Consider the logistic function $P = \dfrac{L}{1 + 100e^{-kt}}$.
(a) Let $k = 1$. Graph P for several values for L. Explain the effect of the parameter L.
(b) Now let $L = 1$. Graph P for several values for k. Explain the effect of the parameter k.

Solution   (a) See Figure 4.72. Notice that the graph levels off at the value L. The parameter L determines the horizontal asymptote and the upper bound for P.

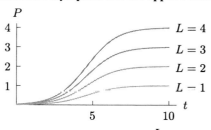
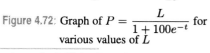

Figure 4.72: Graph of $P = \dfrac{L}{1 + 100e^{-t}}$ for various values of L

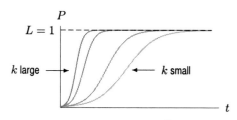

Figure 4.73: Graph of $P = \dfrac{1}{1 + 100e^{-kt}}$ with various values of k

---

[14]See Appendix A: Fitting Formulas to Data.
[15]Just how small is small enough depends on the values of the parameters C and k.

(b) See Figure 4.73. Notice that as k increases, the curve approaches the asymptote more rapidly. The parameter k affects the steepness of the curve.

The Carrying Capacity and the Point of Diminishing Returns

Example 1 suggests that the parameter L of the logistic function is the value at which P levels off, where

$$P = \frac{L}{1 + Ce^{-kt}}.$$

This value L is called the *carrying capacity* and represents the largest population an environment can support.

One way to estimate the carrying capacity is to find the inflection point. The graph of a logistic curve is concave up at first and then concave down. At the inflection point, where the concavity changes, the slope is largest. To the left of this point, the graph is concave up and the rate of growth is increasing. To the right of this point, the graph is concave down and the rate of growth is diminishing. The inflection point is called the *point of diminishing returns*. Problem 49 in the Review Problems shows that this point is at $P = L/2$. See Figure 4.74. Companies sometimes watch for this concavity change in the sales of a new product and use it to estimate the maximum potential sales.

Properties of the logistic function $P = \dfrac{L}{1 + Ce^{-kt}}$:

- The limiting value L represents the carrying capacity for P.
- The point of diminishing returns is the inflection point where P is growing the fastest. It occurs where $P = L/2$.
- The logistic function is approximately exponential for small values of t, with growth rate k.

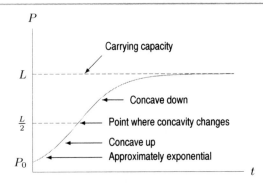

Figure 4.74: Logistic growth

The Years 1790--2000: Another look at the US Population

We used a logistic function to model the US population between 1790 and 1940. How well does this model fit the US population since 1940? We now look at all the population data from 1790 to 2000.

Example 2   If t is in years since 1790 and P is in millions, we used the following logistic function to model the US population between 1790 and 1940:

$$P = \frac{187}{1 + 47e^{-0.0318t}}.$$

According to this function, what is the maximum US population? Is this prediction accurate? How well does this logistic model fit the growth of the US population since 1940?

Table 4.7   *Predicted versus actual US population, in millions, 1940–2000 (logistic model)*

Year	1940	1950	1960	1970	1980	1990	2000
Actual	131.7	150.7	179.3	203.3	226.5	248.7	281.4
Predicted	133.7	145.0	154.4	162.1	168.2	172.9	176.6

Solution   Table 4.7 shows the actual US population between 1940 and 2000 and the predicted values using this logistic model. According to the formula for the logistic function, the upper bound for the population is $L = 187$ million. However, Table 4.7 shows that the actual US population was above this figure by 1970. The fit between the logistic function and the actual population is not a good one beyond 1940. See Figure 4.75.

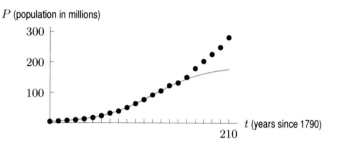

Figure 4.75: The logistic model and the US population, 1790–2000

Despite World War II, which depressed population growth between 1942 and 1945, in the last half of the 1940s the US population surged. The 1950s saw a population growth of 28 million, leaving our logistic model in the dust. This surge in population is referred to as the baby boom.

Once again we have reached a point where our model is no longer useful. This should not lead you to believe that a reasonable mathematical model cannot be found; rather it points out that no model is perfect and that when one model fails, we seek a better one. Just as we abandoned the exponential model in favor of the logistic model for the US population, we could look further.

Sales Predictions

Total sales of a new product often follow a logistic model. For example, when a new compact disc (CD) appears on the market, sales first increase rapidly as word of the CD spreads. Eventually, most of the people who want the CD have already bought it and sales slow down. The graph of total sales against time is concave up at first and then concave down, with the upper bound L equal to the maximum potential sales.

Example 3   Table 4.8 shows the total sales (in thousands) of a new CD since it was introduced.

(a) Find the point where concavity changes in this function. Use it to estimate the maximum potential sales, L.

(b) Using logistic regression, fit a logistic function to this data. What maximum potential sales does this function predict?

Table 4.8   *Total sales of a new CD since its introduction*

t (months)	0	1	2	3	4	5	6	7
P (total sales in 1000s)	0.5	2	8	33	95	258	403	496

Solution

(a) The rate of change of total sales increases until $t = 5$ and decreases after $t = 5$, so the inflection point is at approximately $t = 5$, when $P = 258$. So $L/2 = 258$ and $L = 516$. The maximum potential sales for this CD are estimated to be 516,000.

(b) Logistic regression gives the following function:

$$P = \frac{532}{1 + 869e^{-1.33t}}.$$

Maximum potential sales predicted by this function are $L = 532$, or about 532,000 CDs. See Figure 4.76.

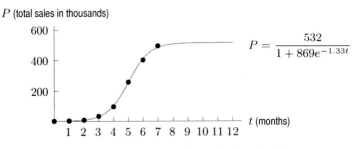

Figure 4.76: Logistic growth: Total sales of a CD

Dose-Response Curves

A *dose-response curve* plots the intensity of physiological response to a drug as a function of the dose administered. As the dose increases, the intensity of the response increases, so a dose-response function is increasing. The intensity of the response is generally scaled as a percentage of the maximum response. The curve cannot go above the maximum response (or 100%), so the curve levels off at a horizontal asymptote. Dose-response curves are generally concave up for low doses and concave down for high doses. A dose-response curve can be modeled by a logistic function with the independent variable being the dose of the drug, not time.

A dose-response curve shows the amount of drug needed to produce the desired effect, as well as the maximum effect attainable and the dose required to obtain it. The slope of the dose-response curve gives information about the therapeutic safety margin of the drug.

Drugs need to be administered in a dose which is large enough to be effective but not so large as to be dangerous. Figure 4.77 shows two different dose-response curves: one with a small slope and one with a large slope. In Figure 4.77(a), there is a broad range of dosages at which the drug is both safe and effective. In Figure 4.77(b), where the slope of the curve is steep, the range of dosages at which the drug is both safe and effective is small. If the slope of the dose-response curve is steep, a small mistake in the dosage can have dangerous results. Administration of such a drug is difficult.

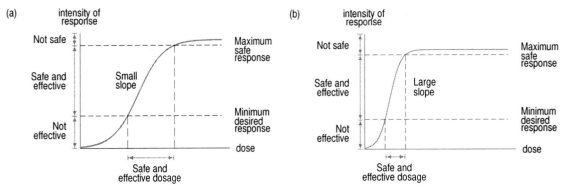

Figure 4.77: What does the slope of the dose-response curve tell us?

Example 4 Figure 4.78 shows dose-response curves for three different drugs used for the same purpose. Discuss the advantages and disadvantages of the three drugs.

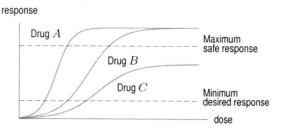

Figure 4.78: What are the advantages and disadvantages of each of these drugs?

Solution Drugs A and B exhibit the same maximum response, while the maximum response of Drug C is significantly less; however all three drugs reach the minimum desired response. The potency of Drugs B and C (the dose required to reach desired effect) is significantly less than the potency of Drug A. (Potency, however, is a relatively unimportant characteristic of a drug, since a less potent drug can simply be given in larger doses.) Drug A has a steeper slope than either of the other two. Both Drugs A and B can exceed the maximum safe response. Thus, Drug C may be the preferred drug despite its lower maximum effect because it is the safest to administer.

Problems for Section 4.7

1. If t is in years since 1990, one model for the population of the world, P, in billions, is

$$P = \frac{40}{1 + 11e^{-0.08t}}.$$

(a) What does this model predict for the maximum sustainable population of the world?
(b) Graph P against t.
(c) According to this model, when will the earth's population reach 20 billion? 39.9 billion?

2. The rate of sales of an automobile anti-theft device are given in the following table.

(a) When is the point of diminishing returns reached?
(b) What are the total sales at this point?
(c) Assuming logistic sales growth, use your answer to part (b) to estimate total potential sales of the device.

Months	1	2	3	4	5	6
Sales per month	140	520	680	750	700	550

3. The following table shows the total sales, in thousands, since a new game was brought to market.

(a) Plot this data and mark on your plot the point of diminishing returns.
(b) Predict total possible sales of this game, using the point of diminishing returns.

Month	0	2	4	6	8	10	12	14
Sales	0	2.3	5.5	9.6	18.2	31.8	42.0	50.8

4. A rumor spreads among a group of 400 people. The number of people, $N(t)$, who have heard the rumor by time t in hours since the rumor started to spread can be approximated by a function of the form

$$N(t) = \frac{400}{1 + 399e^{-0.4t}}.$$

(a) Find $N(0)$ and interpret it.
(b) How many people will have heard the rumor after 2 hours? After 10 hours?
(c) Graph $N(t)$.
(d) Approximately how long will it take until half the people have heard the rumor? Virtually everyone?
(e) Approximately when is the rumor spreading fastest?

5. Investigate the effect of the parameter C on the logistic curve

$$P = \frac{10}{1 + Ce^{-t}}.$$

Substitute several values for C and explain, with a graph and with words, the effect of C on the graph.

6. Write a paragraph explaining why sales of a new product often follow a logistic curve. Explain the benefit to the company of watching for the point of diminishing returns.

7. Figure 4.79 shows the spread of the Code-red computer virus during July 2001. Most of the growth took place starting at midnight on July 19; on July 20, the virus attacked the White House, trying (unsuccessfully) to knock its site off-line. The number of computers infected by the virus is a logistic function of time.

(a) Estimate the limiting value of $f(t)$ as t increases. What does this limiting value represent in terms of Code-red?

(b) Estimate the value of t at which $f''(t) = 0$. Estimate the value of n at this time.

(c) What does the answer to part (b) tell us about Code-red?

(d) How are the answers to parts (a) and (b) related?

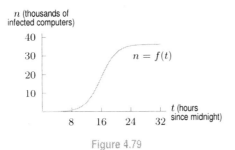

Figure 4.79

8. Find the exact coordinates of the point at which the following curve is steepest:

$$y = \frac{50}{1 + 6e^{-2t}} \qquad \text{for } t \geq 0.$$

9. (a) Draw a logistic curve. Label the carrying capacity L and the point of diminishing returns t_0.

(b) Draw the derivative of the logistic curve. Mark the point t_0 on the horizontal axis.

(c) A company keeps track of the rate of sales (for example, sales per week) rather than total sales. Explain how the company can tell on a graph of rate of sales when the point of diminishing returns is reached.

10. The Tojolobal Mayan Indian community in Southern Mexico has available a fixed amount of land.[16] The proportion, P, of land in use for farming t years after 1935 is modeled with the logistic function

$$P = \frac{1}{1 + 3e^{-0.0275t}}.$$

(a) What proportion of the land was in use for farming in 1935?

(b) What is the long-run prediction of this model?

(c) When was half the land in use for farming?

(d) When is the proportion of land used for farming increasing most rapidly?

11. In the spring of 2003, SARS (Severe Acute Respiratory Syndrome) spread rapidly in several Asian countries and Canada. Table 4.9 gives the total number, P, of SARS cases reported in Hong Kong[17] by day t, where $t = 0$ is March 17, 2003.

(a) Find the average rate of change of P for each interval in Table 4.9.

(b) In early April 2003, there was fear that the disease would spread at an ever-increasing rate for a long time. What is the earliest date by which epidemiologists had evidence to indicate that the rate of new cases had begun to slow?

(c) Explain why an exponential model for P is not appropriate.

(d) It turns out that a logistic model fits the data well. Estimate the value of t at the inflection point. What limiting value of P does this point predict?

(e) The best-fitting logistic function for this data turns out to be

$$P = \frac{1760}{1 + 17.53e^{-0.1408t}}.$$

What limiting value of P does this function predict?

Table 4.9 *Total number of SARS cases in Hong Kong by day t (where t = 0 is March 17, 2003)*

t	P	t	P	t	P	t	P
0	95	26	1108	54	1674	75	1739
5	222	33	1358	61	1710	81	1750
12	470	40	1527	68	1724	87	1755
19	800	47	1621				

12. Substitute $t = 0, 10, 20, \ldots, 70$ into the exponential function used in this section to model the US population 1790–1860. Compare the predicted values of the population with the actual values.

13. On page 209, a logistic function was used to model the US population. Use this function to predict the US population in each of the census years from 1790–1940. Compare the predicted and actual values.

---

[16]Adapted from J. S. Thomas and M. C. Robbins, "The Limits to Growth in a Tojolobal Maya Ejido," *Geoscience and Man 26*, pp. 9–16 (Baton Rouge: Geoscience Publications, 1988).

[17]www.who.int/csr/country/en, accessed July 13, 2003.

14. A curve representing the total number of people, P, infected with a virus often has the shape of a logistic curve of the form

$$P = \frac{L}{1 + Ce^{-kt}},$$

with time t in weeks. Suppose that 10 people originally have the virus and that in the early stages the number of people infected is increasing approximately exponentially, with a continuous growth rate of 1.78. It is estimated that, in the long run, approximately 5000 people will become infected.

(a) What should we use for the parameters k and L?
(b) Use the fact that when $t = 0$, we have $P = 10$, to find C.
(c) Now that you have estimated L, k, and C, what is the logistic function you are using to model the data? Graph this function.
(d) Estimate the length of time until the rate at which people are becoming infected starts to decrease. What is the value of P at this point?

15. If R is percent of maximum response and x is dose in mg, the dose-response curve for a drug is given by

$$R = \frac{100}{1 + 100e^{-0.1x}}.$$

(a) Graph this function.
(b) What dose corresponds to a response of 50% of the maximum? This is the inflection point, at which the response is increasing the fastest.
(c) For this drug, the minimum desired response is 20% and the maximum safe response is 70%. What range of doses is both safe and effective for this drug?

16. Dose-response curves for three different products are given in Figure 4.80.

(a) For the desired response, which drug requires the largest dose? The smallest dose?
(b) Which drug has the largest maximum response? The smallest?
(c) Which drug is the safest to administer? Explain.

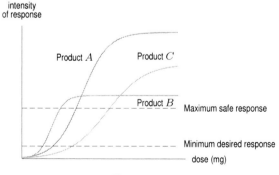

Figure 4.80

17. A dose-response curve is given by $R = f(x)$, where R is percent of maximum response and x is the dose of the drug in mg. The curve has the shape shown in Figure 4.77. The inflection point is at $(15, 50)$ and $f'(15) = 11$.

(a) Explain what $f'(15)$ tells you in terms of dose and response for this drug.
(b) Is $f'(10)$ greater than or less than 11? Is $f'(20)$ greater than or less than 11? Explain.

18. Explain why it is safer to use a drug for which the derivative of the dose-response curve is smaller.

There are two kinds of dose-response curves. One type, discussed in this section, plots the intensity of response against the dose of the drug. We now consider a dose-response curve in which the percentage of subjects showing a specific response is plotted against the dose of the drug. In Problems 19–20, the curve on the left shows the percentage of subjects exhibiting the desired response at the given dose, and the curve on the right shows the percentage of subjects for which the given dose is lethal.

19. In Figure 4.81, what range of doses appears to be both safe and effective for 99% of all patients?

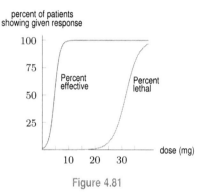

Figure 4.81

20. In Figure 4.82, discuss the possible outcomes and what percent of patients fall in each outcome when 50 mg of the drug is administered.

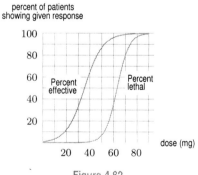

Figure 4.82

21. A population, P, growing logistically is given by

$$P = \frac{L}{1 + Ce^{-kt}}.$$

(a) Show that

$$\frac{L - P}{P} = Ce^{-kt}.$$

(b) Explain why part (a) shows that the ratio of the additional population the environment can support to the existing population decays exponentially.

22. Cell membranes contain ion channels. The fraction, f, of channels that are open is a function of the membrane potential V (the voltage inside the cell minus voltage outside), in millivolts (mV), given by

$$f(V) = \frac{1}{1 + e^{-(V+25)/2}}.$$

(a) Find the values of L, k, and C in the logistic formula for f:

$$f(V) = \frac{L}{1 + Ce^{-kV}}.$$

(b) At what voltages V are 10%, 50% and 90% of the channels open?

4.8 THE SURGE FUNCTION AND DRUG CONCENTRATION

Nicotine in the Blood

When a person smokes a cigarette, the nicotine from the cigarette enters the body through the lungs, is absorbed into the blood, and spreads throughout the body. Most cigarettes contain between 0.5 and 2.0 mg of nicotine; approximately 20% (between 0.1 and 0.4 mg) is actually inhaled and absorbed into the person's bloodstream. As the nicotine leaves the blood, the smoker feels the need for another cigarette. The half-life of nicotine in the bloodstream is about two hours. The lethal dose is considered to be about 60 mg.

The nicotine level in the blood rises as a person smokes, and tapers off when smoking ceases. Table 4.10 shows blood nicotine concentration (in ng/ml) during and after the use of cigarettes. (Smoking occurred during the first ten minutes and the experimental data shown represent average values for ten people.)[18]

The points in Table 4.10 are plotted in Figure 4.83. Functions with this behavior are called *surge functions*. They have equations of the form $y = ate^{-bt}$, where a and b are positive constants.

Table 4.10   *Blood nicotine concentrations during and after the use of cigarettes*

t (minutes)	0	5	10	15	20	25	30	45	60	75	90	105	120
C (ng/ml)	4	12	17	14	13	12	11	9	8	7.5	7	6.5	6

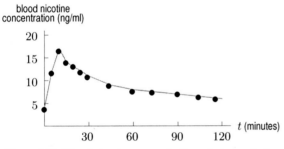

Figure 4.83: Blood nicotine concentrations during and after the use of cigarettes

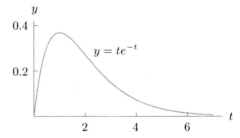

Figure 4.84: One member of the family $y = ate^{-bt}$, with $a = 1$ and $b = 1$

The Family of Functions $y = ate^{-bt}$

What effect do the parameters a and b have on the shape of the graph of $y = ate^{-bt}$? Start by looking at the graph with $a = 1$ and $b = 1$. See Figure 4.84. We consider the effect of the parameter b on the graph of $y = ate^{-bt}$ now; the parameter a is considered in Problem 2 of this section.

---

[18]Benowitz, Porched, Skeiner, Jacog, "Nicotine Absorption and Cardiovascular Effects with Smokeless Tobacco Use: Comparison with Cigarettes and Nicotine Gum," *Clinical Pharmacology and Therapeutics* 44 (1988): 24.

The Effect of the Parameter b on $y = te^{-bt}$

Graphs of $y = te^{-bt}$ for different positive values of b are shown in Figure 4.85. The general shape of the curve does not change as b changes, but as b decreases, the curve rises for a longer period of time and to a higher value.

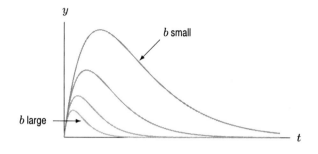

Figure 4.85: Graph of $y = te^{-bt}$, with b varying

Figure 4.86: How does the maximum depend on b?

We see in Figure 4.86 that, when $b = 1$, the maximum occurs at about $t = 1$. When $b = 2$, it occurs at about $t = \frac{1}{2}$, and when $b = 3$, it occurs at about $t = \frac{1}{3}$. The next example shows that the maximum of the function $y = te^{-bt}$ occurs at $t = 1/b$.

Example 1   For $b > 0$, show that the maximum value of $y = te^{-bt}$ occurs at $t = 1/b$ and increases as b decreases.

Solution   The maximum occurs at a critical point where $dy/dt = 0$. Differentiating gives

$$\frac{dy}{dt} = 1 \cdot e^{-bt} + t\left(-be^{-bt}\right) = e^{-bt} - bte^{-bt} = e^{-bt}(1 - bt).$$

So $dy/dt = 0$ where

$$1 - bt = 0$$
$$t = \frac{1}{b}$$

Substituting $t = 1/b$ shows that at the maximum,

$$y = \frac{1}{b}e^{-b(1/b)} = \frac{e^{-1}}{b}.$$

So, for $b > 0$, as b increases, the maximum value of y decreases and vice versa.

> The **surge function** $y = ate^{-bt}$ increases rapidly and then decreases toward zero with a maximum at $t = 1/b$.

Drug Concentration Curves

When the concentration, C, of a drug in the body is plotted against the time, t, since the drug was administered, the curve generally has the shape shown in Figure 4.87. This is called a *drug concentration curve*, and is modeled using a function of the form $C = ate^{-bt}$. Figure 4.87 shows the peak concentration (the maximum concentration of the drug in the body) and the length of time until peak concentration is reached.

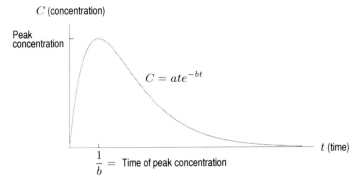

Figure 4.87: Curve showing drug concentration as a function of time

Factors Affecting Drug Absorption

Drug interactions and the age of the patient can affect the drug concentration curve. In Problems 11 and 13, we see that food intake can also affect the rate of absorption of a drug, and (perhaps most surprising) that drug concentration curves can vary markedly between different commercial versions of the same drug.

Example 2   Figure 4.88 shows the drug concentration curves for paracetamol (acetaminophen) alone and for paracetamol taken in conjunction with propantheline. Figure 4.89 shows drug concentration curves for patients known to be slow absorbers of the drug, for paracetamol alone and for paracetamol in conjunction with metoclopramide. Discuss the effects of the additional drugs on peak concentration and the time to reach peak concentration.[19]

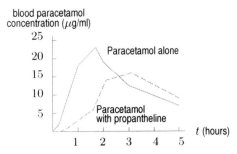

Figure 4.88: Drug concentration curves for paracetamol, normal patients

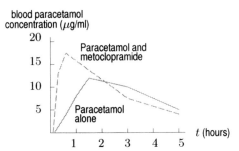

Figure 4.89: Drug concentration curves for paracetamol, patients with slow absorption

Solution   Figure 4.88 shows it takes about 1.5 hours for the paracetamol to reach its peak concentration, and that the maximum concentration reached is about 23 μg of paracetamol per ml of blood. However, if propantheline is administered with the paracetamol, it takes much longer to reach the peak concentration (about three hours, or approximately double the time), and the peak concentration is much lower, at about 16 μg/ml.

Comparing the curves for paracetamol alone in Figures 4.88 and 4.89 shows that the time to reach peak concentration is the same (about 1.5 hours), but the maximum concentration is lower for patients with slow absorption. When metoclopramide is given with paracetamol in Figure 4.89, the peak concentration is reached faster and is higher.

[19]Graeme S. Avery, ed. *Drug Treatment: Principle and Practice of Clinical Pharmacology and Therapeutics* (Sydney: Adis Press, 1976).

Minimum Effective Concentration

The minimum effective concentration of a drug is the blood concentration necessary to achieve a pharmacological response. The time at which this concentration is reached is referred to as onset; termination occurs when the drug concentration falls below this level. See Figure 4.90.

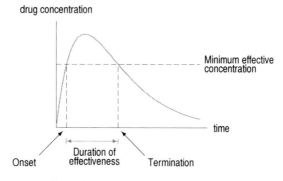

Figure 4.90: When is the drug effective?

Example 3 Depo-Provera was approved for use in the US in 1992 as a contraceptive. Figure 4.91 shows the drug concentration curve for a dose of 150 mg given intramuscularly.[20] The minimum effective concentration is about 4 ng/ml. How often should the drug be administered?

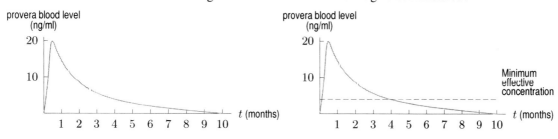

Figure 4.91: Drug concentration curve for Depo-Provera   Figure 4.92: When should the next dose be administered?

Solution The minimum effective concentration on the drug concentration curve is plotted as a dotted horizontal line at 4 ng/ml. See Figure 4.92. We see that the drug becomes effective almost immediately and ceases to be effective after about four months. Doses should be given about every four months.

Although the dosage interval is four months, notice that it takes ten months after injections are discontinued for Depo-Provera to be entirely eliminated from the body. Fertility during that period is unpredictable.

Problems for Section 4.8

1. If time, t, is in hours and concentration, C, is in ng/ml, the drug concentration curve for a drug is given by

$$C = 12.4te^{-0.2t}.$$

(a) Graph this curve.

(b) How many hours does it take for the drug to reach its peak concentration? What is the concentration at that time?

(c) If the minimum effective concentration is 10 ng/ml, during what time period is the drug effective?

(d) Complications can arise whenever the level of the drug is above 4 ng/ml. How long must a patient wait before being safe from complications?

2. Let $b = 1$, and graph $C = ate^{-bt}$ using different values for a. Explain the effect of the parameter a.

3. Figure 4.93 shows drug concentration curves for anhydrous ampicillin for newborn babies and adults.[21] Discuss the differences between newborns and adults in the absorption of this drug.

[20] Robert M. Julien, *A Primer of Drug Action* (W. H. Freeman and Co, 1995).
[21] *Pediatrics*, 1973, 51, 578.

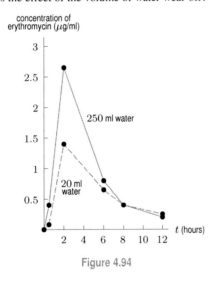

drug concentration
(μg/ml)

Newborns

Adults

t (hours)

Figure 4.93

4. If t is in hours, the drug concentration curve for a drug is given by $C = 17.2te^{-0.4t}$ ng/ml. The minimum effective concentration is 10 ng/ml.

(a) If the second dose of the drug is to be administered when the first dose becomes ineffective, when should the second dose be given?

(b) If you want the onset of effectiveness of the second dose to coincide with termination of effectiveness of the first dose, when should the second dose be given?

5. Absorption of different forms of the antibiotic erythromycin may be increased, decreased, delayed or not affected by food. Figure 4.94 shows the drug concentration levels of erythromycin in healthy, fasting human volunteers who received single oral doses of 500 mg erythromycin tablets, together with either large (250 ml) or small (20 ml) accompanying volumes of water.[22] Discuss the effect of the water on the concentration of erythromycin in the blood. How are the peak concentration and the time to reach peak concentration affected? When does the effect of the volume of water wear off?

concentration of
erythromycin (μg/ml)

250 ml water

20 ml
water

t (hours)

Figure 4.94

6. Hydrocodone bitartrate is a cough suppressant usually administered in a 10 mg oral dose. The peak concentration of the drug in the blood occurs 1.3 hours after consumption and the peak concentration is 23.6 ng/ml. Draw the drug concentration curve for hydrocodone bitartrate.

7. Figure 4.83 shows the concentration of nicotine in the blood during and after smoking a cigarette. Figure 4.95 shows the concentration of nicotine in the blood during and after using chewing tobacco or nicotine gum. (The chewing occurred during the first 30 minutes and the experimental data shown represent the average values for ten patients.)[23] Compare the three nicotine concentration curves (for cigarettes, chewing tobacco and nicotine gum) in terms of peak concentration, the time until peak concentration, and the rate at which the nicotine is eliminated from the bloodstream.

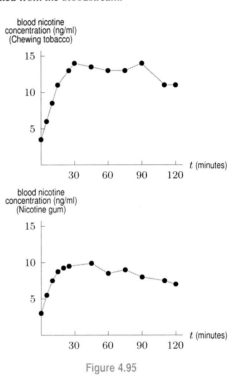

blood nicotine
concentration (ng/ml)
(Chewing tobacco)

t (minutes)

blood nicotine
concentration (ng/ml)
(Nicotine gum)

t (minutes)

Figure 4.95

8. If t is in minutes since the drug was administered, the concentration, $C(t)$ in ng/ml, of a drug in a patient's bloodstream is given by

$$C(t) = 20te^{-0.03t}.$$

(a) How long does it take for the drug to reach peak concentration? What is the peak concentration?

(b) What is the concentration of the drug in the body after 15 minutes? After an hour?

(c) If the minimum effective concentration is 10 ng/ml, when should the next dose be administered?

---

[22]J. W. Bridges and L.F. Chasseaud, *Progress in Drug Metabolism* (New York: John Wiley and Sons, 1980).

[23]Benowitz, Porchet, Skeiner, Jacob, "Nicotine Absorption and Cardiovascular Effects with Smokeless Tobacco Use: Comparison with Cigarettes and Nicotine Gum," *Clinical Pharmacology and Therapeutics* 44 (1988): 24.

9. For time $t \geq 0$, the function $C = ate^{-bt}$ with positive constants a and b gives the concentration, C, of a drug in the body. Figure 4.96 shows the maximum concentration reached (in nanograms per milliliter, ng/ml) after a 10 mg dose of the cough medicine hydrocodone bitartrate. Find the values of a and b.

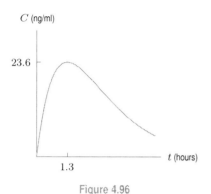

Figure 4.96

10. The method of administering a drug can have a strong influence on the drug concentration curve. Figure 4.97 shows drug concentration curves for penicillin following various routes of administration. Three milligrams per kilogram of body weight were dissolved in water and administered intravenously (IV), intramuscularly (IM), subcutaneously (SC), and orally (PO). The same quantity of penicillin dissolved in oil was administered intramuscularly (P-IM). The minimum effective concentration (MEC) is labeled on the graph.[24]

(a) Which method reaches peak concentration the fastest? The slowest?
(b) Which method has the largest peak concentration? The smallest?
(c) Which method wears off the fastest? The slowest?
(d) Which method has the longest effective duration? The shortest?
(e) When penicillin is administered orally, for approximately what time interval is it effective?

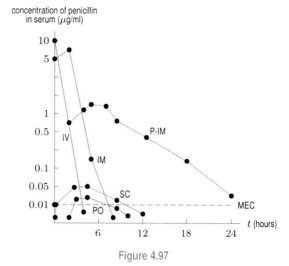

Figure 4.97

11. Figure 4.98 shows the plasma levels of canrenone in a healthy volunteer after a single oral dose of spironolactone given on a fasting stomach and together with a standardized breakfast. (Spironolactone is a diuretic agent that is partially converted into canrenone in the body.)[25] Discuss the effect of food on peak concentration and time to reach peak concentration. Is the effect of the food strongest during the first 8 hours, or after 8 hours?

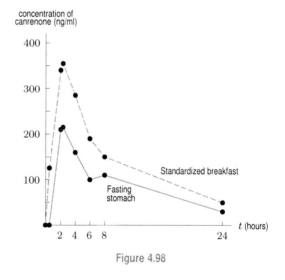

Figure 4.98

12. This problem shows how a surge can be modeled with a difference of exponential decay functions.

(a) Using graphs of e^{-t} and e^{-2t}, explain why the graph of $f(t) = e^{-t} - e^{-2t}$ has the shape of a surge.
(b) Find the critical point and inflection point of f.

[24] J. W. Bridges and L. F. Chasseaud, *Progress in Drug Metabolism* (New York: John Wiley and Sons, 1980).
[25] Welling & Tse, *Pharmacokinetics of Cardiovascular, Central Nervous System, and Antimicrobial Drugs* (The Royal Society of Chemistry, 1985).

13. Figure 4.99 shows drug concentration curves after oral administration of 0.5 mg of four digoxin products. All the tablets met current USP standards of potency, disintegration time, and dissolution rate.[26]

(a) Discuss differences and similarities in the peak concentration and the time to reach peak concentration.

(b) Give possible values for minimum effective concentration and maximum safe concentration that would make Product C or Product D the preferred drug.

(c) Give possible values for minimum effective concentration and maximum safe concentration that would make Product A the preferred drug.

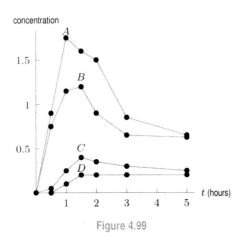

Figure 4.99

14. Figure 4.100 shows a graph of the percentage of drug dissolved against time for four tetracycline products A, B, C, and D. Figure 4.101 shows the drug concentration curves for the same four tetracycline products.[27] Discuss the effect of dissolution rate on peak concentration and time to reach peak concentration.

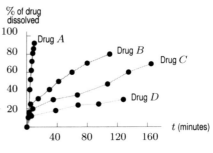

Figure 4.100

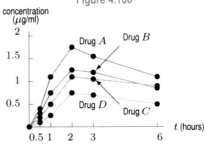

Figure 4.101

CHAPTER SUMMARY

- **Using the first derivative**
  Critical points, local maxima and minima

- **Using the second derivative**
  Inflection points, concavity

- **Optimization**
  Global maxima and minima

- **Maximizing profit and revenue**

- **Average cost**
  Minimizing average cost

- **Elasticity**

- **Families of functions**
  Parameters. The surge function, drug concentration curves. The logistic function, carrying capacity, point of diminishing returns.

REVIEW PROBLEMS FOR CHAPTER FOUR

For Problems 1–2, indicate all critical points on the given graphs. Determine which correspond to local minima, local maxima, global minima, global maxima, or none of these. (Note that the graphs are on closed intervals.)

1.
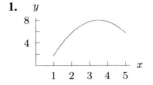

2.
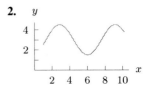

[26]Graeme S. Avery, ed. *Drug Treatment: Principles and Practice of Clinical Pharmacology and Therapeutics* (Sydney: Adis Press, 1976).

[27]J. W. Bridges and L.F. Chasseaud, *Progress in Drug Metabolism* (New York: John Wiley and Sons, 1980).

In Problems 3–4, find the value(s) of x for which:

(a) $f(x)$ has a local maximum or local minimum. Indicate which ones are maxima and which are minima.

(b) $f(x)$ has a global maximum or global minimum.

3. $f(x) = x^{10} - 10x$, and $0 \le x \le 2$

4. $f(x) = x - \ln x$, and $0.1 \le x \le 2$

In Problems 5–7, find the exact global maximum and minimum values of the function.

5. $h(z) = \dfrac{1}{z} + 4z^2$ for $z > 0$

6. $g(t) = \dfrac{1}{t^3 + 1}$ for $t \ge 0$

7. $f(x) = \dfrac{1}{(x - 1)^2 + 2}$

8. On July 1, the price of a stock had a critical point. How could the price have been changing during the time around July 1?

9. Let $C = ate^{-bt}$ represent a drug concentration curve.

(a) Discuss the effect on peak concentration and time to reach peak concentration of varying the parameter a while keeping b fixed.

(b) Discuss the effect on peak concentration and time to reach peak concentration of varying the parameter b while keeping a fixed.

(c) Suppose $a = b$, so $C = ate^{-at}$. Discuss the effect on peak concentration and time to reach peak concentration of varying the parameter a.

For the graphs of f' in Problems 10–13, decide:

(a) Over what intervals is f increasing? Decreasing?

(b) Does f have local maxima or minima? If so, which, and where?

10.
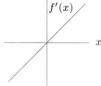

11.
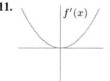

12.
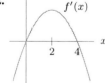

13.
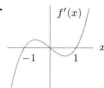

14. The marginal revenue and marginal cost for a certain item are graphed in Figure 4.102. Do the following quantities maximize profit for the company? Explain your answer.

(a) $q = a$ \qquad\qquad (b) $q = b$

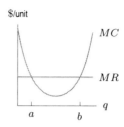

Figure 4.102

Problems 15–18 concern $f(t)$ in Figure 4.103, which gives the length of a human fetus as a function of its age.

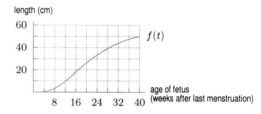

Figure 4.103

15. (a) What are the units of $f'(24)$?

(b) What is the biological meaning of $f'(24) = 1.6$?

16. (a) Which is greater, $f'(20)$ or $f'(36)$?

(b) What does your answer say about fetal growth?

17. (a) At what time does the inflection point occur?

(b) What is the biological significance of this point?

18. Estimate

(a) $f'(20)$ \qquad\qquad (b) $f'(36)$

(c) The average rate of change of length over the 40 weeks shown.

19. Find constants a and b in the function $f(x) = axe^{bx}$ such that $f(\frac{1}{3}) = 1$ and the function has a local maximum at $x = \frac{1}{3}$.

20. Suppose f has a continuous derivative. From the values of $f'(\theta)$ in the following table, estimate the θ values with $1 < \theta < 2.1$ at which $f(\theta)$ has a local maximum or minimum. Identify which is which.

θ	1.0	1.1	1.2	1.3	1.4	1.5
$f'(\theta)$	2.4	0.3	−2.0	−3.5	−3.3	−1.7
θ	1.6	1.7	1.8	1.9	2.0	2.1
$f'(\theta)$	0.8	2.8	3.6	2.8	0.7	−1.6

21. (a) Find the derivative of $f(x) = x^5 + x + 7$. What is its sign?
    (b) How many real roots does the equation $x^5 + x + 7 = 0$ have? How do you know?
    [Hint: How many critical points does this function have?]

22. For the function, f, graphed in Figure 4.104:

    (a) Sketch $f'(x)$.
    (b) Where does $f'(x)$ change its sign?
    (c) Where does $f'(x)$ have local maxima or minima?

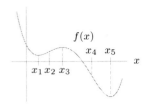

Figure 4.104

23. Using your answer to Problem 22 as a guide, write a short paragraph (using complete sentences) which describes the relationships between the following features of a function f:

    • The local maxima and minima of f.
    • The points at which the graph of f changes concavity.
    • The sign changes of f'.
    • The local maxima and minima of f'.

24. The function $y = t(x)$ is positive and continuous with a global maximum at the point $(3, 3)$. Graph $t(x)$ if $t'(x)$ and $t''(x)$ have the same sign for $x < 3$, but opposite signs for $x > 3$.

25. Each of the graphs in Figure 4.105 belongs to one of the following families of functions. In each case, identify which family is most likely:

    an exponential function,

    a logarithmic function,

    a polynomial (What is the degree? Is the leading coefficient positive or negative?),

    a periodic function,

    a logistic function,

    a surge function.

(a)   (b)

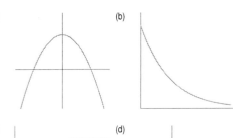

(c)   (d)

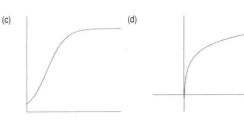

(e)   (f)

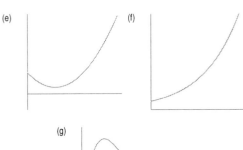

(g)

Figure 4.105

26. The total cost of producing q units of a product is given by $C(q) = q^3 - 60q^2 + 1400q + 1000$ for $0 \le q \le 50$; the product sells for \$788 per unit. What production level maximizes profit? Find the total cost, total revenue, and total profit at this production level. Graph the cost and revenue functions on the same axes, and label the production level at which profit is maximized, and the corresponding cost, revenue, and profit. [Hint: Costs can go as high as \$46,000.]

27. The total cost $C(q)$ of producing q goods is given by:

$$C(q) = 0.01q^3 - 0.6q^2 + 13q.$$

    (a) What is the fixed cost?
    (b) What is the maximum profit if each item is sold for \$7? (Assume you sell everything you produce.)
    (c) Suppose exactly 34 goods are produced. They all sell when the price is \$7 each, but for each \$1 increase in price, 2 fewer goods are sold. Should the price be raised, and if so by how much?

28. The demand equation for a product is $p = b_1 - a_1 q$ and the cost function is $C(q) = b_2 + a_2 q$, where p is the price of the product and q is the quantity sold. Find the value of q, in terms of the positive constants b_1, a_1, b_2, a_2, that maximizes profit.

29. A manufacturer's cost of producing a product is given in Figure 4.106. The manufacturer can sell the product for a price p each (regardless of the quantity sold), so that the total revenue from selling a quantity q is $R(q) = pq$.

(a) The difference $\pi(q) = R(q) - C(q)$ is the total profit. For which quantity q_0 is the profit a maximum? Mark your answer on a sketch of the graph.

(b) What is the relationship between p and $C'(q_0)$? Explain your result both graphically and analytically. What does this mean in terms of economics? (Note that p is the slope of the line $R(q) = pq$. Note also that $\pi(q)$ has a maximum at $q = q_0$, so $\pi'(q_0) = 0$.)

(c) Graph $C'(q)$ and p (as a horizontal line) on the same axes. Mark q_0 on the q-axis.

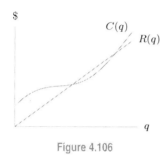

Figure 4.106

30. Let $C(q) = 0.04q^3 - 3q^2 + 75q + 96$ be the total cost of producing q items.

(a) Find the average cost per item as a function of q.

(b) Use a graphing calculator or computer to graph average cost against q.

(c) For what values of q is the average cost per item decreasing? Increasing?

(d) For what value of q is the average cost per item smallest? What is the smallest average cost per item at that point?

31. Graph a cost function where the minimum average cost of $25 per unit is achieved by producing 15,000 units.

32. Figure 4.107 shows the cost, $C(q)$, and the revenue, $R(q)$, for a quantity q. Label the following points on the graph:

(a) The point F representing the fixed costs.

(b) The point B representing the break-even level of production.

(c) The point M representing the level of production at which marginal cost is a minimum.

(d) The point A representing the level of production at which average cost $a(q) = C(q)/q$ is a minimum.

(e) The point P representing the level of production at which profit is a maximum.

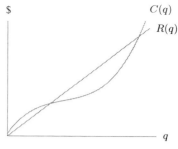

Figure 4.107

In Problems 33–38, cost, $C(q)$, is a positive, increasing, concave down function of quantity produced, q. Which one of the two numbers is the larger?

33. $C'(2)$ and $C'(3)$

34. $C'(5)$ and $\dfrac{C(5) - C(3)}{5 - 3}$

35. $\dfrac{C(100) - C(50)}{50}$ and $\dfrac{C(75) - C(50)}{25}$

36. $\dfrac{C(100) - C(50)}{50}$ and $\dfrac{C(75) - C(25)}{50}$

37. $C'(3)$ and $C(3)/3$

38. $C(10)/10$ and $C(25)/25$

39. The demand for a product is $q = 2000 - 5p$ where q is units sold at a price of p dollars. Find the elasticity if the price is $20, and interpret your answer in terms of demand.

40. Show analytically that if elasticity of demand satisfies $E > 1$, then the derivative of revenue with respect to price satisfies $dR/dp < 0$.

41. Show analytically that if elasticity of demand satisfies $E < 1$, then the derivative of revenue with respect to price satisfies $dR/dp > 0$.

42. The following table gives the percentage, P, of households with cable television.[28]

Year	1977	1978	1979	1980	1981	1982	1983
P	16.6	17.9	19.4	22.6	28.3	35.0	40.5
Year	1984	1985	1986	1987	1988	1989	1990
P	43.7	46.2	48.1	50.5	53.8	57.1	59.0
Year	1991	1992	1993	1994	1995	1996	1997
P	60.6	61.5	62.5	63.4	65.7	66.7	67.3
Year	1998	1999	2000	2001	2002	2003	
P	67.4	68.0	67.8	69.2	68.9	68.0	

(a) Explain why a logistic model is reasonable for this data.

[28] *The World Almanac and Book of Facts 2005*, p. 310 (New York).

(b) Estimate the point of diminishing returns. What limiting value L does this point predict? Does this limiting value appear to be accurate, given the percentages for 2002 and 2003?

(c) If t is in years since 1977, the best fitting logistic function for this data turns out to be

$$P = \frac{68.8}{1 + 3.486e^{-0.237t}},$$

What limiting value does this function predict?

(d) Explain in terms of percentages of households what the limiting value is telling you. Do you think your answer to part (c) is an accurate prediction? What do you think will ultimately be the percentage of households with cable television?

43. Find the dimensions of the rectangle with perimeter 200 meters that has the largest area.

44. A square-bottomed box with a top has a fixed volume, V. What dimensions minimize the surface area?

45. **(a)** Find the critical points of $p(1 - p)^4$.

(b) Classify the critical points as local maxima, local minima, or neither.

(c) What are the maximum and minimum values of $p(1 - p)^4$ on $0 \le x \le 1$?

46. A right triangle has one vertex at the origin and one vertex on the curve $y = e^{-x/3}$ for $1 \le x \le 5$. One of the two perpendicular sides is along the x-axis; the other is parallel to the y-axis. Find the maximum and minimum areas for such a triangle.

47. A pigeon is released from a boat (point B in Figure 4.108) floating on a lake. Because of falling air over the cool water, the energy required to fly one meter over the lake is twice the corresponding energy e required for flying over the bank ($e = 3$ joule/meter). To minimize the energy required to fly from B to the loft, L, the pigeon heads to a point P on the bank and then flies along the bank to L. The distance $\overline{AL}$ is 2000 m, and $\overline{AB}$ is 500 m. The angle at A is a right angle.

(a) Express the energy required to fly from B to L via P as a function of the angle θ (the angle BPA).

(b) What is the optimal angle θ?

(c) Does your answer change if $\overline{AL}$, $\overline{AB}$, and e have different numerical values?

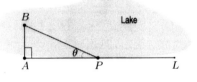

Figure 4.108

48. The bell-shaped curve of statistics has formula

$$p(x) = \frac{1}{\sigma\sqrt{2\pi}} e^{-(x-\mu)^2/(2\sigma^2)}$$

where μ is the mean and σ is the standard deviation.

(a) Where does $p(x)$ have a maximum?

(b) Does $p(x)$ have a point of inflection? If so, where?

49. Consider a population P satisfying the *logistic equation*

$$\frac{dP}{dt} = kP\left(1 - \frac{P}{L}\right).$$

(a) Use the chain rule to find d^2P/dt^2.

(b) Show that the point of diminishing returns, where $d^2P/dt^2 = 0$, occurs where $P = L/2$.

50. Figure 4.109 shows the concentration of bemetizide (a diuretic) in the blood after single oral doses of 25 mg alone, or 25 mg bemetizide and 50 mg triamterene in combination.[29] If the minimum effective concentration in a patient is 40 ng/ml, compare the effect of combining bemetizide with triamterene on peak concentration, time to reach peak concentration, time until onset of effectiveness, and duration of effectiveness. Under what circumstances might it be wise to use the triamterene with the bemetizide?

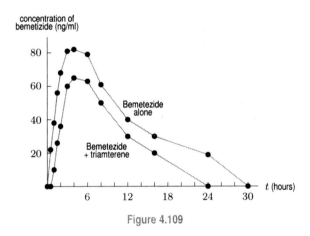

Figure 4.109

51. For $f(x) = \sin(x^2)$ between $x = 0$ and $x = 3$, find the coordinates of all intercepts, critical points, and inflection points to two decimal points.

52. **(a)** Graph $f(x) = x + a \sin x$ for $a = 0.5$ and $a = 3$.

(b) For what values of a is $f(x)$ increasing for all x?

53. **(a)** Graph $f(x) = x^2 + a \sin x$ for $a = 1$ and $a = 20$.

(b) For what values of a is $f(x)$ concave up for all x?

[29] Welling & Tse, *Pharmacokinetics of Cardiovascular, Central Nervous System, and Antimicrobial Drugs* (The Royal Society of Chemistry, 1985).

54. The distance, s, traveled by a cyclist, who starts at 1 pm, is given in Figure 4.110. Time, t, is in hours since noon.

(a) Explain why the quantity s/t is represented by the slope of a line from the origin to the point (t, s) on the graph.

(b) Estimate the time at which the quantity s/t is a maximum.

(c) What is the relationship between the quantity s/t and the instantaneous speed of the cyclist at the time you found in part (b)?

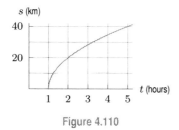

Figure 4.110

55. A bird such as a starling feeds worms to its young. To collect worms, the bird flies to a site where worms are to be found, picks up several in its beak, and flies back to its nest. The *loading curve* in Figure 4.111 shows how the number of worms (the load) a starling collects depends on the time it has been searching for them.[30] The curve is concave down because the bird can pick up worms more efficiently when its beak is empty; when its beak is partly full, the bird becomes much less efficient. The traveling time (from nest to site and back) is represented by the distance PO in Figure 4.111. The bird wants to maximize the rate at which it brings worms to the nest, where

$$\text{Rate worms arrive} = \frac{\text{Load}}{\text{Traveling time} + \text{Searching time}}$$

(a) Draw a line in Figure 4.111 whose slope is this rate.

(b) Using the graph, estimate the load which maximizes this rate.

(c) If the traveling time is increased, does the optimal load increase or decrease? Why?

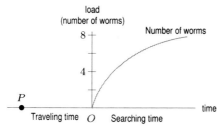

Figure 4.111

56. Table 4.11 shows the total number of cars in a University of Arizona parking lot at 30-minute intervals.[31]

(a) Graph the total number of cars in the parking lot as a function of time. Estimate the capacity of the parking lot, and when it was full.

(b) Construct a table, and then plot a graph, of the rate of arrival of cars as a function of time.

(c) From your graph in part (b), estimate when rush hour occurred.

(d) Explain the relationship between the points on the graphs in parts (a) and (b) where rush hour occurred.

Table 4.11   *Total number of cars, C, at time t*

t	5:00	5:30	6:00	6:30	7:00	7:30	8:00	8:30	9:00
C	4	5	8	18	50	110	170	200	200

57. A line goes through the origin and a point on the curve $y = x^2 e^{-3x}$, for $x \geq 0$. Find the maximum slope of such a line. At what x-value does it occur?

58. A rectangle has one side on the x-axis, one side on the y-axis, one vertex at the origin and one on the curve $y = e^{-2x}$ for $x \geq 0$. Find the

(a) Maximum area   (b) Minimum perimeter

59. A single cell of a bee's honey comb has the shape shown in Figure 4.112. The surface area of this cell is given by

$$A = 6hs + \frac{3}{2}s^2 \left(\frac{-\cos\theta}{\sin\theta} + \frac{\sqrt{3}}{\sin\theta} \right)$$

where h, s, θ are as shown in the picture.

(a) Keeping h and s fixed, for what angle, θ, is the surface area a minimum?

(b) Measurements on bee's cells have shown that the angle actually used by bees is about $\theta = 55°$. Comment.

---

[30] Alex Kacelnik (1984). Reported by J. R. Krebs and N. B. Davis, *An Introduction to Behavioural Ecology* (Oxford: Blackwell, 1987).

[31] Adapted from Nancy Roberts et al., *Introduction to Computer Simulation*, p. 93 (Reading: Addison-Wesley, 1983).

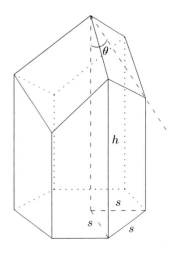

Figure 4.112

60. An organism has size W at time t. For positive constants A, b, and c, the Gompertz growth function gives

$$W = Ae^{-e^{b-ct}}, \quad t \geq 0.$$

(a) Find the intercepts and asymptotes.
(b) Find the critical points and inflection points.
(c) Graph W for various values of A, b, and c.
(d) A certain organism grows fastest when it is about 1/3 of its final size. Would the Gompertz growth function be useful in modeling its growth? Explain.

CHECK YOUR UNDERSTANDING

In Problems 1–80, indicate whether the statement is true or false.

1. The function f has a local maximum at p if $f(p) \leq f(x)$ for points x near p.

2. If $f'(p) = 0$ then p is a critical point of f.

3. If p is a critical point of f then $f'(p) = 0$.

4. If f is increasing at all points to the left of p and decreasing at all points to the right of p then f has a local maximum at p.

5. If $f'(p) = 0$ and $f''(p) > 0$ then f has a local maximum at p.

6. If $f'(p) = 0$ and $f''(p) > 0$ then f has a local minimum at p.

7. If $f''(p) > 0$ then f has a local minimum at p.

8. Every critical point of f is either a local maximum or local minimum of f.

9. A function f must have at least one critical point.

10. If f has a local minimum at $x = p$ then $f'(p) = 0$.

11. A point at which a graph changes concavity is called an inflection point.

12. If $f''(p) = 0$ then p is an inflection point of f.

13. A point p can be both a critical point and an inflection point of a function f.

14. The function $f(x) = x^3$ has an inflection point at $x = 0$.

15. The function $H(x) = x^4$ has an inflection point at $x = 0$.

16. A function f can have one inflection point and no critical points.

17. A function f can have 2 critical points and 3 inflection points.

18. If $f''(x) = x(x + 1)$ then f has an inflection point at $x = -1$.

19. If $f''(x) = x(x + 1)^2$ then f has an inflection point at $x = -1$.

20. If $f''(x) = e^x(x + 1)$ then f has an inflection point at $x = -1$. is an inflection point.

21. A local maximum of f can also be a global maximum of f.

22. A global maximum of f on the interval $1 \leq x \leq 2$ always occurs at a critical point.

23. Every function has a global minimum.

24. If function $y = f(x)$ is increasing on the interval $a \leq x \leq b$, then the global maximum of f on this interval occurs at $x = b$.

25. If a function $y = f(x)$ has $f'(x) < 0$ for all x in the interval $a \leq x \leq b$, then the global maximum of f on this interval occurs at $x = b$.

26. A function $S(x)$ could have a different global maximum on each of the intervals $1 \leq x \leq 2$ and $2 \leq x \leq 3$.

27. The function $k(x) = 1/x$ has a global maximum when $x > 0$.

28. The function f has a global maximum on the interval $-5 \leq x \leq 5$ at p if $f(p) \leq f(x)$ for all $-5 \leq x \leq 5$.

29. The function f has a global minimum on the interval $-5 \leq x \leq 5$ at p if $f(p) \leq f(x)$ for all $-5 \leq x \leq 5$.

30. If $f'(p) = 0$ and $f''(p) > 0$ then p is a global minimum of f.

31. If marginal cost is less than marginal revenue, then increasing quantity sold will increase profit.

32. If marginal cost is greater than marginal revenue, then decreasing quantity sold will increase profit.

33. When marginal revenue equals marginal cost there is maximum profit.

34. If revenue is greater than cost, profit is the vertical distance between the cost and revenue curves.

35. Maximum profit occurs at a critical point of the cost function.

36. Maximum profit can occur when marginal profit is zero.

37. Profit is maximized where the cost and revenue curves cross.

38. Profit is zero where the graphs of the cost and revenue curves cross.

39. If prices are constant, the graph of marginal revenue is a horizontal line.

40. If prices are constant, the graph of revenue is a horizontal line.

41. Average cost is total cost for q items divided by q.

42. Marginal cost and average cost have the same units.

43. Marginal cost is the same thing as average cost.

44. The average cost of q items is the slope of the tangent line to the cost curve at q.

45. If marginal cost is less than average cost, increasing production increases average cost.

46. Marginal cost equals average cost at critical points of marginal cost.

47. If marginal cost is greater than average cost, then increasing production increases average cost.

48. Average cost has a critical point at a quantity where the line from the origin to the cost curve is also tangent to the cost curve.

49. Average cost is a decreasing function of quantity.

50. Marginal and average cost functions are both minimized at the same quantity.

51. The elasticity of demand is given by $E = |q/p \cdot dq/dp|$.

52. If elasticity, E, is greater than one, then demand is inelastic.

53. If elasticity, E, is such that $0 \leq E < 1$ then demand is elastic.

54. Elasticity is a measure of the effect on demand of a change in price.

55. If a product is considered a necessity, the demand is generally elastic.

56. An increase in price always causes an increase in revenue.

57. If elasticity $E > 1$ then revenue increases when price increases.

58. At a critical point of the revenue function, demand is neither elastic nor inelastic.

59. If elasticity $E < 1$ then profit increases with an increase in price.

60. An increase in price always causes an increase in profit.

61. If $P = 1000/(1 + 2e^{-3t})$ then P is a logistic function.

62. If $P = 1000/(1 + 2e^{3t})$ then P is a logistic function.

63. The function $P = 1000/(1 + 2e^{-3t})$ has an inflection point when $t = 500$.

64. The function $P = 1000/(1 + 2e^{-3t})$ has an inflection point when $P = 500$.

65. The logistic function $P = L/(1 + Ce^{-kt})$ approaches L as t increases.

66. If the slope of the dose/response curve is large, a small change in the dose will have a large effect on the response.

67. The carrying capacity of a population growing logistically is the largest population the environment can support.

68. The parameter k in the logistic function $P = L/(1 + Ce^{-kt})$ affects the rate at which the curve approaches its asymptote L.

69. The logistic function $P = L/(1 + Ce^{-kt})$ is concave up for $P < L/2$.

70. The logistic function $P = L/(1 + Ce^{-kt})$ is always increasing.

71. The surge function is an exponential decay function.

72. The surge function has one critical point, which is a global maximum.

73. The surge function is always concave down.

74. If minimum effective concentration is positive and less than peak concentration, then the surge function will intersect the minimum effective concentration line at two points.

75. If minimum effective concentration is greater than peak concentration, then the drug will be effective only for the first half of the time interval.

76. The surge function has exactly one inflection point, and it is always at a t-value greater than the t-value of the critical point.

77. The minimum effective concentration of a drug is the blood concentration necessary to achieve a specific pharmacologic response.

78. The function $y = t/(12e^{3t})$ is a surge function.

79. The function $y = t/(12e^{-3t})$ is a surge function.

80. The surge function $y = 5te^{-3t}$ has maximum at $t = 1/3$.

PROJECTS FOR CHAPTER FOUR

1. Average and Marginal Costs

The total cost of producing a quantity q is $C(q)$. The average cost $a(q)$ is given in Figure 4.113. The following rule is used by economists to determine the marginal cost $C'(q_0)$, for any q_0:

- Construct the tangent line t_1 to $a(q)$ at q_0.
- Let t_2 be the line with the same vertical intercept as t_1 but with twice the slope of t_1.

Then $C'(q_0)$ is the vertical distance shown in Figure 4.113. Explain why this rule works.

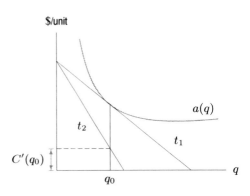

Figure 4.113

2. Firebreaks

The summer of 2000 was devastating for forests in the western US: over 3.5 million acres of trees were lost to fires, making this the worst fire season in 30 years. This project studies a fire management technique called *firebreaks*, which reduce the damage done by forest fires. A firebreak is a strip where trees have been removed in a forest so that a fire started on one side of the strip will not spread to the other side. Having many firebreaks helps confine a fire to a small area. On the other hand, having too many firebreaks involves removing large swaths of trees.[32]

(a) A forest in the shape of a 50 km by 50 km square has firebreaks in rectangular strips 50 km by 0.01 km. The trees between two firebreaks are called a stand of trees. All firebreaks in this forest are parallel to each other and to one edge of the forest, with the first firebreak at the edge of the forest. The firebreaks are evenly spaced throughout the forest. (For example, Figure 4.114 shows four firebreaks.) The total area lost in the case of a fire is the area of the stand of trees in which the fire started plus the area of all the firebreaks.

[32] Adapted from D. Quinney and R. Harding, *Calculus Connections* (New York: John Wiley & Sons, 1996).

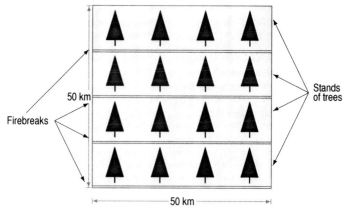

Figure 4.114

    (i) Find the number of firebreaks that minimizes the total area lost to the forest in the case of a fire.

    (ii) If a firebreak is 50 km by b km, find the optimal number of firebreaks as a function of b. If the width, b, of a firebreak is quadrupled, how does the optimal number of firebreaks change?

(b) Now suppose firebreaks are arranged in two equally spaced sets of parallel lines, as shown in Figure 4.115. The forest is a 50 km by 50 km square, and each firebreak is a rectangular strip 50 km by 0.01 km. Find the number of firebreaks in each direction that minimizes the total area lost to the forest in the case of a fire.

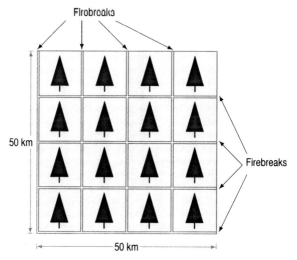

Figure 4.115

3. Production and the Price of Raw Materials

The production function $f(x)$ gives the number of units of an item that a manufacturing company can produce from x units of raw material. The company buys the raw material at price w dollars per unit and sells all it produces at a price of p dollars per unit. The quantity of raw material that maximizes profit is denoted by x^*.

(a) Do you expect the derivative $f'(x)$ to be positive or negative? Justify your answer.

(b) Explain why the formula $\pi(x) = pf(x) - wx$ gives the profit $\pi(x)$ that the company earns as a function of the quantity x of raw materials that it uses.

231

(c) Evaluate $f'(x^*)$.

(d) Assuming it is nonzero, is $f''(x^*)$ positive or negative?

(e) If the supplier of the raw materials is likely to change the price w, then it is appropriate to treat x^* as a function of w. Find a formula for the derivative dx^*/dw and decide whether it is positive or negative.

(f) If the price w goes up, should the manufacturing company buy more or less of the raw material?

Chapter Five

ACCUMULATED CHANGE: THE DEFINITE INTEGRAL

Contents

5.1 DISTANCE AND ACCUMULATED CHANGE

In Chapter 2, we used the derivative to find the rate of change of a function. Here we see how to go in the other direction. If we know the rate of change, can we find the original function? We start by finding the distance traveled from the velocity.

How Do We Measure Distance Traveled?

The rate of change of distance with respect to time is velocity. If we are given the velocity, can we find the distance traveled? Suppose the velocity was 50 miles per hour throughout a four-hour trip. What is the total distance traveled? Since

$$\text{Distance} = \text{Velocity} \times \text{Time},$$

we have

$$\text{Distance traveled} = (50 \text{ miles/hour}) \times (4 \text{ hours})$$
$$= 200 \text{ miles}.$$

The graph of velocity against time is the horizontal line in Figure 5.1. Notice that the distance traveled is represented by the shaded area under the graph.

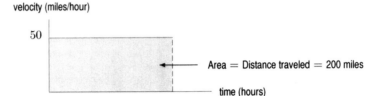

velocity (miles/hour)

50

Area = Distance traveled = 200 miles

time (hours)

4

Figure 5.1: Area shaded represents distance traveled in 4 hours at 50 mph

Now let's see what happens if the velocity is not constant.

Example 1 Suppose that you travel 30 miles/hour for 2 hours, then 40 miles/hour for 1/2 hour, then 20 miles/hour for 4 hours. What is the total distance you traveled?

Solution We compute the distances traveled for each of the three legs of the trip and add them to find the total distance traveled:

$$\text{Distance} = (30 \text{ miles/hour})(2 \text{ hours}) + (40 \text{ miles/hour})(1/2 \text{ hour}) + (20 \text{ miles/hour})(4 \text{ hours})$$
$$= 60 \text{ miles} + 20 \text{ miles} + 80 \text{ miles}$$
$$= 160 \text{ miles}.$$

You travel 160 miles on this trip.

A Thought Experiment: How Far Did the Car Go?

In Example 1, the velocity was constant over intervals. Of course, this is not always the case; we now look at an example where the velocity is continually changing.

Velocity Data Every Two Seconds

Suppose a car is moving with increasing velocity and suppose we measure the car's velocity every two seconds, obtaining the data in Table 5.1:

Table 5.1 *Velocity of car every two seconds*

Time (sec)	0	2	4	6	8	10
Velocity (ft/sec)	20	30	38	44	48	50

234

How far has the car traveled? Since we don't know how fast the car is moving at every moment, we can't calculate the distance exactly, but we can make an estimate. The velocity is increasing, so the car is going at least 20 ft/sec for the first two seconds. Since Distance = Velocity × Time, the car goes at least $20 \cdot 2 = 40$ feet during the first two seconds. Likewise, it goes at least $30 \cdot 2 = 60$ feet during the next two seconds, and so on. During the ten-second period it goes at least

$$20 \cdot 2 + 30 \cdot 2 + 38 \cdot 2 + 44 \cdot 2 + 48 \cdot 2 = 360 \text{ feet.}$$

Thus, 360 feet is an underestimate of the total distance traveled during the ten seconds.

To get an overestimate, we can reason in a similar way: During the first two seconds, the car's velocity is at most 30 ft/sec, so it moves at most $30 \cdot 2 = 60$ feet. In the next two seconds it moves at most $38 \cdot 2 = 76$ feet, and so on. Therefore, over the ten-second period it moves at most

$$30 \cdot 2 + 38 \cdot 2 + 44 \cdot 2 + 48 \cdot 2 + 50 \cdot 2 = 420 \text{ feet.}$$

Therefore,

$$360 \text{ feet} \leq \text{Total distance traveled} \leq 420 \text{ feet.}$$

There is a difference of 60 feet between the upper and lower estimates.

Velocity Data Every One Second

What if we want a more accurate estimate? We could make more frequent velocity measurements, say every second. The data is in Table 5.2.

As before, we get a lower estimate for each second by using the velocity at the beginning of that second. During the first second the velocity is at least 20 ft/sec, so the car travels at least $20 \cdot 1 = 20$ feet. During the next second the car moves at least 26 feet, and so on. So we say

$$\begin{aligned}\text{New lower estimate} &= 20 \cdot 1 + 26 \cdot 1 + 30 \cdot 1 + 35 \cdot 1 + 38 \cdot 1 \\ &\quad + 42 \cdot 1 + 44 \cdot 1 + 46 \cdot 1 + 48 \cdot 1 + 49 \cdot 1 \\ &= 378 \text{ feet.}\end{aligned}$$

Notice that this is greater than the old lower estimate of 360 feet.

We get a new upper estimate by considering the velocity at the end of each second. During the first second the velocity is at most 26 ft/sec, and so the car moves at most $26 \cdot 1 = 26$ feet; in the next second it moves at most 30 feet, and so on.

$$\begin{aligned}\text{New upper estimate} &= 26 \cdot 1 + 30 \cdot 1 + 35 \cdot 1 + 38 \cdot 1 + 42 \cdot 1 \\ &\quad + 44 \cdot 1 + 46 \cdot 1 + 48 \cdot 1 + 49 \cdot 1 + 50 \cdot 1 \\ &= 408 \text{ feet.}\end{aligned}$$

This is less than the old upper estimate of 420 feet. Now we know that

$$378 \text{ feet} \leq \text{Total distance traveled} \leq 408 \text{ feet.}$$

Notice that the difference between the new upper and lower estimates is now 30 feet, half of what it was before. By halving the interval of measurement, we have halved the difference between the upper and lower estimates.

Table 5.2   *Velocity of car every second*

Time (sec)	0	1	2	3	4	5	6	7	8	9	10
Velocity (ft/sec)	20	26	30	35	38	42	44	46	48	49	50

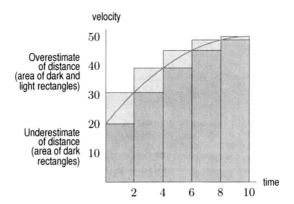

Figure 5.2: Shaded area estimates distance traveled. Velocity measured every 2 seconds

Visualizing Distance on the Velocity Graph

Consider the two-second data in Table 5.1. We can represent both upper and lower estimates on a graph of the velocity against time. The velocity can be graphed by plotting these data and drawing a smooth curve through the data points. (See Figure 5.2.)

We use the fact that for a rectangle, Area = Height × Width. The area of the first dark rectangle is $20 \cdot 2 = 40$, the lower estimate of the distance moved during the first two seconds. The area of the second dark rectangle is $30 \cdot 2 = 60$, the lower estimate for the distance moved in the next two seconds. The total area of the dark rectangles represents the lower estimate for the total distance moved during the ten seconds.

If the dark and light rectangles are considered together, the first area is $30 \cdot 2 = 60$, the upper estimate for the distance moved in the first two seconds. The second area is $38 \cdot 2 = 76$, the upper estimate for the next two seconds. Continuing this calculation suggests that the upper estimate for the total distance is represented by the sum of the areas of the dark and light rectangles. Therefore, the area of the light rectangles alone represents the difference between the two estimates.

Figure 5.3 shows a graph of the one-second data. The area of the dark rectangles again represents the lower estimate, and the area of the dark and light rectangles together represents the upper estimate. The total area of the light rectangles is smaller in Figure 5.3 than in Figure 5.2, so the underestimate and overestimate are closer for the one-second data than for the two-second data.

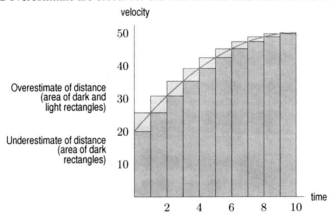

Figure 5.3: Shaded area estimates distance traveled. Velocity measured every second

Visualizing Distance on the Velocity Graph: Area Under Curve

As we make more frequent velocity measurements, the rectangles used to estimate the distance traveled fit the curve more closely. See Figures 5.4 and 5.5. In the limit, as the number of subdivisions increases, we see that the distance traveled is given by the area between the velocity curve and the horizontal axis. See Figure 5.6. In general:

> If the velocity is positive, the total distance traveled is the area under the velocity curve.

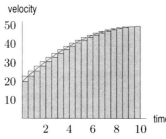

Figure 5.4: Velocity measured every 1/2 second

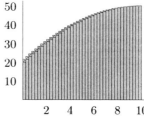

Figure 5.5: Velocity measured every 1/4 second

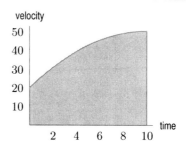

Figure 5.6: Distance traveled is area under curve

Example 2 With time t in seconds, the velocity of a bicycle, in feet per second, is given by $v(t) = 5t$. How far does the bicycle travel in 3 seconds?

Solution The velocity is linear. See Figure 5.7. The distance traveled is the area between the line $v(t) = 5t$ and the t-axis. Since this region is a triangle of height 15 and base 3

$$\text{Distance traveled} = \text{Area of triangle} = \frac{1}{2} \cdot 15 \cdot 3 = 22.5 \text{ feet.}$$

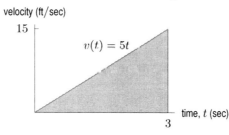

Figure 5.7: Shaded area represents distance traveled

Approximating Total Change from Rate of Change

We have seen how to use the rate of change of distance (the velocity) to calculate the total distance traveled. We can use the same method to find total change from rate of change of other quantities.

Example 3 A city's population grows at the rate of 5000 people/year for 3 years and then grows at the rate of 3000 people/year for the next 4 years. What is the total change in the population of the city during this 7-year period?

Solution The units of people/year remind us that we are given a rate of change of the population (people) with respect to time (years). If the rate of change is constant, we know that

$$\text{Total change in population} = \text{Rate of change per year} \times \text{Number of years.}$$

Thus, the total change in this population is

$$\begin{aligned}
\text{Total change} &= (5000 \text{ people/year})(3 \text{ years}) + (3000 \text{ people/year})(4 \text{ years}) \\
&= 15{,}000 \text{ people} + 12{,}000 \text{ people} \\
&= 27{,}000 \text{ people.}
\end{aligned}$$

The change in population is 27,000 people. Notice that this does not tell us the population of the city at the end of 7 years; it tells us the change in the population. For example, if the population were initially 100,000, it would be 127,000 at the end of the 7-year period.

Example 4   The rate of sales (in games per week) of a new video game is shown in Table 5.3. Assuming that the rate of sales increased throughout the 20-week period, estimate the total number of games sold during this period.

Table 5.3   *Weekly sales of a video game*

Time (weeks)	0	5	10	15	20
Rate of sales (games per week)	0	585	892	1875	2350

Solution   If the rate of sales is constant, we have

$$\text{Total sales } = \text{ Rate of sales per week } \times \text{ Number of weeks.}$$

How many games were sold during the first five weeks? During this time, sales went from 0 to 585 games per week. If we assume that 585 games were sold every week, we get an overestimate for the sales in the first five weeks of (585 games/week)(5 weeks) = 2925 games. Similar overestimates for each of the five-week periods gives an overestimate for the entire 20-week period:

$$\text{Overestimate for total sales } = 585 \cdot 5 + 892 \cdot 5 + 1875 \cdot 5 + 2350 \cdot 5 = 28{,}510 \text{ games.}$$

We underestimate the total sales by taking the lower value for rate of sales during each of the five-week periods:

$$\text{Underestimate for total sales } = 0 \cdot 5 + 585 \cdot 5 + 892 \cdot 5 + 1875 \cdot 5 = 16{,}760 \text{ games.}$$

Thus, the total sales of the game during the 20-week period is between 16,760 and 28,510 games. A good single estimate of total sales is the average of these two numbers:

$$\text{Total sales } \approx \frac{16{,}760 + 28{,}510}{2} = 22{,}635 \text{ games.}$$

Problems for Section 5.1

1. (a) Sketch a graph of the velocity function for the trip described in Example 1 on page 234.
    (b) Represent the total distance traveled on this graph.

2. Graph the rate of sales against time for the video game data in Example 4. Represent graphically the overestimate and the underestimate calculated in that example.

3. A car comes to a stop six seconds after the driver applies the brakes. While the brakes are on, the velocities recorded are in Table 5.4.

    Table 5.4

Time since brakes applied (sec)	0	2	4	6
Velocity (ft/sec)	88	45	16	0

    (a) Give lower and upper estimates for the distance the car traveled after the brakes were applied.
    (b) On a sketch of velocity against time, show the lower and upper estimates of part (a).

4. A car starts moving at time $t = 0$ and goes faster and faster. Its velocity is shown in the following table. Estimate how far the car travels during the 12 seconds.

t (seconds)	0	3	6	9	12
Velocity (ft/sec)	0	10	25	45	75

5. The velocity of a car is $f(t) = 5t$ meters/sec. Use a graph of $f(t)$ to find the exact distance traveled by the car, in meters, from $t = 0$ to $t = 10$ seconds.

6. Figure 5.8 shows the velocity, v, of an object (in meters/sec). Estimate the total distance the object traveled between $t = 0$ and $t = 6$.

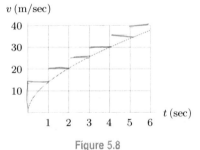

Figure 5.8

7. A car accelerates smoothly from 0 to 60 mph in 10 seconds with the velocity given in Figure 5.9. Estimate how far the car travels during the 10-second period.

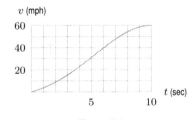

Figure 5.9

8. The following table gives world oil consumption, in billions of barrels per year.[1] Estimate total oil consumption during this 25-year period.

Year	1980	1985	1990	1995	2000	2005
Oil (bn barrels/yr)	22.3	21.3	23.9	24.9	27.0	29.3

9. Filters at a water treatment plant become less effective over time. The rate at which pollution passes through the filters into a nearby lake is given in the following table.

  (a) Estimate the total quantity of pollution entering the lake during the 30-day period.

  (b) Your answer to part (a) is only an estimate. Give bounds (lower and upper estimates) between which the true quantity of pollution must lie. (Assume the rate of pollution is continually increasing.)

Day	0	6	12	18	24	30
Rate (kg/day)	7	8	10	13	18	35

10. The rate of change of the world's population, in millions of people per year, is given in the following table.

  (a) Use this data to estimate the total change in the world's population between 1950 and 2000.

  (b) The world population was 2555 million people in 1950 and 6085 million people in 2000. Calculate the true value of the total change in the population. How does this compare with your estimate in part (a)?

Year	1950	1960	1970	1980	1990	2000
Rate of change	37	41	78	77	86	79

11. A village wishes to measure the quantity of water that is piped to a factory during a typical morning. A gauge on the water line gives the flow rate (in cubic meters per hour) at any instant. The flow rate is about 100 m³/hr at 6 am and increases steadily to about 280 m³/hr at 9 am. Using only this information, give your best estimate of the total volume of water used by the factory between 6 am and 9 am.

12. Roger runs a marathon. His friend Jeff rides behind him on a bicycle and clocks his speed every 15 minutes. Roger starts out strong, but after an hour and a half he is so exhausted that he has to stop. Jeff's data follow:

Time since start (min)	0	15	30	45	60	75	90
Speed (mph)	12	11	10	10	8	7	0

  (a) Assuming that Roger's speed is never increasing, give upper and lower estimates for the distance Roger ran during the first half hour.

  (b) Give upper and lower estimates for the distance Roger ran in total during the entire hour and a half.

13. A car initially going 50 ft/sec brakes at a constant rate (constant negative acceleration), coming to a stop in 5 seconds.

  (a) Graph the velocity from $t = 0$ to $t = 5$.

  (b) How far does the car travel?

  (c) How far does the car travel if its initial velocity is doubled, but it brakes at the same constant rate?

14. Figure 5.10 shows the rate of change of a fish population. Estimate the total change in the population during this 12-month period.

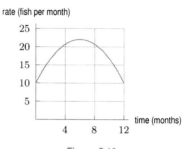

Figure 5.10

15. Two cars start at the same time and travel in the same direction along a straight road. Figure 5.11 gives the velocity, v, of each car as a function of time, t. Which car:

  (a) Attains the larger maximum velocity?

  (b) Stops first?

  (c) Travels farther?

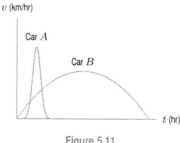

Figure 5.11

[1] www.bp.com/centres/energy/world_stat_rev/oil/reserves.asp. and https://www.cia.gov/library/publications/the-world-factbook/print/xx.html.

16. Two cars travel in the same direction along a straight road. Figure 5.12 shows the velocity, v, of each car at time t. Car B starts 2 hours after car A and car B reaches a maximum velocity of 50 km/hr.

(a) For approximately how long does each car travel?

(b) Estimate car A's maximum velocity.

(c) Approximately how far does each car travel?

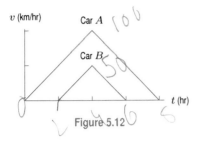

v (km/hr)

Car A

Car B

t (hr)

Figure 5.12

17. Your velocity is given by $v(t) = t^2 + 1$ in m/sec, with t in seconds. Estimate the distance, s, traveled between $t = 0$ and $t = 5$. Explain how you arrived at your estimate.

18. An old rowboat has sprung a leak. Water is flowing into the boat at a rate, $r(t)$, given in the following table.

t minutes	0	5	10	15
$r(t)$ liters/min	12	20	24	16

(a) Compute upper and lower estimates for the volume of water that has flowed into the boat during the 15 minutes.

(b) Draw a graph to illustrate the lower estimate.

19. The value of a mutual fund increases at a rate of $R = 500e^{0.04t}$ dollars per year, with t in years since 2010.

(a) Using $t = 0, 2, 4, 6, 8, 10$, make a table of values for R.

(b) Use the table to estimate the total change in the value of the mutual fund between 2010 and 2020.

20. A car speeds up at a constant rate from 10 to 70 mph over a period of half an hour. Its fuel efficiency (in miles per gallon) increases with speed; values are in the table. Make lower and upper estimates of the quantity of fuel used during the half hour.

Speed (mph)	10	20	30	40	50	60	70
Fuel efficiency (mpg)	15	18	21	23	24	25	26

5.2 THE DEFINITE INTEGRAL

In Section 5.1 we saw how to approximate total change given the rate of change. We now see how to make the approximation more accurate.

Improving the Approximation: The Role of n and Δt

To approximate total change, we construct a sum. We use the notation Δt for the size of the t-intervals used. We use n to represent the number of subintervals of length Δt. In the following example, we see how decreasing Δt (and increasing n) improves the accuracy of the approximation.

Example 1 If t is in hours since the start of a 20-hour period, a bacteria population increases at a rate given by

$$f(t) = 3 + 0.1t^2 \text{ millions of bacteria per hour.}$$

Make an underestimate of the total change in the number of bacteria over this period using

(a) $\Delta t = 4$ hours   **(b)** $\Delta t = 2$ hours   **(c)** $\Delta t = 1$ hour

Solution **(a)** The rate of change is $f(t) = 3 + 0.1t^2$. If we use $\Delta t = 4$, we measure the rate every 4 hours and $n = 20/4 = 5$. See Table 5.5. An underestimate for the population change during the first 4 hours is (3.0 million/hour)(4 hours) = 12 million. Combining the contributions from all the subintervals gives the underestimate:

$$\text{Total change} \approx 3.0 \cdot 4 + 4.6 \cdot 4 + 9.4 \cdot 4 + 17.4 \cdot 4 + 28.6 \cdot 4 = 252.0 \text{ million bacteria.}$$

The rate of change is graphed in Figure 5.13 (a); the area of the shaded rectangles represents this underestimate. Notice that $n = 5$ is the number of rectangles in the graph.

Table 5.5   *Rate of change with $\Delta t = 4$ using $f(t) = 3 + 0.1t^2$ million bacteria/hour*

t (hours)	0	4	8	12	16	20
$f(t)$	3.0	4.6	9.4	17.4	28.6	43.0

(b) If we use $\Delta t = 2$, we measure $f(t)$ every 2 hours and $n = 20/2 = 10$. See Table 5.6. The underestimate is

$$\text{Total change} \approx 3.0 \cdot 2 + 3.4 \cdot 2 + 4.6 \cdot 2 + \cdots + 35.4 \cdot 2 = 288.0 \text{ million bacteria.}$$

Figure 5.13 (b) suggests that this estimate is more accurate than the estimate made in part (a).

Table 5.6   *Rate of change with $\Delta t = 2$ using $f(t) = 3 + 0.1t^2$ million bacteria/hour*

t (hours)	0	2	4	6	8	10	12	14	16	18	20
$f(t)$	3.0	3.4	4.6	6.6	9.4	13.0	17.4	22.6	28.6	35.4	43.0

(c) If we use $\Delta t = 1$, then $n = 20$ and a similar calculation shows that we have

$$\text{Total change} \approx 307.0 \text{ million bacteria.}$$

The shaded area in Figure 5.13 (c) represents this estimate; it is the most accurate of the three.

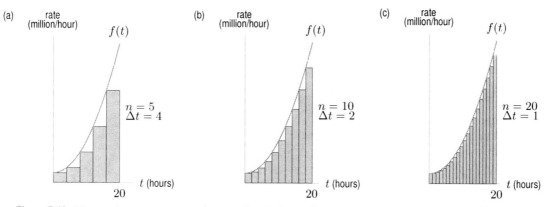

Figure 5.13: More and more accurate estimates of total change from rate of change. In each case, $f(t)$ is the rate of change, and the shaded area approximates total change. Largest n and smallest Δt give the best estimate.

Notice that as n gets larger, the estimate improves and the area of the shaded rectangles approaches the area under the curve.

Left- and Right-Hand Sums

Suppose we have a function $f(t)$ that is continuous for $a \le t \le b$. We divide the interval from a to b into n equal subdivisions, each of width Δt, so

$$\Delta t = \frac{b - a}{n}.$$

We let $t_0, t_1, t_2, \ldots, t_n$ be endpoints of the subdivisions, as in Figures 5.14 and 5.15. We construct two sums, similar to the overestimates and underestimates in Section 3.1. For a *left-hand sum*, we use the values of the function from the left end of the interval. For a *right-hand sum*, we use the values of the function from the right end of the interval. We have:

$$\text{Left-hand sum} = f(t_0)\Delta t + f(t_1)\Delta t + \cdots + f(t_{n-1})\Delta t$$

and

$$\text{Right-hand sum} = f(t_1)\Delta t + f(t_2)\Delta t + \cdots + f(t_n)\Delta t.$$

These sums represent the shaded areas in Figures 5.14 and 5.15, provided $f(t) \geq 0$. In Figure 5.14, the first rectangle has width Δt and height $f(t_0)$, since the top of its left edge just touches the curve, and hence it has area $f(t_0)\Delta t$. The second rectangle has width Δt and height $f(t_1)$, and hence has area $f(t_1)\Delta t$, and so on. The sum of all these areas is the left-hand sum. The right-hand sum, shown in Figure 5.15, is constructed in the same way, except that each rectangle touches the curve on its right edge instead of its left.

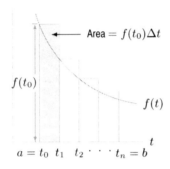

Figure 5.14: Left-hand sum: Area of rectangles   Figure 5.15: Right-hand sum: Area of rectangles

Writing Left- and Right-Hand Sums Using Sigma Notation

Both the left-hand and right-hand sums can be written more compactly using *sigma*, or summation, notation. The symbol $\sum$ is a capital sigma, or Greek letter "S." We write

$$\text{Right-hand sum} = \sum_{i=1}^{n} f(t_i)\Delta t = f(t_1)\Delta t + f(t_2)\Delta t + \cdots + f(t_n)\Delta t.$$

The $\sum$ tells us to add terms of the form $f(t_i)\Delta t$. The "$i = 1$" at the base of the sigma sign tells us to start at $i = 1$, and the "n" at the top tells us to stop at $i = n$.

In the left-hand sum we start at $i = 0$ and stop at $i = n - 1$, so we write

$$\text{Left-hand sum} = \sum_{i=0}^{n-1} f(t_i)\Delta t = f(t_0)\Delta t + f(t_1)\Delta t + \cdots + f(t_{n-1})\Delta t.$$

Taking the Limit to Obtain the Definite Integral

If f is a rate of change of some quantity, then the left-hand sum and the right-hand sum approximate the total change in the quantity. For most functions f, the approximation is improved by increasing the value of n. To find the total change exactly, we take larger and larger values of n and look at the values approached by the left and right sums. This is called taking the *limit* of these sums as n goes to infinity and is written $\lim_{n \to \infty}$. If f is continuous for $a \leq t \leq b$, the limits of the left- and right-hand sums exist and are equal. The *definite integral* is the common limit of these sums.

Suppose f is continuous for $a \leq t \leq b$. The **definite integral** of f from a to b, written

$$\int_a^b f(t)\, dt,$$

is the limit of the left-hand or right-hand sums with n subdivisions of $[a, b]$ as n gets arbitrarily large. In other words, if $t_0, t_1, \ldots t_n$ are the endpoints of the subdivisions,

$$\int_a^b f(t)\, dt = \lim_{n \to \infty} (\text{Left-hand sum}) = \lim_{n \to \infty} \left(\sum_{i=0}^{n-1} f(t_i) \Delta t \right)$$

and

$$\int_a^b f(t)\, dt = \lim_{n \to \infty} (\text{Right-hand sum}) = \lim_{n \to \infty} \left(\sum_{i=1}^{n} f(t_i) \Delta t \right).$$

Each of these sums is called a *Riemann sum*, f is called the *integrand*, and a and b are called the *limits of integration*.

The "$\int$" notation comes from an old-fashioned "S," which stands for "sum" in the same way that $\sum$ does. The "dt" in the integral comes from the factor Δt. Notice that the limits on the $\sum$ symbol are 0 and $n - 1$ for the left-hand sum, and 1 and n for the right-hand sum, whereas the limits on the $\int$ sign are a and b.

When $f(t)$ is positive, the left- and right-hand sums are represented by the sums of areas of rectangles, so the definite integral is represented graphically by an area.

Computing a Definite Integral

In practice, we often approximate definite integrals numerically using a calculator or computer. They compute sums for larger and larger values of n, and eventually give a value for the integral. Different calculators and computers may give slightly different estimates, owing to round-off error and the fact that they may use different approximation methods.

Example 2   Compute $\displaystyle\int_1^3 t^2\, dt$ and represent this integral as an area.

Solution   Using a calculator, we find

$$\int_1^3 t^2\, dt = 8.667.$$

The integral represents the area between $t = 1$ and $t = 3$ under the curve $f(t) = t^2$. See Figure 5.16.

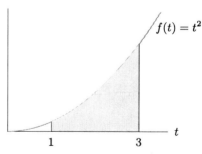

Figure 5.16: Shaded area $= \int_1^3 t^2\, dt$

Estimating a Definite Integral from a Table or Graph

If we have a formula for the integrand, $f(x)$, we can calculate the integral $\int_a^b f(x)\,dx$ using a calculator or computer. If, however, we have only a table of values or a graph of $f(x)$, we can still estimate the integral.

Example 3   Values for a function $f(t)$ are in the following table. Estimate $\int_{20}^{30} f(t)dt$.

t	20	22	24	26	28	30
$f(t)$	5	7	11	18	29	45

Solution   Since we have only a table of values, we use left- and right-hand sums to approximate the integral. The values of $f(t)$ are spaced 2 units apart, so $\Delta t = 2$ and $n = (30-20)/2 = 5$. Calculating the left-hand and right-hand sums gives

$$\begin{aligned}
\text{Left-hand sum} &= f(20)\cdot 2 + f(22)\cdot 2 + f(24)\cdot 2 + f(26)\cdot 2 + f(28)\cdot 2 \\
&= 5\cdot 2 + 7\cdot 2 + 11\cdot 2 + 18\cdot 2 + 29\cdot 2 \\
&= 10 + 14 + 22 + 36 + 58 \\
&= 140.
\end{aligned}$$

$$\begin{aligned}
\text{Right-hand sum} &= f(22)\cdot 2 + f(24)\cdot 2 + f(26)\cdot 2 + f(28)\cdot 2 + f(30)\cdot 2 \\
&= 7\cdot 2 + 11\cdot 2 + 18\cdot 2 + 29\cdot 2 + 45\cdot 2 \\
&= 14 + 22 + 36 + 58 + 90 \\
&= 220.
\end{aligned}$$

Both left- and right-hand sums approximate the integral. We generally get a better estimate by averaging the two:

$$\int_{20}^{30} f(t)dt \approx \frac{140 + 220}{2} = 180.$$

Example 4   The function $f(x)$ is graphed in Figure 5.17. Estimate $\int_0^6 f(x)\,dx$.

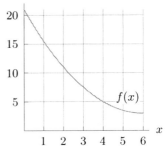

Figure 5.17: Estimate $\int_0^6 f(x)\,dx$

Solution   We approximate the integral using left- and right-hand sums with $n = 3$, so $\Delta x = 2$. Figures 5.18 and 5.19 give

$$\text{Left-hand sum} = f(0) \cdot 2 + f(2) \cdot 2 + f(4) \cdot 2 = 21 \cdot 2 + 11 \cdot 2 + 5 \cdot 2 = 74,$$
$$\text{Right-hand sum} = f(2) \cdot 2 + f(4) \cdot 2 + f(6) \cdot 2 = 11 \cdot 2 + 5 \cdot 2 + 3 \cdot 2 = 38.$$

We estimate the integral by taking the average:

$$\int_0^6 f(x)\,dx \approx \frac{74 + 38}{2} = 56.$$

Alternatively, since the integral equals the area under the curve between $x = 0$ and $x = 6$, we can estimate it by counting grid squares. Each grid square has area $5 \cdot 1 = 5$, and the region under $f(x)$ includes about 10.5 grid squares, so the area is about $10.5 \cdot 5 = 52.5$.

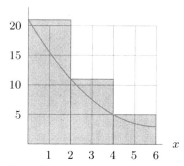

Figure 5.18: Area of shaded region is left-hand sum with $n = 3$

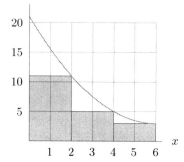

Figure 5.19: Area of shaded region is right-hand sum with $n = 3$

Rough Estimates of a Definite Integral

When calculating an integral using a calculator or computer, it is useful to have a rough idea of the value you expect. This helps detect errors in entering the integral.

Example 5   Three people calculated $\int_1^3 \frac{1}{t}\,dt$ on a calculator and got values 0.023, 11.984, and 1.526. Explain how you can be sure that none of these values is correct.

Solution   Figure 5.20 shows left- and right-hand approximations to $\int_1^3 \frac{1}{t}\,dt$ with $n = 1$. We see that the right-hand sum $2(1/3)$ is an underestimate of $\int_1^3 \frac{1}{t}\,dt$. Since 0.023 is less than $2/3$, this value must be wrong. Similarly, we see that the left-hand sum $2(1)$ is an overestimate of the integral. Since 11.984 is larger than 2, this value is wrong. Since the graph is concave up, the value of the integral is closer to the smaller value of the two sums, $2/3$, than to the larger value, 2. Thus, the integral is less than the average of the two sums, $4/3$. Since $1.526 > 4/3$, this value is wrong too.

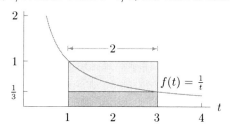

Figure 5.20: Left- and right-hand sums with $n = 1$

Problems for Section 5.2

1. Estimate $\int_0^6 2^x \, dx$ using a left-hand sum with $n = 2$.

2. Estimate $\int_0^{12} \dfrac{1}{x+1} \, dx$ using a left-hand sum with $n = 3$.

3. Use the following table to estimate $\int_0^{25} f(x) \, dx$.

x	0	5	10	15	20	25
$f(x)$	100	82	69	60	53	49

4. Use the following table to estimate $\int_3^4 W(t) \, dt$. What are n and Δt?

t	3.0	3.2	3.4	3.6	3.8	4.0
$W(t)$	25	23	20	15	9	2

5. Use the following table to estimate $\int_0^{15} f(x) \, dx$.

x	0	3	6	9	12	15
$f(x)$	50	48	44	36	24	8

6. Use the table to estimate $\int_0^{40} f(x) \, dx$. What values of n and Δx did you use?

x	0	10	20	30	40
$f(x)$	350	410	435	450	460

7. Using Figure 5.21, draw rectangles representing each of the following Riemann sums for the function f on the interval $0 \le t \le 8$. Calculate the value of each sum.

(a) Left-hand sum with $\Delta t = 4$
(b) Right-hand sum with $\Delta t = 4$
(c) Left-hand sum with $\Delta t = 2$
(d) Right-hand sum with $\Delta t = 2$

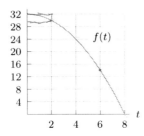

Figure 5.21

8. Use Figure 5.22 to estimate $\int_0^{20} f(x) \, dx$.

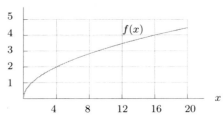

Figure 5.22

9. Use Figure 5.23 to estimate $\int_{-10}^{15} f(x) \, dx$.

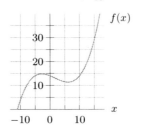

Figure 5.23

Use the graphs in Problems 10–11 to estimate $\int_0^3 f(x) \, dx$.

10.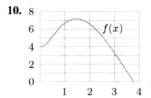

11.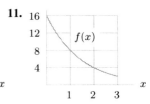

For Problems 12–15:

(a) Use a graph of the integrand to make a rough estimate of the integral. Explain your reasoning.
(b) Use a computer or calculator to find the value of the definite integral.

12. $\displaystyle\int_0^1 x^3 \, dx$

13. $\displaystyle\int_0^3 \sqrt{x} \, dx$

14. $\displaystyle\int_0^1 3^t \, dt$

15. $\displaystyle\int_1^2 x^x \, dx$

16. The rate of change of a quantity is given by $f(t) = t^2 + 1$. Make an underestimate and an overestimate of the total change in the quantity between $t = 0$ and $t = 8$ using

(a) $\Delta t = 4$ (b) $\Delta t = 2$ (c) $\Delta t = 1$

What is n in each case? Graph $f(t)$ and shade rectangles to represent each of your six answers.

17. **(a)** Use a calculator or computer to find $\int_0^6 (x^2 + 1)\, dx$. Represent this value as the area under a curve.

(b) Estimate $\int_0^6 (x^2 + 1)\, dx$ using a left-hand sum with $n = 3$. Represent this sum graphically on a sketch of $f(x) = x^2 + 1$. Is this sum an overestimate or underestimate of the true value found in part (a)?

(c) Estimate $\int_0^6 (x^2 + 1)\, dx$ using a right-hand sum with $n = 3$. Represent this sum on your sketch. Is this sum an overestimate or underestimate?

18. Using Figure 5.24, find the value of $\int_1^6 f(x)\, dx$.

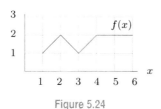

Figure 5.24

19. The graph of a function $f(t)$ is given in Figure 5.25. Which of the following four numbers could be an estimate of $\int_0^1 f(t)\,dt$ accurate to two decimal places? Explain how you chose your answer.

(a) ~~−98.35~~   **(b)** 71.84

(c) ~~100.12~~   **(d)** 93.47

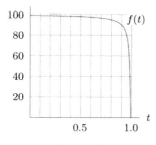

Figure 5.25

In Problems 20–29, use a calculator or computer to evaluate the integral.

20. $\displaystyle\int_0^5 x^2\, dx$ 21. $\displaystyle\int_1^5 (3x + 1)^2\, dx$

22. $\displaystyle\int_1^4 \frac{1}{\sqrt{1 + x^2}}\, dx$ 23. $\displaystyle\int_{-1}^1 \frac{1}{e^t}\, dt$

24. $\displaystyle\int_{1.1}^{1.7} 10(0.85)^t\, dt$ 25. $\displaystyle\int_1^2 2^x\, dx$

26. $\displaystyle\int_1^2 (1.03)^t\, dt$ 27. $\displaystyle\int_1^3 \ln x\, dx$

28. $\displaystyle\int_{1.1}^{1.7} e^t \ln t\, dt$ 29. $\displaystyle\int_{-3}^3 e^{-t^2}\, dt$

30. Use the expressions for left and right sums on page 242 and Table 5.7.

(a) If $n = 4$, what is Δt? What are t_0, t_1, t_2, t_3, t_4? What are $f(t_0), f(t_1), f(t_2), f(t_3), f(t_4)$?

(b) Find the left and right sums using $n = 4$.

(c) If $n = 2$, what is Δt? What are t_0, t_1, t_2? What are $f(t_0), f(t_1), f(t_2)$?

(d) Find the left and right sums using $n = 2$.

Table 5.7

t	15	17	19	21	23
$f(t)$	10	13	18	20	30

31. Use the expressions for left and right sums on page 242 and Table 5.8.

(a) If $n = 4$, what is Δt? What are t_0, t_1, t_2, t_3, t_4? What are $f(t_0), f(t_1), f(t_2), f(t_3), f(t_4)$?

(b) Find the left and right sums using $n = 4$.

(c) If $n = 2$, what is Δt? What are t_0, t_1, t_2? What are $f(t_0), f(t_1), f(t_2)$?

(d) Find the left and right sums using $n = 2$.

Table 5.8

t	0	4	8	12	16
$f(t)$	25	23	22	20	17

5.3 THE DEFINITE INTEGRAL AS AREA

The Definite Integral as an Area: When $f(x)$ is Positive

If $f(x)$ is continuous and positive, each term $f(x_0)\Delta x, f(x_1)\Delta x, \ldots$ in a left- or right-hand Riemann sum represents the area of a rectangle. See Figure 5.26. As the width Δx of the rectangles approaches zero, the rectangles fit the curve of the graph more exactly, and the sum of their areas gets closer to the area under the curve shaded in Figure 5.27. In other words:

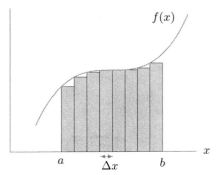

Figure 5.26: Area of rectangles approximating the area under the curve

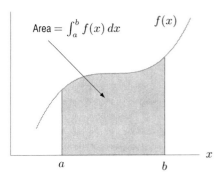

Figure 5.27: Shaded area is the definite integral $\int_a^b f(x)\,dx$

When $f(x)$ is positive and $a < b$:

$$\text{Area under graph of } f \text{ between } a \text{ and } b = \int_a^b f(x)\,dx.$$

Example 1   Find the area under the graph of $y = 10x(3^{-x})$ between $x = 0$ and $x = 3$.

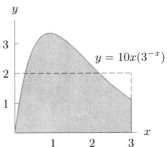

Figure 5.28: Area shaded $= \int_0^3 10x(3^{-x})\,dx$

Solution   The area we want is shaded in Figure 5.28. A rough estimate of this area is 6, since it has about the same area as a rectangle of width 3 and height 2. To find the area more accurately, we say

$$\text{Area shaded } = \int_0^3 10x(3^{-x})\,dx.$$

Using a calculator or computer to evaluate the integral, we obtain

$$\text{Area shaded } = \int_0^3 10x(3^{-x})\,dx = 6.967 \approx 7 \text{ square units.}$$

Relationship Between Definite Integral and Area: When $f(x)$ is Not Positive

We assumed in drawing Figure 5.27 that the graph of $f(x)$ lies above the x-axis. If the graph lies below the x-axis, then each value of $f(x)$ is negative, so each $f(x)\Delta x$ is negative, and the area gets counted negatively. In that case, the definite integral is the negative of the area between the graph of f and the horizontal axis.

Example 2   What is the relation between the definite integral $\int_{-1}^{1} (x^2 - 1)\,dx$ and the area between the parabola $y = x^2 - 1$ and the x-axis?

Solution   The parabola lies below the x-axis between $x = -1$ and $x = 1$. (See Figure 5.29.) So,

$$\int_{-1}^{1} (x^2 - 1)\, dx = -\text{Area} = -1.33.$$

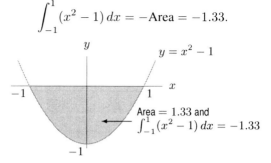

Figure 5.29: Integral $\int_{-1}^{1} (x^2 - 1)\, dx$ is negative of shaded area

Summarizing, assuming $f(x)$ is continuous, we have:

> **When $f(x)$ is positive for some x-values and negative for others**, and $a < b$:
>
> $$\int_{a}^{b} f(x)\, dx$$ is the sum of the areas above the x-axis, counted positively, and the areas below the x-axis, counted negatively.

In the following example, we break up the integral. The properties that allow us to do this are discussed in the Focus on Theory section at the end of this chapter.

Example 3   Interpret the definite integral $\int_{0}^{4} (x^3 - 7x^2 + 11x)\, dx$ in terms of areas.

Solution   Figure 5.30 shows the graph of $f(x) = x^3 - 7x^2 + 11x$ crossing below the x-axis at about $x = 2.38$. The integral is the area above the x-axis, A_1, minus the area below the x-axis, A_2. Computing the integral with a calculator or computer shows

$$\int_{0}^{4} (x^3 - 7x^2 + 11x)\, dx = 2.67.$$

Breaking the integral into two parts and calculating each one separately gives

$$\int_{0}^{2.38} (x^3 - 7x^2 + 11x)\, dx = 7.72 \quad \text{and} \quad \int_{2.38}^{4} (x^3 - 7x^2 + 11x)\, dx = -5.05,$$

so $A_1 = 7.72$ and $A_2 = 5.05$. Then, as we would expect,

$$\int_{0}^{4} (x^3 - 7x^2 + 11x)\, dx = A_1 - A_2 = 7.72 - 5.05 = 2.67.$$

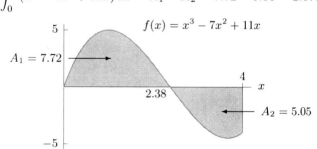

Figure 5.30: Integral $\int_{0}^{4} (x^3 - 7x^2 + 11x)\, dx = A_1 - A_2$

Example 4   Find the total area of the shaded regions in Figure 5.30.

Solution   We saw in Example 3 that $A_1 = 7.72$ and $A_2 = 5.05$. Thus we have

$$\text{Total shaded area} = A_1 + A_2 = 7.72 + 5.05 = 12.77.$$

Example 5   For each of the functions graphed in Figure 5.31, decide whether $\int_0^5 f(x)\,dx$ is positive, negative or approximately zero.

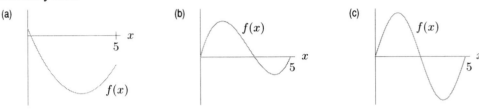

Figure 5.31: Is $\int_0^5 f(x)dx$ positive, negative or zero?

Solution   (a) The graph lies almost entirely below the x-axis, so the integral is negative.

(b) The graph lies partly below the x-axis and partly above the x-axis. However, the area above the x-axis is larger than the area below the x-axis, so the integral is positive.

(c) The graph lies partly below the x-axis and partly above the x-axis. Since the areas above and below the x-axis appear to be approximately equal in size, the integral is approximately zero.

Area Between Two Curves

We can use rectangles to approximate the area between two curves. If $g(x) \leq f(x)$, as in Figure 5.32, the height of a rectangle is $f(x) - g(x)$. The area of the rectangle is $(f(x) - g(x))\Delta x$, and we have the following result:

If $g(x) \leq f(x)$ for $a \leq x \leq b$:

$$\text{Area between graphs of } f(x) \text{ and } g(x) \text{ for } a \leq x \leq b = \int_a^b (f(x) - g(x))\,dx.$$

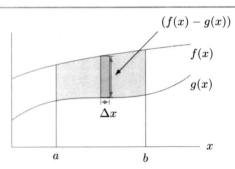

Figure 5.32: Area between two curves $= \int_a^b (f(x) - g(x))\,dx$

Example 6   Graphs of $f(x) = 4x - x^2$ and $g(x) = \frac{1}{2}x^{3/2}$ for $x \geq 0$ are shown in Figure 5.33. Use a definite integral to estimate the area enclosed by the graphs of these two functions.

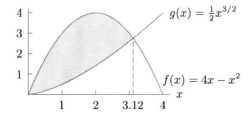

Figure 5.33: Find the area between $f(x) = 4x - x^2$ and $g(x) = \frac{1}{2}x^{3/2}$ using an integral

Solution   The region enclosed by the graphs of the two functions is shaded in Figure 5.33. The two graphs cross at $x = 0$ and at $x \approx 3.12$. Between these values, the graph of $f(x) = 4x - x^2$ lies above the graph of $g(x) = \frac{1}{2}x^{3/2}$. Using a calculator or computer to evaluate the integral, we get

$$\text{Area between graphs} = \int_0^{3.12} \left((4x - x^2) - \frac{1}{2}x^{3/2} \right) dx = 5.906.$$

Problems for Section 5.3

1. Find the area under $y = x^3 + 2$ between $x = 0$ and $x = 2$. Sketch this area.

2. Find the area under $P = 100(0.6)^t$ between $t = 0$ and $t = 8$.

3. Find the area between $y = x+5$ and $y = 2x+1$ between $x = 0$ and $x = 2$.

4. Find the area enclosed by $y = 3x$ and $y = x^2$.

5. (a) What is the area between the graph of $f(x)$ in Figure 5.34 and the x-axis, between $x = 0$ and $x = 5$?
(b) What is $\int_0^5 f(x)\,dx$?

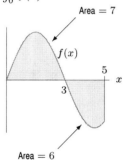

Figure 5.34

In Problems 6–9, decide whether $\int_{-3}^{3} f(x)\,dx$ is positive, negative, or approximately zero.

6.

7.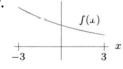

8.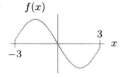

9.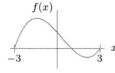

10. (a) Estimate (by counting the squares) the total area shaded in Figure 5.35.
(b) Using Figure 5.35, estimate $\int_0^8 f(x)\,dx$.
(c) Why are your answers to parts (a) and (b) different?

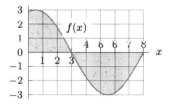

Figure 5.35

11. Using Figure 5.36, estimate $\int_{-3}^{5} f(x)\,dx$.

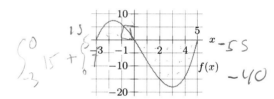

Figure 5.36

12. Given $\int_{-1}^{0} f(x)\,dx = 0.25$ and Figure 5.37, estimate:
(a) $\int_0^1 f(x)\,dx$ **(b)** $\int_{-1}^{1} f(x)\,dx$
(c) The total shaded area.

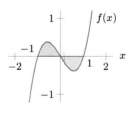

Figure 5.37

13. Given $\int_{-2}^{0} f(x)dx = 4$ and Figure 5.38, estimate:

   (a) $\int_{0}^{2} f(x)dx$ **(b)** $\int_{-2}^{2} f(x)dx$

   (c) The total shaded area.

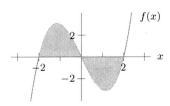

Figure 5.38

In Problems 14–17, match the graph with one of the following possible values for the integral $\int_{0}^{5} f(x)\,dx$:

   I. $- 10.4$ II. $- 2.1$ III. 5.2 IV. 10.4

14.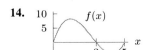

15.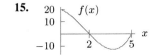

16.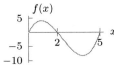

17.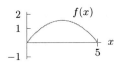

18. Use Figure 5.39 to find the values of

   (a) $\int_{a}^{b} f(x)\,dx$ **(b)** $\int_{b}^{c} f(x)\,dx$

   (c) $\int_{a}^{c} f(x)\,dx$ **(d)** $\int_{a}^{c} |f(x)|\,dx$

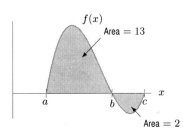

Figure 5.39

19. Using Figure 5.40, list the following integrals in increasing order (from smallest to largest). Which integrals are negative, which are positive? Give reasons.

   I. $\int_{a}^{b} f(x)\,dx$ II. $\int_{a}^{c} f(x)\,dx$ III. $\int_{a}^{e} f(x)\,dx$

   IV. $\int_{b}^{e} f(x)\,dx$ V. $\int_{b}^{c} f(x)\,dx$

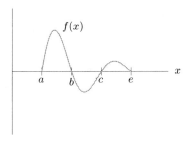

Figure 5.40

20. (a) Graph $f(x) = x(x + 2)(x - 1)$.

   (b) Find the total area between the graph and the x-axis between $x = -2$ and $x = 1$.

   (c) Find $\int_{-2}^{1} f(x)\,dx$ and interpret it in terms of areas.

21. (a) Using Figure 5.41, find $\int_{-3}^{0} f(x)\,dx$.

   (b) If the area of the shaded region is A, estimate $\int_{-3}^{4} f(x)\,dx$.

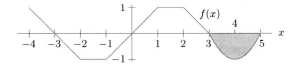

Figure 5.41

22. Use the following table to estimate the area between $f(x)$ and the x-axis on the interval $0 \le x \le 20$.

x	0	5	10	15	20
$f(x)$	15	18	20	16	12

For Problems 23–24, compute the definite integral and interpret the result in terms of areas.

23. $\int_{1}^{4} \dfrac{x^2 - 3}{x}\,dx$.   **24.** $\int_{1}^{4} (x - 3\ln x)\,dx$.

25. Compute the definite integral $\int_{0}^{4} \cos \sqrt{x}\,dx$ and interpret the result in terms of areas.

26. Find the area between the graph of $y = x^2 - 2$ and the x-axis, between $x = 0$ and $x = 3$.

In Problems 27–34, use an integral to find the specified area.

27. Under $y = 6x^3 - 2$ for $5 \le x \le 10$.

28. Under $y = 2\cos(t/10)$ for $1 \le t \le 2$.

29. Under $y = 5\ln(2x)$ and above $y = 3$ for $3 \le x \le 5$.

30. Between $y = \sin x + 2$ and $y = 0.5$ for $6 \le x \le 10$.

31. Between $y = \cos x + 7$ and $y = \ln(x - 3)$, $5 \le x \le 7$.

32. Above the curve $y = x^4 - 8$ and below the x-axis.

33. Above the curve $y = -e^x + e^{2(x-1)}$ and below the x-axis, for $x \ge 0$.

34. Between $y = \cos t$ and $y = \sin t$ for $0 \le t \le \pi$.

5.4 INTERPRETATIONS OF THE DEFINITE INTEGRAL

The Notation and Units for the Definite Integral

Just as the Leibniz notation dy/dx for the derivative reminds us that the derivative is the limit of a quotient of differences, the notation for the definite integral,

$$\int_a^b f(x)\,dx,$$

reminds us that an integral is a limit of a sum. The integral sign is a misshapen S. Since the terms being added are products of the form "$f(x)$ times a difference in x," we have the following result:

> The unit of measurement for $\int_a^b f(x)\,dx$ is the product of the units for $f(x)$ and the units for x.

For example, if x and $f(x)$ have the same units, then the integral $\int_a^b f(x)\,dx$ is measured in square units, say cm $\times$ cm $=$ cm^2. This is what we would expect, since the integral represents an area.

Similarly, if $f(t)$ is velocity in meters/second and t is time in seconds, then the integral

$$\int_a^b f(t)\,dt$$

has units of (meters/sec) $\times$ (sec) $=$ meters, which is what we expect since the integral represents change in position. We saw in Section 5.1 that total change could be approximated by a Riemann sum formed using the rate of change. In the limit, we have:

> If $f(t)$ is a rate of change of a quantity, then
>
> $$\text{Total change in quantity between } t = a \text{ and } t = b = \int_a^b f(t)\,dt$$

The units correspond: If $f(t)$ is a rate of change, with units of quantity/time, then $f(t)\Delta t$ and the definite integral have units of (quantity/time) $\times$ (time) $=$ quantity.

Example 1 A bacteria colony initially has a population of 14 million bacteria. Suppose that t hours later the population is growing at a rate of $f(t) = 2^t$ million bacteria per hour.

(a) Give a definite integral that represents the total change in the bacteria population during the time from $t = 0$ to $t = 2$.

(b) Find the population at time $t = 2$.

Solution (a) Since $f(t) = 2^t$ gives the rate of change of population, we have

$$\text{Change in population between } t = 0 \text{ and } t = 2 = \int_0^2 2^t\,dt.$$

(b) Using a calculator, we find $\int_0^2 2^t\,dt = 4.328$. The bacteria population was 14 million at time $t = 0$ and increased 4.328 million between $t = 0$ and $t = 2$. Therefore, at time $t = 2$,

$$\text{Population} = 14 + 4.328 = 18.328 \text{ million bacteria.}$$

Example 2 Suppose that $C(t)$ represents the cost per day to heat your home in dollars per day, where t is time measured in days and $t = 0$ corresponds to January 1, 2010. Interpret $\int_0^{90} C(t)\,dt$.

Solution   The units for the integral $\int_0^{90} C(t)\, dt$ are (dollars/day) × (days) = dollars. The integral represents the cost in dollars to heat your house for the first 90 days of 2010, namely the months of January, February, and March.

Example 3   A man starts 50 miles away from his home and takes a trip in his car. He moves on a straight line, and his home lies on this line. His velocity is given in Figure 5.42.

(a) When is the man closest to his home? Approximately how far away is he then?

(b) When is the man farthest from his home? How far away is he then?

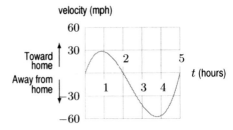

Figure 5.42: Velocity of trip starting 50 miles from home

Solution   What happens on this trip? The velocity function is positive the first two hours and negative between $t = 2$ and $t = 5$. So the man moves toward his home during the first two hours, then turns around at $t = 2$ and moves away from his home. The distance he travels is represented by the area between the graph of velocity and the t-axis; since the area below the axis is greater than the area above the axis, we see that he ends up farther away from home than when he started. Thus he is closest to home at $t = 2$ and farthest from home at $t = 5$. We can estimate how far he went in each direction by estimating areas.

(a) The man starts out 50 miles from home. The distance the man travels during the first two hours is the area under the curve between $t = 0$ and $t = 2$. This area corresponds to about one grid square. Since each grid square has area (30 miles/hour)(1 hour) = 30 miles, the man travels about 30 miles toward home. He is closest to home after 2 hours, and he is about 20 miles away at that time.

(b) Between $t = 2$ and $t = 5$, the man moves away from his home. Since this area is equal to about 3.5 grid squares, which is $(3.5)(30) = 105$ miles, he has moved 105 miles farther from home. He was already 20 miles from home, so at $t = 5$ he is about 125 miles from home. He is farthest from home at $t = 5$.

Notice that the man has covered a total distance of $30 + 105 = 135$ miles. However, he went toward his home for 30 miles and away from his home for 105 miles. His *net* change in position is 75 miles.

Example 4   The rates of growth of the populations of two species of plants (measured in new plants per year) are shown in Figure 5.43. Assume that the populations of the two species are equal at time $t = 0$.

(a) Which population is larger after one year? After two years?

(b) How much does the population of species 1 increase during the first two years?

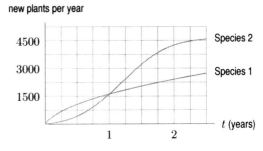

Figure 5.43: Population growth rates for two species of plants

Solution
(a) The rate of growth of the population of species 1 is higher than that of species 2 throughout the first year, so the population of species 1 is larger after one year. After two years, the situation is less clear, since the population of species 1 increased faster for the first year and that of species 2 for the second. However, if $r(t)$ is the rate of growth of a population, we have

$$\text{Total change in population during first two years} = \int_0^2 r(t)dt.$$

This integral is the area under the graph of $r(t)$. For $t = 0$ to $t = 2$, the area under the species 1 graph in Figure 5.43 is smaller than the area under the species 2 graph, so the population of species 2 is larger after two years.

(b) The population change for species 1 is the area of the region under the graph of $r(t)$ between $t = 0$ and $t = 2$ in Figure 5.43. The region consists of about 16.5 grid squares, each of area (750 plants/year)(0.25 year) = 187.5 plants, giving a total of (16.5)(187.5) = 3093.75 plants. The population of species 1 increases by about 3100 plants during the two years.

Bioavailability of Drugs

In pharmacology, the definite integral is used to measure *bioavailability*; that is, the overall presence of a drug in the bloodstream during the course of a treatment. Unit bioavailability represents 1 unit concentration of the drug in the bloodstream for 1 hour. For example, a concentration of 3 μg/cm^3 in the blood for 2 hours has bioavailability of $3 \cdot 2 = 6$ (μg/cm^3)-hours.

Ordinarily the concentration of a drug in the blood is not constant. Typically, the concentration in the blood increases as the drug is absorbed into the bloodstream, and then decreases as the drug is broken down and excreted.[2] (See Figure 5.44.)

Suppose that we want to calculate the bioavailability of a drug that is in the blood with concentration $C(t)\mu$g/cm^3 at time t for the time period $0 \leq t \leq T$. Over a small interval Δt, we estimate

$$\text{Bioavailability} \approx \text{Concentration} \times \text{Time} = C(t)\Delta t.$$

Summing over all subintervals gives

$$\text{Total bioavailability} \approx \sum C(t)\Delta t.$$

In the limit as $n \to \infty$, where n is the number of intervals of width Δt, the sum becomes an integral. So for $0 \leq t \leq T$, we have

$$\text{Bioavailability} = \int_0^T C(t)dt.$$

That is, the total bioavailability of a drug is equal to the area under the drug concentration curve.

concentration of drug in blood stream

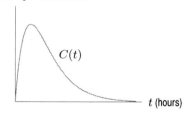

Figure 5.44: Curve showing drug concentration as a function of time

[2]*Drug Treatment*, Graeme S. Avery (Ed.) (Sydney: Adis Press, 1976).

Example 5   Blood concentration curves[3] of two drugs are shown in Figure 5.45. Describe the differences and similarities between the two drugs in terms of peak concentration, speed of absorption into the bloodstream, and total bioavailability.

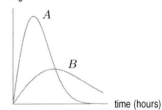

concentration of drug in blood stream

Figure 5.45: Concentration curves of two drugs

Solution   Drug A has a peak concentration more than twice as high as that of drug B. Because drug A achieves peak concentration sooner than drug B, drug A appears to be absorbed more rapidly into the blood stream than drug B. Finally, drug A has greater total bioavailability, since the area under the graph of the concentration function for drug A is greater than the area under the graph for drug B.

Problems for Section 5.4

1. The following table gives the emissions, E, of nitrogen oxides in millions of metric tons per year in the US.[4] Let t be the number of years since 1970 and $E = f(t)$.

   (a) What are the units and meaning of $\int_0^{30} f(t)dt$?

   (b) Estimate $\int_0^{30} f(t)dt$.

Year	1970	1975	1980	1985	1990	1995	2000
E	26.9	26.4	27.1	25.8	25.5	25.0	22.6

2. Annual coal production in the US (in quadrillion BTU per year) is given in the table.[5] Estimate the total amount of coal produced in the US between 1960 and 1990. If $r = f(t)$ is the rate of coal production t years since 1960, write an integral to represent the 1960–1990 coal production.

Year	1960	1965	1970	1975	1980	1985	1990
Rate	10.82	13.06	14.61	14.99	18.60	19.33	22.46

In Problems 3–6, explain in words what the integral represents and give units.

3. $\int_1^3 v(t)\,dt$, where $v(t)$ is velocity in meters/sec and t is time in seconds.

4. $\int_0^6 a(t)\,dt$, where $a(t)$ is acceleration in km/hr² and t is time in hours.

5. $\int_{2000}^{2004} f(t)\,dt$, where $f(t)$ is the rate at which the world's population is growing in year t, in billion people per year.

6. $\int_0^5 s(x)\,dx$, where $s(x)$ is rate of change of salinity (salt concentration) in gm/liter per cm in sea water, and where x is depth below the surface of the water in cm.

7. Oil leaks out of a tanker at a rate of $r = f(t)$ gallons per minute, where t is in minutes. Write a definite integral expressing the total quantity of oil which leaks out of the tanker in the first hour.

8. A cup of coffee at 90°C is put into a 20°C room when $t = 0$. The coffee's temperature is changing at a rate of $r(t) = -7(0.9^t)$ °C per minute, with t in minutes. Estimate the coffee's temperature when $t = 10$.

9. After a foreign substance is introduced into the blood, the rate at which antibodies are made is given by

$$r(t) = \frac{t}{t^2 + 1} \text{ thousands of antibodies per minute,}$$

where time, t, is in minutes. Assuming there are no antibodies present at time $t = 0$, find the total quantity of antibodies in the blood at the end of 4 minutes.

---

[3]*Drug Treatment*, Graeme S. Avery (Ed.) (Sydney: Adis Press, 1976).
[4]*The World Almanac and Book of Facts 2005*, p. 177 (New York: World Almanac Books).
[5]*World Almanac*, 1995.

10. World annual natural gas[6] consumption, N, in millions of metric tons of oil equivalent, is approximated by $N = 1770 + 53t$, where t is in years since 1990.

   (a) How much natural gas was consumed in 1990? In 2010?
   (b) Estimate the total amount of natural gas consumed during the 20-year period from 1990 to 2010.

11. Solar photovoltaic (PV) cells are the world's fastest-growing energy source.[7] Annual solar PV production, S, in megawatts, is approximated by $S = 277e^{0.368t}$, where t is in years since 2000. Estimate the total solar PV production between 2000 and 2010.

12. The velocity of a car (in miles per hour) is given by $v(t) = 40t - 10t^2$, where t is in hours.

   (a) Write a definite integral for the distance the car travels during the first three hours.
   (b) Sketch a graph of velocity against time and represent the distance traveled during the first three hours as an area on your graph.
   (c) Use a computer or calculator to find this distance.

13. Your velocity is $v(t) = \ln(t^2 + 1)$ ft/sec for t in seconds, $0 \le t \le 3$. Estimate the distance traveled during this time.

14. Figure 5.46 shows the rate of change of the quantity of water in a water tower, in liters per day, during the month of April. If the tower had 12,000 liters of water in it on April 1, estimate the quantity of water in the tower on April 30.

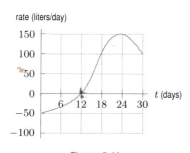

Figure 5.46

15. Figure 5.47 shows the weight growth rate of a human fetus.

   (a) What property of a graph of weight as a function of age corresponds to the fact that the function in Figure 5.47 is increasing?
   (b) Estimate the weight of a baby born in week 40.

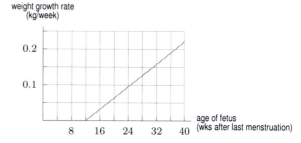

Figure 5.47: Rate of increase of fetal weight

Problems 16–18 concern the future of the US Social Security Trust Fund, out of which pensions are paid. Figure 5.48 shows the rates (billions of dollars per year) at which income, $I(t)$, from taxes and interest is projected to flow into the fund and at which expenditures, $E(t)$, flow out of the fund. Figure 5.49 shows the value of the fund as a function of time.[8]

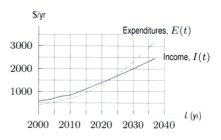

Figure 5.48

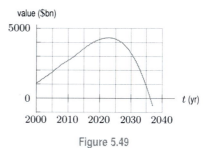

Figure 5.49

16. (a) Write each of the following areas in Figure 5.48 from 2000 to 2015 as an integral and explain its significance for the fund.
   (i) Under the income curve
   (ii) Under the expenditure curve
   (iii) Between the income and expenditure curves
   (b) Use Figure 5.49 to estimate the area between the two curves from 2000 to 2015.

17. Decide when the value of the fund is projected to be a maximum using
   (a) Figure 5.48   **(b)** Figure 5.49

---

[6] *BP Statistical Review of World Energy 2009*, http://www.bp.com/ (accessed 9/6/2009).
[7] *Vital Signs 2007-2008*, The Worldwatch Institute, W.W. Norton & Company, 2007, p. 38.
[8] The 2009 OASDI Trustees Report, www.ssa.gov/OACT/TR/2009.

18. Express the projected increase in value of the fund from 2000 to 2030 as an integral.

19. A forest fire covers 2000 acres at time $t = 0$. The fire is growing at a rate of $8\sqrt{t}$ acres per hour, where t is in hours. How many acres are covered 24 hours later?

20. Water is pumped out of a holding tank at a rate of $5 - 5e^{-0.12t}$ liters/minute, where t is in minutes since the pump is started. If the holding tank contains 1000 liters of water when the pump is started, how much water does it hold one hour later?

21. Figure 5.50 shows the rate of growth of two trees. If the two trees are the same height at time $t = 0$, which tree is taller after 5 years? After 10 years?

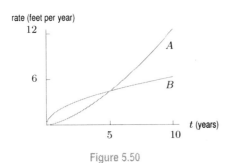

Figure 5.50

22. Figure 5.51 shows the number of sales per month made by two salespeople. Which person has the most total sales after 6 months? After the first year? At approximately what times (if any) have they sold roughly equal total amounts? Approximately how many total sales has each person made at the end of the first year?

Figure 5.51

23. Height velocity graphs are used by endocrinologists to follow the progress of children with growth deficiencies. Figure 5.52 shows the height velocity curves of an average boy and an average girl between ages 3 and 18.

(a) Which curve is for girls and which is for boys? Explain how you can tell.

(b) About how much does the average boy grow between ages 3 and 10?

(c) The growth spurt associated with adolescence and the onset of puberty occurs between ages 12 and 15 for the average boy and between ages 10 and 12.5 for the average girl. Estimate the height gained by each average child during this growth spurt.

(d) When fully grown, about how much taller is the average man than the average woman? (The average boy and girl are about the same height at age 3.)

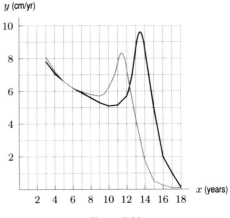

Figure 5.52

24. The birth rate, B, in births per hour, of a bacteria population is given in Figure 5.53. The curve marked D gives the death rate, in deaths per hour, of the same population.

(a) Explain what the shape of each of these graphs tells you about the population.

(b) Use the graphs to find the time at which the net rate of increase of the population is at a maximum.

(c) At time $t = 0$ the population has size N. Sketch the graph of the total number born by time t. Also sketch the graph of the number alive at time t. Estimate the time at which the population is a maximum.

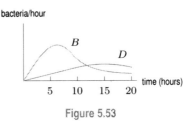

Figure 5.53

25. The rates of consumption of stores of protein and fat in the human body during 8 weeks of starvation are shown in Figure 5.54. Does the body burn more fat or more protein during this period?

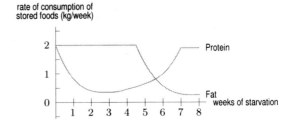

Figure 5.54

A healthy human heart pumps about 5 liters of blood per minute. Problems 26–27 refer to Figure 5.55, which shows the response of the heart to bleeding. The pumping rate drops and then returns to normal if the person recovers fully, or drops to zero if the person dies.

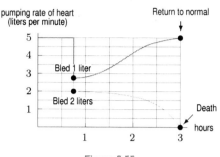

Figure 5.55

26. **(a)** If the body is bled 2 liters, how much blood is pumped during the three hours leading to death?
   (b) If $f(t)$ is the pumping rate in liters per minute at time t hours, express your answer to part (a) as a definite integral.
   (c) How much more blood would have been pumped during the same time period if there had been no bleeding? Illustrate your answer on the graph.

27. **(a)** If the body is bled 1 liter, how much blood is pumped during the three hours leading to full recovery?
   (b) If $g(t)$ is the pumping rate in liters per minute at time t hours, express your answer to part (a) as a definite integral.
   (c) How much more blood would have been pumped during the same time period if there had been no bleeding? Show your answer as an area on the graph.

28. The amount of waste a company produces, W, in tons per week, is approximated by $W = 3.75e^{-0.008t}$, where t is in weeks since January 1, 2005. Waste removal for the company costs \$15/ton. How much does the company pay for waste removal during the year 2005?

Problems 29–32 show the velocity, in cm/sec, of a particle moving along the x-axis. Compute the particle's change in position, left (negative) or right (positive), between times $t = 0$ and $t = 5$ seconds.

29.
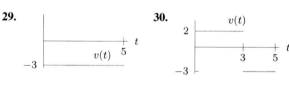

30.

31.
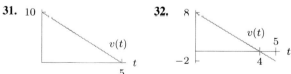

32.

33. Figure 5.56 gives your velocity during a trip starting from home. Positive velocities take you away from home and negative velocities take you toward home. Where are you at the end of the 5 hours? When are you farthest from home? How far away are you at that time?

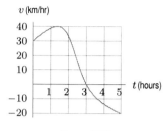

Figure 5.56

34. A bicyclist is pedaling along a straight road for one hour with a velocity v shown in Figure 5.57. She starts out five kilometers from the lake and positive velocities take her toward the lake. [Note: The vertical lines on the graph are at 10 minute (1/6 hour) intervals.]

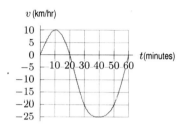

Figure 5.57

   (a) Does the cyclist ever turn around? If so, at what time(s)?
   (b) When is she going the fastest? How fast is she going then? Toward the lake or away?
   (c) When is she closest to the lake? Approximately how close to the lake does she get?
   (d) When is she farthest from the lake? Approximately how far from the lake is she then?

35. Figure 5.58 shows plasma concentration curves for two drugs used to slow a rapid heart rate. Compare the two products in terms of level of peak concentration, time until peak concentration, and overall bioavailability.

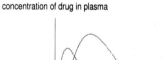

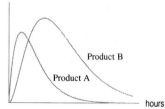

Figure 5.58

36. Figure 5.59 compares the concentration in blood plasma for two pain relievers. Compare the two products in terms of level of peak concentration, time until peak concentration, and overall bioavailability.

concentration of drug in plasma

Product B

Product A

hours

Figure 5.59

37. Draw plasma concentration curves for two drugs A and B if product A has the highest peak concentration, but product B is absorbed more quickly and has greater overall bioavailability.

38. A two-day environmental cleanup started at 9 am on the first day. The number of workers fluctuated as shown in Figure 5.60. If the workers were paid $10 per hour, how much was the total personnel cost of the cleanup?

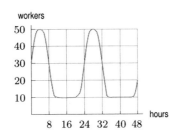

workers

50
40
30
20
10

8 16 24 32 40 48   hours

Figure 5.60

39. Suppose in Problem 38 that the workers were paid $10 per hour for work during the time period 9 am to 5 pm and were paid $15 per hour for work during the rest of the day. What would the total personnel costs of the clean up have been under these conditions?

40. At the site of a spill of radioactive iodine, radiation levels were four times the maximum acceptable limit, so an evacuation was ordered. If R_0 is the initial radiation level

(at $t = 0$) and t is the time in hours, the radiation level $R(t)$, in millirems/hour, is given by

$$R(t) = R_0(0.996)^t.$$

(a) How long does it take for the site to reach the acceptable level of radiation of 0.6 millirems/hour?
(b) How much total radiation (in millirems) has been emitted by that time?

41. If you jump out of an airplane and your parachute fails to open, your downward velocity (in meters per second) t seconds after the jump is approximated by

$$v(t) = 49(1 - (0.8187)^t).$$

(a) Write an expression for the distance you fall in T seconds.
(b) If you jump from 5000 meters above the ground, estimate, using trial and error, how many seconds you fall before hitting the ground.

42. The Montgolfier brothers (Joseph and Etienne) were eighteenth-century pioneers in the field of hot-air ballooning. Had they had the appropriate instruments, they might have left us a record, like that shown in Figure 5.61, of one of their early experiments. The graph shows their vertical velocity, v, with upward as positive.

(a) Over what intervals was the acceleration positive? Negative?
(b) What was the greatest altitude achieved, and at what time?
(c) This particular flight ended on top of a hill. How do you know that it did, and what was the height of the hill above the starting point?

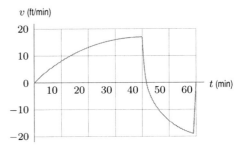

v (ft/min)

20

10

0   10   20   30   40   50   60 t (min)

−10

−20

Figure 5.61

5.5 THE FUNDAMENTAL THEOREM OF CALCULUS

In Section 5.4, we saw that the total change of a quantity can be obtained by integrating its rate of change. Since the change in F between a and b is $F(b) - F(a)$ and the rate of change is $F'(t)$ we have the following result:

The Fundamental Theorem of Calculus

If $F'(t)$ is continuous for $a \leq t \leq b$, then

$$\int_a^b F'(t)\, dt = F(b) - F(a).$$

In words:

The definite integral of the derivative of a function gives the total change in the function.

In Section 7.3 we see how to use the Fundamental Theorem to compute a definite integral. The Focus on Theory Section gives another version of the Fundamental Theorem.

Example 1 Figure 5.62 shows $F'(t)$, the rate of change of the value, $F(t)$, of an investment over a 5-month period.

(a) When is the value of the investment increasing in value and when is it decreasing?

(b) Does the investment increase or decrease in value during the 5 months?

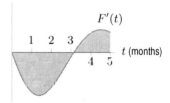

rate of change of value
of investment ($/month)

Figure 5.62: Did the investment increase or decrease in value over these 5 months?

Solution (a) The investment decreased in value during the first 3 months, since the rate of change of value is negative then. The value rose during the last 2 months.

(b) We want to find the total change in the value of the investment between $t = 0$ and $t = 5$. Since the total change is the integral of the rate of change, $F'(t)$, we are looking for

$$\text{Total change in value} = \int_0^5 F'(t)\,dt.$$

The integral equals the shaded area above the t-axis minus the shaded area below the t-axis. Since in Figure 5.62 the area below the axis is greater than the area above the axis, the integral is negative. The total change in value of the investment during this time is negative, so it decreased in value.

Marginal Cost and Change in Total Cost

Suppose $C(q)$ represents the cost of producing q items. The derivative, $C'(q)$, is the marginal cost. Since marginal cost $C'(q)$ is the rate of change of the cost function with respect to quantity, by the Fundamental Theorem, the integral

$$\int_a^b C'(q)\,dq$$

represents the total change in the cost function between $q = a$ and $q = b$. In other words, the integral gives the amount it costs to increase production from a units to b units.

The cost of producing 0 units is the fixed cost $C(0)$. The area under the marginal cost curve between $q = 0$ and $q = b$ is the total increase in cost between a production of 0 and a production of b. This is called the *total variable cost*. Adding this to the fixed cost gives the total cost to produce b units. In summary,

---

If $C'(q)$ is a marginal cost function and $C(0)$ is the fixed cost,

$$\text{Cost to increase production from } a \text{ units to } b \text{ units} = C(b) - C(a) = \int_a^b C'(q)\, dq$$

$$\text{Total variable cost to produce } b \text{ units} = \int_0^b C'(q)\, dq$$

$$\text{Total cost of producing } b \text{ units} = \text{Fixed cost} + \text{Total variable cost}$$

$$= C(0) + \int_0^b C'(q)\, dq$$

Example 2   A marginal cost curve is given in Figure 5.63. If the fixed cost is $1000, estimate the total cost of producing 250 items.

Solution   The total cost of production is Fixed cost + Variable cost. The variable cost of producing 250 items is represented by the area under the marginal cost curve. The area in Figure 5.63 between $q = 0$ and $q = 250$ is about 20 grid squares. Each grid square has area (2 dollars/item)(50 items) = 100 dollars, so

$$\text{Total variable cost} = \int_0^{250} C'(q)\, dq \approx 20(100) = 2000.$$

The total cost to produce 250 items is given by :

$$\text{Total cost} = \text{Fixed cost} + \text{Total variable cost}$$

$$\approx \$1000 + \$2000 = \$3000.$$

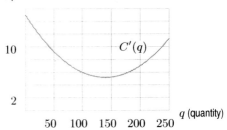

Figure 5.63: A marginal cost curve

Problems for Section 5.5

1. If the marginal cost function $C'(q)$ is measured in dollars per ton, and q gives the quantity in tons, what are the units of measurement for $\int_{800}^{900} C'(q)\, dq$? What does this integral represent?

2. The marginal cost function of a product, in dollars per unit, is $C'(q) = q^2 - 50q + 700$. If fixed costs are $500, find the total cost to produce 50 items.

3. The total cost in dollars to produce q units of a product is $C(q)$. Fixed costs are $20,000. The marginal cost is

$$C'(q) = 0.005q^2 - q + 56.$$

   (a) On a graph of $C'(q)$, illustrate graphically the total variable cost of producing 150 units.
   (b) Estimate $C(150)$, the total cost to produce 150 units.
   (c) Find the value of $C'(150)$ and interpret your answer in terms of costs of production.
   (d) Use parts (b) and (c) to estimate $C(151)$.

4. A marginal cost function $C'(q)$ is given in Figure 5.64. If the fixed costs are $10,000, estimate:

  (a) The total cost to produce 30 units.
  (b) The additional cost if the company increases production from 30 units to 40 units.
  (c) The value of $C'(25)$. Interpret your answer in terms of costs of production.

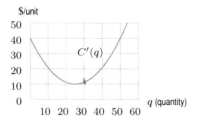

Figure 5.64

5. The population of Tokyo grew at the rate shown in Figure 5.65. Estimate the change in population between 1970 and 1990.

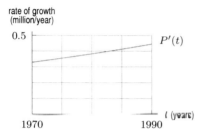

Figure 5.65

6. The marginal cost function for a company is given by

$$C'(q) = q^2 - 16q + 70 \text{ dollars/unit},$$

where q is the quantity produced. If $C(0) = 500$, find the total cost of producing 20 units. What is the fixed cost and what is the total variable cost for this quantity?

7. The marginal cost $C'(q)$ (in dollars per unit) of producing q units is given in the following table.

  (a) If fixed cost is $10,000, estimate the total cost of producing 400 units.
  (b) How much would the total cost increase if production were increased one unit, to 401 units?

q	0	100	200	300	400	500	600
$C'(q)$	25	20	18	22	28	35	45

8. The marginal cost function of producing q mountain bikes is

$$C'(q) = \frac{600}{0.3q + 5}.$$

  (a) If the fixed cost in producing the bicycles is $2000, find the total cost to produce 30 bicycles.
  (b) If the bikes are sold for $200 each, what is the profit (or loss) on the first 30 bicycles?
  (c) Find the marginal profit on the 31st bicycle.

9. The marginal revenue function on sales of q units of a product is $R'(q) = 200 - 12\sqrt{q}$ dollars per unit.

  (a) Graph $R'(q)$.
  (b) Estimate the total revenue if sales are 100 units.
  (c) What is the marginal revenue at 100 units? Use this value and your answer to part (b) to estimate the total revenue if sales are 101 units.

10. Figure 5.66 shows $P'(t)$, the rate of change of the price of stock in a certain company at time t.

  (a) At what time during this five-week period was the stock at its highest value? At its lowest value?
  (b) If $P(t)$ represents the price of the stock, arrange the following quantities in increasing order:

$$P(0), \ P(1), \ P(2), \ P(3), \ P(4), \ P(5).$$

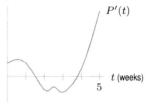

Figure 5.66

11. Ice is forming on a pond at a rate given by

$$\frac{dy}{dt} = \frac{\sqrt{t}}{2} \text{ inches per hour,}$$

where y is the thickness of the ice in inches at time t measured in hours since the ice started forming.

  (a) Estimate the thickness of the ice after 8 hours.
  (b) At what rate is the thickness of the ice increasing after 8 hours?

12. The net worth, $f(t)$, of a company is growing at a rate of $f'(t) = 2000 - 12t^2$ dollars per year, where t is in years since 2005. How is the net worth of the company expected to change between 2005 and 2015? If the company is worth $40,000 in 2005, what is it worth in 2015?

13. The graph of a derivative $f'(x)$ is shown in Figure 5.67. Fill in the table of values for $f(x)$ given that $f(0) = 2$.

x	0	1	2	3	4	5	6
$f(x)$	2						

Figure 5.67: Graph of f', not f

CHAPTER SUMMARY

- **Definite integral as limit of right-hand or left-hand sums**
- **Interpretations of the definite integral**
  Total change from rate of change, change in position given velocity, area, bioavailability, total variable cost.
- **Working with the definite integral**
  Estimate definite integral from graph, table of values, or formula.
- **Fundamental Theorem of Calculus**

REVIEW PROBLEMS FOR CHAPTER FIVE

1. The velocity $v(t)$ in Table 5.9 is decreasing, $2 \leq t \leq 12$. Using $n = 5$ subdivisions to approximate the total distance traveled, find

   (a) An upper estimate   (b) A lower estimate

   Table 5.9

t	2	4	6	8	10	12
$v(t)$	44	42	41	40	37	35

2. The velocity $v(t)$ in Table 5.10 is increasing, $0 \leq t \leq 12$.

   (a) Find an upper estimate for the total distance traveled using
       (i) $n = 4$ (ii) $n = 2$
   (b) Which of the two answers in part (a) is more accurate? Why?
   (c) Find a lower estimate of the total distance traveled using $n = 4$.

   Table 5.10

t	0	3	6	9	12
$v(t)$	34	37	38	40	45

3. Use the following table to estimate $\int_{10}^{26} f(x)\, dx$.

x	10	14	18	22	26
$f(x)$	100	88	72	50	28

4. If $f(t)$ is measured in miles per hour and t is measured in hours, what are the units of $\int_a^b f(t)\, dt$?

5. If $f(t)$ is measured in meters/second^2 and t is measured in seconds, what are the units of $\int_a^b f(t)\, dt$?

6. If $f(t)$ is measured in dollars per year and t is measured in years, what are the units of $\int_a^b f(t)\, dt$?

7. If $f(x)$ is measured in pounds and x is measured in feet, what are the units of $\int_a^b f(x)\, dx$?

8. As coal deposits are depleted, it becomes necessary to strip-mine larger areas for each ton of coal. Figure 5.68 shows the number of acres of land per million tons of coal that will be defaced during strip-mining as a function of the number of million tons removed, starting from the present day.

   (a) Estimate the total number of acres defaced in extracting the next 4 million tons of coal (measured from the present day). Draw four rectangles under the curve, and compute their area.
   (b) Reestimate the number of acres defaced using rectangles above the curve.
   (c) Use your answers to parts (a) and (b) to get a better estimate of the actual number of acres defaced.

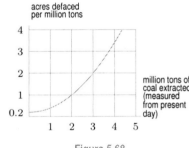

Figure 5.68

For Problems 9–14, use a calculator or computer to evaluate the integral.

9. $\int_0^{10} 2^{-x}\, dx$

10. $\int_1^5 (x^2 + 1)\, dx$

11. $\int_0^1 \sqrt{1 + t^2}\, dt$

12. $\int_{-1}^1 \frac{x^2 + 1}{x^2 - 4}\, dx$

13. $\int_2^3 \frac{-1}{(r + 1)^2}\, dr$

14. $\int_1^3 \frac{z^2 + 1}{z}\, dz$

15. Find the area under the graph of $f(x) = x^2 + 2$ between $x = 0$ and $x = 6$.

In Problems 16–19, find the given area.

16. Between $y = x^2$ and $y = x^3$ for $0 \le x \le 1$.

17. Between $y = x^{1/2}$ and $y = x^{1/3}$ for $0 \le x \le 1$.

18. Between $y = 3x$ and $y = x^2$.

19. Between $y = x$ and $y = \sqrt{x}$.

20. Coal gas is produced at a gasworks. Pollutants in the gas are removed by scrubbers, which become less and less efficient as time goes on. The following measurements, made at the start of each month, show the rate at which pollutants are escaping (in tons/month) in the gas:

Time (months)	0	1	2	3	4	5	6
Rate pollutants escape	5	7	8	10	13	16	20

(a) Make an overestimate and an underestimate of the total quantity of pollutants that escape during the first month.

(b) Make an overestimate and an underestimate of the total quantity of pollutants that escape during the six months.

21. A student is speeding down Route 11 in his fancy red Porsche when his radar system warns him of an obstacle 400 feet ahead. He immediately applies the brakes, starts to slow down, and spots a skunk in the road directly ahead of him. The "black box" in the Porsche records the car's speed every two seconds, producing the following table. The speed decreases throughout the 10 seconds it takes to stop, although not necessarily at a uniform rate.

Time since brakes applied (sec)	0	2	4	6	8	10
Speed (ft/sec)	100	80	50	25	10	0

(a) What is your best estimate of the total distance the student's car traveled before coming to rest?

(b) Which one of the following statements can you justify from the information given?

  (i) The car stopped before getting to the skunk.

  (ii) The "black box" data is inconclusive. The skunk may or may not have been hit.

  (iii) The skunk was hit by the car.

22. The velocity of a particle moving along the x-axis is given by $f(t) = 6 - 2t$ cm/sec. Use a graph of $f(t)$ to find the exact change in position of the particle from time $t = 0$ to $t = 4$ seconds.

23. A baseball thrown directly upward at 96 ft/sec has velocity $v(t) = 96 - 32t$ ft/sec at time t seconds.

(a) Graph the velocity from $t = 0$ to $t = 6$.

(b) When does the baseball reach the peak of its flight? How high does it go?

(c) How high is the baseball at time $t = 5$?

24. A news broadcast in early 1993 said the typical American's annual income is changing at a rate of $r(t) = 40(1.002)^t$ dollars per month, where t is in months from January 1, 1993. How much did the typical American's income change during 1993?

25. Two species of plants have the same populations at time $t = 0$ and the growth rates shown in Figure 5.69.

(a) Which species has a larger population at the end of 5 years? At the end of 10 years?

(b) Which species do you think has the larger population after 20 years? Explain.

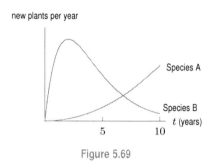

Figure 5.69

26. Figure 5.70 represents your velocity, v, on a bicycle trip along a straight road which starts 10 miles from home. Write a paragraph describing your trip: Do you start out going toward or away from home? How long do you continue in that direction and how far are you from home when you turn around? How many times do you change direction? Do you ever get home? Where are you at the end of the four-hour bike ride?

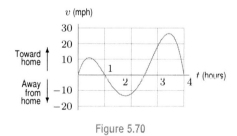

Figure 5.70

27. Figure 5.71 shows the length growth rate of a human fetus.

    (a) What feature of a graph of length as a function of age corresponds to the maximum in Figure 5.71?
    (b) Estimate the length of a baby born in week 40.

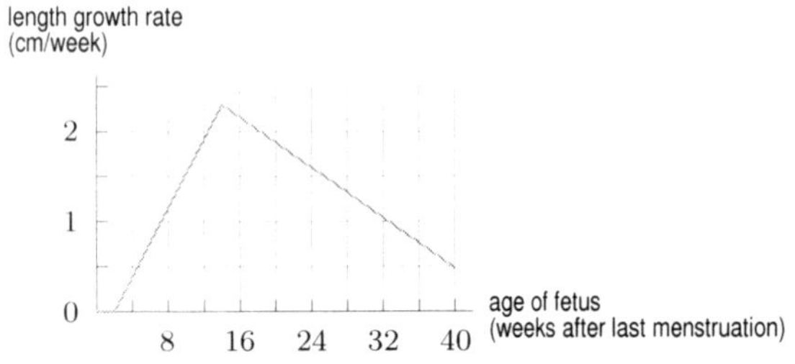

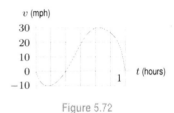

Figure 5.71

28. A bicyclist pedals along a straight road with velocity, v, given in Figure 5.72. She starts 5 miles from a lake; positive velocities take her away from the lake and negative velocities take her toward the lake. When is the cyclist farthest from the lake, and how far away is she then?

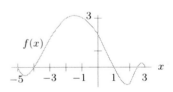

Figure 5.72

29. Using Figure 5.73, decide whether each of the following definite integrals is positive or negative.

    (a) $\int_{-5}^{-4} f(x)\,dx$ (b) $\int_{-4}^{1} f(x)\,dx$
    (c) $\int_{1}^{3} f(x)\,dx$ (d) $\int_{-5}^{3} f(x)\,dx$

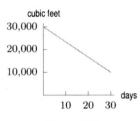

Figure 5.73

30. Using Figure 5.73, arrange the following definite integrals in ascending order:
    $\int_{-5}^{-3} f(x)\,dx,\ \int_{-5}^{-1} f(x)\,dx,\ \int_{-5}^{1} f(x)\,dx,\ \int_{-5}^{3} f(x)\,dx.$

31. Use a graph of $y = 2^{-x^2}$ to explain why $\int_{-1}^{1} 2^{-x^2}\,dx$ must be between 0 and 2.

32. Without computation, show that $2 \le \int_{0}^{2} \sqrt{1 + x^3}\,dx \le 6.$

33. With t in seconds, the velocity of an object is $v(t) = 10 + 8t - t^2$ m/sec.

    (a) Represent the distance traveled during the first 5 seconds as a definite integral and as an area.
    (b) Estimate the distance traveled by the object during the first 5 seconds by estimating the area.
    (c) Calculate the distance traveled.

34. The world's oil is being consumed at a continuously increasing rate, $f(t)$ (in billions of barrels per year), where t is in years since the start of 1995.

    (a) Write a definite integral which represents the total quantity of oil used between the start of 1995 and the start of 2010.
    (b) Suppose $r = 32(1.05)^t$. Using a left-hand sum with five subdivisions, find an approximate value for the total quantity of oil used between the start of 1995 and the start of 2010.
    (c) Interpret each of the five terms in the sum from part (b) in terms of oil consumption.

35. A car moves along a straight line with velocity, in feet/second, given by

    $$v(t) = 6 - 2t \quad \text{for } t \ge 0.$$

    (a) Describe the car's motion in words. (When is it moving forward, backward, and so on?)
    (b) The car's position is measured from its starting point. When is it farthest forward? Backward?

36. The marginal cost of drilling an oil well depends on the depth at which you are drilling; drilling becomes more expensive, per meter, as you dig deeper into the earth. The fixed costs are 1,000,000 riyals (the riyal is the unit of currency of Saudi Arabia), and, if x is the depth in meters, the marginal costs are

    $$C'(x) = 4000 + 10x \quad \text{riyals/meter.}$$

    Find the total cost of drilling a 500-meter well.

37. A warehouse charges its customers $5 per day for every 10 cubic feet of space used for storage. Figure 5.74 records the storage used by one company over a month. How much will the company have to pay?

Figure 5.74

38. One of the earliest pollution problems brought to the attention of the Environmental Protection Agency (EPA) was the case of the Sioux Lake in eastern South Dakota. For years a small paper plant located nearby had been discharging waste containing carbon tetrachloride (CCl_4) into the waters of the lake. At the time the EPA learned of the situation, the chemical was entering at a rate of 16 cubic yards/year.

The agency immediately ordered the installation of filters designed to slow (and eventually stop) the flow of CCl_4 from the mill. Implementation of this program took exactly three years, during which the flow of pollutant was steady at 16 cubic yards/year. Once the filters were installed, the flow declined. From the time the filters were installed until the time the flow stopped, the rate of flow was well approximated by

$$\text{Rate (in cubic yards/year)} = t^2 - 14t + 49,$$

where t is time measured in years since the EPA learned of the situation (thus, $t \geq 3$).

(a) Draw a graph showing the rate of CCl_4 flow into the lake as a function of time, beginning at the time the EPA first learned of the situation.

(b) How many years elapsed between the time the EPA learned of the situation and the time the pollution flow stopped entirely?

(c) How much CCl_4 entered the waters during the time shown in the graph in part (a)?

39. (a) Graph $x^3 - 5x^2 + 4x$, marking $x = 1, 2, 3, 4, 5$.

(b) Use your graph and the area interpretation of the definite integral to decide which of the five numbers

$$I_n = \int_0^n (x^3 - 5x^2 + 4x)\,dx \quad \text{for } n = 1, 2, 3, 4, 5$$

is largest. Which is smallest? How many of the numbers are positive? (Don't calculate the integrals.)

40. A mouse moves back and forth in a straight tunnel, attracted to bits of cheddar cheese alternately introduced to and removed from the ends (right and left) of the tunnel. The graph of the mouse's velocity, v, is given in Figure 5.75, with positive velocity corresponding to motion toward the right end. Assuming that the mouse starts

($t = 0$) at the center of the tunnel, use the graph to estimate the time(s) at which:

(a) The mouse changes direction.

(b) The mouse is moving most rapidly to the right; to the left.

(c) The mouse is farthest to the right of center; farthest to the left.

(d) The mouse's speed (i.e., the magnitude of its velocity) is decreasing.

(e) The mouse is at the center of the tunnel.

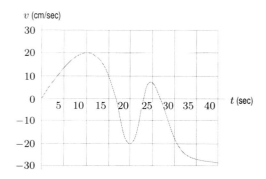

Figure 5.75

41. Pollution is being dumped into a lake at a rate which is increasing at a constant rate from 10 kg/year to 50 kg/year until a total of 270 kg has been dumped. Sketch a graph of the rate at which pollution is being dumped in the lake against time. How long does it take until 270 kg has been dumped?

For Problems 42–44, suppose $F(0) = 0$ and $F'(x) = 4 - x^2$.

42. Calculate $F(b)$ for $b = 0,\ 0.5,\ 1,\ 1.5,\ 2,\ 2.5$.

43. Using a graph of F', decide where F is increasing and where F is decreasing for $0 \leq x \leq 2.5$.

44. Does F have a maximum value for $0 \leq x \leq 2.5$? If so, what is it, and at what value of x does it occur?

CHECK YOUR UNDERSTANDING

In Problems 1–50, indicate whether the statement is true or false.

1. If the velocity is positive, then the total distance traveled is the area under the velocity curve.

2. If velocity is constant and positive, then distance traveled during a time interval is the velocity multiplied by the length of the interval.

3. If velocity is positive and increasing, using the velocity at the beginning of each subinterval gives an overestimate

of the distance traveled.

4. A car traveling with increasing velocity (in feet per second) given in the following table travels more than 40 feet between $t = 4$ and $t = 6$ seconds.

t	4	6
$v(t)$	10	20

5. A car traveling with increasing velocity (in feet per second) given in the following table travels less than 40 feet between $t = 4$ and $t = 6$ seconds.

t	4	6
$v(t)$	10	20

6. If water flows out of a tank at an increasing rate starting at 20 gallons per minute at the start of a 10-minute period and ending at 50 gallons per minute at the end of the 10-minute period, then at least 200 gallons have flowed out of the tank.

7. If water flows out of a tank at an increasing rate starting at 20 gallons per minute at the start of a 10-minute period and ending at 50 gallons per minute at the end of the 10-minute period, then at least 500 gallons have flowed out of the tank.

8. At a construction site, money is spent at a continuous, decreasing rate of $r(t)$ dollars per day, where t is in days. If the rates at two particular instants, $t = 3$ and $t = 5$, are given in the table, then at least \$3000 is spent between $t = 3$ and $t = 5$.

t	3	5
$r(t)$	1000	800

9. An object traveling with velocity given by $v(t) = 3t$ feet per second travels exactly 150 feet when $0 \leq t \leq 10$ seconds.

10. The units of total change and rate of change of a quantity are the same.

11. If $f(t)$ is increasing, then the left-hand sum gives an overestimate of $\int_a^b f(t)\, dt$.

12. If $f(t)$ is concave up, then the left-hand sum gives an overestimate of $\int_a^b f(t)\, dt$.

13. A left-hand sum estimate for $\int_3^5 f(t)\, dt$ is 3000, if $f(t)$ is shown in the following table.

t	3	5
$f(t)$	1000	800

14. A right-hand sum estimate for $\int_3^5 f(t)\, dt$ is 1600, if $f(t)$ is shown in the following table.

t	3	5
$f(t)$	1000	800

15. If the number of terms in a Riemann sum increases, the quantity Δt decreases.

16. A 4-term Riemann sum on the interval $4 \leq t \leq 6$ has $\Delta t = 2$.

17. The right-hand sum for $\int_a^b f(t)\, dt$ is the total area of rectangles that are above the graph of $f(t)$.

18. If $f(t) \geq 0$ then $\int_a^b f(t)\, dt$ is the area between the t-axis and the graph of $f(t)$, for $a \leq t \leq b$.

19. If $f(t) = t^2$ then the left-hand sum gives an overestimate for $\int_{-2}^{-1} f(t)\, dt$.

20. The right-hand sum for $\int_5^{25} t^2\, dt$ using $n = 5$ rectangles has $\Delta t = 5$.

21. The integral $\int_0^2 f(x)\, dx$ gives the area between the x-axis and the graph of $f(x)$ between $x = 0$ and $x = 2$.

22. If $\int_0^2 f(x)\, dx = 0$ then $f(x) = 0$ for all $0 \leq x \leq 2$.

23. If $\int_0^2 f(x)\, dx > 0$ then $f(x) > 0$ for all $0 \leq x \leq 2$.

24. If $f(x) > 0$ for all $0 \leq x \leq 2$ then $\int_0^2 f(x)\, dx > 0$.

25. The integral $\int_{-1}^1 (x^2 - 1)\, dx$ is the negative of the area between the graph of $f(x) = x^2 - 1$ and the x-axis for $-1 \leq x \leq 1$.

26. If $f(x)$ is both positive and negative when $0 \leq x \leq 2$ then $\int_0^2 f(x)\, dx$ is the sum of the areas between the graph and the x-axis for $0 \leq x \leq 2$.

27. The integral $\int_0^3 (1 - x)\, dx$ is the area between the graph of $f(x) = 1 - x$ and the x-axis between $x = 0$ and $x = 3$.

28. The integral $\int_{-1}^2 e^x\, dx$ is the area between the graph of $f(x) = e^x$ and the x-axis between $x = -1$ and $x = 2$.

29. If $f(x) \geq g(x)$ for all $a \leq x \leq b$ then $\int_a^b f(x) - g(x)\, dx$ gives the area between the graphs of $f(x)$ and $g(x)$ for $a \leq x \leq b$.

30. If the graph of $f(x)$ has more area below the x-axis than above the x-axis when $1 \leq x \leq 10$ then $\int_1^{10} f(x)\, dx$ is negative.

31. If flow rate $r(t)$ has units gallons per hour, and t has units in hours, then $\int_0^{100} r(t)\, dt$ has units of gallons.

32. If expense rate $s(t)$ has units dollars per day, and t has units in days, then $\int_{20}^{35} s(t)\, dt$ has units of days.

33. If acceleration $a(t)$ has units feet/(second)^2, and t has units in seconds, then $\int_{2.1}^{7.2} a(t)\, dt$ has units of feet/second.

34. If $w(t)$ is the number of workers on a building construction site as a function of t, the number of days since the start of construction, then $\int_0^{60} w(t)\, dt$ has units of workers per day.

35. If $C(t)$ is the rate, in dollars per day, for electricity to cool and heat a home on day t, where t is measured in days since January 1, then $\int_0^{365} C(t)\,dt$ has units of dollars.

36. If electricity costs $f(t)$ dollars per day, and t is measured in days, then $\int_0^{30} f(t)\,dt$ gives the total cost of electricity, in dollars, for the 30-day period from $t = 0$ to $t = 30$.

37. If $f(t)$ gives the rate of change of water in a reservoir, in gallons per day, and t is measured in days, then $\int_0^{30} f(t)\,dt$ gives the total volume of water, in gallons, in the reservoir on day $t = 30$.

38. If $f(t)$ gives the rate of change of water in a reservoir, in gallons per day, and t is measured in days, then $\int_0^{30} f(t)\,dt$ gives the total change in the volume of water, in gallons, in the reservoir during the period $t = 0$ to $t = 30$.

39. Bioavailability is the integral of blood concentration as a function of time.

40. If the peak concentration of drug A is greater than the peak concentration of drug B then the bioavailability of drug A is greater than the bioavailability of drug B.

41. The Fundamental Theorem of Calculus says if $F(t)$ is continuous, then $\int_a^b F(t)\,dt = F(b) - F(a)$.

42. The Fundamental Theorem of Calculus says if $g'(t)$ is continuous, then $\int_a^b g'(t)\,dt = g(b) - g(a)$.

43. The Fundamental Theorem of Calculus says if $F'(t)$ is continuous, then $\int_a^b F'(t)\,dt = F'(b) - F'(a)$.

44. If $C(q)$ is the cost to produce q units, then $\int_{100}^{200} C'(q)\,dq$ is the amount it costs to increase production from 100 to 200 units.

45. If $C(q)$ is the cost to produce q units, then $\int_0^{1000} C'(q)\,dq$ is the total amount it costs to produce 1000 units.

46. If the marginal cost function of a product is $C'(q) = 0.03q + 0.1$ dollars, and the fixed costs are \$1500, then the total cost to produce 1000 items is \$15,100.

47. The integral of the marginal cost $C'(q)$ function can be used to find the fixed costs.

48. If $\int_0^5 F'(t)\,dt = 0$, then F is constant on the interval between $t = 0$ and $t = 5$.

49. If $\int_0^5 F'(t)\,dt < 0$, then $F(0) > F(5)$.

50. If $F'(t)$ is constant, then $\int_0^5 F'(t)\,dt = 0$.

PROJECTS FOR CHAPTER FIVE

1. Carbon Dioxide in Pond Water Biological activity in a pond is reflected in the rate at which carbon dioxide, CO_2, is added to or withdrawn from the water. Plants take CO_2 out of the water during the day for photosynthesis and put CO_2 into the water at night. Animals put CO_2 into the water all the time as they breathe. Biologists are interested in how the net rate at which CO_2 enters a pond varies during the day. Figure 5.76 shows this rate as a function of time of day.[9] The rate is measured in millimoles (mmol) of CO_2 per liter of water per hour; time is measured in hours past dawn. At dawn, there were 2.600 mmol of CO_2 per liter of water.

   (a) What can be concluded from the fact that the rate is negative during the day and positive at night?

   (b) Some scientists have suggested that plants respire (breathe) at a constant rate at night, and that they photosynthesize at a constant rate during the day. Does Figure 5.76 support this view?

   (c) When was the CO_2 content of the water at its lowest? How low did it go?

   (d) How much CO_2 was released into the water during the 12 hours of darkness? Compare this quantity with the amount of CO_2 withdrawn from the water during the 12 hours of daylight. How can you tell by looking at the graph whether the CO_2 in the pond is in equilibrium?

   (e) Estimate the CO_2 content of the water at three-hour intervals throughout the day. Use your estimates to plot a graph of CO_2 content throughout the day.

---

[9]Data from R. J. Beyers, *The Pattern of Photosynthesis and Respiration in Laboratory Microsystems* (Mem. 1st Ital. Idrobiol., 1965).

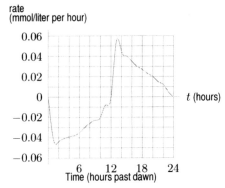

Figure 5.76: Rate at which CO_2 is entering the pond

2. Flooding in the Grand Canyon

The Glen Canyon Dam at the top of the Grand Canyon prevents natural flooding. In 1996, scientists decided an artificial flood was necessary to restore the environmental balance. Water was released through the dam at a controlled rate[10] shown in Figure 5.77. The figure also shows the rate of flow of the last natural flood in 1957.

(a) At what rate was water passing through the dam in 1996 before the artificial flood?

(b) At what rate was water passing down the river in the pre-flood season in 1957?

(c) Estimate the maximum rates of discharge for the 1996 and 1957 floods.

(d) Approximately how long did the 1996 flood last? How long did the 1957 flood last?

(e) Estimate how much additional water passed down the river in 1996 as a result of the artificial flood.

(f) Estimate how much additional water passed down the river in 1957 as a result of the flood.

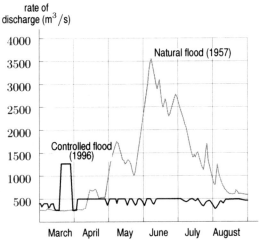

Figure 5.77

---

[10]Adapted from M. Collier, R. Webb, E. Andrews, "Experimental Flooding in Grand Canyon" in *Scientific American* (January 1997).

FOCUS ON THEORY

THEOREMS ABOUT DEFINITE INTEGRALS

The Second Fundamental Theorem of Calculus

The Fundamental Theorem of Calculus tells us that if we have a function F whose derivative is a continuous function f, then the definite integral of f is given by

$$\int_a^b f(t)\, dt = F(b) - F(a).$$

We now take a different point of view. If a is fixed and the upper limit is x, then the value of the integral is a function of x. We define a new function G on the interval by

$$G(x) = \int_a^x f(t)\, dt.$$

To visualize G, suppose that f is positive and $x > a$. Then $G(x)$ is the area under the graph of f in Figure 5.78. If f is continuous on an interval containing a, then it can be shown that G is defined for all x on that interval.

We now consider the derivative of G. Using the definition of the derivative,

$$G'(x) = \lim_{h \to 0} \frac{G(x+h) - G(x)}{h}.$$

Suppose f and h are positive. Then we can visualize

$$G(x) = \int_a^x f(t)\, dt$$

and

$$G(x + h) = \int_a^{x+h} f(t)\, dt$$

as areas, which leads to representing

$$G(x + h) - G(x) = \int_x^{x+h} f(t)\, dt$$

as a difference of two areas.

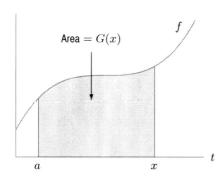

Figure 5.78: Representing $G(x)$ as an area

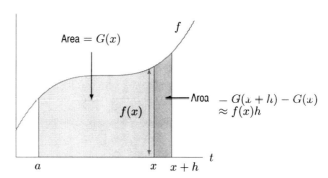

Figure 5.79: $G(x+h) - G(x)$ is the area of a roughly rectangular region

From Figure 5.79, we see that, if h is small, $G(x+h) - G(x)$ is roughly the area of a rectangle of height $f(x)$ and width h (shaded darker in Figure 5.79), so we have

$$G(x+h) - G(x) \approx f(x)h,$$

hence

$$\frac{G(x+h) - G(x)}{h} \approx f(x).$$

The same result holds when h is negative, suggesting that

$$G'(x) = \lim_{h \to 0} \frac{G(x+h) - G(x)}{h} = f(x).$$

This result is another form of the Fundamental Theorem of Calculus. It is usually stated as follows:

Second Fundamental Theorem of Calculus

If f is a continuous function on an interval, and if a is any number in that interval, then the function G defined on the interval by

$$G(x) = \int_a^x f(t)\, dt$$

has derivative f; that is, $G'(x) = f(x)$.

Properties of the Definite Integral

In this chapter, we have used the following properties to break up definite integrals.

Sums and Multiples of Definite Integrals

If a, b, and c are any numbers and f and g are continuous functions, then

1. $\displaystyle\int_a^c f(x)\, dx + \int_c^b f(x)\, dx = \int_a^b f(x)\, dx.$

2. $\displaystyle\int_a^b (f(x) \pm g(x))\, dx = \int_a^b f(x)\, dx \pm \int_a^b g(x)\, dx.$

3. $\displaystyle\int_a^b cf(x)\, dx = c\int_a^b f(x)\, dx.$

In words:
1. The integral from a to c plus the integral from c to b is the integral from a to b.
2. The integral of the sum (or difference) of two functions is the sum (or difference) of their integrals.
3. The integral of a constant times a function is that constant times the integral of the function.

These properties can best be visualized by thinking of the integrals as areas or as the limit of the sum of areas of rectangles.

Problems on the Second Fundamental Theorem of Calculus

For Problems 1–4, find $G'(x)$.

1. $G(x) = \int_a^x t^3 \, dt$

2. $G(x) = \int_a^x 3^t \, dt$

3. $G(x) = \int_a^x te^t \, dt$

4. $G(x) = \int_a^x \ln y \, dy$

5. Let $F(b) = \int_0^b 2^x \, dx$.

    (a) What is $F(0)$?

    (b) Does the value of F increase or decrease as b increases? (Assume $b \geq 0$.)

    (c) Estimate $F(1)$, $F(2)$, and $F(3)$.

6. For $x = 0, 0.5, 1.0, 1.5,$ and 2.0, make a table of values for $I(x) = \int_0^x \sqrt{t^4 + 1} \, dt$.

7. Assume that $F'(t) = \sin t \cos t$ and $F(0) = 1$. Find $F(b)$ for $b = 0, 0.5, 1, 1.5, 2, 2.5,$ and 3.

Let $\int_a^b f(x) \, dx = 8$, $\int_a^b (f(x))^2 \, dx = 12$, $\int_a^b g(t) \, dt = 2$, and $\int_a^b (g(t))^2 \, dt = 3$. Find the integrals in Problems 8–11.

8. $\int_a^b (f(x) + g(x)) \, dx$

9. $\int_a^b \left((f(x))^2 - (g(x))^2 \right) dx$

10. $\int_a^b (f(x))^2 \, dx - \left(\int_a^b f(x) \, dx \right)^2$

11. $\int_a^b cf(z) \, dz$

Chapter Six

USING THE DEFINITE INTEGRAL

Contents

6.1 AVERAGE VALUE

In this section we show how to interpret the definite integral as the average value of a function.

The Definite Integral as an Average

We know how to find the average of n numbers: Add them and divide by n. But how do we find the average value of a continuously varying function? Let us consider an example. Suppose $f(t)$ is the temperature at time t, measured in hours since midnight, and that we want to calculate the average temperature over a 24-hour period. One way to start would be to average the temperatures at n equally spaced times, $t_1, t_2, \ldots, t_n$, during the day.

$$\text{Average temperature} \approx \frac{f(t_1) + f(t_2) + \cdots + f(t_n)}{n}.$$

The larger we make n, the better the approximation. We can rewrite this expression as a Riemann sum over the interval $0 \le t \le 24$ if we use the fact that $\Delta t = 24/n$, so $n = 24/\Delta t$:

$$\text{Average temperature} \approx \frac{f(t_1) + f(t_2) + \cdots + f(t_n)}{24/\Delta t}$$

$$= \frac{f(t_1)\Delta t + f(t_2)\Delta t + \cdots + f(t_n)\Delta t}{24}$$

$$= \frac{1}{24} \sum_{i=1}^{n} f(t_i)\Delta t.$$

As $n \to \infty$, the Riemann sum tends toward an integral, and the approximation gets better. We expect that

$$\text{Average temperature} = \lim_{n \to \infty} \frac{1}{24} \sum_{i=1}^{n} f(t_i)\Delta t$$

$$= \frac{1}{24} \int_{0}^{24} f(t)\, dt.$$

Generalizing for any function f, if $a < b$, we have

$$\text{Average value of } f \text{ on the interval from } a \text{ to } b = \frac{1}{b-a} \int_{a}^{b} f(x)\, dx.$$

The units of $f(x)$ are the same as the units of the average value of $f(x)$.

How to Visualize the Average on a Graph

The definition of average value tells us that

$$(\text{Average value of } f) \cdot (b - a) = \int_{a}^{b} f(x)\, dx.$$

Let's interpret the integral as the area under the graph of f. If $f(x)$ is positive, then the average value of f is the height of a rectangle whose base is $(b - a)$ and whose area is the same as the area between the graph of f and the x-axis. (See Figure 6.1.)

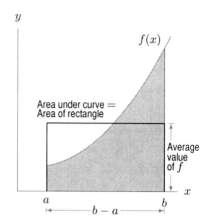

Figure 6.1: Area and average value

Example 1   Suppose that $C(t)$ represents the daily cost of heating your house, in dollars per day, where t is time in days and $t = 0$ corresponds to January 1, 2010. Interpret $\dfrac{1}{90 - 0} \displaystyle\int_0^{90} C(t)\, dt$.

Solution   The units for the integral $\int_0^{90} C(t)\, dt$ are (dollars/day)×(days) = dollars. The integral represents the total cost in dollars to heat your house for the first 90 days of 2010, namely the months of January, February, and March. The expression $\frac{1}{90-0}\int_0^{90} C(t)\, dt$ represents the average cost per day to heat your house during the first 90 days of 2010. It is measured in (1/days) × (dollars) = dollars/day, the same units as $C(t)$.

Example 2   The population of McAllen, Texas can be modeled by the function

$$P = f(t) = 570(1.037)^t,$$

where P is in thousands of people and t is in years since 2000. Use this function to predict the average population of McAllen between the years 2020 and 2040.

Solution   We want the average value of $f(t)$ between $t = 20$ and $t = 40$. Using a calculator to evaluate the integral, we get

$$\text{Average population} = \frac{1}{40 - 20} \int_{20}^{40} f(t)\, dt = \frac{1}{20}(34{,}656.2) = 1732.81.$$

The average population of McAllen between 2020 and 2040 is predicted to be about 1733 thousand people.

Example 3   (a) For the function $f(x)$ graphed in Figure 6.2, evaluate $\displaystyle\int_0^5 f(x)\, dx$.

(b) Find the average value of $f(x)$ on the interval $x = 0$ to $x = 5$. Check your answer graphically.

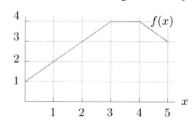

Figure 6.2: Estimate $\int_0^5 f(x)dx$

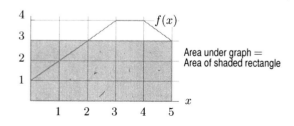

Figure 6.3: Average value of $f(x)$ is 3

Solution   (a) Since $f(x) \geq 0$, the definite integral is the area of the region under the graph of $f(x)$ between $x = 0$ and $x = 5$. Figure 6.2 shows that this region consists of 13 full grid squares and 4 half grid squares, each grid square of area 1, for a total area of 15, so

$$\int_0^5 f(x)\, dx = 15.$$

(b) The average value of $f(x)$ on the interval from 0 to 5 is given by

$$\text{Average value} = \frac{1}{5 - 0}\int_0^5 f(x)\, dx = \frac{1}{5}(15) = 3.$$

To check the answer graphically, draw a horizontal line at $y = 3$ on the graph of $f(x)$. (See Figure 6.3.) Then observe that, between $x = 0$ and $x = 5$, the area under the graph of $f(x)$ is equal to the area of the rectangle with height 3.

Problems for Section 6.1

1. What is the average value of the function f in Figure 6.4 over the interval $1 \leq x \leq 6$?

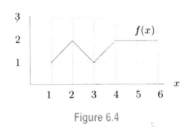

Figure 6.4

2. Use Figure 6.5 to estimate the following:

   (a) The integral $\int_0^5 f(x)\, dx$.
   (b) The average value of f between $x = 0$ and $x = 5$ by estimating visually the average height.
   (c) The average value of f between $x = 0$ and $x = 5$ by using your answer to part (a).

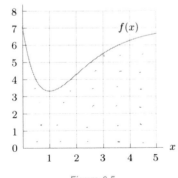

Figure 6.5

3. Find the average value of $g(t) = 1 + t$ over the interval $[0, 2]$

4. Find the average value of the function $f(x) = 5 + 4x - x^2$ between $x = 0$ and $x = 3$.

In Problems 5–6, estimate the average value of the function between $x = 0$ and $x = 7$.

5.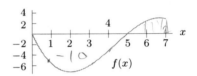

6.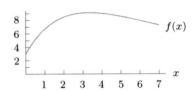

In Problems 7–8, estimate the average value of the function $f(x)$ on the interval from $x = a$ to $x = b$.

7.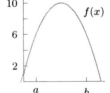

8.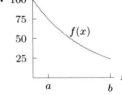

In Problems 9–10 annual income for ages 25 to 85 is given graphically. People sometimes spend less than their income (to save for retirement) or more than their income (taking out a loan). The process of spreading out spending over a lifetime is called consumption smoothing.

(a) Find the average annual income for these years.

(b) Assuming that the person spends at a constant rate equal to their average income, when are they spending less than they earn, and when are they spending more?

9.

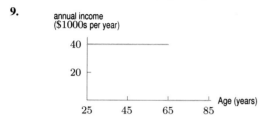

10.

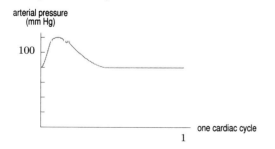

11. The value, V, of a Tiffany lamp, worth \$225 in 1975, increases at 15% per year. Its value in dollars t years after 1975 is given by

$$V = 225(1.15)^t.$$

Find the average value of the lamp over the period 1975–2010.

12. If t is measured in days since June 1, the inventory $I(t)$ for an item in a warehouse is given by

$$I(t) = 5000(0.9)^t.$$

(a) Find the average inventory in the warehouse during the 90 days after June 1.

(b) Graph $I(t)$ and illustrate the average graphically.

Problems 13–14 refer to Figure 6.6, which shows human arterial blood pressure during the course of one heartbeat.

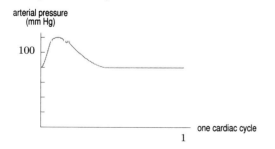

Figure 6.6

13. (a) Estimate the maximum blood pressure, called the systolic pressure.

(b) Estimate the minimum blood pressure, called the diastolic pressure.

(c) Calculate the average of the systolic and diastolic pressures.

(d) Is the average arterial pressure over the entire cycle greater than, less than, or equal to the answer for part (c)?

14. Estimate the average arterial blood pressure over one cardiac cycle.

15. Figure 6.7 shows the rate, $f(x)$, in thousands of algae per hour, at which a population of algae is growing, where x is in hours.

(a) Estimate the average value of the rate over the interval $x = -1$ to $x = 3$.

(b) Estimate the total change in the population over the interval $x = -3$ to $x = 3$.

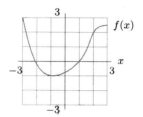

Figure 6.7

16. The population of the world t years after 2000 is predicted to be $P = 6.1e^{0.0125t}$ billion.

(a) What population is predicted in 2010?

(b) What is the predicted average population between 2000 and 2010?

17. The number of hours, H, of daylight in Madrid as a function of date is approximated by the formula

$$H = 12 + 2.4 \sin[0.0172(t - 80)],$$

where t is the number of days since the start of the year. Find the average number of hours of daylight in Madrid:

(a) in January   (b) in June   (c) over a year

(d) Explain why the relative magnitudes of your answers to parts (a), (b), and (c) are reasonable.

18. A bar of metal is cooling from $1000°$C to room temperature, $20°$C. The temperature, H, of the bar t minutes after it starts cooling is given, in $°$C, by

$$H = 20 + 980e^{-0.1t}.$$

(a) Find the temperature of the bar at the end of one hour.

(b) Find the average value of the temperature over the first hour.

(c) Is your answer to part (b) greater or smaller than the average of the temperatures at the beginning and the end of the hour? Explain this in terms of the concavity of the graph of H.

19. The rate of sales (in sales per month) of a company is given, for t in months since January 1, by

$$r(t) = t^4 - 20t^3 + 118t^2 - 180t + 200.$$

(a) Graph the rate of sales per month during the first year ($t = 0$ to $t = 12$). Does it appear that more sales were made during the first half of the year, or during the second half?

(b) Estimate the total sales during the first 6 months of the year and during the last 6 months of the year.

(c) What are the total sales for the entire year?

(d) Find the average sales per month during the year.

20. Throughout much of the 20th century, the yearly consumption of electricity in the US increased exponentially at a continuous rate of 7% per year. Assume this trend continues and that the electrical energy consumed in 1900 was 1.4 million megawatt-hours.

(a) Write an expression for yearly electricity consumption as a function of time, t, in years since 1900.

(b) Find the average yearly electrical consumption throughout the 20th century.

(c) During what year was electrical consumption closest to the average for the century?

(d) Without doing the calculation for part (c), how could you have predicted which half of the century the answer would be in?

21. Using Figure 6.8, list the following numbers from least to greatest:

(a) $f'(1)$

(b) The average value of f on $0 \leq x \leq 4$

(c) $\int_0^1 f(x)dx$

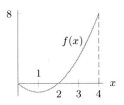

Figure 6.8

22. Using Figure 6.9, list from least to greatest,

(a) $f'(1)$.

(b) The average value of $f(x)$ on $0 \leq x \leq a$.

(c) The average value of the rate of change of $f(x)$, for $0 \leq x \leq a$.

(d) $\int_0^a f(x)\,dx$.

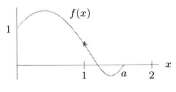

Figure 6.9

6.2   CONSUMER AND PRODUCER SURPLUS

Supply and Demand Curves

As we saw in Chapter 1, the quantity of a certain item produced and sold can be described by the supply and demand curves of the item. The *supply curve* shows what quantity, q, of the item the producers supply at different prices, p. The consumers' behavior is reflected in the *demand curve*, which shows what quantity of goods are bought at various prices. See Figure 6.10.

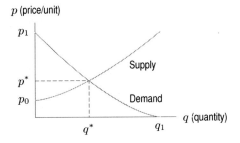

Figure 6.10: Supply and demand curves

It is assumed that the market settles at the *equilibrium price* p^* and *equilibrium quantity* q^* where the graphs cross. At equilibrium, a quantity q^* of an item is produced and sold for a price of p^* each.

Consumer and Producer Surplus

Notice that at equilibrium, a number of consumers have bought the item at a lower price than they would have been willing to pay. (For example, there are some consumers who would have been willing to pay prices up to p_1.) Similarly, there are some suppliers who would have been willing to produce the item at a lower price (down to p_0, in fact). We define the following terms:

- The **consumer surplus** measures the consumers' gain from trade. It is the total amount gained by consumers by buying the item at the current price rather than at the price they would have been willing to pay.
- The **producer surplus** measures the suppliers' gain from trade. It is the total amount gained by producers by selling at the current price, rather than at the price they would have been willing to accept.

In the absence of price controls, the current price is assumed to be the equilibrium price.

Both consumers and producers are richer for having traded. The consumer and producer surplus measure how much richer they are.

Suppose that all consumers buy the good at the maximum price they are willing to pay. Subdivide the interval from 0 to q^* into intervals of length Δq. Figure 6.11 shows that a quantity Δq of items are sold at a price of about p_1, another Δq are sold for a slightly lower price of about p_2, the next Δq for a price of about p_3, and so on. Thus, the consumers' total expenditure is about

$$p_1 \Delta q + p_2 \Delta q + p_3 \Delta q + \cdots = \sum p_i \Delta q.$$

If the demand curve has equation[1] $p = f(q)$, and if all consumers who were willing to pay more than p^* paid as much as they were willing, then as $\Delta q \to 0$, we would have

$$\text{Consumer expenditure} = \int_0^{q^*} f(q)\, dq = \text{Area under demand curve from 0 to } q^*.$$

If all goods are sold at the equilibrium price, the consumers' actual expenditure is only $p^* q^*$, which is the area of the rectangle between the q-axis and the line $p = p^*$ from $q = 0$ to $q = q^*$. The consumer surplus is the difference between the total consumer expenditure if all consumers pay the maximum they are willing to pay and the actual consumer expenditure if all consumers pay the current price. The consumer surplus is represented by the area in Figure 6.12. Similarly, the producer surplus is represented by the area in Figure 6.13. (See Problems 14 and 15.) Thus:

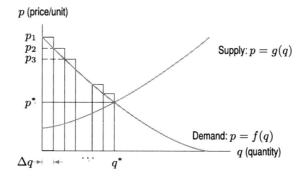

Figure 6.11: Calculation of consumer surplus

---

[1]Note that here p is written as a function of q.

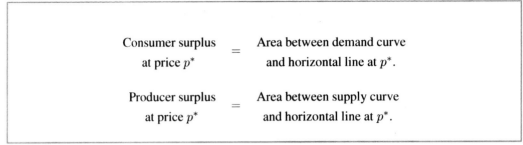

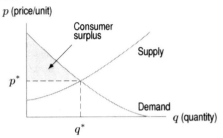

Figure 6.12: Consumer surplus

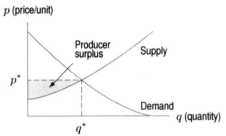

Figure 6.13: Producer surplus

Example 1   The supply and demand curves for a product are given in Figure 6.14.

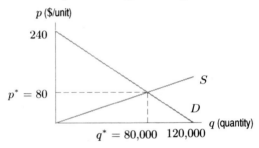

Figure 6.14: Supply and demand curves for a product

(a) What are the equilibrium price and quantity?

(b) At the equilibrium price, calculate and interpret the consumer and producer surplus.

Solution   (a) The equilibrium price is $p^* = \$80$ and the equilibrium quantity is $q^* = 80,000$ units.

(b) The consumer surplus is the area under the demand curve and above the line $p = 80$. (See Figure 6.15.) We have

$$\text{Consumer surplus} = \text{Area of triangle} = \frac{1}{2}\text{Base} \cdot \text{Height} = \frac{1}{2}80,000 \cdot 160 = \$6,400,000.$$

This tells us that consumers gain $6,400,000 in buying goods at the equilibrium price instead of at the price they would have been willing to pay.

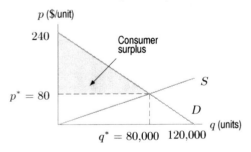

Figure 6.15: Consumer surplus

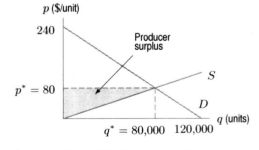

Figure 6.16: Producer surplus

The producer surplus is the area above the supply curve and below the line $p = 80$. (See Figure 6.16.) We have

$$\text{Producer surplus} = \text{Area of triangle} = \frac{1}{2}\text{Base} \cdot \text{Height} = \frac{1}{2} \cdot 80{,}000 \cdot 80 = \$3{,}200{,}000.$$

So, producers gain \$3,200,000 by supplying goods at the equilibrium price instead of the price at which they would have been willing to provide the goods.

Wage and Price Controls

In a free market, the price of a product generally moves to the equilibrium price, unless outside forces keep the price artificially high or artificially low. Rent control, for example, keeps prices below market value, whereas cartel pricing or the minimum wage law raise prices above market value. What happens to consumer and producer surplus at non-equilibrium prices?

Example 2   The dairy industry has cartel pricing: the government has set milk prices artificially high. What effect does raising the price to p^+ from the equilibrium price have on:

(a)   Consumer surplus?   (b)   Producer surplus?

(c)   Total gains from trade (that is, Consumer surplus + Producer surplus)?

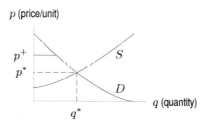

Figure 6.17: What is the effect of the artificially high price, p^+, on consumer and producer surplus? (q^* and p^* are equilibrium values)

Solution   (a) A graph of possible supply and demand curves for the milk industry is given in Figure 6.17. Suppose that the price is fixed at p^+, above the equilibrium price. Consumer surplus is the difference between the amount the consumers paid (p^+) and the amount they would have been willing to pay (given on the demand curve). This is the area shaded in Figure 6.18. This consumer surplus is less than the consumer surplus at the equilibrium price, shown in Figure 6.19.

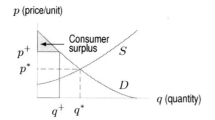

Figure 6.18: Consumer surplus: Artificial price

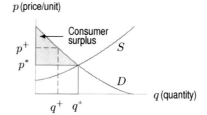

Figure 6.19: Consumer surplus: Equilibrium price

(b) At a price of p^+, the quantity sold, q^+, is less than it would have been at the equilibrium price. The producer surplus is represented by the area between p^+ and the supply curve at this reduced demand. This area is shaded in Figure 6.20. Compare this producer surplus (at the artificially high price) to the producer surplus in Figure 6.21 (at the equilibrium price). In this case, producer surplus appears to be greater at the artificial price than at the equilibrium price. (However, different supply and demand curves might lead to a different answer.)

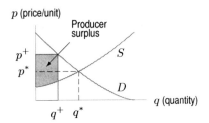

Figure 6.20: Producer surplus: Artificial price

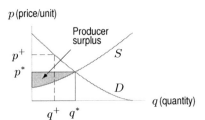

Figure 6.21: Producer surplus: Equilibrium price

(c) The total gains from trade (Consumer surplus + Producer surplus) at the price of p^+ is represented by the area shaded in Figure 6.22. The total gains from trade at the equilibrium price of p^* is represented by the area shaded in Figure 6.23. Under artificial price conditions, the total gains from trade decrease. The total financial effect of the artificially high price on all producers and consumers combined is negative.

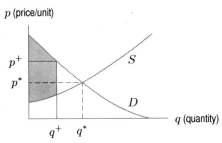

Figure 6.22: Total gains from trade: Artificial price

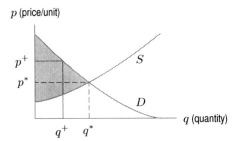

Figure 6.23: Total gains from trade: Equilibrium price

Problems for Section 6.2

1. (a) What are the equilibrium price and quantity for the supply and demand curves in Figure 6.24?
   (b) Shade the areas representing the consumer and producer surplus and estimate them.

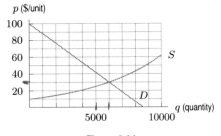

Figure 6.24

2. The supply and demand curves for a product are given in Figure 6.25. Estimate the equilibrium price and quantity and the consumer and producer surplus. Shade areas representing the consumer surplus and the producer surplus.

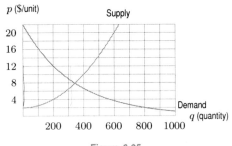

Figure 6.25

3. Find the consumer surplus for the demand curve $p = 100 - 3q^2$ when 5 units are sold.

4. Given the demand curve $p = 35 - q^2$ and the supply curve $p = 3 + q^2$, find the producer surplus when the market is in equilibrium.

5. Find the consumer surplus for the demand curve $p = 100 - 4q$ when $q = 10$.

6. **(a)** Estimate the equilibrium price and quantity for the supply and demand curves in Figure 6.26.
   (b) Estimate the consumer and producer surplus.
   (c) The price is set artificially low at $p^- = 4$ dollars per unit. Estimate the consumer and producer surplus at this price. Compare your answers to the consumer and producer surplus at the equilibrium price.

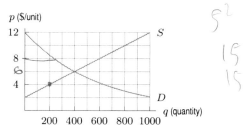

Figure 6.26

7. Supply and demand curves for a product are in Figure 6.27.

   (a) Estimate the equilibrium price and quantity.
   (b) Estimate the consumer and producer surplus. Shade them.
   (c) What are the total gains from trade for this product?

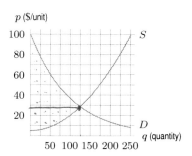

Figure 6.27

8. Supply and demand curves are in Figure 6.27. A price of $40 is artificially imposed.

   (a) At the $40 price, estimate the consumer surplus, the producer surplus, and the total gains from trade.
   (b) Compare your answers in this problem to your answers in Problem 7. Discuss the effect of price controls on the consumer surplus, producer surplus, and gains from trade in this case.

9. The demand curve for a product has equation $p = 20e^{-0.002q}$ and the supply curve has equation $p = 0.02q + 1$ for $0 \le q \le 1000$, where q is quantity and p is price in \$/unit.

   (a) Which is higher, the price at which 300 units are supplied or the price at which 300 units are demanded? Find both prices.
   (b) Sketch the supply and demand curves. Find the equilibrium price and quantity.
   (c) Using the equilibrium price and quantity, calculate and interpret the consumer and producer surplus.

10. Show graphically that the maximum total gains from trade occurs at the equilibrium price. Do this by showing that if outside forces keep the price artificially high or low, the total gains from trade (consumer surplus + producer surplus) are lower than at the equilibrium price.

11. Rent controls on apartments are an example of price controls on a commodity. They keep the price artificially low (below the equilibrium price). Sketch a graph of supply and demand curves, and label on it a price p^- below the equilibrium price. What effect does forcing the price down to p^- have on:

   (a) The producer surplus?
   (b) The consumer surplus?
   (c) The total gains from trade (Consumer surplus + Producer surplus)?

12. Supply and demand data are in Tables 6.1 and 6.2.

   (a) Which table shows supply and which shows demand?
   (b) Estimate the equilibrium price and quantity.
   (c) Estimate the consumer and producer surplus.

Table 6.1

q (quantity)	0	100	200	300	400	500	600
p (\$/unit)	60	50	41	32	25	20	17

Table 6.2

q (quantity)	0	100	200	300	400	500	600
p (\$/unit)	10	14	18	22	25	28	34

13. The total gains from trade (consumer surplus + producer surplus) is largest at the equilibrium price. What about the consumer surplus and producer surplus separately?

   (a) Suppose a price is artificially high. Can the consumer surplus at the artificial price be larger than the consumer surplus at the equilibrium price? What about the producer surplus? Sketch possible supply and demand curves to illustrate your answers.
   (b) Suppose a price is artificially low. Can the consumer surplus at the artificial price be larger than the consumer surplus at the equilibrium price? What about the producer surplus? Sketch possible supply and demand curves to illustrate your answers.

In Problems 14–16, the supply and demand curves have equations $p = S(q)$ and $p = D(q)$, respectively, with equilibrium at (q^*, p^*).

14. Using Riemann sums, explain the economic significance of $\int_0^{q^*} S(q)\, dq$ to the producers.

15. Using Riemann sums, give an interpretation of producer surplus, $\int_0^{q^*} (p^* - S(q))\, dq$ analogous to the interpretation of consumer surplus.

16. Referring to Figures 6.12 and 6.13 on page 282, mark the regions representing the following quantities and explain their economic meaning:

(a) $p^* q^*$

(b) $\int_0^{q^*} D(q)\, dq$

(c) $\int_0^{q^*} S(q)\, dq$

(d) $\int_0^{q^*} D(q)\, dq - p^* q^*$

(e) $p^* q^* - \int_0^{q^*} S(q)\, dq$ **(f)** $\int_0^{q^*} (D(q) - S(q))\, dq$

6.3 PRESENT AND FUTURE VALUE

In Chapter 1 on page 54, we introduced the present and future value of a single payment. In this section we see how to calculate the present and future value of a continuous stream of payments.

Income Stream

When we consider payments made to or by an individual, we usually think of *discrete* payments, that is, payments made at specific moments in time. However, we may think of payments made by a company as being *continuous*. The revenues earned by a huge corporation, for example, come in essentially all the time, and therefore they can be represented by a continuous *income stream*. Since the rate at which revenue is earned may vary from time to time, the income stream is described by

$$S(t) \text{ dollars/year.}$$

Notice that $S(t)$ is a *rate* at which payments are made (its units are dollars per year, for example) and that the rate depends on the time, t, usually measured in years from the present.

Present and Future Values of an Income Stream

Just as we can find the present and future values of a single payment, so we can find the present and future values of a stream of payments. As before, the future value represents the total amount of money that you would have if you deposited an income stream into a bank account as you receive it and let it earn interest until that future date. The present value represents the amount of money you would have to deposit today (in an interest-bearing bank account) in order to match what you would get from the income stream by that future date.

When we are working with a continuous income stream, we will assume that interest is compounded continuously. If the interest rate is r, the present value, P, of a deposit, B, made t years in the future is

$$P = Be^{-rt}.$$

Suppose that we want to calculate the present value of the income stream described by a rate of $S(t)$ dollars per year, and that we are interested in the period from now until M years in the future. In order to use what we know about single deposits to calculate the present value of an income stream, we divide the stream into many small deposits, and imagine each deposited at one instant. Dividing the interval $0 \leq t \leq M$ into subintervals of length Δt:

Assuming Δt is small, the rate, $S(t)$, at which deposits are being made does not vary much within one subinterval. Thus, between t and $t + \Delta t$:

$$\text{Amount paid} \approx \text{Rate of deposits} \times \text{Time}$$
$$\approx (S(t) \text{ dollars/year})(\Delta t \text{ years})$$
$$= S(t)\Delta t \text{ dollars}.$$

The deposit of $S(t)\Delta t$ is made t years in the future. Thus, assuming a continuous interest rate r,

$$\begin{array}{c} \text{Present value of money} \\ \text{deposited in interval } t \text{ to } t + \Delta t \end{array} \approx S(t)\Delta t e^{-rt}.$$

Summing over all subintervals gives

$$\text{Total present value} \approx \sum S(t)e^{-rt}\Delta t \text{ dollars}.$$

In the limit as $\Delta t \to 0$, we get the following integral:

$$\text{Present value} = \int_0^M S(t)e^{-rt}dt.$$

As in Section 1.7, the value M years in the future is given by

$$\text{Future value} = \text{Present value} \cdot e^{rM}.$$

Example 1 Find the present and future values of a constant income stream of $1000 per year over a period of 20 years, assuming an interest rate of 6% compounded continuously.

Solution Using $S(t) = 1000$ and $r = 0.06$, we have

$$\text{Present value} = \int_0^{20} 1000 e^{-0.06t} dt = \$11{,}647.$$

We can get the future value, B, from the present value, P, using $B = Pe^{rt}$, so

$$\text{Future value} = 11{,}647 e^{0.06(20)} = \$38{,}669.$$

Notice that since money was deposited at a rate of $1000 a year for 20 years, the total amount deposited was $20,000. The future value is $38,669, so the money has almost doubled because of the interest.

Example 2 Suppose you want to have $50,000 in 8 years' time in a bank account earning 2% interest, compounded continuously.

(a) If you make one lump sum deposit now, how much should you deposit?
(b) If you deposit money continuously throughout the 8-year period, at what rate should you deposit it?

Solution (a) If you deposit a lump sum of P, then P is the present value of $50,000. So, using $B = Pe^{rt}$ with $B = 50{,}000$ and $r = 0.02$ and $t = 8$:

$$50000 = Pe^{0.02(8)}$$
$$P = \frac{50000}{e^{0.02(8)}} = 42{,}607.$$

If you deposit $42,607 into the account now, you will have $50,000 in 8 years' time.

(b) Suppose you deposit money at a constant rate of $\$S$ per year. Then

$$\text{Present value of deposits} = \int_0^8 Se^{-0.02t}\, dt$$

Since S is a constant, we can take it out in front of the integral sign:

$$\text{Present value} - S\int_0^8 e^{-0.02t}\, dt \approx S(7.3928).$$

But the present value of the continuous deposit must be the same as the present value of the lump sum deposit; that is, $\$42,607$. So

$$42,607 \approx S(7.3928)$$
$$S \approx \$5763.$$

To meet your goal of $\$50,000$, you need to deposit money at a continuous rate of $\$5,763$ per year, or about $\$480$ per month.

Problems for Section 6.3

1. Find the present and future values of an income stream of $\$3000$ per year over a 15-year period, assuming a 6% annual interest rate compounded continuously.

2. Draw a graph, with time in years on the horizontal axis, of what an income stream might look like for a company that sells sunscreen in the northeast United States.

3. (a) Find the present and future value of an income stream of $\$6000$ per year for a period of 10 years if the interest rate, compounded continuously, is 5%.
   (b) How much of the future value is from the income stream? How much is from interest?

4. Find the present and future values of an income stream of $\$12,000$ a year for 20 years. The interest rate is 6%, compounded continuously.

5. A small business expects an income stream of $\$5000$ per year for a four-year period.
   (a) Find the present value of the business if the annual interest rate, compounded continuously, is
      (i) 3%   (ii) 10%
   (b) In each case, find the value of the business at the end of the four-year period.

6. A bond is guaranteed to pay $100 + 10t$ dollars per year for 10 years, where t is in years from the present. Find the present value of this income stream, given an interest rate of 5%, compounded continuously.

7. A recently-installed machine earns the company revenue at a continuous rate of $60,000t + 45,000$ dollars per year during the first six months of operation and at the continuous rate of 75,000 dollars per year after the first six months. The cost of the machine is $\$150,000$, the interest rate is 7% per year, compounded continuously, and t is time in years since the machine was installed.

(a) Find the present value of the revenue earned by the machine during the first year of operation.
(b) Find how long it will take for the machine to pay for itself; that is, how long it will take for the present value of the revenue to equal the cost of the machine?

8. At what constant, continuous rate must money be deposited into an account if the account is to contain $\$20,000$ in 5 years? The account earns 6% interest compounded continuously.

9. Your company needs $\$500,000$ in two years' time for renovations and can earn 9% interest on investments.
   (a) What is the present value of the renovations?
   (b) If your company deposits money continuously at a constant rate throughout the two-year period, at what rate should the money be deposited so that you have the $\$500,000$ when you need it?

10. A company is expected to earn $\$50,000$ a year, at a continuous rate, for 8 years. You can invest the earnings at an interest rate of 7%, compounded continuously. You have the chance to buy the rights to the earnings of the company now for $\$350,000$. Should you buy? Explain.

11. Sales of Version 6.0 of a computer software package start out high and decrease exponentially. At time t, in years, the sales are $s(t) = 50e^{-t}$ thousands of dollars per year. After two years, Version 7.0 of the software is released and replaces Version 6.0. You can invest earnings at an interest rate of 6%, compounded continuously. Calculate the total present value of sales of Version 6.0 over the two-year period.

12. Intel Corporation is a leading manufacturer of integrated circuits. In 2004, Intel generated profits at a continuous rate of 7.5 billion dollars per year.[2] Assume the interest rate was 8.5% per year compounded continuously.

    (a) What was the present value of Intel's profits over the 2004 one-year time period?
    (b) What was the value at the end of the year of Intel's profits over the 2004 one-year time period?

13. Harley-Davidson Inc. manufactures motorcycles. During the years following 2003 (the company's 100th anniversary), the company's net revenue can be approximated[3] by $4.6 + 0.4t$ billion dollars per year, where t is time in years since January 1, 2003. Assume this rate holds through January 1, 2013, and assume a continuous interest rate of 3.5% per year.

    (a) What was the net revenue of the Harley-Davidson Company in 2003? What is the projected net revenue in 2013?
    (b) What was the present value, on January 1, 2003, of Harley-Davidson's net revenue for the ten years from January 1, 2003 to January 1, 2013?
    (c) What is the future value, on January 1, 2013, of net revenue for the preceding 10 years?

14. McDonald's Corporation licenses and operates a chain of 31,377 fast-food restaurants throughout the world. Between 2005 and 2008, McDonald's has been generating revenue at continuous rates between 17.9 and 22.8 billion dollars per year.[4] Suppose that McDonald's rate of revenue stays within this range. Use an interest rate of 4.5% per year compounded continuously. Fill in the blanks:

    (a) The present value of McDonald's revenue over a five-year time period is between _____

and _____ billion dollars.

    (b) The present value of McDonald's revenue over a twenty-five-year time period is between _____ and _____ billion dollars.

15. Your company is considering buying new production machinery. You want to know how long it will take for the machinery to pay for itself; that is, you want to find the length of time over which the present value of the profit generated by the new machinery equals the cost of the machinery. The new machinery costs $130,000 and earns profit at the continuous rate of $80,000 per year. Use an interest rate of 8.5% per year compounded continuously.

16. An oil company discovered an oil reserve of 100 million barrels. For time $t > 0$, in years, the company's extraction plan is a linear declining function of time as follows:

$$q(t) = a - bt,$$

where $q(t)$ is the rate of extraction of oil in millions of barrels per year at time t and $b = 0.1$ and $a = 10$.

    (a) How long does it take to exhaust the entire reserve?
    (b) The oil price is a constant $20 per barrel, the extraction cost per barrel is a constant $10, and the market interest rate is 10% per year, compounded continuously. What is the present value of the company's profit?

17. The value of good wine increases with age. Thus, if you are a wine dealer, you have the problem of deciding whether to sell your wine now, at a price of $$P$ a bottle, or to sell it later at a higher price. Suppose you know that the amount a wine-drinker is willing to pay for a bottle of this wine t years from now is $$P(1 + 20\sqrt{t})$. Assuming continuous compounding and a prevailing interest rate of 5% per year, when is the best time to sell your wine?

6.4 INTEGRATING RELATIVE GROWTH RATES

Population Growth Rates

In Chapter 5, we saw how to calculate change in a population, P, from its derivative, dP/dt, using the Fundamental Theorem of Calculus. However, population growth rates are often given by the relative rate of change $(1/P)(dP/dt)$, rather than by the derivative. For example, we saw that the population of Nevada is growing at 3.6% per year.

We can use the relative growth rate to find the percentage change in a population. Recall that

$$\text{Relative growth rate} = \frac{1}{P}\frac{dP}{dt} = \frac{d}{dt}(\ln P).$$

By the Fundamental Theorem of Calculus, the integral of the relative growth rate gives the total change in $\ln(P)$:

$$\int_a^b \frac{P'(t)}{P(t)}\, dt = \int_a^b \frac{d}{dt}(\ln P(t))\, dt = \ln(P(b)) - \ln(P(a)) = \ln\left(\frac{P(b)}{P(a)}\right).$$

[2] www.intel.com, accessed May 28, 2005.
[3] 2007 Harley-Davidson Annual Report, accessed from www.harley-davidson.com.
[4] McDonald's Annual Report 2007 accessed at www.mcdonalds.com.

Example 1   The relative rate of growth $P'(t)/P(t)$ of a population $P(t)$ over a 50-year period is given in Figure 6.28. By what factor did the population increase during the period?

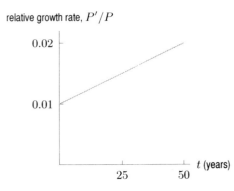

relative growth rate, P'/P

Figure 6.28: The relative growth rate of population

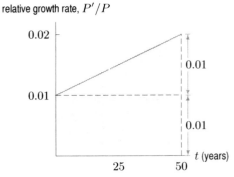

relative growth rate, P'/P

Figure 6.29: Calculating the percentage change in a population from the relative growth rate

Solution   We have

$$\ln\left(\frac{P(50)}{P(0)}\right) = \int_0^{50} \frac{P'(t)}{P(t)}\,dt.$$

This integral equals the area under the graph of $P'(t)/P(t)$ between $t = 0$ and $t = 50$. See Figure 6.29. The area of a rectangle and triangle gives

$$\text{Area} = 50(0.01) + \frac{1}{2}\cdot 50(0.01) = 0.75.$$

Thus,

$$\ln\left(\frac{P(50)}{P(0)}\right) = 0.75$$
$$\frac{P(50)}{P(0)} = e^{0.75} = 2.1$$
$$P(50) = 2.1P(0).$$

The population more than doubled during the 50 years, increasing by a factor of about 2.1. We cannot determine the amount by which the population increased unless we know how large the population was initially.

If everything else remains constant, the relative growth rate increases if either the relative birth rate increases or the relative death rate decreases. Even if the relative birth rate decreases, we can still see an increase in the relative growth rate if the relative death rate decreases faster. This is the case with the population of the world today. The difference between the birth rate and the death rate is an important variable.

Example 2   Figure 6.30 shows relative birth rates and relative death rates for developed and developing countries.[5]

(a) Which is changing faster, the birth rate or the death rate? What does this tell you about how the populations are changing?

---

[5] *Food and Population: A Global Concern*, Elaine Murphy, Washington, DC: Population Reference Bureau, Inc., 1984, p. 2.

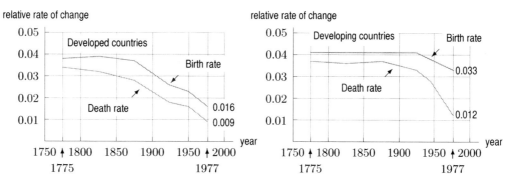

relative rate of change

relative rate of change

Figure 6.30: Birth and death rates in developed and developing countries, 1775-1977

(b) By what percentage did the population of developing countries increase between 1800 and 1900?

(c) By what percentage did the population of developing countries increase between 1950 and 1977?

Solution

(a) In developed countries, the birth rate is higher than the death rate, so the population is increasing. The birth rate and the death rate are both decreasing, and at approximately the same rate, so the population of developed countries is increasing at a constant relative rate.

   In developing countries, the birth rate is higher than the death rate and both have been decreasing since about 1925. In recent years, the death rate has been decreasing faster than the birth rate, so the relative rate of population growth is increasing. The decline in the death rate has contributed significantly to the recent population growth in developing countries.

(b) The relative rate of growth of the population is the difference between the relative birth rate and the relative death rate, so it is represented in Figure 6.30 by the vertical distance between the birth-rate curve and the death-rate curve. The area of the region between the two curves from 1800 to 1900, shown in Figure 6.31, gives the change in $\ln P(t)$. The region has area approximately equal to a rectangle of height 0.005 and width 100, so has area 0.5. We have

$$\ln\left(\frac{P(1900)}{P(1800)}\right) = \int_{1800}^{1900} \frac{P'(t)}{P(t)}\, dt \approx 0.5$$
$$\frac{P(1900)}{P(1800)} \approx e^{0.5} = 1.65$$

During the nineteenth century the population of developing countries increased by about 65%, a factor of 1.65.

(c) The shaded region between the birth rate and death rate curves from 1950 to 1977 in Figure 6.31 consists of approximately 1.5 rectangles, each of which has area $(0.01)(27) = 0.27$. The area of this region is approximately $(1.5)(0.27) = 0.405$. We have

$$\ln\left(\frac{P(1977)}{P(1950)}\right) = \int_{1950}^{1977} \frac{P'(t)}{P(t)}\, dt \approx 0.405,$$

so

$$\frac{P(1977)}{P(1950)} \approx e^{0.405} = 1.50$$

Between 1950 and 1977, the population of developing countries increased by about 50%, a factor of 1.5.

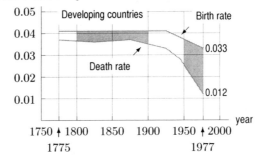

Figure 6.31: Relative growth rate of population = Relative birth rate − Relative death rate

Problems for Section 6.4

1. A town has a population of 1000. Fill in the table assuming that the town's population grows by

(a)   50 people per year   (b)   5% per year

Year	0	1	2	3	4	...	10
Population	1000					...	

2. Birth and death rates are often reported as births or deaths per thousand members of the population. What is the relative rate of growth of a population with a birth rate of 30 births per 1000 and a death rate of 20 deaths per 1000?

3. The relative growth rate of a population is given in Figure 6.32. By what percentage does the population change over the 10-year period?

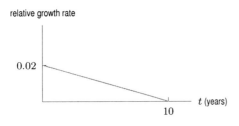

Figure 6.32

4. The article "Job Scene: What's Hot, What's Not in the Next 10 Years," appearing in the *Chicago Tribune* in July 1996, contained the following quote about predicted change in jobs between 1995 and 2005:

- Employment is projected to increase to 144.7 million from 127 million – only **14** percent. The growth rate was **24** percent from 1983 to 1994.

[6]www.unaids.org, May 2009.

- Jobs in services and retail trades are expected to increase by **16.2** million. Business services, health and education will account for **9.1** million new jobs.
- Manufacturing will lose **1.3** million jobs, a continuation of its decline.
- Jobs for those with master's degrees will grow the most, **28** percent.

(a) The first number in bold is 14. Is this an absolute change, a relative change, an absolute rate of change, or a relative rate of change? Find each of the other three measures of change, corresponding to 14.

(b) Identify each of the numbers in bold as an absolute change, a relative change, an absolute rate of change, or a relative rate of change.

5. Table 6.3 shows the cumulative number of AIDS deaths worldwide.[6] Find the absolute increase in AIDS deaths between 2003 and 2004 and between 2006 and 2007. Find the relative increase between 2003 and 2004 and between 2006 and 2007.

Table 6.3   *Cumulative AIDS deaths worldwide, in millions*

Year	2003	2004	2005	2006	2007
Cases	30.2	33.3	35.5	37.6	39.6

6. A population is 100 at time $t = 0$, with t in years.

(a) If the population has a constant absolute growth rate of 10 people per year, find a formula for the size of the population at time t.

(b) If the population has a constant relative growth rate of 10% per year, find a formula for the size of the population at time t.

(c) Graph both functions on the same axes.

7. The size of a bacteria population is 4000. Find a formula for the size, P, of the population t hours later if the population is decreasing by

(a)   100 bacteria per hour (b)   5% per hour

In which case does the bacteria population reach 0 first?

8. The relative growth rate of a population $P(t)$ is given in Figure 6.33. By what percentage does the population change over the 8-year period? If the population was 10,000 at time $t = 0$, what is the population 8 years later?

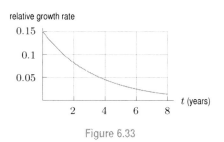

Figure 6.33

In Problems 9–12, a graph of the relative rate of change of a population is given with t in years. By approximately what percentage does the population change over the 10-year period?

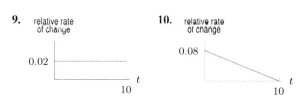

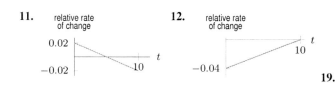

Problems 13–16 show the relative rate of change of f for $0 \le t \le 10$. Give the intervals on which f is increasing and on which f is decreasing.

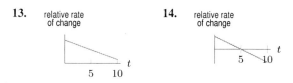

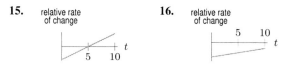

17. The population, P, in millions, of Nicaragua was 5.78 million in 2008 and growing at an annual rate of 1.8%.

(a) Write a formula for P as a function of t, where t is years since 2008.

(b) Find the projected average rate of change (or absolute growth rate) in Nicaragua between 2008 and 2009, and between 2009 and 2010. Explain why your answers are different.

(c) Use your answers to part (b) to confirm that the relative rate of change (or relative growth rate) over both time intervals was 1.8%.

In Problems 18–19, the graph shows relative birth and death rates for a population $P(t)$. Determine whether the population is increasing or decreasing, and find the percentage by which the population changes during the 10-year period.

18.

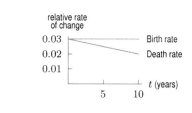

19.

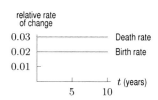

CHAPTER SUMMARY

- **Average value**
- **Consumer and producer surplus**
  Wage and price controls

- **Present and future value**
  Income stream
- **Relative growth rates**

REVIEW PROBLEMS FOR CHAPTER SIX

1. **(a)** Use Figure 6.34 to find $\int_0^6 f(x)\,dx$.
   (b) What is the average value of f on the interval $x = 0$ to $x = 6$?

Figure 6.34

2. Find the average value of $g(t) = e^t$ over the interval $0 \le t \le 10$.

3. **(a)** What is the average value of $f(x) = \sqrt{1 - x^2}$ over the interval $0 \le x \le 1$?
   (b) How can you tell whether this average value is more or less than 0.5 without doing any calculations?

4. A service station orders 100 cases of motor oil every 6 months. The number of cases of oil remaining t months after the order arrives is modeled by

$$f(t) = 100e^{-0.5t}.$$

   (a) How many cases are there at the start of the six-month period? How many cases are left at the end of the six-month period?
   (b) Find the average number of cases in inventory over the six-month period.

5. The quantity of a radioactive substance at time t is

$$Q(t) = 4(0.96)^t \text{ grams.}$$

   (a) Find $Q(10)$ and $Q(20)$.
   (b) Find the average of $Q(10)$ and $Q(20)$.
   (c) Find the average value of $Q(t)$ over the interval $10 \le t \le 20$.
   (d) Use the graph of $Q(t)$ to explain the relative sizes of your answers in parts (b) and (c).

In Problems 6–7, estimate the average value of $f(x)$ from $x = a$ to $x = b$.

6.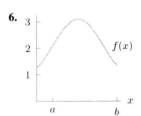

7.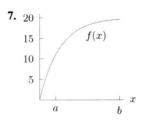

8. The function f in Figure 6.35 is symmetric about the y-axis. Consider the average value of f over the following intervals:

   I. $0 \le x \le 1$ II. $0 \le x \le 2$
   III. $0 \le x \le 5$ IV. $-2 \le x \le 2$

   (a) For which interval is the average value of f least?
   (b) For which interval is the average value of f greatest?
   (c) For which pair of intervals are the average values equal?

9. For the function f in Figure 6.35, write an expression involving one or more definite integrals that denotes:

   (a) The average value of f for $0 \le x \le 5$.
   (b) The average value of $|f|$ for $0 \le x \le 5$.

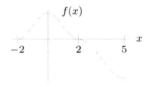

Figure 6.35

10. For a product, the supply curve is $p = 5 + 0.02q$ and the demand curve is $p = 30e^{-0.003q}$, where p is the price and q is the quantity sold at that price. Find:

   (a) The equilibrium price and quantity
   (b) The consumer and producer surplus

11. Sketch possible supply and demand curves where the consumer surplus at the equilibrium price is

   (a) Greater than the producer surplus.
   (b) Less than the producer surplus.

12. In May 1991, *Car and Driver* described a Jaguar that sold for \$980,000. At that price only 50 have been sold. It is estimated that 350 could have been sold if the price had been \$560,000. Assuming that the demand curve is a straight line, and that \$560,000 and 350 are the equilibrium price and quantity, find the consumer surplus at the equilibrium price.

13. For a product, the demand curve is $p = 100e^{-0.008q}$ and the supply curve is $p = 4\sqrt{q} + 10$ for $0 \le q \le 500$, where q is quantity and p is price in dollars per unit.

   (a) At a price of \$50, what quantity are consumers willing to buy and what quantity are producers willing to supply? Will the market push prices up or down?
   (b) Find the equilibrium price and quantity. Does your answer to part (a) support the observation that market forces tend to push prices closer to the equilibrium price?
   (c) At the equilibrium price, calculate and interpret the consumer and producer surplus.

14. Calculate the present value of a continuous revenue stream of $1000 per year for 5 years at an interest rate of 9% per year compounded continuously.

15. (a) A bank account earns 10% interest compounded continuously. At what constant, continuous rate must a parent deposit money into such an account in order to save $100,000 in 10 years for a child's college expenses?
    (b) If the parent decides instead to deposit a lump sum now in order to attain the goal of $100,000 in 10 years, how much must be deposited now?

16. The Hershey Company is the largest US producer of chocolate. Between 2005 and 2008, Hershey generated net sales at a rate approximated by $4.8 + 0.1t$ billion dollars per year, where t is the time in years since January 1, 2005.[7] Assume this rate continues through the year 2010 and that the interest rate is 2% per year compounded continuously.

    (a) Find the value, on January 1, 2005, of Hershey's net sales during the five-year period from from January 1, 2005 to January 1, 2010.
    (b) Find the value, on January 1, 2010, of Hershey's net sales during the same five-year period.

17. Determine the constant income stream that needs to be invested over a period of 10 years at an interest rate of 5% per year compounded continuously to provide a present value of $5000.

18. In 1980 West Germany made a loan of 20 billion Deutsche Marks to the Soviet Union, to be used for the construction of a natural gas pipeline connecting Siberia to Western Russia, and continuing to West Germany (Urengoi–Uschgorod–Berlin). Assume that the deal was as follows: In 1985, upon completion of the pipeline, the Soviet Union would deliver natural gas to West Germany, at a constant rate, for all future times. Assuming a constant price of natural gas of 0.10 Deutsche Mark per cubic meter, and assuming West Germany expects 10% annual interest on its investment (compounded continuously), at what rate does the Soviet Union have to deliver the gas, in billions of cubic meters per year? Keep in mind that delivery of gas could not begin until the pipeline was completed. Thus, West Germany received no return on its investment until after five years had passed. (Note: A more complex deal of this type was actually made between the two countries.)

19. Figure 6.36 gives the relative growth rate of a population.

    (a) On what interval is the population increasing? By what percentage does the population increase during this interval?
    (b) On what interval is the population decreasing? By what percentage does the population decrease during this interval?

(c) By what percentage does the population change during the 15-year period shown?

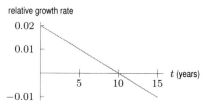

Figure 6.36

20. The number of reported burglary offenses in the US has been mostly decreasing since 1990.[8] If $P(t)$ is the number of reported burglaries as a function of year t, the relative rate of change per year, $P'(t)/P(t)$, is given in the table.

    (a) Estimate the integral of the relative rate of change:
    $$\int_{1990}^{2002} \frac{P'(t)}{P(t)}\, dt.$$
    (b) By what percent did the number of burglaries change during the period 1990–2002?

Year	1990	1991	1992	1993	1994	1995	1996
Rel. rate	−0.03	0.03	−0.06	−0.05	−0.04	−0.04	−0.03

Year	1997	1998	1999	2000	2001	2002	
Rel. rate	−0.02	−0.05	−0.10	−0.02	0.03	0.02	

21. In 1990 humans generated $1.4 \cdot 10^{20}$ joules of energy from petroleum. At the time, it was estimated that all of the earth's petroleum would generate approximately 10^{22} joules. Assuming the use of energy generated by petroleum increases by 2% each year, how long will it be before all of our petroleum resources are used up?

22. In everyday language, exponential growth means very fast growth. In this problem, you will see that any exponentially growing function eventually grows faster than any power function.

    (a) Show that the relative growth rate of the function $f(x) = x^n$, for fixed $n > 0$ and for $x > 0$, decreases as x increases.
    (b) Assume $k > 0$ is fixed. Explain why, for large x, the relative growth rate of the function $g(x) = e^{kx}$ is larger than the relative growth rate of $f(x)$.

---

[7] 2008 Annual Report, accessed at www.thehersheycompany.com.
[8] *The World Almanac and Book of Facts 2005*, p. 162 (New York).

CHECK YOUR UNDERSTANDING

In Problems 1–40, indicate whether the statement is true or false.

1. The average value of $f(x)$ on the interval $0 \leq x \leq 10$ is $\int_0^{10} f(x)\, dx$.

2. If $f(0) = 0$ and $f(10) = 50$, then the average value of f on the interval $0 \leq x \leq 10$ is between 0 and 50.

3. If f is a continuous increasing function and $f(0) = 0$ and $f(10) = 50$, then the average value of f on the interval $0 \leq x \leq 10$ is between 0 and 50.

4. The units of the average value of f are the same as the units of f.

5. The average value of f over the interval $0 \leq x \leq 20$ is twice the average value of f over the interval $0 \leq x \leq 10$.

6. The average value of $2f(x)$ over the interval $0 \leq x \leq 10$ is twice the average value of $f(x)$ over the interval $0 \leq x \leq 10$.

7. If the average value of f over the interval $0 \leq x \leq 10$ is 0, then $f(x) = 0$ for all $0 \leq x \leq 10$.

8. If $f(x) > g(x) \geq 0$, then the average value of $f(x)$ for $0 \leq x \leq 10$ is greater than the average value of $g(x)$ for $0 \leq x \leq 10$.

9. The average value of f on the interval $5 \leq x \leq 7$ is one half of $\int_5^7 f(x)\, dx$.

10. The average value of the cost function $C(q)$ for $0 \leq q \leq 10$ is the same as the average cost for 10 items.

11. The supply curve shows the quantity of an item that a producer supplies at different prices.

12. The demand curve shows the quantity of an item that consumers will purchase at various prices.

13. The equilibrium price p^* is where the demand curve crosses the q-axis.

14. Consumer surplus is the total amount gained by consumers by purchasing items at prices higher than they are willing to pay.

15. The units of consumer surplus are the same as the units of producer surplus.

16. The units of producer surplus are price times quantity.

17. Total gains from trade is the difference between consumer and producer surplus.

18. The area under the demand curve from 0 to the equilibrium quantity q^* is the amount consumers spend at the equilibrium price p^*.

19. Consumer surplus at price p^* is the area between the demand curve and the horizontal line at p^*.

20. Producer surplus at price p^* is the area between the demand curve and the horizontal line at p^*.

21. Assuming a positive interest rate, the present value of an income stream is always less than its future value.

22. The future value of an income stream of $1000 per year for 5 years is more than $5000 if the interest rate is 2%.

23. The present value of an income stream of $1000 per year for 5 years is more than $5000 if the interest rate is 2%.

24. A single payment at the beginning of the year of $1000 has the same future value as an income stream of $1000 per year if we assume a continuous interest rate of 3% for one year for both.

25. If $S(t)$ is an income stream, then $S(t)$ has units of dollars.

26. With a continuous interest rate of r, the future value of a payment of $3000 today in 5 years is $3000e^{-5r}$.

27. With a continuous interest rate of r, we would have to deposit $3000e^{-5r}$ today to grow to $3000 in 5 years.

28. If an income stream of 2000 dollars per year grows to $11,361 in five years, then the amount of interest earned over the five-year period is $1361.

29. The present value of an income stream of 2000 dollars per year that starts now and pays out over 6 years with a continuous interest rate of 2% is $\int_0^6 2000 e^{-2t}\, dt$.

30. The future value of an income stream of 2000 dollars per year that starts now and accumulates over 9 years with a positive interest rate is greater than $18,000.

31. The relative rate of change of population P with respect to time t is dP/dt.

32. Percent rate of change has the same units as rate of change.

33. If a population P is growing at a continuous rate of 2% per year, then $dP/dt = 0.02$.

34. The relative growth rate of P is the derivative of $\ln(P(t))$.

35. If the percent birth rate is higher than the percent death rate in a country, then the country's population is increasing.

36. If the percent birth rate and the percent death rate in a country are both decreasing, then the country's population is decreasing.

37. The integral of the relative growth rate of a population P gives the total change in P.

38. If the integral of the relative growth rate of a population P over 30 years is 0.5 then the population has decreased by 50% over the 30-year period.

39. If the integral of the relative growth rate of a population is negative on an interval, then the population at the end of the interval is less than the population at the beginning of the interval.

40. If the integral of the relative growth rate of a population is equal to A on an interval, then the population at the end of the interval is e^A times the population at the beginning of the interval.

PROJECTS FOR CHAPTER SIX

1. Distribution of Resources

Whether a resource is distributed evenly among members of a population is often an important political or economic question. How can we measure this? How can we decide if the distribution of wealth in this country is becoming more or less equitable over time? How can we measure which country has the most equitable income distribution? This problem describes a way of making such measurements. Suppose the resource is distributed evenly. Then any 20% of the population will have 20% of the resource. Similarly, any 30% will have 30% of the resource and so on. If, however, the resource is not distributed evenly, the poorest $p\%$ of the population (in terms of this resource) will not have $p\%$ of the goods. Suppose $F(x)$ represents the fraction of the resources owned by the poorest fraction x of the population. Thus $F(0.4) = 0.1$ means that the poorest 40% of the population owns 10% of the resource.

(a) What would F be if the resource were distributed evenly?

(b) What must be true of any such F? What must $F(0)$ and $F(1)$ equal? Is F increasing or decreasing? Is the graph of F concave up or concave down?

(c) Gini's index of inequality, G, is one way to measure how evenly the resource is distributed. It is defined by

$$G = 2 \int_0^1 [x - F(x)] \, dx.$$

Show graphically what G represents.

(d) Graphical representations of Gini's index for two countries are given in Figures 6.37 and 6.38. Which country has the more equitable distribution of wealth? Discuss the distribution of wealth in each of the two countries.

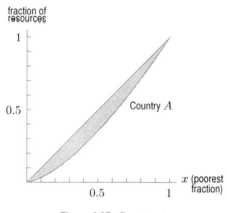

Figure 6.37: Country A

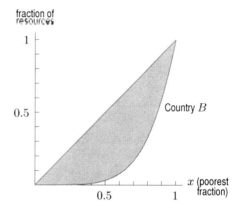

Figure 6.38: Country B

(e) What is the maximum possible value of Gini's index of inequality, G? What is the minimum possible value? Sketch graphs in each case. What is the distribution of resources in each case?

2. Yield from an Apple Orchard

Figure 6.39 is a graph of the annual yield, $y(t)$, in bushels per year, from an orchard t years after planting. The trees take about 10 years to get established, but for the next 20 years they give a substantial yield. After about 30 years, however, age and disease start to take their toll, and the annual yield falls off.[9]

(a) Represent on a sketch of Figure 6.39 the total yield, $F(M)$, up to M years, with $0 \leq M \leq 60$. Write an expression for $F(M)$ in terms of $y(t)$.

---

[9] From Peter D. Taylor, *Calculus: The Analysis of Functions* (Toronto: Wall & Emerson, Inc., 1992).

(b) Sketch a graph of $F(M)$ against M for $0 \le M \le 60$.

(c) Write an expression for the average annual yield, $a(M)$, up to M years.

(d) When should the orchard be cut down and replanted? Assume that we want to maximize average revenue per year, and that fruit prices remain constant, so that this is achieved by maximizing average annual yield. Use the graph of $y(t)$ to estimate the time at which the average annual yield is a maximum. Explain your answer geometrically and symbolically.

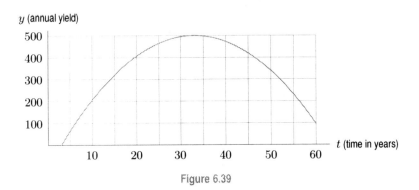

Figure 6.39

Chapter Nine

FUNCTIONS OF SEVERAL VARIABLES

Contents

9.1 UNDERSTANDING FUNCTIONS OF TWO VARIABLES

In business, science, and politics, outcomes are rarely determined by a single variable. For example, consider an airline's ticket pricing. To avoid flying planes with many empty seats, it sells some tickets at full price and some at a discount. For a particular route, the airline's revenue, R, earned in a given time period is determined by the number of full-price tickets, x, and the number of discount tickets, y, sold. We say that R is a function of x and y, and we write

$$R = f(x, y).$$

This is just like the function notation of one-variable calculus. The variable R is the dependent variable and the variables x and y are the independent variables. The letter f stands for the *function* or rule that gives the value, or output, of R corresponding to given values of x and y. The collection of all possible inputs, (x, y), is called the *domain* of f. We say a function is an *increasing* (*decreasing*) function of one of its variables if it increases (decreases) as that variable increases while the other independent variables are held constant.

A function of two variables can be represented numerically by a table of values, algebraically by a formula, or pictorially by a contour diagram. In this section we give numerical and algebraic examples; contour diagrams are introduced in Section 9.2.

Functions Given Numerically

The revenue, R, (in dollars) from a particular airline route is shown in Table 9.1 as a function of the number of full-price tickets and the number of discount tickets sold.

Table 9.1   *Revenue from ticket sales as a function of x and y*

Number of full-price tickets, x

		100	200	300	400
	200	75,000	110,000	145,000	180,000
Number of discount tickets, y	400	115,000	150,000	185,000	220,000
	600	155,000	190,000	225,000	260,000
	800	195,000	230,000	265,000	300,000
	1000	235,000	270,000	305,000	340,000

Values of x are shown across the top, values of y are down the left side, and the corresponding values of $f(x, y)$ are in the table. For example, to find the value of $f(300, 600)$, we look in the column corresponding to $x = 300$ at the row $y = 600$, where we find the number 225,000. Thus,

$$f(300, 600) = 225,000.$$

This means that the revenue from 300 full-price tickets and 600 discount tickets is $225,000. We see in Table 9.1 that f is an increasing function of x and an increasing function of y.

Notice how this differs from the table of values of a one-variable function, where one row or one column is enough to list the values of the function. Here many rows and columns are needed because the function has an output for every pair of values of the independent variables.

Functions Given Algebraically

The function given in Table 9.1 can be represented by a formula. Looking across the rows, we see that each additional 100 full-price tickets sold raises the revenue by $35,000, so each full-price ticket must cost $350. Similarly, looking down a column shows that an additional 200 discount tickets sold increases the revenue by $40,000, so each discount ticket must cost $200. Thus, the revenue function is given by the formula

$$R = 350x + 200y.$$

Example 1   Give a formula for the function $M = f(B, t)$ where M is the amount of money in a bank account t years after an initial investment of B dollars, if interest accrues at a rate of 5% per year compounded
(a) Annually   (b) Continuously.

Solution   (a) Annual compounding means that M increases by a factor of 1.05 every year, so
$$M = f(B,t) = B(1.05)^t.$$

(b) Continuous compounding means that M grows according to the function e^{kt}, with $k = 0.05$, so
$$M = f(B,t) = Be^{0.05t}.$$

Example 2   A car rental company charges \$40 a day and 15 cents a mile for its cars.
(a) Write a formula for the cost, C, of renting a car as a function of the number of days, d, and the number of miles driven, m.
(b) If $C = f(d, m)$, find $f(5, 300)$ and interpret it.

Solution   (a) The total cost in dollars of renting a car is 40 times the numbers of days plus 0.15 times the number of miles, so
$$C = 40d + 0.15m.$$

(b) We have
$$f(5, 300) = 40(5) + 0.15(300)$$
$$= 200 + 45$$
$$= 245.$$

We see that $f(5, 300) = 245$. This tells us that if we rent a car for 5 days and drive it 300 miles, it costs us \$245.

Strategy to Investigate Functions of Two Variables: Vary One Variable at a Time

We can learn a great deal about a function of two variables by letting one variable vary while holding the other fixed. This gives a function of one variable, called a *cross-section* of the original function.

Concentration of a Drug in the Blood

When a drug is injected into muscle tissue, it diffuses into the bloodstream. The concentration of the drug in the blood increases until it reaches a maximum, and then decreases. The concentration, C (in mg per liter), of the drug in the blood is a function of two variables: x, the amount (in mg) of the drug given in the injection, and t, the time (in hours) since the injection was administered. We are told that
$$C = f(x,t) = te^{-t(5-x)} \qquad \text{for } 0 \le x \le 4 \text{ and } t \ge 0.$$

Example 3   In terms of the drug concentration in the blood, explain the significance of the cross-sections:
(a) $f(4, t)$ (b) $f(x, 1)$

Solution   (a) Holding x fixed at 4 means that we are considering an injection of 4 mg of the drug; letting t vary means we are watching the effect of this dose as time passes. Thus the function $f(4, t)$ describes the concentration of the drug in the blood resulting from a 4-mg injection as a function of time. Figure 9.1 shows the graph of $f(4, t) = te^{-t}$. Notice that the concentration in the blood from this dose is at a maximum at 1 hour after injection, and that the concentration in the blood eventually approaches zero.

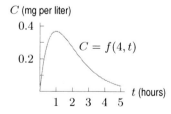

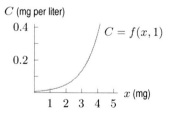

Figure 9.1: The function $f(4, t)$ shows the concentration in the blood resulting from a 4-mg injection

Figure 9.2: The function $f(x, 1)$ shows the concentration in the blood 1 hour after the injection

(b) Holding t fixed at 1 means that we are focusing on the blood 1 hour after the injection; letting x vary means we are considering the effect of different doses at that instant. Thus, the function $f(x, 1)$ gives the concentration of the drug in the blood 1 hour after injection as a function of the amount injected. Figure 9.2 shows the graph of $f(x, 1) = e^{-(5-x)} = e^{x-5}$. Notice that $f(x, 1)$ is an increasing function of x. This makes sense: If we administer more of the drug, the concentration in the bloodstream is higher.

Example 4   Continue with $C = f(x, t) = te^{-t(5-x)}$. Graph the cross-sections of $f(a, t)$ for $a = 1, 2, 3,$ and 4 on the same axes. Describe how the graph changes for larger values of a and explain what this means in terms of drug concentration in the blood.

Solution   The one-variable function $f(a, t)$ represents the effect of an injection of a mg at time t. Figure 9.3 shows the graphs of the four functions $f(1, t) = te^{-4t}$, $f(2, t) = te^{-3t}$, $f(3, t) = te^{-2t}$, and $f(4, t) = te^{-t}$ corresponding to injections of 1, 2, 3, and 4 mg of the drug. The general shape of the graph is the same in every case: The concentration in the blood is zero at the time of injection $t = 0$, then increases to a maximum value, and then decreases toward zero again. We see that if a larger dose of the drug is administered, the peak of the graph is later and higher. This makes sense, since a larger dose will take longer to diffuse fully into the bloodstream and will produce a higher concentration when it does.

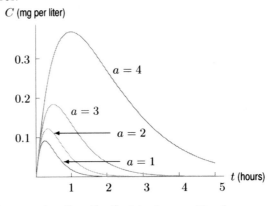

Figure 9.3: Concentration $C = f(a, t)$ of the drug resulting from an a-mg injection

Problems for Section 9.1

Problems 1–2 concern the cost, C, of renting a car from a company which charges \$40 a day and 15 cents a mile, so $C = f(d, m) = 40d + 0.15m$, where d is the number of days, and m is the number of miles.

1. Make a table of values for C, using $d = 1, 2, 3, 4$ and $m = 100, 200, 300, 400$. You should have 16 values in your table.

2. (a) Find $f(3, 200)$ and interpret it.
   (b) Explain the significance of $f(3, m)$ in terms of rental car costs. Graph this function, with C as a function of m.
   (c) Explain the significance of $f(d, 100)$ in terms of rental car costs. Graph this function, with C as a function of d.

Problems 3–7 refer to Table 9.2, which shows[1] the weekly beef consumption, C, (in lbs) of an average household as a function of p, the price of beef (in \$/lb) and I, annual household income (in \$1000s).

Table 9.2   *Quantity of beef bought (lbs/household/week)*

		p		
	3.00	3.50	4.00	4.50
20	2.65	2.59	2.51	2.43
40	4.14	4.05	3.94	3.88
I 60	5.11	5.00	4.97	4.84
80	5.35	5.29	5.19	5.07
100	5.79	5.77	5.60	5.53

3. Give tables for beef consumption as a function of p, with I fixed at $I = 20$ and $I = 100$. Give tables for beef consumption as a function of I, with p fixed at $p = 3.00$ and $p = 4.00$. Comment on what you see in the tables.

4. How does beef consumption vary as a function of household income if the price of beef is held constant?

5. Make a table showing the amount of money, M, that the average household spends on beef (in dollars per household per week) as a function of the price of beef and household income.

[1] From Richard G. Lipsey, *An Introduction to Positive Economics*, 3rd Ed. (London: Weidenfeld and Nicolson, 1971).

6. Make a table of the proportion, P, of household income spent on beef per week as a function of price and income. (Note that P is the fraction of income spent on beef.)
7. Express P, the proportion of household income spent on beef per week, in terms of the original function $f(I, p)$ which gave consumption as a function of p and I.
8. Graph the bank-account function f in Example 1 on page 350, holding B fixed at $B = 10, 20, 30$ and letting t vary. Then graph f, holding t fixed at $t = 0, 5, 10$ and letting B vary. Explain what you see.
9. The total sales of a product, S, can be expressed as a function of the price p charged for the product and the amount, a, spent on advertising, so $S = f(p, a)$. Do you expect f to be an increasing or decreasing function of p? Do you expect f to be an increasing or decreasing function of a? Why?
10. The number, n, of new cars sold in a year is a function of the price of new cars, c, and the average price of gas, g.
    - **(a)** If c is held constant, is n an increasing or decreasing function of g? Why?
    - **(b)** If g is held constant, is n an increasing or decreasing function of c? Why?
11. The heat index is a temperature which tells you how hot it feels as a result of the combination of temperature and humidity. See Table 9.3. Heat exhaustion is likely to occur when the heat index reaches 105°F.
    - **(a)** If the temperature is 80°F and the humidity is 50%, how hot does it feel?
    - **(b)** At what humidity does 90°F feel like 90°F?
    - **(c)** Make a table showing the approximate temperature at which heat exhaustion becomes a danger, as a function of humidity.
    - **(d)** Explain why the heat index is sometimes above the actual temperature and sometimes below it.

Table 9.3 *Heat index (°F) as a function of humidity (H%) and temperature (T°F)*

H \ T	70	75	80	85	90	95	100	105	110	115
0	64	69	73	78	83	87	91	95	99	103
10	65	70	75	80	85	90	95	100	105	111
20	66	72	77	82	87	93	99	105	112	120
30	67	73	78	84	90	96	104	113	123	135
40	68	74	79	86	93	101	110	123	137	151
50	69	75	81	88	96	107	120	135	150	
60	70	76	82	90	100	114	132	149		

12. Using Table 9.3, graph heat index as a function of humidity with temperature fixed at 70°F and at 100°F. Explain the features of each graph and the difference between them in common-sense terms.
13. The monthly payments, P dollars, on a mortgage in which A dollars were borrowed at an annual interest rate of $r\%$ for t years is given by $P = f(A, r, t)$. Is f an increasing or decreasing function of A? Of r? Of t?
14. You are planning a trip whose principal cost is gasoline.
    - **(a)** Make a table showing how the daily fuel cost varies as a function of the price of gasoline (in dollars per gallon) and the number of gallons you buy each day.
    - **(b)** If your car goes 30 miles on each gallon of gasoline, make a table showing how your daily fuel cost varies as a function of your daily travel distance and the price of gas.

In Problems 15–16, the fallout, V (in kilograms per square kilometer), from a volcanic explosion depends on the distance, d, from the volcano and the time, t, since the explosion:

$$V = f(d, t) = (\sqrt{t})e^{-d}.$$

15. On the same axes, graph cross-sections of f with $t = 1$, and $t = 2$. As distance from the volcano increases, how does the fallout change? Look at the relationship between the graphs: how does the fallout change as time passes? Explain your answers in terms of volcanoes.
16. On the same axes, graph cross-sections of f with $d = 0$, $d = 1$, and $d = 2$. As time passes since the explosion, how does the fallout change? Look at the relationship between the graphs: how does fallout change as a function of distance? Explain your answers in terms of volcanoes.
17. An airport can be cleared of fog by heating the air. The amount of heat required, $H(T, w)$ (in calories per cubic meter of fog), depends on the temperature of the air, T (in °C), and the wetness of the fog, w (in grams per cubic meter of fog). Figure 9.4 shows several graphs of H against T with w fixed.
    - **(a)** Estimate $H(20, 0.3)$ and explain what information it gives us.
    - **(b)** Make a table of values for $H(T, w)$. Use $T = 0, 10, 20, 30, 40$, and $w = 0.1, 0.2, 0.3, 0.4$.

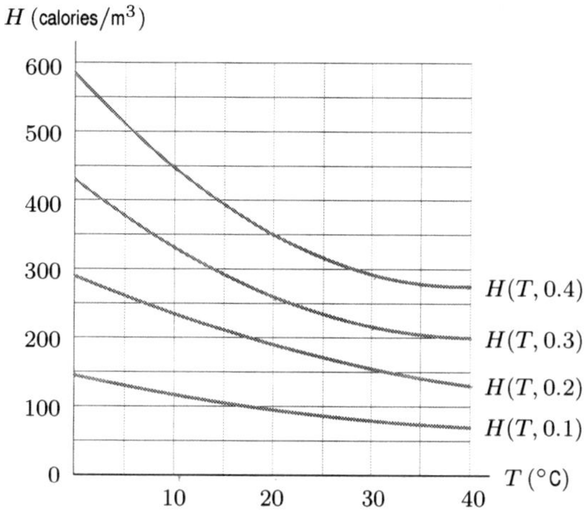

Figure 9.4

9.2   CONTOUR DIAGRAMS

How can we visualize a function of two variables? Just as a function of one variable can be represented by a graph, a function of two variables can be represented by a surface in space or by a *contour diagram* in the plane. Numerical information is more easily obtained from contour diagrams, so we concentrate on their use.

Weather Maps

Figure 9.5 shows a weather map from a newspaper. This contour diagram shows the predicted high temperature, T, in degrees Fahrenheit (°F), throughout the US on that day. The curves on the map, called *isotherms*, separate the country into zones, according to whether T is in the 60s, 70s, 80s, 90s, or 100s. (*Iso* means same and *therm* means heat.) Notice that the isotherm separating the 80s and 90s zones connects all the points where the temperature is predicted to be exactly 90°F.

If the function $T = f(x, y)$ gives the predicted high temperature (in °F) on this particular day as a function of latitude x and longitude y, then the isotherms are graphs of the equations

$$f(x, y) = c$$

where c is a constant. In general, such curves are called *contours*, and a graph showing selected contours of a function is called a contour diagram.

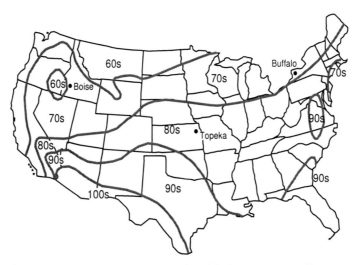

Figure 9.5: Weather map showing predicted high temperatures, T, on a summer day

Example 1   Estimate the predicted value of T in Boise, Idaho; Topeka, Kansas; and Buffalo, New York.

Solution   Boise and Buffalo are in the 70s region, and Topeka is in the 80s region. Thus, the predicted temperature in Boise and Buffalo is between 70 and 80 while the predicted temperature in Topeka is between 80 and 90.

In fact, we can say more. Although both Boise and Buffalo are in the 70s, Boise is quite close to the $T = 70$ isotherm, whereas Buffalo is quite close to the $T = 80$ isotherm. So we estimate that the temperature will be in the low 70s in Boise and in the high 70s in Buffalo. Topeka is about halfway between the $T = 80$ isotherm and the $T = 90$ isotherm. Thus, we guess that the temperature in Topeka will be in the mid-80s. In fact, the actual high temperatures for that day were 71°F for Boise, 79°F for Buffalo, and 86°F for Topeka.

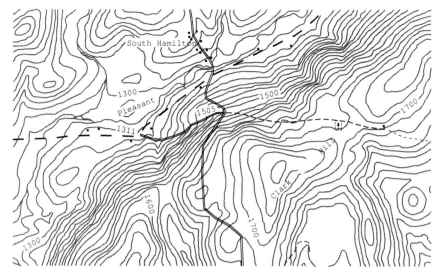

Figure 9.6: A topographical map showing the region around South Hamilton, NY

Topographical Maps

Another common example of a contour diagram is a topographical map like that shown in Figure 9.6. Here, the contours separate regions of lower elevation from regions of higher elevation, and give an overall picture of the nature of the terrain. Such topographical maps are frequently colored green at the lower elevations and brown, red, or even white at the higher elevations.

Example 2   Explain why the topographical map shown in Figure 9.7 corresponds to the surface of the terrain shown in Figure 9.8.

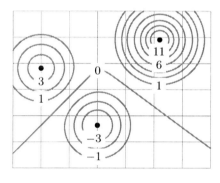

Figure 9.7: A topographical map

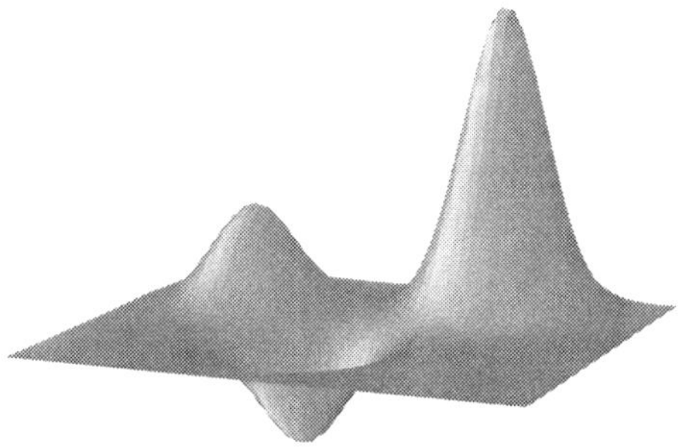

Figure 9.8: Terrain corresponding to the topographical map in Figure 9.7

Solution   We see from the topographical map in Figure 9.7 that there are two hills, one with height about 12, and the other with height about 4. Most of the terrain is around height 0, and there is one valley with height about -4. This matches the surface of the terrain in Figure 9.8 since there are two hills (one taller than the other) and one valley.

The contours on a topographical map outline the contour or shape of the land. Because every point along the same contour has the same elevation, contours are also called *level curves* or *level sets*. We usually draw contours for equally spaced values of the function. The more closely spaced the contours, the steeper the terrain; the more widely spaced the contours, the flatter the terrain (provided, of course, that the elevation between contours varies by a constant amount). Certain

305

features have distinctive characteristics. A mountain peak is typically surrounded by contours like those in Figure 9.9. A pass in a range of mountains may have contours that look like Figure 9.10. A long valley has parallel contours indicating the rising elevations on both sides of the valley (see Figure 9.11); a long ridge of mountains has the same type of contours, only the elevations decrease on both sides of the ridge. Notice that the elevation numbers on the contours are as important as the curves themselves.

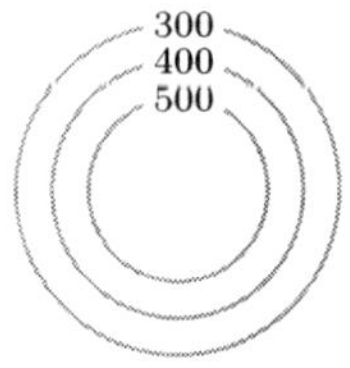

Figure 9.9: Mountain peak

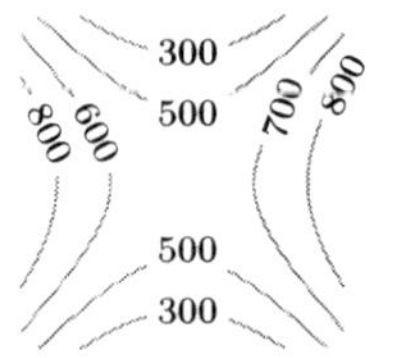

Figure 9.10: Pass between two mountains

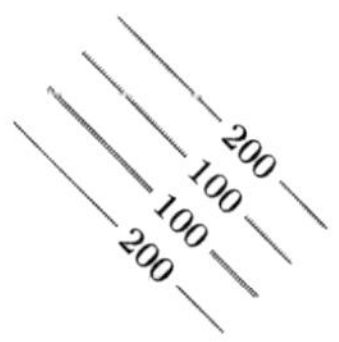

Figure 9.11: Long valley

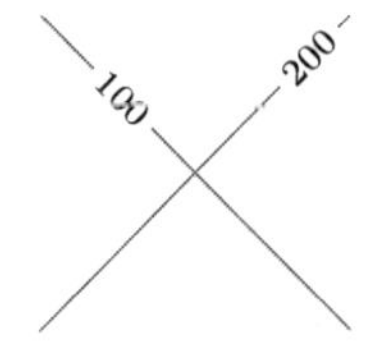

Figure 9.12: Impossible contour lines

There are some things contours cannot do. Two contours corresponding to different elevations cannot cross each other as shown in Figure 9.12. If they did, the point of intersection of the two curves would have two different elevations, which is impossible (assuming the terrain has no overhangs).

Using Contour Diagrams

Consider the effect of different weather conditions on US corn production. What would happen if the average temperature were to increase (due to global warming, for example) or if the rainfall were to decrease (due to a drought)? One way of estimating the effect of these climatic changes is to use Figure 9.13. This map is a contour diagram giving the corn production $C = f(R, T)$ in the US as a function of the total rainfall, R, in inches, and average temperature, T, in degrees Fahrenheit, during the growing season.[2] Suppose at the present time, $R = 15$ inches and $T = 76°$F. Production is measured as a percentage of the present production; thus, the contour through $R = 15, T = 76$ is $C = 100$, that is, $C = f(15, 76) = 100$.

Example 3   Use Figure 9.13 to evaluate $f(18, 78)$ and $f(12, 76)$ and explain the answers in terms of corn production.

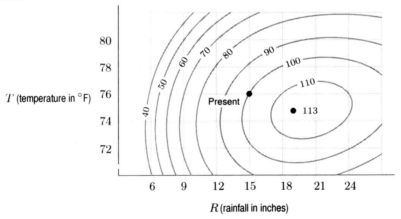

Figure 9.13: Corn production, C, as a function of rainfall and temperature

Solution   The point with R-coordinate 18 and T-coordinate 78 is on the contour with value $C = 100$, so $f(18, 78) = 100$. This means that if the annual rainfall were 18 inches and the temperature were 78°F, the country would produce about the same amount of corn as at present, although it would be wetter and warmer than it is now. The point with R-coordinate 12 and T-coordinate 76 is about

[2]Adapted from S. Beaty and R. Healy, *The Future of American Agriculture*, Scientific American, Vol. 248, No. 2, February, 1983.

halfway between the $C = 80$ and $C = 90$ contours, so $f(12, 76) \approx 85$. This means that if the rainfall dropped to 12 inches and the temperature stayed at 76°F, then corn production would drop to about 85% of what it is now.

Example 4   Describe how corn production changes as a function of rainfall if temperature is fixed at the present value in Figure 9.13. Describe how corn production changes as a function of temperature if rainfall is held constant at the present value. Give common-sense explanations for your answers.

Solution   To see what happens to corn production if the temperature stays fixed at 76°F but the rainfall changes, look along the horizontal line $T = 76$. Starting from the present and moving left along the line $T = 76$, the values on the contours decrease. In other words, if there is a drought, corn production decreases. Conversely, as rainfall increases, that is, as we move from the present to the right along the line $T = 76$, corn production increases, reaching a maximum of more than 110% when $R = 21$, and then decreases (too much rainfall floods the fields). If, instead, rainfall remains at the present value and temperature increases, we move up the vertical line $R = 15$. Under these circumstances corn production decreases; a $2°$ increase causes a 10% drop in production. This makes sense since hotter temperatures lead to greater evaporation and hence drier conditions, even with rainfall constant at 15 inches. Similarly, a decrease in temperature leads to a very slight increase in production, reaching a maximum of around 102% when $T = 74$, followed by a decrease (the corn won't grow if it is too cold).

Cobb-Douglas Production Functions

Suppose you are running a small printing business, and decide to expand because you have more orders than you can handle. How should you expand? Should you start a night shift and hire more workers? Should you buy more expensive but faster computers which will enable the current staff to keep up with the work? Or should you do some combination of the two?

Obviously, the way such a decision is made in practice involves many other considerations—such as whether you could get a suitably trained night shift, or whether there are any faster computers available. Nevertheless, you might model the quantity, P, of work produced by your business as a function of two variables: your total number, N, of workers, and the total value, V, of your equipment.

How would you expect such a production function to behave? In general, having more equipment and more workers enables you to produce more. However, increasing equipment without increasing the number of workers will increase production a bit, but not beyond a point. (If equipment is already lying idle, having more of it won't help.) Similarly, increasing the number of workers without increasing equipment will increase production, but not past the point where the equipment is fully utilized, as any new workers would have no equipment available to them.

Example 5   Explain why the contour diagram in Figure 9.14 does not model the behavior expected of the production function, whereas the contour diagram in Figure 9.15 does.

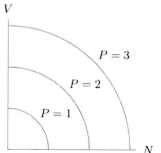

Figure 9.14: Incorrect contours for printing production

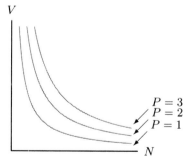

Figure 9.15: Correct contours for printing production

Solution   Production P should be an increasing function of N and an increasing function of V. We see that both contour diagrams (in Figure 9.14 and Figure 9.15) satisfy this condition. Which of the contour diagrams has production increasing in the correct way? First look at the contour diagram in Figure 9.14. Fixing V at a particular value and letting N increase means moving to the right on the contour diagram. As we do so, we cross contours with larger and larger P values, meaning that production increases indefinitely. On the other hand, in Figure 9.15, as we move in the same direction we eventually find ourselves moving nearly parallel to the contours, crossing them less and less frequently. Therefore, production increases more and more slowly as N increases while V is held fixed. Similarly, if we hold N fixed and let V increase, the contour diagram in Figure 9.14 shows production increasing at a steady rate, whereas Figure 9.15 shows production increasing, but at a decreasing rate. Thus, Figure 9.15 fits the expected behavior of the production function best.

The Cobb-Douglas Production Model

In 1928, Cobb and Douglas used a simple formula to model the production of the entire US economy in the first quarter of the 20th century. Using government estimates of P, the total yearly production between 1899 and 1922, and of K, the total capital investment over the same period, and of L, the total labor force, they found that P was well approximated by the function:

$$P = 1.01 L^{0.75} K^{0.25}.$$

This function turned out to model the US economy surprisingly accurately, both for the period on which it was based, and for some time afterward. The contour diagram of this function is similar to that in Figure 9.15. In general, production is often modeled by a function of the following form:

> ## Cobb-Douglas Production Function
>
> $$P = f(N, V) = c N^{\alpha} V^{\beta}$$
>
> where P is the total quantity produced and c, α, and β are positive constants with $0 < \alpha < 1$ and $0 < \beta < 1$.

Contour Diagrams and Tables

Table 9.4 shows the heat index as a function of temperature and humidity. The heat index is a temperature which tells you how hot it feels as a result of the combination of the two. We can also display this function using a contour diagram. Scales for the two independent variables (temperature and humidity) go on the axes. The heat indices shown range from 64 to 151, so we will draw contours at values of 70, 80, 90, 100, 110, 120, 130, 140, and 150. How do we know where the contour for 70 goes? Table 9.4 shows that, when humidity is 0%, a heat index of 70 occurs between 75°F and 80°F, so the contour will go approximately through the point $(76, 0)$. It also goes through the point $(75, 10)$. Continuing in this way, we can approximate the 70 contour. See Figure 9.16. You can construct all the contours in Figure 9.17 in a similar way.

Table 9.4   *Heat index (°F)*

		Temperature (°F)									
		70	75	80	85	90	95	100	105	110	115
	0	64	69	73	78	83	87	91	95	99	103
	10	65	70	75	80	85	90	95	100	105	111
	20	66	72	77	82	87	93	99	105	112	120
Humidity (%)	30	67	73	78	84	90	96	104	113	123	135
	40	68	74	79	86	93	101	110	123	137	151
	50	69	75	81	88	96	107	120	135	150	
	60	70	76	82	90	100	114	132	149		

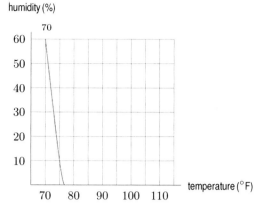

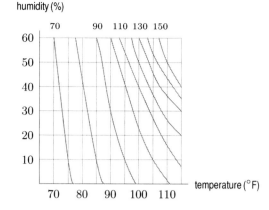

Figure 9.16: The contour for a heat index of 70

Figure 9.17: Contour diagram for the heat index

Example 6 Heat exhaustion is likely to occur where the heat index is 105 or higher. On the contour diagram in Figure 9.17, shade in the region where heat exhaustion is likely to occur.

Solution The shaded region in Figure 9.18 shows the values of temperature and humidity at which the heat index is above 105.

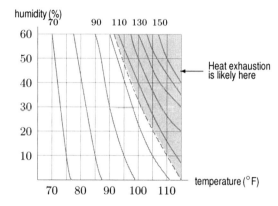

Figure 9.18: Shaded region shows conditions under which heat exhaustion is likely

Finding Contours Algebraically

Algebraic equations for the contours of a function f are easy to find if we have a formula for $f(x, y)$. A contour consists of all the points (x, y) where $f(x, y)$ has a constant value, c. Its equation is

$$f(x, y) = c.$$

Example 7 Draw a contour diagram for the airline revenue function $R = 350x + 200y$. Include contours for $R = 4000, 8000, 12000, 16000$.

Solution The contour for $R = 4000$ is given by

$$350x + 200y = 4000.$$

This is the equation of a line with intercepts $x = 4000/350 = 11.43$ and $y = 4000/200 = 20$. (See Figure 9.19.) The contour for $R = 8000$ is given by

$$350x + 200y = 8000.$$

This is the equation of a parallel line with intercepts $x = 8000/350 = 22.86$ and $y = 8000/200 = 40$. The contours for $R = 12000$ and $R = 16000$ are parallel lines drawn similarly. (See Figure 9.19.)

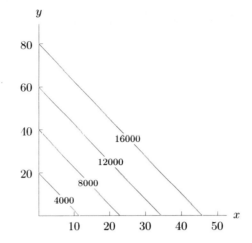

Figure 9.19: A contour diagram for $R = 350x + 200y$

Problems for Section 9.2

1. Figure 9.20 shows contours for the function $z = f(x, y)$. Is z an increasing or a decreasing function of x? Is z an increasing or a decreasing function of y?

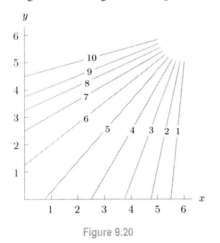

Figure 9.20

2. Figure 9.21 is a contour diagram for the demand for orange juice as a function of the price of orange juice and the price of apple juice. Which axis corresponds to orange juice? Which to apple juice? Explain.

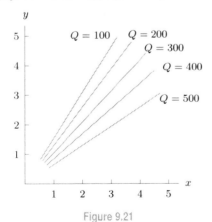

Figure 9.21

3. Figure 9.22 shows contour diagrams of temperature in °C in a room at three different times. Describe the heat flow in the room. What could be causing this?

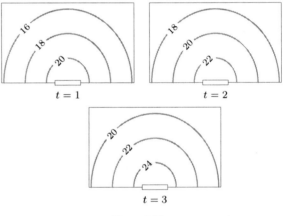

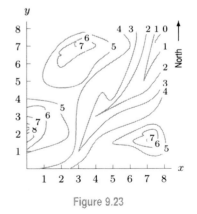

Figure 9.22

4. A topographic map is given in Figure 9.23. How many hills are there? Estimate the x- and y-coordinates of the tops of the hills. Which hill is the highest? A river runs through the valley; in which direction is it flowing?

Figure 9.23

5. Draw a contour diagram for the function $C = 40d + 0.15m$. Include contours for $C = 50, 100, 150, 200$.

Problems 6–8 refer to the map in Figure 9.5 on page 354.

6. Give the range of daily high temperatures for:

    (a) Pennsylvania   **(b)** North Dakota

    (c) California

7. Sketch a possible graph of the predicted high temperature T on a line north-south through Topeka.

8. Sketch possible graphs of the predicted high temperature on a north-south line and an east-west line through Boise.

9. Maple syrup production is highest when the nights are cold and the days are warm. Make a possible contour diagram for maple syrup production as a function of the high (daytime) temperature and the low (nighttime) temperature. Label the contours with 10, 20, 30, and 40 (in liters of maple syrup).

10. A manufacturer sells two products, one at a price of $3000 a unit and the other at a price of $12,000 a unit. A quantity q_1 of the first product and q_2 of the second product are sold at a total cost of $4000 to the manufacturer.

    (a) Express the manufacturer's profit, π, as a function of q_1 and q_2.
    (b) Sketch contours of π for $\pi = 10,000$, $\pi = 20,000$, and $\pi = 30,000$ and the break-even curve $\pi = 0$.

11. Sketch a contour diagram for $z = y - \sin x$. Include at least four labeled contours. Describe the contours in words and how they are spaced.

12. The contour diagram in Figure 9.24 shows your happiness as a function of love and money.

    (a) Describe in words your happiness as a function of:

        (i) Money, with love fixed.

        (ii) Love, with money fixed.

    (b) Graph two different cross-sections with love fixed and two different cross-sections with money fixed.

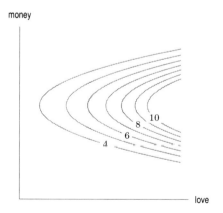

Figure 9.24

13. The concentration, C, of a drug in the blood is given by $C = f(x, t) = te^{-t(5-x)}$, where x is the amount of

drug injected (in mg) and t is the number of hours since the injection. The contour diagram of $f(x, t)$ is given in Figure 9.25. Explain the diagram by varying one variable at a time: describe f as a function of x if t is held fixed, and then describe f as a function of t if x is held fixed.

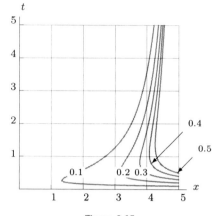

Figure 9.25

14. The cornea is the front surface of the eye. Corneal specialists use a TMS, or Topographical Modeling System, to produce a "map" of the curvature of the eye's surface. A computer analyzes light reflected off the eye and draws level curves joining points of constant curvature. The regions between these curves are colored different colors.

    The first two pictures in Figure 9.26 are cross-sections of eyes with constant curvature, the smaller being about 38 units and the larger about 50 units. For contrast, the third eye has varying curvature.

    (a) Describe in words how the TMS map of an eye of constant curvature will look.
    (b) Draw the TMS map of an eye with the cross-section in Figure 9.27. Assume the eye is circular when viewed from the front, and the cross-section is the same in every direction. Put reasonable numeric labels on your level curves.

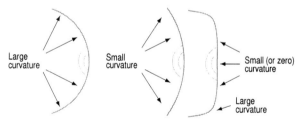

Figure 9.26: Pictures of eyes with different curvature

Figure 9.27

In Problems 15–20, sketch a contour diagram for the function with at least four labeled contours. Describe in words the contours and how they are spaced.

15. $f(x, y) = x + y$ **16.** $f(x, y) = 3x + 3y$

17. $f(x, y) = x + y + 1$ **18.** $f(x, y) = 2x - y$

19. $f(x, y) = -x - y$ **20.** $f(x, y) = y - x^2$

21. Figure 9.28 shows contours of the function giving the species density of breeding birds at each point in the US, Canada, and Mexico.[3] Are the following statements true or false? Explain your answers.

(a) Moving from south to north across Canada, the species density increases.

(b) In general, peninsulas (for example, Florida, Baja California, the Yucatan) have lower species densities than the areas around them.

(c) The species density around Miami is over 100.

(d) The greatest rate of change in species density with distance is in Mexico. If this is true, mark the point and direction giving the maximum rate.

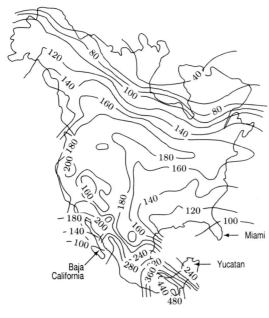

Figure 9.28

22. Each of the contour diagrams in Figure 9.29 shows population density in a certain region of a city. Choose the contour diagram that best corresponds to each of the following situations. Many different matchings are possible. Pick a reasonable one and justify your choice.

(a) The middle contour is a highway.

(b) The middle contour is an open sewage canal.

(c) The middle contour is a railroad line.

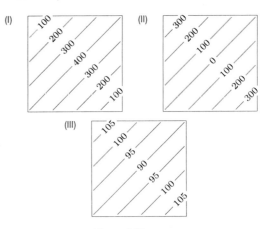

Figure 9.29

23. In a small printing business, $P = 2N^{0.6}V^{0.4}$, where N is the number of workers, V is the value of the equipment, and P is production, in thousands of pages per day.

(a) If this company has a labor force of 300 workers and 200 units worth of equipment, what is production?

(b) If the labor force is doubled (to 600 workers), how does production change?

(c) If the company purchases enough equipment to double the value of its equipment (to 400 units), how does production change?

(d) If both N and V are doubled from the values given in part (a), how does production change?

24. Figure 9.30 shows a contour map of a hill with two paths, A and B.

(a) On which path, A or B, will you have to climb more steeply?

(b) On which path, A or B, will you probably have a better view of the surrounding countryside? (Assume trees do not block your view.)

(c) Alongside which path is there more likely to be a stream?

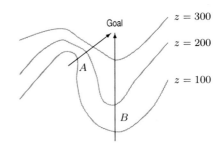

Figure 9.30

---
[3]From the undergraduate senior thesis of Professor Robert Cook, Director of Harvard's Arnold Arboretum.

25. Match the following descriptions of a company's success with the contour diagrams of success as a function of money and work in Figure 9.31.

(a) Our success is measured in dollars, plain and simple. More hard work won't hurt, but it also won't help.

(b) No matter how much money or hard work we put into the company, we just can't make a go of it.

(c) Although we are not always totally successful, it seems that the amount of money invested does not matter. As long as we put hard work into the company, our success increases.

(d) The company's success is based on both hard work and investment.

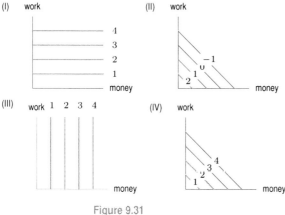

Figure 9.31

26. Figure 9.32 shows cardiac output (in liters per minute) in patients suffering from shock as a function of blood pressure in the central veins (in mm Hg) and the time in hours since the onset of shock.[4]

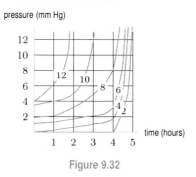

Figure 9.32

(a) In a patient with blood pressure of 4 mm Hg, what is cardiac output when the patient first goes into shock? Estimate cardiac output three hours later. How much time has passed when cardiac output is reduced to 50% of the initial value?

(b) In patients suffering from shock, is cardiac output an increasing or decreasing function of blood pressure?

(c) Is cardiac output an increasing or decreasing function of time, t, where t represents the elapsed time since the patient went into shock?

(d) If blood pressure is 3 mm Hg, explain how cardiac output changes as a function of time. In particular, does it change rapidly or slowly during the first two hours of shock? During hours 2 to 4? During the last hour of the study? Explain why this information is useful to a physician treating a patient for shock.

27. Antibiotics can be toxic in large doses. If repeated doses of an antibiotic are to be given, the rate at which the medicine is excreted through the kidneys should be monitored by a physician. One measure of kidney function is the glomerular filtration rate, or GFR, which measures the amount of material crossing the outer (or glomerular) membrane of the kidney, in milliliters per minute. A normal GFR is about 125 ml/min. Figure 9.33 gives a contour diagram of the percent, P, of a dose of mezlocillin (an antibiotic) excreted, as a function of the patient's GFR and the time, t, in hours since the dose was administered.[5]

(a) In a patient with a GFR of 50, approximately how long will it take for 30% of the dose to be excreted?

(b) In a patient with a GFR of 60, approximately what percent of the dose has been excreted after 5 hours?

(c) Explain how we can tell from the graph that, for a patient with a fixed GFR, the amount excreted changes very little after 12 hours.

(d) Is the percent excreted an increasing or decreasing function of time? Explain why this makes sense.

(e) Is the percent excreted an increasing or decreasing function of GFR? Explain what this means to a physician giving antibiotics to a patient with kidney disease.

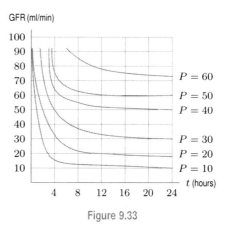

Figure 9.33

[4]Arthur C. Guyton and John E. Hall, *Textbook of Medical Physiology, Ninth Edition*, p. 288 (Philadelphia: W. B. Saunders, 1996).

[5]Peter G. Welling and Francis L. S. Tse, *Pharmacokinetics of Cardiovascular, Central Nervous System, and Antimicrobial Drugs*, The Royal Society of Chemistry, 1985, p. 316.

9.3 PARTIAL DERIVATIVES

In one-variable calculus we saw how the derivative measures the rate of change of a function. We begin by reviewing this idea.

Rate of Change of Airline Revenue

In Section 9.1 we saw a two-variable function which gives an airline's revenue, R, as a function of the number of full price tickets, x, and the number of discount tickets, y, sold:

$$R = f(x, y) = 350x + 200y.$$

If we fix the number of discount tickets at $y = 10$, we have a one-variable function

$$R = f(x, 10) = g(x) = 350x + 2000.$$

The rate of the change of revenue with respect to x is given by the one-variable derivative

$$g'(x) = 350.$$

This tells us that, if y is fixed at 10, then the revenue increases by \$350 for each additional full price ticket sold. We call $g'(x)$ the *partial derivative of R with respect to x* at the point $(x, 10)$. If $R = f(x, y)$, we write

$$\frac{\partial R}{\partial x} = f_x(x, 10) = g'(x) = 350.$$

Example 1   Find the rate of change of revenue, R, as y increases with x fixed at $x = 20$.

Solution   Substituting $x = 20$ into $R = 350x + 200y$ gives the one-variable function

$$R = h(y) = 350(20) + 200y = 7000 + 200y.$$

The rate of change of R as y increases with x fixed is

$$\frac{\partial R}{\partial y} = f_y(20, y) = h'(y) = 200.$$

We call $\partial R/\partial y = f_y(20, y)$ the *partial derivative of R with respect to y* at the point $(20, y)$. The fact that both partial derivatives of R are positive corresponds to the fact that the revenue is increasing as more of either type of ticket is sold.

Definition of the Partial Derivative

For any function $f(x, y)$ we study the influence of x and y separately on the value $f(x, y)$ by keeping one fixed and letting the other vary. The method of the previous example allows us to calculate the rates of change of $f(x, y)$ with respect to x and y. For all points (a, b) at which the limits exist, we make the following definitions:

Partial Derivatives of f with Respect to x and y

The *partial derivative of f with respect to x* at (a, b) is the derivative of f with y constant:

$$f_x(a, b) = \begin{array}{c} \text{Rate of change of } f \text{ with } y \text{ fixed} \\ \text{at } b, \text{ at the point } (a, b) \end{array} = \lim_{h \to 0} \frac{f(a + h, b) - f(a, b)}{h}.$$

The *partial derivative of f with respect to y* at (a, b) is the derivative of f with x constant:

$$f_y(a, b) = \begin{array}{c} \text{Rate of change of } f \text{ with } x \text{ fixed} \\ \text{at } a, \text{ at the point } (a, b) \end{array} = \lim_{h \to 0} \frac{f(a, b + h) - f(a, b)}{h}.$$

If we think of a and b as variables, $a = x$ and $b = y$, we have the **partial derivative functions** $f_x(x, y)$ and $f_y(x, y)$.

Just as with ordinary derivatives, there is an alternative notation:

Alternative Notation for Partial Derivatives

If $z = f(x, y)$ we can write

$$f_x(x, y) = \frac{\partial z}{\partial x} \quad \text{and} \quad f_y(x, y) = \frac{\partial z}{\partial y}$$

$$f_x(a, b) = \frac{\partial z}{\partial x}\Big|_{(a, b)} \quad \text{and} \quad f_y(a, b) = \frac{\partial z}{\partial y}\Big|_{(a, b)}$$

We use the symbol ∂ to distinguish partial derivatives from ordinary derivatives. In cases where the independent variables have names different from x and y, we adjust the notation accordingly. For example, the partial derivatives of $f(u, v)$ are denoted by f_u and f_v.

Estimating Partial Derivatives from a Table

Example 2 An experiment[6] done on rats to measure the toxicity of formaldehyde yielded the data shown in Table 9.5. The values in the table show the percent, P, of rats that survived an exposure with concentration c (in parts per million) after t months, so $P = f(t, c)$. Using Table 9.5, estimate $f_t(18, 6)$ and $f_c(18, 6)$. Interpret your answers in terms of formaldehyde toxicity.

Table 9.5 *Percent, P, of rat population surviving after exposure to formaldehyde vapor*

		\multicolumn{13}{c}{Time t (months)}
		0	2	4	6	8	10	12	14	16	18	20	22	24
Conc. c (ppm)	0	100	100	100	100	100	100	100	100	100	100	99	97	95
	2	100	100	100	100	100	100	100	100	99	98	97	95	92
	6	100	100	100	99	99	98	96	96	95	93	90	86	80
	15	100	100	100	99	99	99	99	96	93	82	70	58	36

Solution For $f_t(18, 6)$, we fix c at 6 ppm, and find the rate of change of percent surviving, P, with respect to t. We have

$$f_t(18, 6) \approx \frac{\Delta P}{\Delta t} = \frac{f(20, 6) - f(18, 6)}{20 - 18} = \frac{90 - 93}{20 - 18} \approx -1.5 \text{ \% per month.}$$

[6] James E. Gibson, *Formaldehyde Toxicity*, p. 125 (Hemisphere Publishing Company, McGraw-Hill, 1983).

This is the rate of change of percent surviving, P, *in the time t direction* at the point $(18, 6)$. The fact that it is negative means that P is decreasing as we read across the $c = 6$ row of the table in the direction of increasing t (that is, horizontally from left to right in Table 9.5). For $f_c(18, 6)$, we fix t at 18, and calculate the rate of change of P as we move in the direction of increasing c (that is, from top to bottom in Table 9.5). We have

$$f_c(18, 6) \approx \frac{\Delta P}{\Delta c} = \frac{f(18, 15) - f(18, 6)}{15 - 6} = \frac{82 - 93}{15 - 6} = -1.22\% \text{ per ppm.}$$

The rate of change of P as c increases is about -1.22% per ppm. This means that as the concentration increases by 1 ppm from 6 ppm, the percent surviving 18 months decreases by about 1.22% per unit increase ppm. The partial derivative is negative because fewer rats survive this long when the concentration of formaldehyde increases. (That is, P goes down as c goes up.)

Using Partial Derivatives to Estimate Values of the Function

Example 3   Use Table 9.5 and partial derivatives to estimate the percent of rats surviving if they are exposed to formaldehyde with a concentration of

(a)   6 ppm for 18.5 months   (b)   18 ppm for 24 months   (c)   9 ppm for 20.5 months

Solution   (a) Since $t = 18.5$ and $c = 6$, we want to evaluate $P = f(18.5, 6)$. Table 9.5 tells us that $f(18, 6) = 93\%$ and we have just calculated

$$\left.\frac{\partial P}{\partial t}\right|_{(18,6)} = f_t(18, 6) = -1.5\% \text{ per month.}$$

This partial derivative tells us that after 18 months of exposure to formaldehyde at a concentration of 6 ppm, P decreases by 1.5% for every additional month of exposure. Therefore after an additional 0.5 month, we have

$$P \approx 93 - 1.5(0.5) = 92.25\%.$$

(b) Now we wish to evaluate $f(24, 18)$. The closest entry to this in Table 9.5 is $f(24, 15) = 36$. We keep t fixed at 24 and increase c from 15 to 18. We estimate the rate of change in P as c changes; this is $\partial P / \partial c$. We see from Table 9.5 that

$$\left.\frac{\partial P}{\partial c}\right|_{(24,15)} \approx \frac{\Delta P}{\Delta c} = \frac{36 - 80}{15 - 6} = -4.89\% \text{ per ppm.}$$

The percent surviving 24 months goes down from 36% by about 4.89% for every unit increase in the formaldehyde concentration above 15 ppm. We have:

$$f(24, 18) \approx 36 - 4.89(3) = 21.33\%.$$

We estimate that only about 21% of the rats would survive for 24 months if they were exposed to formaldehyde as strong as 18 ppm. Since this figure is an extrapolation from the available data, we should use it with caution.

(c) To estimate $f(20.5, 9)$, we use the closest entry $f(20, 6) = 90$. As we move from $(20, 6)$ to $(20.5, 9)$, the percentage, P, changes both due to the change in t and due to the change in c. We estimate the two partial derivatives at $t = 20$, $c = 6$:

$$\left.\frac{\partial P}{\partial t}\right|_{(20,6)} \approx \frac{\Delta P}{\Delta t} = \frac{86 - 90}{22 - 20} = -2\% \text{ per month,}$$

$$\left.\frac{\partial P}{\partial c}\right|_{(20,6)} \approx \frac{\Delta P}{\Delta c} = \frac{70 - 90}{15 - 6} = -2.22\% \text{ per month.}$$

The change in P due to a change of $\Delta t = 0.5$ month and $\Delta c = 3$ ppm is

$$\begin{aligned} \Delta P &\approx \text{Change due to } \Delta t + \text{Change due to } \Delta c \\ &= -2(0.5) - 2.22(3) \\ &= -7.66. \end{aligned}$$

So for $t = 20.5$, $c = 9$ we have

$$f(20.5, 9) \approx f(20, 6) - 7.66 = 82.34\%.$$

In Example 3 part (c), we used the relationship between ΔP, Δt, and Δc. In general, the relationship between the change Δf, in function value $f(x, y)$ and the changes Δx and Δy is as follows:

Local Linearity

$$\begin{array}{ccc} \text{Change} \\ \text{in } f \end{array} \approx \begin{array}{c} \text{Rate of change} \\ \text{in } x\text{-direction} \end{array} \cdot \Delta x + \begin{array}{c} \text{Rate of change} \\ \text{in } y\text{-direction} \end{array} \cdot \Delta y$$

$$\Delta f \approx f_x \cdot \Delta x + f_y \cdot \Delta y$$

Estimating Partial Derivatives from a Contour Diagram

If we move parallel to one of the axes on a contour diagram, the partial derivative is the rate of change of the value of the function on the contours. For example, if the values on the contours are increasing in the direction of positive change, then the partial derivative must be positive.

Example 4   Figure 9.34 shows the contour diagram for the temperature $H(x, t)$ (in °F) in a room as a function of distance x (in feet) from a heater and time t (in minutes) after the heater has been turned on. What are the signs of $H_x(10, 20)$ and $H_t(10, 20)$? Estimate these partial derivatives and explain the answers in practical terms.

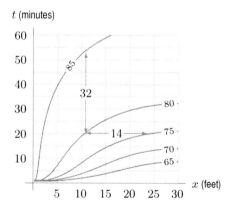

Figure 9.34: Temperature in a heated room

Solution   The point $(10, 20)$ is on the $H = 80$ contour. As x increases, we move toward the $H = 75$ contour, so H is decreasing and $H_x(10, 20)$ is negative. This makes sense because as we move further from the heater, the temperature drops. On the other hand, as t increases, we move toward the $H = 85$ contour, so H is increasing and $H_t(10, 20)$ is positive. This also makes sense, because it says that as time passes, the room warms up.

To estimate the partial derivatives, use a difference quotient. Looking at the contour diagram, we see there is a point on the $H = 75$ contour about 14 units to the right of $(10, 20)$. Hence, H decreases by 5 when x increases by 14, so the rate of change of H with respect to x is about $\Delta H/\Delta x = -5/14 \approx -0.36$. Thus, we find

$$H_x(10, 20) \approx -0.36°\text{F/ft.}$$

This means that near the point 10 feet from the heater, after 20 minutes the temperature drops about 1/3 of a degree for each foot we move away from the heater.

To estimate $H_t(10, 20)$, we look again at the contour diagram and notice that the $H = 85$ contour is about 32 units directly above the point $(10, 20)$. So H increases by 5 when t increases by 32. Hence,

$$H_t(10, 20) \approx \frac{\Delta H}{\Delta t} = \frac{5}{32} \approx 0.16°\text{F/min.}$$

This means that after 20 minutes the temperature is going up about 1/6 of a degree each minute at the point 10 ft from the heater.

Using Units to Interpret Partial Derivatives

The units of the independent and dependent variables can often be helpful in explaining the meaning of a partial derivative.

Example 5   Suppose that your weight w in pounds is a function $f(c, n)$ of the number c of calories you consume daily and the number n of minutes you exercise daily. Using the units for w, c and n, interpret in everyday terms the statements

$$\left.\frac{\partial w}{\partial c}\right|_{(2000,15)} = 0.02 \quad \text{and} \quad \left.\frac{\partial w}{\partial n}\right|_{(2000,15)} = -0.025.$$

Solution   The units of $\partial w/\partial c$ are pounds per calorie. The statement

$$\left.\frac{\partial w}{\partial c}\right|_{(2000,15)} = 0.02$$

means that if you are presently consuming 2000 calories daily and exercising 15 minutes daily, you will weigh 0.02 pounds more for each extra calorie you consume daily, or about 2 pounds for each extra 100 calories per day. The units of $\partial w/\partial n$ are pounds per minute. The statement

$$\left.\frac{\partial w}{\partial n}\right|_{(2000,15)} = -0.025$$

means that for the same calorie consumption and number of minutes of exercise, you will weigh 0.025 pounds less for each extra minute you exercise daily, or about 1 pound less for each extra 40 minutes per day. So if you eat an extra 100 calories each day and exercise about 80 minutes more each day, your weight should remain roughly steady.

Problems for Section 9.3

1. Using the contour diagram for $f(x, y)$ in Figure 9.35, decide whether each of these partial derivatives is positive, negative, or approximately zero.

(a) $f_x(4, 1)$ (b) $f_y(4, 1)$
(c) $f_x(5, 2)$ (d) $f_y(5, 2)$

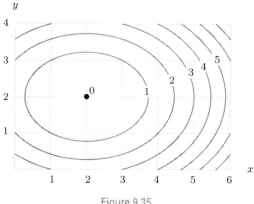

Figure 9.35

2. According to the contour diagram for $f(x, y)$ in Figure 9.35, which is larger: $f_x(3, 1)$ or $f_x(5, 2)$? Explain.

For Problems 3–4 refer to Table 9.4 on page 358 giving the heat index, I, in °F, as a function $f(H, T)$ of the relative humidity, H, and the temperature, T, in °F. The heat index is a temperature which tells you how hot it feels as a result of the combination of humidity and temperature.

3. Estimate $\partial I/\partial H$ and $\partial I/\partial T$ for typical weather conditions in Tucson in summer ($H = 10$, $T = 100$). What do your answers mean in practical terms for the residents of Tucson?

4. Answer the question in Problem 3 for Boston in summer ($H = 50$, $T = 80$).

5. The demand for coffee, Q, in pounds sold per week, is a function of the price of coffee, c, in dollars per pound and the price of tea, t, in dollars per pound, so $Q = f(c, t)$.

(a) Do you expect f_c to be positive or negative? What about f_t? Explain.
(b) Interpret each of the following statements in terms of the demand for coffee:

$f(3, 2) = 780$ $f_c(3, 2) = -60$ $f_t(3, 2) = 20$

6. The quantity Q (in pounds) of beef that a certain community buys during a week is a function $Q = f(b, c)$ of the prices of beef, b, and chicken, c, during the week. Do you expect $\partial Q/\partial b$ to be positive or negative? What about $\partial Q/\partial c$?

[7]From the August 28, 1994, issue of *Parade Magazine*.

7. A drug is injected into a patient's blood vessel. The function $c = f(x, t)$ represents the concentration of the drug at a distance x mm in the direction of the blood flow measured from the point of injection and at time t seconds since the injection. What are the units of the following partial derivatives? What are their practical interpretations? What do you expect their signs to be?

(a) $\partial c/\partial x$ (b) $\partial c/\partial t$

8. Table 9.6 gives the number of calories burned per minute, $B = f(s, w)$, for someone roller-blading,[7] as a function of the person's weight, w, and speed, s.

(a) Is f_w positive or negative? Is f_s positive or negative? What do your answers tell us about the effect of weight and speed on calories burned per minute?
(b) Estimate $f_w(160, 10)$ and $f_s(160, 10)$. Interpret your answers.

Table 9.6   *Calories burned per minute*

$w \backslash s$	8 mph	9 mph	10 mph	11 mph
120 lbs	4.2	5.8	7.4	8.9
140 lbs	5.1	6.7	8.3	9.9
160 lbs	6.1	7.7	9.2	10.8
180 lbs	7.0	8.6	10.2	11.7
200 lbs	7.9	9.5	11.1	12.6

9. Estimate $z_x(1, 0)$ and $z_x(0, 1)$ and $z_y(0, 1)$ from the contour diagram for $z(x, y)$ in Figure 9.36.

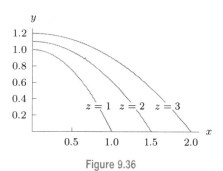

Figure 9.36

10. The monthly mortgage payment in dollars, P, for a house is a function of three variables:

$$P = f(A, r, N),$$

where A is the amount borrowed in dollars, r is the interest rate, and N is the number of years before the mortgage is paid off.

(a) $f(92000, 14, 30) = 1090.08$. What does this tell you, in financial terms?

(b) $\left.\dfrac{\partial P}{\partial r}\right|_{(92000,14,30)} = 72.82$. What is the financial significance of the number 72.82?

(c) Would you expect $\partial P/\partial A$ to be positive or negative? Why?

(d) Would you expect $\partial P/\partial N$ to be positive or negative? Why?

11. The sales of a product, $S = f(p, a)$, are a function of the price, p, of the product (in dollars per unit) and the amount, a, spent on advertising (in thousands of dollars).

  (a) Do you expect f_p to be positive or negative? Why?

  (b) Explain the meaning of the statement $f_a(8, 12) = 150$ in terms of sales.

12. Figure 9.37 shows a contour diagram for the monthly payment P as a function of the interest rate, $r\%$, and the amount, L, of a 5-year loan. Estimate $\partial P/\partial r$ and $\partial P/\partial L$ at the point where $r = 8$ and $L = 5000$. Give the units and the financial meaning of your answers.

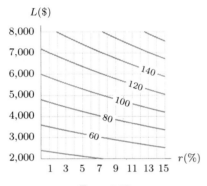

$L(\$)$

Figure 9.37

13. Figure 9.13 on page 356 gives a contour diagram of corn production as a function of rainfall, R, in inches and temperature, T, in $°F$. Corn production, C, is measured as a percentage of the present production, and $C = f(R, T)$. Estimate the following quantities. Give units and interpret your answers in terms of corn production:

  (a) $f_R(15, 76)$ **(b)** $f_T(15, 76)$

14. Use the diagram from Problem 17 in Section 9.1, to estimate $H_T(T, w)$ for $T = 10, 20, 30$ and $w = 0.1, 0.2, 0.3$. What is the practical meaning of these partial derivatives?

15. People commuting to a city can choose to go either by bus or by train. The number of people who choose either method depends in part upon the price of each. Let $f(P_1, P_2)$ be the number of people who take the bus when P_1 is the price of a bus ride and P_2 is the price of a train ride. What can you say about the signs of $\partial f/\partial P_1$ and $\partial f/\partial P_2$? Explain your answers.

16. In the 1940s the quantity, q, of beer sold each year in Britain was found to depend on I (the aggregate personal income, adjusted for taxes and inflation), p_1 (the average price of beer), and p_2 (the average price of all other goods and services). Would you expect $\partial q/\partial I, \partial q/\partial p_1, \partial q/\partial p_2$ to be positive or negative? Give reasons for your answers.

17. An airline's revenue, R, is a function of the number of full price tickets, x, and the number of discount tickets, y, sold. Values of $R = f(x, y)$ are in Table 9.1 on page 350.

  (a) Evaluate $f(200, 400)$, and interpret your answer.

  (b) Is $f_x(200, 400)$ positive or negative? Is $f_y(200, 400)$ positive or negative? Explain.

  (c) Estimate the partial derivatives in part (b). Give units and interpret your answers in terms of revenue.

18. In Problem 17 the revenue is \$150,000 when 200 full-price tickets and 400 discount tickets are sold; that is, $f(200, 400) = 150,000$. Use this fact and the partial derivatives $f_x(200, 400) = 350$ and $f_y(200, 400) = 200$ to estimate the revenue when

  (a) $x = 201$ and $y = 400$ **(b)** $x = 200$ and $y = 405$

  (c) $x = 203$ and $y = 406$

19. Table 9.5 on page 365 gives the percent of rats surviving, P, as a function of time, t, in months and concentration of formaldehyde, c, in ppm, so $P = f(t, c)$. Use partial derivatives to estimate the percent surviving after 26 months when the concentration is 15.

20. For a function $f(x, y)$, we are given $f(100, 20) = 2750$, and $f_x(100, 20) = 4$, and $f_y(100, 20) = 7$. Estimate $f(105, 21)$.

21. For a function $f(r, s)$, we are given $f(50, 100) = 5.67$, and $f_r(50, 100) = 0.60$, and $f_s(50, 100) = -0.15$. Estimate $f(52, 108)$.

22. The cardiac output, represented by c, is the volume of blood flowing through a person's heart, per unit time. The systemic vascular resistance (SVR), represented by s, is the resistance to blood flowing through veins and arteries. Let p be a person's blood pressure. Then p is a function of c and s, so $p = f(c, s)$.

  (a) What does $\partial p/\partial c$ represent?

  Suppose now that $p = kcs$, where k is a constant.

  (b) Sketch the level curves of p. What do they represent? Label your axes.

  (c) For a person with a weak heart, it is desirable to have the heart pumping against less resistance, while maintaining the same blood pressure. Such a person may be given the drug nitroglycerine to decrease the SVR and the drug Dopamine to increase the cardiac output. Represent this on a graph showing level curves. Put a point A on the graph representing the person's state before drugs are given and a point B for after.

(d) Right after a heart attack, a patient's cardiac output drops, thereby causing the blood pressure to drop. A common mistake made by medical residents is to get the patient's blood pressure back to normal by using drugs to increase the SVR, rather than by increasing the cardiac output. On a graph of the level curves of p, put a point D representing the patient before the heart attack, a point E representing the patient right after the heart attack, and a third point F representing the patient after the resident has given the drugs to increase the SVR.

23. In each case, give a possible contour diagram for the function $f(x, y)$ if
    (a) $f_x > 0$ and $f_y > 0$ (b) $f_x > 0$ and $f_y < 0$
    (c) $f_x < 0$ and $f_y > 0$ (d) $f_x < 0$ and $f_y < 0$

24. Figure 9.38 shows contours of $f(x, y)$ with values of f on the contours omitted. If $f_x(P) > 0$, find the sign of

(a) $f_y(P)$ (b) $f_y(Q)$ (c) $f_x(Q)$

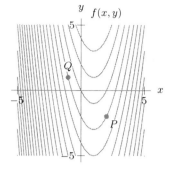

Figure 9.38

9.4 COMPUTING PARTIAL DERIVATIVES ALGEBRAICALLY

The partial derivative $f_x(x, y)$ is the ordinary derivative of the function $f(x, y)$ with respect to x with y fixed, and the partial derivative $f_y(x, y)$ is the ordinary derivative of $f(x, y)$ with respect to y with x fixed. Thus, we can use all the techniques for differentiation from single-variable calculus to find partial derivatives.

Example 1   Let $f(x, y) = x^2 + 5y^2$. Find $f_x(3, 2)$ and $f_y(3, 2)$ algebraically.

Solution   We use the fact that $f_x(3, 2)$ is the derivative of $f(x, 2)$ at $x = 3$. To find f_x, we fix y at 2:
$$f(x, 2) = x^2 + 5(2^2) = x^2 + 20.$$
Differentiating with respect to x gives
$$f_x(x, 2) = 2x   \text{so}   f_x(3, 2) = 2(3) = 6.$$
Similarly, $f_y(3, 2)$ is the derivative of $f(3, y)$ at $y = 2$. To find f_y, we fix x at 3:
$$f(3, y) = 3^2 + 5y^2 = 9 + 5y^2.$$
Differentiating with respect to y, we have
$$f_y(3, y) = 10y   \text{so}   f_y(3, 2) = 10(2) = 20.$$

Example 2   Let $f(x, y) = x^2 + 5y^2$ as in Example 1. Find f_x and f_y as functions of x and y.

Solution   To find f_x, we treat y as a constant. Thus $5y^2$ is a constant and the derivative with respect to x of this term is 0. We have
$$f_x(x, y) = 2x + 0 = 2x.$$
To find f_y, we treat x as a constant and so the derivative of x^2 with respect to y is zero. We have
$$f_y(x, y) = 0 + 10y = 10y.$$

Example 3   Find both partial derivatives of each of the following functions:
    (a) $f(x, y) = 3x + e^{-5y}$ (b) $f(x, y) = x^2 y$ (c) $f(u, v) = u^2 e^{2v}$

Solution   (a) To find f_x, we treat y as a constant, so the term e^{-5y} is a constant, and the derivative of this term is zero. Likewise, to find f_y, we treat x as a constant. We have

$$f_x(x, y) = 3 + 0 = 3 \quad \text{and} \quad f_y(x, y) = 0 + (-5)e^{-5y} = -5e^{-5y}.$$

   (b) To find f_x, we treat y as a constant, so the function is treated as a constant times x^2. The derivative of a constant times x^2 is the constant times $2x$, and so we have

$$f_x(x, y) = (2x)y = 2xy \quad \text{Similarly,} \quad f_y(x, y) = (x^2)(1) = x^2.$$

   (c) To find f_u, we treat v as a constant, and to find f_v, we treat u as a constant. We have

$$f_u(u, v) = (2u)(e^{2v}) = 2ue^{2v} \quad \text{and} \quad f_v(u, v) = u^2(2e^{2v}) = 2u^2e^{2v}.$$

Example 4   The concentration C of bacteria in the blood (in millions of bacteria/ml) following the injection of an antibiotic is a function of the dose x (in gm) injected and the time t (in hours) since the injection. Suppose we are told that $C = f(x, t) = te^{-xt}$. Evaluate the following quantities and explain what each one means in practical terms: (a) $f_x(1, 2)$ (b) $f_t(1, 2)$

Solution   (a) To find f_x, we treat t as a constant and differentiate with respect to x, giving

$$f_x(x, t) = -t^2e^{-xt}.$$

Substituting $x = 1, t = 2$ gives

$$f_x(1, 2) = -4e^{-2} \approx -0.54.$$

To see what $f_x(1, 2)$ means, think about the function $f(x, 2)$ of which it is the derivative. The graph of $f(x, 2)$ in Figure 9.39 gives the concentration of bacteria as a function of the dose two hours after the injection. The derivative $f_x(1, 2)$ is the slope of this graph at the point $x = 1$; it is negative because a larger dose reduces the bacteria population. More precisely, the partial derivative $f_x(1, 2)$ gives the rate of change of bacteria concentration with respect to the dose injected, namely a decrease in bacteria concentration of 0.54 million/ml per gram of additional antibiotic injected.

   (b) To find f_t, treat x as a constant and differentiate using the product rule:

$$f_t(x, t) = 1 \cdot e^{-xt} - xte^{-xt}.$$

Substituting $x = 1, t = 2$ gives

$$f_t(1, 2) = e^{-2} - 2e^{-2} \approx -0.14.$$

To see what $f_t(1, 2)$ means, think about the function $f(1, t)$ of which it is the derivative. The graph of $f(1, t)$ in Figure 9.40 gives the concentration of bacteria at time t if the dose of antibiotic is 1 gm. The derivative $f_t(1, 2)$ is the slope of the graph at the point $t = 2$; it is negative because after 2 hours the concentration of bacteria is decreasing. More precisely, the partial derivative $f_t(1, 2)$ gives the rate at which the bacteria concentration is changing with respect to time, namely a decrease in bacteria concentration of 0.14 million/ml per hour.

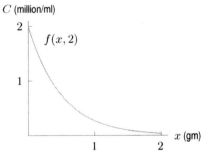

Figure 9.39: Bacteria concentration after 2 hours as a function of the quantity of antibiotic injected

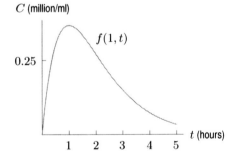

Figure 9.40: Bacteria concentration as a function of time if 1 unit of antibiotic is injected

Example 5   Let's consider a small printing business where N is the number of workers, V is the value of the equipment (in units of $\$25,000$), and P is the production, measured in thousands of pages per day. Suppose the production function for this company is given by

$$P = f(N, V) = 2N^{0.6}V^{0.4}.$$

(a) If this company has a labor force of 100 workers and 200 units' worth of equipment, what is the production output of the company?

(b) Find $f_N(100, 200)$ and $f_V(100, 200)$. Interpret your answers in terms of production.

Solution   (a) We have $N = 100$ and $V = 200$, so

$$\text{Production} = 2(100)^{0.6}(200)^{0.4} = 263.9 \text{ thousand pages per day.}$$

(b) To find f_N, we treat V as a constant and differentiate with respect to N:

$$f_N(N, V) = 2(0.6)N^{-0.4}V^{0.4}.$$

Substituting $N = 100$, $V = 200$ gives

$$f_N(100, 200) = 1.2(100^{-0.4})(200^{0.4}) \approx 1.583 \text{ thousand pages/worker.}$$

This tells us that if we have 200 units of equipment and increase the number of workers by 1 from 100 to 101, the production output will go up by about 1.58 units, or 1580 pages per day. Similarly, to find $f_V(100, 200)$, we treat N as a constant and differentiate with respect to V:

$$f_V(N, V) = 2(0.4)N^{0.6}V^{-0.6}.$$

Substituting $N = 100$, $V = 200$ gives

$$f_V(100, 200) = 0.8(100^{0.6})(200^{-0.6}) \approx 0.53 \text{ thousand pages/unit of equipment.}$$

This tells us that if we have 100 workers and increase the value of the equipment by 1 unit ($\$25,000$) from 200 units to 201 units, the production goes up by about 0.53 units, or 530 pages per day.

Second-Order Partial Derivatives

Since the partial derivatives of a function are themselves functions, we can usually differentiate them, giving *second-order partial derivatives*. A function $z = f(x, y)$ has two first-order partial derivatives, f_x and f_y, and four second-order partial derivatives.

The Second-Order Partial Derivatives of $z = f(x, y)$

$$\frac{\partial^2 z}{\partial x^2} = f_{xx} = (f_x)_x, \qquad \frac{\partial^2 z}{\partial x \partial y} = f_{yx} = (f_y)_x,$$

$$\frac{\partial^2 z}{\partial y \partial x} = f_{xy} = (f_x)_y, \qquad \frac{\partial^2 z}{\partial y^2} = f_{yy} = (f_y)_y.$$

It is usual to omit the parentheses, writing f_{xy} instead of $(f_x)_y$ and $\dfrac{\partial^2 z}{\partial y\, \partial x}$ instead of $\dfrac{\partial}{\partial y}\left(\dfrac{\partial z}{\partial x}\right)$.

Example 6   Use the values of the function $f(x, y)$ in Table 9.7 to estimate $f_{xy}(1, 2)$ and $f_{yx}(1, 2)$.

Table 9.7   *Values of $f(x, y)$*

		\multicolumn{3}{c}{x}		
		0.9	1.0	1.1
	1.8	4.72	5.83	7.06
y	2.0	6.48	8.00	9.60
	2.2	8.62	10.65	12.88

Solution   Since $f_{xy} = (f_x)_y$, we first estimate f_x:

$$f_x(1, 2) \approx \frac{f(1.1, 2) - f(1, 2)}{0.1} = \frac{9.60 - 8.00}{0.1} = 16.0,$$

$$f_x(1, 2.2) \approx \frac{f(1.1, 2.2) - f(1, 2.2)}{0.1} = \frac{12.88 - 10.65}{0.1} = 22.3.$$

Thus,

$$f_{xy}(1, 2) \approx \frac{f_x(1, 2.2) - f_x(1, 2)}{0.2} = \frac{22.3 - 16.0}{0.2} = 31.5.$$

Similarly,

$$f_{yx}(1, 2) \approx \frac{f_y(1.1, 2) - f_y(1, 2)}{0.1} \approx \frac{1}{0.1} \left(\frac{f(1.1, 2.2) - f(1.1, 2)}{0.2} - \frac{f(1, 2.2) - f(1, 2)}{0.2} \right)$$

$$= \frac{1}{0.1} \left(\frac{12.88 - 9.60}{0.2} - \frac{10.65 - 8.00}{0.2} \right) = 31.5.$$

Observe that in this example, $f_{xy} = f_{yx}$ at the point $(1, 2)$.

Example 7   Compute the four second-order partial derivatives of $f(x, y) = xy^2 + 3x^2 e^y$.

Solution   From $f_x(x, y) = y^2 + 6xe^y$ we get

$$f_{xx}(x, y) = \frac{\partial}{\partial x}(y^2 + 6xe^y) = 6e^y \quad \text{and} \quad f_{xy}(x, y) = \frac{\partial}{\partial y}(y^2 + 6xe^y) = 2y + 6xe^y.$$

From $f_y(x, y) = 2xy + 3x^2 e^y$ we get

$$f_{yx}(x, y) = \frac{\partial}{\partial x}(2xy + 3x^2 e^y) = 2y + 6xe^y \quad \text{and} \quad f_{yy}(x, y) = \frac{\partial}{\partial y}(2xy + 3x^2 e^y) = 2x + 3x^2 e^y.$$

Observe that $f_{xy} = f_{yx}$ in this example.

The Mixed Partial Derivatives Are Equal

It is not an accident that the estimates for $f_{xy}(1, 2)$ and $f_{yx}(1, 2)$ are equal in Example 6, because the same values of the function are used to calculate each one. The fact that $f_{xy} = f_{yx}$ in Example 7 corroborates the following general result:

> If f_{xy} and f_{yx} are continuous at (a, b), then
>
> $$f_{xy}(a, b) = f_{yx}(a, b).$$

Most of the functions we will encounter not only have f_{xy} and f_{yx} continuous, but all their higher-order partial derivatives (such as f_{xxy} or f_{xyyy}) will be continuous. We call such functions *smooth*.

Problems for Section 9.4

Find the partial derivatives in Problems 1–13. The variables are restricted to a domain on which the function is defined.

1. f_x and f_y if $f(x, y) = 2x^2 + 3y^2$
2. f_x and f_y if $f(x, y) = 100x^2y$
3. f_x and f_y if $f(x, y) = x^2 + 2xy + y^3$
4. z_x if $z = x^2y + 2x^5y$
5. f_u and f_v if $f(u, v) = u^2 + 5uv + v^2$
6. $\dfrac{\partial z}{\partial x}$ if $z = x^2e^y$
7. $\dfrac{\partial Q}{\partial p}$ if $Q = 5a^2p - 3ap^3$
8. f_t if $f(t, a) = 5a^2t^3$
9. f_x and f_y if $f(x, y) = 5x^2y^3 + 8xy^2 - 3x^2$
10. f_x and f_y if $f(x, y) = 10x^2e^{3y}$
11. $\dfrac{\partial P}{\partial r}$ if $P = 100e^{rt}$
12. $\dfrac{\partial A}{\partial h}$ if $A = \frac{1}{2}(a + b)h$
13. $\dfrac{\partial}{\partial m}\left(\frac{1}{2}mv^2\right)$
14. If $f(x, y) = x^3 + 3y^2$, find $f(1, 2)$, $f_x(1, 2)$, $f_y(1, 2)$.
15. If $f(u, v) = 5uv^2$, find $f(3, 1)$, $f_u(3, 1)$, and $f_v(3, 1)$.
16. (a) Let $f(x, y) = x^2 + y^2$. Estimate $f_x(2, 1)$ and $f_y(2, 1)$ using the contour diagram for f in Figure 9.41.
    (b) Estimate $f_x(2, 1)$ and $f_y(2, 1)$ from a table of values for f with $x = 1.9, 2, 2.1$ and $y = 0.9, 1, 1.1$.
    (c) Compare your estimates in parts (a) and (b) with the exact values of $f_x(2, 1)$ and $f_y(2, 1)$ found algebraically.

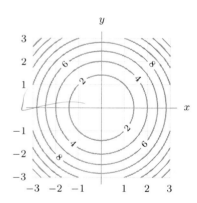

Figure 9.41

17. The amount of money, B, in a bank account earning interest at a continuous rate, r, depends on the amount deposited, P, and the time, t, it has been in the bank, where
$$B = Pe^{rt}.$$
Find $\partial B/\partial t$, $\partial B/\partial r$ and $\partial B/\partial P$ and interpret each in financial terms.

18. The cost of renting a car from a certain company is \$40 per day plus 15 cents per mile, and so we have
$$C = 40d + 0.15m.$$
Find $\partial C/\partial d$ and $\partial C/\partial m$. Give units and explain why your answers make sense.

19. A company's production output, P, is given in tons, and is a function of the number of workers, N, and the value of the equipment, V, in units of \$25,000. The production function for the company is
$$P = f(N, V) = 5N^{0.75}V^{0.25}.$$
The company currently employs 80 workers, and has equipment worth \$750,000. What are N and V? Find the values of f, f_N, and f_V at these values of N and V. Give units and explain what each answer means in terms of production.

For Problems 20–31, calculate all four second-order partial derivatives and confirm that the mixed partials are equal.

20. $f(x, y) = x^2y$
21. $f(x, y) = x^2 + 2xy + y^2$
22. $f(x, y) = xe^y$
23. $f(x, y) = \dfrac{2x}{y}, \quad y \neq 0$
24. $f = 5 + x^2y^2$
25. $f = e^{xy}$
26. $Q = 5p_1^2p_2^{-1}, \quad p_2 \neq 0$
27. $V = \pi r^2h$
28. $P = 2KL^2$
29. $B = 5xe^{-2t}$
30. $f(x, t) = t^3 - 4x^2t$
31. $f = 100e^{rt}$

32. Is there a function f which has the following partial derivatives? If so what is it? Are there any others?
$$f_x(x, y) = 4x^3y^2 - 3y^4,$$
$$f_y(x, y) = 2x^4y - 12xy^3.$$

33. Show that the Cobb-Douglas function
$$Q = bK^\alpha L^{1-\alpha} \quad \text{where} \quad 0 < \alpha < 1$$
satisfies the equation
$$K\frac{\partial Q}{\partial K} + L\frac{\partial Q}{\partial L} = Q.$$

Problems 34–36 are about the money supply, M, which is the total value of all the cash and checking account balances in an economy. It is determined by the value of all the cash, B, the ratio, c, of cash to checking deposits, and the fraction, r, of checking account deposits that banks hold as cash:
$$M = \frac{c + 1}{c + r}B.$$

(a) Find the partial derivative.
(b) Give its sign.
(c) Explain the significance of the sign in practical terms.

34. $\partial M/\partial B$ 35. $\partial M/\partial r$ 36. $\partial M/\partial c$

9.5 CRITICAL POINTS AND OPTIMIZATION

To optimize a function means to find the largest or smallest value of the function. If the function represents profit, we may want to find the conditions that maximize profit. On the other hand, if the function represents cost, we may want to find the conditions that minimize cost. In Chapter 4, we saw how to optimize a function of one variable by investigating critical points. In this section, we see how to extend the notions of critical points and local extrema to a function of more than one variable.

Local and Global Maxima and Minima for Functions of Two Variables

Functions of several variables, like functions of one variable, can have *local* and *global extrema* (that is, local and global maxima and minima). A function has a local extremum at a point where it takes on the largest or smallest value in a small region around the point. Global extrema are the largest or smallest value anywhere. For a function f defined on a domain R, we say:

- f has a **local maximum** at P_0 if $f(P_0) \geq f(P)$ for all points P near P_0
- f has a **local minimum** at P_0 if $f(P_0) \leq f(P)$ for all points P near P_0
- f has a **global maximum** at P_0 if $f(P_0) \geq f(P)$ for all points P in R
- f has a **global minimum** at P_0 if $f(P_0) \leq f(P)$ for all points P in R

Example 1   Table 9.8 gives a table of values for a function $f(x, y)$. Estimate the location and value of any global maxima or minima for $0 \leq x \leq 1$ and $0 \leq y \leq 20$.

Table 9.8   *Where are the extreme points of this function $f(x, y)$?*

		x				
		0	0.2	0.4	0.6	0.8	1.0
	0	80	84	82	76	71	65
	5	86	90	88	73	77	71
y	10	91	95	93	88	82	76
	15	87	91	89	84	78	72
	20	82	86	84	79	73	67

Solution   The global maximum value of the function appears to be 95 at the point $(0.2, 10)$. Since the table only gives certain values, we cannot be sure that this is exactly the maximum. (The function might have a larger value at, for example, $(0.3, 11)$.) The global minimum value of this function on the points given is 65 at the point $(1, 0)$.

Example 2   Figure 9.42 gives a contour diagram for a function $f(x, y)$. Estimate the location and value of any local maxima or minima. Are any of these global maxima or minima on the square shown?

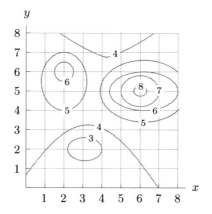

Figure 9.42: Where are the local and global extreme points of this function?

Solution   There is a local maximum of above 8 near the point $(6, 5)$, a local maximum of above 6 near the point $(2, 6)$, and a local minimum of below 3 near the point $(3, 2)$. The value above 8 is the global maximum and the value below 3 is the global minimum on the given domain.

In Example 1 and Example 2, we can estimate the location and value of extreme points, but we do not have enough information to find them exactly. This is usually true when we are given a table of values or a contour diagram. To find local or global extrema exactly, we usually need to have a formula for the function.

Finding a Local Maximum or Minimum Analytically

In one-variable calculus, the local extrema of a function occur at points where the derivative is zero or undefined. How does this generalize to the case of functions of two or more variables? Suppose that a function $f(x, y)$ has a local maximum at a point (x_0, y_0) which is not on the boundary of the domain of f. If the partial derivative $f_x(x_0, y_0)$ were defined and positive, then we could increase f by increasing x. If $f_x(x_0, y_0) < 0$, then we could increase f by decreasing x. Since f has a local maximum at (x_0, y_0), there can be no direction in which f is increasing, so we must have $f_x(x_0, y_0) = 0$. Similarly, if $f_y(x_0, y_0)$ is defined, then $f_y(x_0, y_0) = 0$. The case in which $f(x, y)$ has a local minimum is similar. Therefore, we arrive at the following conclusion:

> If a function $f(x, y)$ has a local maximum or minimum at a point (x_0, y_0) not on the boundary of the domain of f, then either
> $$f_x(x_0, y_0) = 0 \quad \text{and} \quad f_y(x_0, y_0) = 0$$
> or (at least) one partial derivative is undefined at the point (x_0, y_0). Points where each of the partial derivatives is either zero or undefined are called **critical points**.

As in the single-variable case, the fact that (x_0, y_0) is a critical point for f does not necessarily mean that f has a maximum or a minimum there.

How Do We Find Critical Points?

To find critical points of a function f, we find the points where both partial derivatives of f are zero or undefined.

Example 3   Find and analyze the critical points of $f(x, y) = x^2 - 2x + y^2 - 4y + 5$.

Solution   To find the critical points, we set both partial derivatives equal to zero:

$$f_x(x, y) = 2x - 2 = 0,$$
$$f_y(x, y) = 2y - 4 = 0.$$

Solving these equations gives $x = 1$ and $y = 2$. Hence, f has only one critical point, namely $(1, 2)$. What is the behavior of f near $(1, 2)$? The values of the function in Table 9.9 suggest that the function has a local minimum value of 0 at the point $(1, 2)$.

Table 9.9   *Values of $f(x, y)$ near the point $(1, 2)$*

		x				
		0.8	0.9	1.0	1.1	1.2
	1.8	0.08	0.05	0.04	0.05	0.08
	1.9	0.05	0.02	0.01	0.02	0.05
y	2.0	0.04	0.01	0.00	0.01	0.04
	2.1	0.05	0.02	0.01	0.02	0.05
	2.2	0.08	0.05	0.04	0.05	0.08

Example 4   A manufacturing company produces two products which are sold in two separate markets. The company's economists analyze the two markets and determine that the quantities, q_1 and q_2, demanded by consumers and the prices, p_1 and p_2 (in dollars), of each item are related by the equations

$$p_1 = 600 - 0.3q_1 \quad \text{and} \quad p_2 = 500 - 0.2q_2.$$

Thus, if the price for either item increases, the demand for it decreases. The company's total production cost is given by

$$C = 16 + 1.2q_1 + 1.5q_2 + 0.2q_1q_2.$$

If the company wants to maximize its total profits, how much of each product should it produce? What is the maximum profit?[8]

Solution   The total revenue R is the sum of the revenues, p_1q_1 and p_2q_2, from each market. Substituting for p_1 and p_2, we get

$$R = p_1q_1 + p_2q_2$$
$$= (600 - 0.3q_1)q_1 + (500 - 0.2q_2)q_2$$
$$= 600q_1 - 0.3q_1^2 + 500q_2 - 0.2q_2^2.$$

Thus the total profit π is given by

$$\pi = R - C$$
$$= 600q_1 - 0.3q_1^2 + 500q_2 - 0.2q_2^2 - (16 + 1.2q_1 + 1.5q_2 + 0.2q_1q_2)$$
$$= -16 + 598.8q_1 - 0.3q_1^2 + 498.5q_2 - 0.2q_2^2 - 0.2q_1q_2.$$

To maximize π, we compute partial derivatives:

$$\frac{\partial \pi}{\partial q_1} = 598.8 - 0.6q_1 - 0.2q_2,$$

$$\frac{\partial \pi}{\partial q_2} = 498.5 - 0.4q_2 - 0.2q_1.$$

---

[8]Adapted from M. Rosser, *Basic Mathematics for Economists*, p. 316 (New York: Routledge, 1993).

Since the partial derivatives are defined everywhere, the only critical points of π are those where the partial derivatives of π are both equal to zero. Thus, we solve the equations for q_1 and q_2,

$$598.8 - 0.6q_1 - 0.2q_2 = 0,$$
$$498.5 - 0.4q_2 - 0.2q_1 = 0,$$

giving

$$q_1 = 699.1 \approx 699 \quad \text{and} \quad q_2 = 896.7 \approx 897.$$

To see whether this is a maximum, we look at a table of values of profit π around this point. Table 9.10 suggests that profit is greatest at $(699, 897)$. So the company should produce 699 units of the first product priced at \$390.30 per unit, and 897 units of the second product priced at \$320.60 per unit. The maximum profit is then $\pi(699, 897) = \$432,797$.

Table 9.10   *Does this profit function have a maximum at $(699, 897)$?*

			Quantity, q_1	
		698	699	700
Quantity, q_2	896	432,796.4	432,796.9	432,796.8
	897	432,796.7	432,797.0	432,796.7
	898	432,796.6	432,796.7	432,796.2

Is a Critical Point a Local Maximum or a Local Minimum?

We can often see whether a critical point is a local maximum or minimum or neither by looking at a table or contour diagram. The following analytic method may also be useful in distinguishing between local maxima and minima.[9] It is analogous to the Second Derivative Test in Chapter 4.

Second Derivative Test for Functions of Two Variables

Suppose (x_0, y_0) is a critical point where $f_x(x_0, y_0) = f_y(x_0, y_0) = 0$. Let

$$D = f_{xx}(x_0, y_0)f_{yy}(x_0, y_0) - f_{xy}(x_0, y_0)^2.$$

- If $D > 0$ and $f_{xx}(x_0, y_0) > 0$, then f has a local minimum at (x_0, y_0).
- If $D > 0$ and $f_{xx}(x_0, y_0) < 0$, then f has a local maximum at (x_0, y_0).
- If $D < 0$, then f has neither a local maximum or minimum at (x_0, y_0).
- If $D = 0$, the test is inconclusive.

Example 5   Use the second derivative test to confirm that the critical point $q_1 = 699.1$, $q_2 = 896.7$ gives a local maximum of the profit function π of Example 4.

Solution   To see whether or not we have found a maximum point, we compute the second-order partial derivatives:

$$\frac{\partial^2 \pi}{\partial q_1^2} = -0.6, \quad \frac{\partial^2 \pi}{\partial q_2^2} = -0.4, \quad \frac{\partial^2 \pi}{\partial q_1 \partial q_2} = -0.2.$$

Since

$$D = \frac{\partial^2 \pi}{\partial q_1^2}\frac{\partial^2 \pi}{\partial q_2^2} - \left(\frac{\partial^2 \pi}{\partial q_1 \partial q_2}\right)^2 = (-0.6)(-0.4) - (-0.2)^2 = 0.2 > 0,$$

the second derivative test implies that we have found a local maximum point.

[9]An explanation of this test can be found, for example, in *Multivariable Calculus* by W. McCallum et al. (New York: John Wiley, 1997).

Problems for Section 9.5

1. Figure 9.43 shows contours of $f(x,y)$. List x- and y-coordinates and the value of the function at any local maximum and local minimum points, and identify which is which. Are any of these local extrema also global extrema on the region shown? If so, which ones?

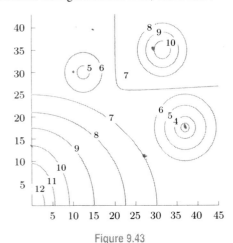

Figure 9.43

2. Figure 9.44 shows contours of $f(x,y)$. List the x- and y-coordinates and the value of the function at any local maximum and local minimum points, and identify which is which. Are any of these local extrema also global extrema on the region shown? If so, which ones?

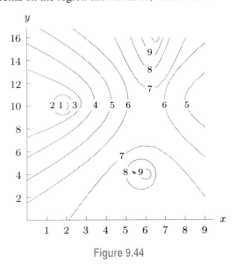

Figure 9.44

In Problems 3–5, estimate the position and approximate value of the global maxima and minima on the region shown.

3.

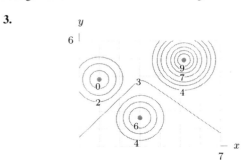

4.

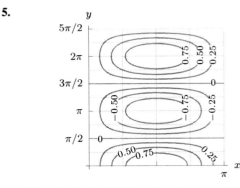

5.

In Problems 6–15, find all the critical points and determine whether each is a local maximum, local minimum, or neither.

6. $f(x,y) = x^2 + 4x + y^2$

7. $f(x,y) = x^2 + xy + 3y$

8. $f(x,y) = x^2 + y^2 + 6x - 10y + 8$

9. $f(x,y) = y^3 - 3xy + 6x$

10. $f(x,y) = x^2 - 2xy + 3y^2 - 8y$

11. $f(x,y) = x^3 - 3x + y^3 - 3y$

12. $f(x,y) = x^3 + y^2 - 3x^2 + 10y + 6$

13. $f(x,y) = x^3 + y^3 - 6y^2 - 3x + 9$

14. $f(x,y) = x^3 + y^3 - 3x^2 - 3y + 10$

15. $f(x,y) = 400 - 3x^2 - 4x + 2xy - 5y^2 + 48y$

16. By looking at the weather map in Figure 9.5 on page 354, find the maximum and minimum daily high temperatures in the states of Mississippi, Alabama, Pennsylvania, New York, California, Arizona, and Massachusetts.

17. For $f(x,y) = A - (x^2 + Bx + y^2 + Cy)$, what values of A, B, and C give f a local maximum value of 15 at the point $(-2, 1)$?

18. A missile has a guidance device which is sensitive to both temperature, $t°C$, and humidity, h. The range in km over which the missile can be controlled is given by

$$\text{Range} = 27{,}800 - 5t^2 - 6ht - 3h^2 + 400t + 300h.$$

What are the optimal atmospheric conditions for controlling the missile?

19. The quantity of a product demanded by consumers is a function of its price. The quantity of one product demanded may also depend on the price of other products. For example, the demand for tea is affected by the price of coffee; the demand for cars is affected by the price of gas. The quantities demanded, q_1 and q_2, of two products depend on their prices, p_1 and p_2, as follows:

$$q_1 = 150 - 2p_1 - p_2$$

$$q_2 = 200 - p_1 - 3p_2.$$

(a) What does the fact that the coefficients of p_1 and p_2 are negative tell you? Give an example of two products that might be related this way.

(b) If one manufacturer sells both products, how should the prices be set to generate the maximum possible revenue? What is that maximum possible revenue?

20. Two products are manufactured in quantities q_1 and q_2 and sold at prices of p_1 and p_2, respectively. The cost of producing them is given by

$$C = 2q_1^2 + 2q_2^2 + 10.$$

(a) Find the maximum profit that can be made, assuming the prices are fixed.

(b) Find the rate of change of that maximum profit as p_1 increases.

21. A company operates two plants which manufacture the same item and whose total cost functions are

$$C_1 = 8.5 + 0.03q_1^2 \quad \text{and} \quad C_2 = 5.2 + 0.04q_2^2,$$

where q_1 and q_2 are the quantities produced by each plant. The company is a monopoly. The total quantity demanded, $q = q_1 + q_2$, is related to the price, p, by

$$p = 60 - 0.04q.$$

How much should each plant produce in order to maximize the company's profit?[10]

9.6 CONSTRAINED OPTIMIZATION

Many real optimization problems are constrained by external circumstances. For example, a city wanting to build a public transportation system has a limited number of tax dollars available. A nation trying to maintain its balance of trade must spend less on imports than it carns on exports. In this section, we see how to find an optimum value under such constraints.

A Constrained Optimization Problem

Suppose we want to maximize the production of a company under a budget constraint. Suppose production, f, is a function of two variables, x and y, which are quantities of two raw materials, and

$$f(x,y) = x^{2/3}y^{1/3}.$$

If x and y are purchased at prices of p_1 and p_2 dollars per unit, what is the maximum production f that can be obtained with a budget of c dollars?

To increase f without regard to the budget, we simply increase x and y. However, the budget prevents us from increasing x and y beyond a certain point. Exactly how does the budget constrain us? Suppose that x and y each cost $100 per unit, and suppose that the total budget is $378,000. The amount spent on x and y together is given by $g(x,y) = 100x + 100y$, and since we can't spend more than the budget allows, we must have:

$$g(x,y) = 100x + 100y \le 378,000.$$

The goal is to maximize the function

$$f(x,y) = x^{2/3}y^{1/3}.$$

Since we expect to exhaust the budget, we have

$$100x + 100y = 378,000.$$

---
[10]Adapted from M. Rosser, *Basic Mathematics for Economists*, p. 318 (New York: Routledge, 1993).

Example 1   A company has production function $f(x, y) = x^{2/3}y^{1/3}$ and budget constraint
$100x + 100y = 378,000$.

(a) If \$100,000 is spent on x, how much can be spent on y? What is the production in this case?
(b) If \$200,000 is spent on x, how much can be spent on y? What is the production in this case?
(c) Which of the two options above is the better choice for the company? Do you think this is the best of all possible options?

Solution   (a) If the company spends \$100,000 on x, then it has \$278,000 left to spend on y. In this case, we have $100x = 100,000$, so $x = 1000$, and $100y = 278,000$, so $y = 2780$. Therefore,

$$\text{Production} = f(1000, 2780) = (1000)^{2/3}(2780)^{1/3} = 1406 \text{ units.}$$

(b) If the company spends \$200,000 on x, then it has \$178,000 left to spend on y. Therefore, $x = 2000$ and $y = 1780$, and so

$$\text{Production} = f(2000, 1780) = (2000)^{2/3}(1780)^{1/3} = 1924 \text{ units.}$$

(c) Of these two options, (b) is better since production is larger in this case. This is probably not optimal, since there are many other combinations of x and y that we have not checked.

Graphical Approach: Maximizing Production Subject to a Budget Constraint

How can we find the maximum value of production? We maximize the *objective function*

$$f(x, y) = x^{2/3}y^{1/3}$$

subject to $x \geq 0$ and $y \geq 0$ and the budget constraint

$$g(x, y) = 100x + 100y = 378,000.$$

The constraint is represented by the line in Figure 9.45. Any point on or below the line represents a pair of values of x and y that we can afford. A point on the line completely exhausts the budget, while a point below the line represents values of x and y which can be bought without using up the budget. Any point above the line represents a pair of values that we cannot afford.

Figure 9.45 also shows some contours of the production function f. Since we want to maximize f, we want to find the point which lies on the contour with the largest possible f value *and* which lies within the budget. The point we are looking for must lie on the budget constraint because we should spend all the available money. The key observation is this: The maximum occurs at a point P where the budget constraint is tangent to a production contour. (See Figure 9.45.) The reason is

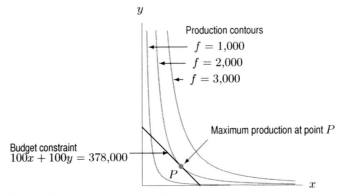

Figure 9.45: At the optimal point P the budget constraint is tangent to a production contour

that if we are on the constraint line to the left of P, moving right on the constraint increases f; if we are on the line to the right of P, moving left increases f. Thus, the maximum value of f on the budget constraint line occurs at the point P.

In general, provided f and g are smooth, we have the following result:

> If $f(x, y)$ has a global maximum or minimum on the constraint $g(x, y) = c$, it occurs at a point where the graph of the constraint is tangent to a contour of f, or at an endpoint of the constraint.[11]

Analytical Approach: The Method of Lagrange Multipliers

Suppose we want to optimize $f(x, y)$ subject to the constraint $g(x, y) = c$. We make the following definition.

> Suppose P_0 is a point satisfying the constraint $g(x, y) = c$.
> - f has a **local maximum** at P_0 **subject to the constraint** if $f(P_0) \geq f(P)$ for all points P near P_0 satisfying the constraint.
> - f has a **global maximum** at P_0 **subject to the constraint** if $f(P_0) \geq f(P)$ for all points P satisfying the constraint.
> Local and global minima are defined similarly.

It can be shown[12] that the constraint is tangent to a contour of f at the point which satisfies the equations laid out in the following method.

> **Method of Lagrange Multipliers** To optimize $f(x, y)$ subject to the constraint $g(x, y) = c$, solve the following system of three equations:
>
> $$f_x(x, y) = \lambda g_x(x, y),$$
> $$f_y(x, y) = \lambda g_y(x, y),$$
> $$g(x, y) = c,$$
>
> for the three unknowns x, y, and λ; the number λ is called the *Lagrange multiplier*. If f has a constrained global maximum or minimum, then it occurs at one of the solutions (x_0, y_0) to this system or at an endpoint of the constraint.

Example 2   Maximize $f(x, y) = x^{2/3}y^{1/3}$ subject to $100x + 100y = 378{,}000$ and $x \geq 0, y \geq 0$.

Solution   Differentiating gives

$$f_x(x, y) = \frac{2}{3}x^{-1/3}y^{1/3} \qquad \text{and} \qquad f_y(x, y) = \frac{1}{3}x^{2/3}y^{-2/3}$$

[11] If the constraint has endpoints.
[12] See W. McCallum, et al., *Multivariable Calculus* (New York: John Wiley, 2009).

and

$$g_x(x, y) = 100 \qquad \text{and} \qquad g_y(x, y) = 100,$$

leading to the equations

$$\frac{2}{3}x^{-1/3}y^{1/3} = \lambda(100)$$

$$\frac{1}{3}x^{2/3}y^{-2/3} = \lambda(100)$$

$$100x + 100y = 378{,}000.$$

The first two equations show that we must have

$$\frac{2}{3}x^{-1/3}y^{1/3} = \frac{1}{3}x^{2/3}y^{-2/3}.$$

Using the fact that $x^{-1/3} = 1/x^{1/3}$, we can rewrite this as

$$\frac{2y^{1/3}}{3x^{1/3}} = \frac{x^{2/3}}{3y^{2/3}}.$$

Multiplying through by the denominators gives

$$2y^{1/3}(3y^{2/3}) = x^{2/3}(3x^{1/3}),$$

and simplifying using $y^{1/3} \cdot y^{2/3} = y^1$ gives

$$6y = 3x$$
$$2y = x.$$

Since we must also satisfy the constraint $100x + 100y = 378{,}000$, we substitute $x = 2y$ and get

$$100(2y) + 100y = 378{,}000$$
$$300y = 378{,}000$$
$$y = 1260.$$

Since $x = 2y$, we have $x = 2520$. The optimum value occurs at $x = 2520$ and $y = 1260$. For these values,

$$f(2520, 1260) = (2520)^{2/3}(1260)^{1/3} \approx 2000.1.$$

The endpoints of the constraint are the points $(3780, 0)$ and $(0, 3780)$. Since

$$f(3780, 0) = f(0, 3780) = 0,$$

we see that the maximum value of f is approximately 2000 and that it occurs at $x = 2520$ and $y = 1260$.

The Meaning of λ

In the previous example, we never found (or needed) the value of λ. However, λ does have a practical interpretation. In the production problem we maximized

$$f(x, y) = x^{2/3}y^{1/3}$$

subject to the constraint

$$g(x, y) = 100x + 100y = 378{,}000.$$

We solved the equations

$$\frac{2}{3}x^{-1/3}y^{1/3} = 100\lambda,$$

$$\frac{1}{3}x^{2/3}y^{-2/3} = 100\lambda,$$

$$100x + 100y = 378,000,$$

to get $x = 2520, y = 1260$. Continuing to find λ gives us

$$\lambda \approx 0.0053.$$

Suppose now we do another, apparently unrelated calculation. Suppose our budget is increased by $1000, from $378,000 to $379,000. The new budget constraint is

$$100x + 100y = 379,000.$$

The corresponding solution is at $x = 2527, y = 1263$ and the new maximum value (instead of $f = 2000.1$) is

$$f = (2527)^{2/3}(1263)^{1/3} \approx 2005.4.$$

The additional $1000 in the budget increased the production level f by 5.3 units. Notice that production increased by $5.3/1000 = 0.0053$ units per dollar, which is our value of λ. The value of λ represents the extra production achieved by increasing the budget by one dollar—in other words, the extra "bang" you get for an extra "buck" of budget.

Solving for λ in either of the equations $f_x = \lambda g_x$ or $f_y = \lambda g_y$ suggests that the Lagrange multiplier is given by the ratio of the changes:

$$\lambda \approx \frac{\Delta f}{\Delta g} = \frac{\text{Change in optimum value of } f}{\text{Change in } g}.$$

These results suggest the following interpretations of the Lagrange multiplier λ:

> - The value of λ is approximately the change in the optimum value of f when the value of the constraint is increased by 1 unit.
> - The value of λ represents the rate of change of the optimum value of f as the constraint increases.

Example 3   The quantity of goods produced according to the function $f(x, y) = x^{2/3}y^{1/3}$ is maximized subject to the budget constraint $100x + 100y = 378,000$. Suppose the budget is increased to allow a small increase in production. What price must the product sell for if it is to be worth the increased budget?

Solution   We know that $\lambda = 0.0053$. Therefore, increasing the budget by $1 increases production by about 0.0053 unit. In order to make the increase in budget profitable, the extra goods produced must sell for more than $1. If the price is p in dollars, we must have $0.0053p > 1$. Thus, we need $p > 1/0.0053 \approx \$189$.

Example 4   If x and y are the amounts of raw materials used, the quantity, Q, of a product manufactured is

$$Q = xy.$$

Assume that x costs $20 per unit, y costs $10 per unit, and the production budget is $10,000.

(a) How many units of x and y should be purchased in order to maximize production? How many units are produced at that point?
(b) Find the value of λ and interpret it.

Solution   (a) We maximize $f(x, y) = xy$ subject to the constraint $g(x, y) = 20x + 10y = 10,000$ and $x \geq 0$, $y \geq 0$. We have the following partial derivatives:

$$f_x = y, \qquad f_y = x, \qquad \text{and} \qquad g_x = 20, \qquad g_y = 10.$$

The method of Lagrange multipliers gives the following equations:

$$y = 20\lambda$$
$$x = 10\lambda$$
$$20x + 10y = 10,000.$$

Substituting values of x and y from the first two equations into the third gives

$$20(10\lambda) + 10(20\lambda) = 10,000$$
$$400\lambda = 10,000$$
$$\lambda = 25.$$

Substituting $\lambda = 25$ in the first two equations above gives $x = 250$ and $y = 500$. The endpoints of the constraint are the point $(500, 0)$ and $(0, 1000)$. Since

$$f(500, 0) = 0 \quad \text{and} \quad f(0, 1000) = 0,$$

the maximum value of the production function is

$$f(250, 500) = 250 \cdot 500 = 125,000 \text{ units.}$$

The company should purchase 250 units of x and 500 units of y, giving 125,000 units of products.

(b) We have $\lambda = 25$. This tells us that if the budget is increased by \$1, we expect production to go up by about 25 units. If the budget goes up by \$1000, maximum production will increase about 25,000 to a total of roughly 150,000 units.

The Lagrangian Function

Constrained optimization problems are frequently solved using a *Lagrangian function*, $\mathcal{L}$. For example, to optimize the function $f(x, y)$ subject to the constraint $g(x, y) = c$, we use the Lagrangian function

$$\mathcal{L}(x, y, \lambda) = f(x, y) - \lambda(g(x, y) - c).$$

To see why the function $\mathcal{L}$ is useful, compute the partial derivatives of $\mathcal{L}$:

$$\frac{\partial \mathcal{L}}{\partial x} = \frac{\partial f}{\partial x} - \lambda \frac{\partial g}{\partial x},$$
$$\frac{\partial \mathcal{L}}{\partial y} = \frac{\partial f}{\partial y} - \lambda \frac{\partial g}{\partial y},$$
$$\frac{\partial \mathcal{L}}{\partial \lambda} = -(g(x, y) - c).$$

Notice that if (x_0, y_0) gives a maximum or minimum value of $f(x, y)$ subject to the constraint $g(x, y) = c$ and λ_0 is the corresponding Lagrange multiplier, then at the point (x_0, y_0, λ_0), we have

$$\frac{\partial \mathcal{L}}{\partial x} = 0 \quad \text{and} \quad \frac{\partial \mathcal{L}}{\partial y} = 0 \quad \text{and} \quad \frac{\partial \mathcal{L}}{\partial \lambda} = 0.$$

In other words, (x_0, y_0, λ_0) is a critical point for the unconstrained problem of optimization of the Lagrangian, $\mathcal{L}(x, y, \lambda)$.

We can therefore attack constrained optimization problems in two steps. First, write down the Lagrangian function $\mathcal{L}$. Second, find the critical points of $\mathcal{L}$.

Problems for Section 9.6

In Problems 1–10, use Lagrange multipliers to find the maximum or minimum values of $f(x, y)$ subject to the constraint.

1. $f(x, y) = x^2 + 4xy, \quad x + y = 100$

2. $f(x, y) = xy, \quad 5x + 2y = 100$

3. $f(x, y) = x^2 + 3y^2 + 100, \quad 8x + 6y = 88$

4. $f(x, y) = 5xy, \quad x + 3y = 24$

5. $f(x, y) = x + y, \quad x^2 + y^2 = 1$

6. $f(x, y) = x^2 + y^2, \quad 4x - 2y = 15$

7. $f(x, y) = 3x - 2y, \quad x^2 + 2y^2 = 44$

8. $f(x, y) = x^2 + y, \quad x^2 - y^2 = 1$

9. $f(x, y) = xy, \quad 4x^2 + y^2 = 8$

10. $f(x, y) = x^2 + y^2, \quad x^4 + y^4 = 2$

11. Figure 9.46 shows contours of $f(x, y)$ and the constraint $g(x, y) = c$. Approximately what values of x and y maximize $f(x, y)$ subject to the constraint? What is the approximate value of f at this maximum?

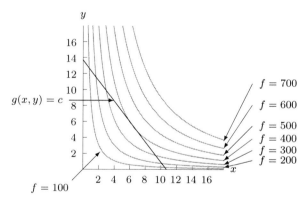

Figure 9.46

12. Figure 9.47 shows contours labeled with values of $f(x, y)$ and a constraint $g(x, y) = c$. Mark the approximate points at which:

   (a) f has a maximum
   (b) f has a maximum on the constraint $g = c$.

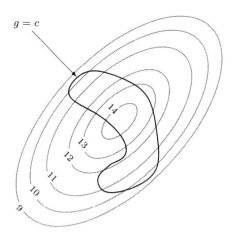

Figure 9.47

13. The quantity, Q, of a good produced depends on the quantities x_1 and x_2 of two raw materials used:

$$Q = x_1^{0.6} x_2^{0.4}.$$

A unit of x_1 costs \$127, and a unit of x_2 costs \$92. We want to minimize the cost, C, of producing 500 units of the good.

   (a) What is the objective function?
   (b) What is the constraint?

14. The quantity, Q, of a certain product manufactured depends on the quantity of labor, L, and of capital, K, used according to the function

$$Q = 900L^{1/2}K^{2/3}.$$

Labor costs \$100 per unit and capital costs \$200 per unit. What combination of labor and capital should be used to produce 36,000 units of the goods at minimum cost? What is that minimum cost?

15. The Cobb-Douglas production function for a product is

$$P = 5L^{0.8}K^{0.2},$$

where P is the quantity produced, L is the size of the labor force, and K is the amount of total equipment. Each unit of labor costs \$300, each unit of equipment costs \$100, and the total budget is \$15,000.

   (a) Make a table of L and K values which exhaust the budget. Find the production level, P, for each.
   (b) Use the method of Lagrange multipliers to find the optimal way to spend the budget.

16. A firm manufactures a commodity at two different factories. The total cost of manufacturing depends on the quantities, q_1 and q_2, supplied by each factory, and is expressed by the *joint cost function*,

$$C = f(q_1, q_2) = 2q_1^2 + q_1 q_2 + q_2^2 + 500.$$

The company's objective is to produce 200 units, while minimizing production costs. How many units should be supplied by each factory?

17. The quantity, Q, of a product manufactured by a company is given by

$$Q = aK^{0.6} L^{0.4},$$

where a is a positive constant, K is the quantity of capital and L is the quantity of labor used. Capital costs are $20 per unit, labor costs are $10 per unit, and the company wants costs for capital and labor combined to be no higher than $150. Suppose you are asked to consult for the company, and learn that 5 units each of capital and labor are being used.

(a) What do you advise? Should the company use more or less labor? More or less capital? If so, by how much?

(b) Write a one-sentence summary that could be used to sell your advice to the board of directors.

18. For a cost function, $f(x, y)$, the minimum cost for a production of 50 is given by $f(33, 87) = 1200$, with $\lambda = 15$. Estimate the cost if the production quota is:

(a) Raised to 51   (b) Lowered to 49

19. A company has the production function $P(x, y)$, which gives the number of units that can be produced for given values of x and y; the cost function $C(x, y)$ gives the cost of production for given values of x and y.

(a) If the company wishes to maximize production at a cost of $50,000, what is the objective function f? What is the constraint equation? What is the meaning of λ in this situation?

(b) If instead the company wishes to minimize the costs at a fixed production level of 2000 units, what is the objective function f? What is the constraint equation? What is the meaning of λ in this situation?

20. You have set aside 20 hours to work on two class projects. You want to maximize your grade (measured in points), which depends on how you divide your time between the two projects.

(a) What is the objective function for this optimization problem and what are its units?

(b) What is the constraint?

(c) Suppose you solve the problem by the method of Lagrange multipliers. What are the units for λ?

(d) What is the practical meaning of the statement $\lambda = 5$?

21. A steel manufacturer can produce $P(K, L)$ tons of steel using K units of capital and L units of labor, with production costs $C(K, L)$ dollars. With a budget of $600,000, the maximum production is 2,500,000 tons, using $400,000 of capital and $200,000 of labor. The Lagrange multiplier is $\lambda = 3.17$.

(a) What is the objective function?
(b) What is the constraint?
(c) What are the units for λ?
(d) What is the practical meaning of the statement $\lambda = 3.17$?

22. The quantity, q, of a product manufactured depends on the number of workers, W, and the amount of capital invested, K, and is given by the Cobb-Douglas function

$$q = 6W^{3/4} K^{1/4}.$$

In addition, labor costs are $10 per worker and capital costs are $20 per unit and the budget is $3000.

(a) What are the optimum number of workers and the optimum number of units of capital?

(b) Recompute the optimum values of W and K when the budget is increased by $1. Check that increasing the budget by $1 allows the production of λ extra units of the product, where λ is the Lagrange multiplier.

23. The terminal velocity (meters/second) that a two-stage rocket achieves is a function of the amount of fuel x_1 and x_2 (measured in liters) loaded into the two stages. You wish to minimize the total quantity of fuel required to achieve a specified terminal velocity, v_0.

(a) What is the objective function and what are its units?
(b) What is the constraint?
(c) Suppose you solve the problem by the method of Lagrange multipliers. What are the units for λ?
(d) What is the practical meaning of the statement $\lambda = 8$ when the terminal velocity of the rocket is 50 meters/second?

24. Each person tries to balance his or her time between leisure and work. The tradeoff is that as you work less your income falls. Therefore each person has *indifference curves* which connect the number of hours of leisure, l, and income, s. If, for example, you are indifferent between 0 hours of leisure and an income of $1125 a week on the one hand, and 10 hours of leisure and an income of $750 a week on the other hand, then the points $l = 0$, $s = 1125$, and $l = 10$, $s = 750$ both lie on the same indifference curve. Table 9.11 gives information on three indifference curves, I, II, and III.

Table 9.11

Weekly income			Weekly leisure hours		
I	II	III	I	II	III
1125	1250	1375	0	20	40
750	875	1000	10	30	50
500	625	750	20	40	60
375	500	625	30	50	70
250	375	500	50	70	90

(a) Graph the three indifference curves.

(b) You have 100 hours a week available for work and leisure combined, and you earn $10/hour. Write an equation in terms of l and s which represents this constraint.

(c) On the same axes, graph this constraint.

(d) Estimate from the graph what combination of leisure hours and income you would choose under these circumstances. Give the corresponding number of hours per week you would work.

25. If x_1 and x_2 are the number of items of two goods bought, a customer's utility is

$$U(x_1, x_2) = 2x_1 x_2 + 3x_1.$$

The unit cost is $1 for the first good and $3 for the second. Use Lagrange multipliers to find the maximum value of U if the consumer's disposable income is $100. Estimate the new optimal utility if the consumer's disposable income increases by $6.

CHAPTER SUMMARY

- **Functions of two variables**
  Represented by: tables, graphs, formulas, cross-sections (one variable fixed), contours (function value fixed).

- **Partial derivatives**
  Definition as a difference quotient, interpreting using units, estimating from a contour diagram or a table, computing from a formula, second-order partial derivatives.

- **Optimization**
  Critical points, local and global maxima and minima.

- **Constrained optimization**
  Geometric interpretation of Lagrange multiplier method, solving Lagrange multiplier problems algebraically, interpreting λ, Lagrangian function.

REVIEW PROBLEMS FOR CHAPTER NINE

1. Use Table 9.12. Is f an increasing or decreasing function of x? Is f an increasing or decreasing function of y?

Table 9.12   *Values of a function $f(x,y)$*

		y				
		0	1	2	3	4	5
	0	102	107	114	123	135	150
	20	96	101	108	117	129	144
x	40	90	95	102	111	123	138
	60	85	90	97	106	118	133
	80	81	86	93	102	114	129

2. The balance, B, in dollars, in a bank account depends on the amount deposited, A dollars, the annual interest rate, $r\%$, and the time, t, in months since the deposit, so $B = f(A, r, t)$.

(a) Is f an increasing or decreasing function of A? Of r? Of t?

(b) Interpret the statement $f(1250, 1, 25) \approx 1276$. Give units.

For each of the functions in Problems 3–4, make a contour

plot in the region $-2 < x < 2$ and $-2 < y < 2$. In each case, what is the equation and the shape of the contour lines?

3. $z = 3x - 5y + 1$ 4. $z = 2x^2 + y^2$

5. Table 9.13 shows the wind-chill factor as a function of wind speed and temperature. Draw a possible contour diagram for this function. Include contours at wind chills of $20°$, $0°$, and $-20°$.

6. The temperature adjusted for wind-chill is a temperature which tells you how cold it feels, as a result of the combination of wind and temperature.[13] See Table 9.13.

(a) If the temperature is $0°$F and the wind speed is 15 mph, how cold does it feel?

(b) If the temperature is $35°$F, what wind speed makes it feel like $24°$F?

(c) If the temperature is $25°$F, what wind speed makes it feel like $12°$F?

(d) If the wind is blowing at 20 mph, what temperature feels like $0°$F?

---

[13]Data from www.nws.noaa.gov/om/windchill, accessed on May 22, 2009.

Table 9.13   *Temperature adjusted for wind-chill (°F) as a function of wind speed and temperature*

| | | | Temperature (°F) | | | | | |
	35	30	25	20	15	10	5	0
5	31	25	19	13	7	1	−5	−11
10	27	21	15	9	3	−4	−10	−16
15	25	19	13	6	0	−7	−13	−19
20	24	17	11	4	−2	−9	−15	−22
25	23	16	9	3	−4	−11	−17	−24

Wind Speed (mph)

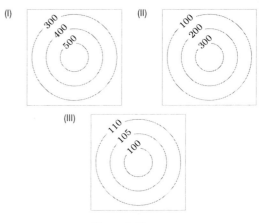

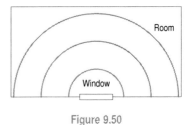

Figure 9.49

7. Using Table 9.13, make tables of the temperature adjusted for wind-chill as a function of wind speed for temperatures of 20°F and 0°F.

8. Using Table 9.13, make tables of the temperature adjusted for wind-chill as a function of temperature for wind speeds of 5 mph and 20 mph.

9. Figure 9.48 is a contour diagram for the sales of a product as a function of the price of the product and the amount spent on advertising. Which axis corresponds to the amount spent on advertising? Explain.

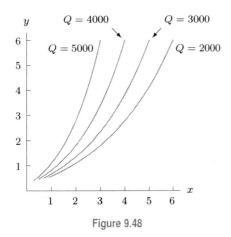

Figure 9.48

11. Figure 9.50 shows the contours of the temperature H in a room near a recently opened window. Label the three contours with reasonable values of H if the house is in the following locations.

(a) Minnesota in winter (where winters are harsh).
(b) San Francisco in winter (where winters are mild).
(c) Houston in summer (where summers are hot).
(d) Oregon in summer (where summers are mild).

Figure 9.50

10. Each of the contour diagrams in Figure 9.49 shows population density in a certain region. Choose the contour diagram that best corresponds to each of the following situations. Many different matchings are possible. Pick any reasonable one and justify your choice.

(a) The center of the diagram is a city.
(b) The center of the diagram is a lake.
(c) The center of the diagram is a power plant.

12. Figure 9.51 is a contour diagram of the monthly payment on a 5-year car loan as a function of the interest rate and the amount you borrow. The interest rate is 8% and you borrow $6000 for a used car.

(a) What is your monthly payment?
(b) If interest rates drop to 6%, how much more can you borrow without increasing your monthly payment?
(c) Make a table of how much you can borrow, without increasing your monthly payment, as a function of the interest rate.

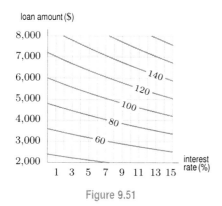

Figure 9.51

13. Match tables (a)–(d) with the contour diagrams (I)–(IV) in Figure 9.52.

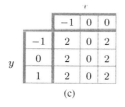

(a)

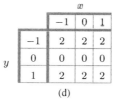

(b)

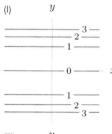

(c)

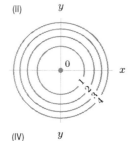

(d)

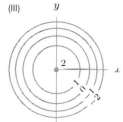

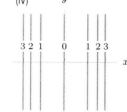

Figure 9.52

14. Figure 9.53 gives contour diagrams for different Cobb-Douglas production functions $F(L, K)$. Match each contour diagram with the correct statement.

(A) Tripling each input triples output.

(B) Quadrupling each input doubles output.

(C) Doubling each input almost triples output.

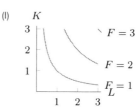
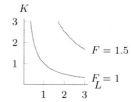
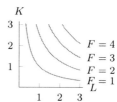

Figure 9.53

15. Figure 9.54 is a contour diagram for $z = f(x, y)$. Is f_x positive or negative? Is f_y positive or negative? Estimate $f(2, 1)$, $f_x(2, 1)$, and $f_y(2, 1)$.

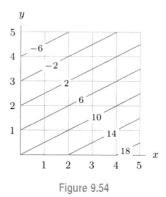

Figure 9.54

16. You borrow $\$A$ at an interest rate of $r\%$ (per month) and pay it off over t months by making monthly payments of $P = g(A, r, t)$ dollars. In financial terms, what do the following statements tell you?

(a) $g(8000, 1, 24) = 376.59$

(b) $\left.\dfrac{\partial g}{\partial A}\right|_{(8000,1,24)} = 0.047$

(c) $\left.\dfrac{\partial g}{\partial r}\right|_{(8000,1,24)} = 44.83$

For Problems 17–19, refer to Table 9.13 on page 390 giving the temperature adjusted for wind-chill, C, in °F, as a function $f(w, T)$ of the wind speed, w, in mph, and the temperature, T, in °F. The temperature adjusted for wind-chill tells you how cold it feels, as a result of the combination of wind and temperature.

17. Estimate $f_w(10, 25)$. What does your answer mean in practical terms?

18. Estimate $f_T(5, 20)$. What does your answer mean in practical terms?

19. From Table 9.13 you can see that when the temperature is 20°F, the temperature adjusted for wind-chill drops by an average of about 0.8°F with every 1 mph increase in wind speed from 5 mph to 10 mph. Which partial derivative is this telling you about?

20. Suppose that x is the price of one brand of gasoline and y is the price of a competing brand. Then q_1, the quantity of the first brand sold in a fixed time period, depends on both x and y, so $q_1 = f(x, y)$. Similarly, if q_2 is the quantity of the second brand sold during the same period, $q_2 = g(x, y)$. What do you expect the signs of the following quantities to be? Explain.

(a) $\partial q_1/\partial x$ and $\partial q_2/\partial y$
(b) $\partial q_1/\partial y$ and $\partial q_2/\partial x$

21. Suppose that x is the average price of a new car and that y is the average price of a gallon of gasoline. Then q_1, the number of new cars bought in a year, depends on both x and y, so $q_1 = f(x, y)$. Similarly, if q_2 is the quantity of gas bought in a year, then $q_2 = g(x, y)$.

(a) What do you expect the signs of $\partial q_1/\partial x$ and $\partial q_2/\partial y$ to be? Explain.
(b) What do you expect the signs of $\partial q_1/\partial y$ and $\partial q_2/\partial x$ to be? Explain.

22. Figure 9.55 shows the density of the fox population P (in foxes per square kilometer) for southern England. Draw two different graphs of the fox population as a function of kilometers north, with kilometers east fixed at two different values, and draw two different graphs of the fox population as a function of kilometers east, with kilometers north fixed at two different values.

23. Figure 9.55 gives a contour diagram for the number n of foxes per square kilometer in southwestern England. Estimate $\partial n/\partial x$ and $\partial n/\partial y$ at the points A, B, and C, where x is kilometers east and y is kilometers north.

Find the partial derivatives in Problems 24–29. The variables are restricted to a domain on which the function is defined.

24. f_x and f_y if $f(x, y) = x^2 + xy + y^2$

25. P_a and P_b if $P = a^2 - 2ab^2$

26. $\dfrac{\partial Q}{\partial p_1}$ and $\dfrac{\partial Q}{\partial p_2}$ if $Q = 50p_1p_2 - p_2^2$

27. $\dfrac{\partial f}{\partial x}$ and $\dfrac{\partial f}{\partial t}$ if $f = 5xe^{-2t}$

28. $\dfrac{\partial P}{\partial K}$ and $\dfrac{\partial P}{\partial L}$ if $P = 10K^{0.7}L^{0.3}$

29. f_x and f_y if $f(x, y) = \sqrt{x^2 + y^2}$

30. A manufacturing company produces two items in quantities q_1 and q_2, respectively. Total production costs are given by

$$\text{Cost} = f(q_1, q_2) = 16 + 1.2q_1 + 1.5q_2 + 0.2q_1q_2.$$

Find $f(500, 1000)$, $f_{q_1}(500, 1000)$, and $f_{q_2}(500, 1000)$. Give units with your answers and interpret each of your answers in terms of production cost.

31. The Cobb-Douglas production function for a product is given by

$$Q = 25K^{0.75}L^{0.25},$$

where Q is the quantity produced for a capital investment of K dollars and a labor investment of L.

(a) Find Q_K and Q_L.
(b) Find the values of Q, Q_K and Q_L given that $K = 60$ and $L = 100$.
(c) Interpret each of the values you found in part (b) in terms of production.

32. Figure 9.56 is a contour diagram of $f(x, y)$. In each of the following cases, list the marked points in the diagram (there may be none or more than one) at which

(a) $f_x < 0$ (b) $f_y > 0$
(c) $f_{xx} > 0$ (d) $f_{yy} < 0$

kilometers north

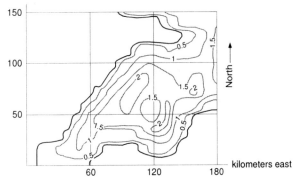

Figure 9.55

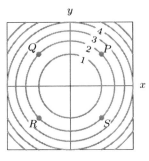

Figure 9.56

33. You are an anthropologist observing a native ritual. Sixteen people arrange themselves with their backs to you along a bench; all but the three on the far left side are seated. The first person on the far left is standing with her hands at her side, the second is standing with his hands raised and the third is standing with her hands at her side. At some unseen signal, the first one sits down, and everyone else copies what his neighbor to the left was doing one second earlier. Every second that passes, this behavior is repeated until all are once again seated.

   (a) Draw graphs at several different times showing how the height depends upon the distance along the bench.
   (b) Graph the location of the raised hands as a function of time.
   (c) What US ritual is most closely related to what you have observed?

34. You are in a stadium doing the wave. This is a ritual in which members of the audience stand up and down in such a way as to create a wave that moves around the stadium. Normally a single wave travels all the way around the stadium, but we assume there is a continuous sequence of waves. Let $h(x, t) = 5 + \cos(0.5x - t)$ be the function describing this stadium wave. The value of $h(x, t)$ gives the height (in feet) of the head of the spectator in seat x at time t seconds. Evaluate $h_x(2, 5)$ and $h_t(2, 5)$ and interpret each in terms of the wave.

35. Find all critical points of $f(x, y) = x^3 - 3x + y^2$. Make a table of values to determine if each critical point is a local minimum, a local maximum, or neither.

36. Find all critical points of $f(x, y) = x^2 + 3y^2 - 4x + 6y + 10$.

37. A company sells two products which are partial substitutes for each other, such as coffee and tea. If the price of one product rises, then the demand for the other product rises. The quantities demanded, q_1 and q_2, are given as a function of the prices, p_1 and p_2, by

$$q_1 = 517 - 3.5p_1 + 0.8p_2, \quad q_2 = 770 - 4.4p_2 + 1.4p_1.$$

   (a) Write total sales revenue as a function of p_1 and p_2.
   (b) What prices should the company charge in order to maximize the total sales revenue? [14]

Figure 9.57 shows contours of f. In Problems 38–40 give an approximate maximum or minimum value of f for $0 \le x \le 300$, $0 \le y \le 300$ subject to the given constraint.

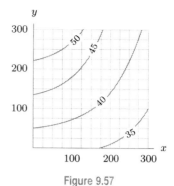

Figure 9.57

38. Minimum, constraint is $y = 100$

39. Maximum, constraint is $y = 100$

40. Maximum, constraint is $y = x$

41. The quantity, Q, of a good produced depends on the quantities x_1 and x_2 of two raw materials used:

$$Q = x_1^{0.3} x_2^{0.7}.$$

A unit of x_1 costs \$10, and a unit of x_2 costs \$25. We want to maximize production with a budget of \$50 thousand for raw materials.

   (a) What is the objective function?
   (b) What is the constraint?

42. Figure 9.58 shows level curves for production $f(x, y)$ as a function of the quantities x and y of two raw materials utilized. The cost of the materials is $15x + 20y$ thousand dollars. What is the maximum production possible with a budget of 300 thousand dollars? How much of each raw material should be purchased to achieve this maximum?

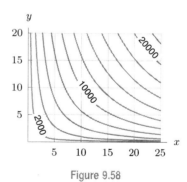

Figure 9.58

43. An automobile manufacturing plant currently employs 1500 workers and has capital investment of 15 million dollars per month. The production function is

$$Q = x^{0.4} y^{0.6},$$

[14] Adapted from M. Rosser, *Basic Mathematics for Economists*, p. 318 (New York: Routledge, 1993).

where Q is the number of cars produced per month, x is the number of workers and y is capital investment. Each worker's salary is $5000 per month and each unit of capital costs $4000 per month.

(a) How many cars is the factory currently assembling each month?

(b) Due to the sluggish economy, the factory decides to decrease production to 2000 cars per month. The factory wants to minimize the cost of producing these 2000 cars. How many workers will need to be laid off? By what amount should monthly investment be decreased?

(c) Give the value of the Lagrange multiplier, λ, and interpret it in terms of the car factory.

44. A company manufactures x units of one item and y units of another. The total cost in dollars, C, of producing these two items is approximated by the function

$$C = 5x^2 + 2xy + 3y^2 + 800.$$

(a) If the production quota for the total number of items (both types combined) is 39, find the minimum production cost.

(b) Estimate the additional production cost or savings if the production quota is raised to 40 or lowered to 38.

45. A monopolistic producer of two goods A and B has a joint total cost function

$$C = 10q_1 + q_1 q_2 + 10q_2,$$

where q_1 and q_2 denote the quantities of A and B respectively. The demand curves for the corresponding prices p_1 and p_2 are

$$p_1 = 50 - q_1 + q_2$$
$$p_2 = 30 + 2q_1 - q_2.$$

(a) Find the maximum profit if the firm produces a total of 15.

(b) Estimate the new optimal profit if the production quota increases by one unit.

CHECK YOUR UNDERSTANDING

In Problems 1–60, indicate whether the statement is true or false.

1. If $Q = f(x, t)$ gives the quantity Q of pollutants, in ppm (parts per million), at a distance x kilometers from a waste incinerator at time t hours after incineration ends, we have $f(1.5, 6) = 4.1$. The units for 1.5 are ppm.

2. If $Q = f(x, t)$ gives the quantity Q of pollutants, in ppm (parts per million), at a distance x kilometers from a waste incinerator at time t hours after incineration ends, we have $f(1.5, 6) = 4.1$. The units for 6 are hours.

3. If $Q = f(x, t)$ gives the quantity Q of pollutants, in ppm (parts per million), at a distance x kilometers from a waste incinerator at time t hours after incineration ends, we have $f(1.5, 6) = 4.1$. The units for 4.1 are kilometers.

4. If $Q = f(x, t)$ gives the quantity Q of pollutants, in ppm (parts per million), at a distance x kilometers from a waste incinerator at time t hours after incineration ends, then Q is most likely an increasing function of x.

5. If $Q = f(x, t)$ gives the quantity Q of pollutants, in ppm (parts per million), at a distance x kilometers from a waste incinerator at time t hours after incineration ends, in the statement $f(2, 3) = 5$, the units of the 2 are km.

6. If $f(x, y)$ is a function of two variables, then f is an increasing function of x if f increases when both x increases and y increases.

7. A function $f(x, y)$ can be both an increasing function of x and a decreasing function of y.

8. The function $f(x, y) = x^2 - y$ is a decreasing function of y.

9. If $f(x, y) = e^{xy} - y^2$, then the cross-section $x = 1$ is $f(1, y) = e^y - 1$.

10. If $f(x, y) = e^{xy} - y^2$, then the cross-section $y = 0$ is $f(x, 0) = 0$.

11. The function $f(x, y) = y + 3x - 1$ has contours which are lines.

12. The function $f(x, y) = x^2 - y$ has contours which are parabolas.

13. For a given function $f(x, y)$ the two contours $f(x, y) = 1$ and $f(x, y) = 2$ can never intersect.

14. Contours of the function $g(x, y) = 3x + 2y$ are parallel lines.

15. The function $P = 3N^{1/2} V^{1/2}$ is a Cobb-Douglas production function.

16. The function $P = 10N^3 V^2$ is a Cobb-Douglas production function.

17. The function $f(x, y)$ has the same value at all points along a specific contour.

18. A contour can consist of a single point.

19. Contours of a Cobb-Douglas production function are decreasing, concave up.

20. Every point in the domain of $g(x, y)$ has a contour that contains it.

21. The partial derivative $f_x(a, b)$ is the rate of change of the function f with respect to x at the point (a, b) with y fixed at b.

22. If $P = f(m, t)$ gives the price P (in dollars) of a used car as a function of its mileage m (in miles) and its age t (in years), then $\partial P/\partial m$ has units miles per year.

23. If $P = f(m, t)$ gives the price P (in dollars) of a used car as a function of its mileage m (in miles) and its age t (in years), then $\partial P/\partial m$ is most likely to be negative.

24. If $W = f(c, t)$ gives the weight W (in pounds) of a person as a function of their daily food intake c (in calories) and daily aerobic exercise time t (in hours), then $\partial W/\partial c$ has units pounds per calorie.

25. If $W = f(c, t)$ gives the weight W (in pounds) of a person as a function of their daily food intake c (in calories) and daily aerobic exercise time t (in hours), then $\partial W/\partial c$ is most likely to be negative.

26. If $W = f(c, t)$ gives the weight W (in pounds) of a person as a function of their daily food intake c (in calories) and daily aerobic exercise time t (in hours), then $\partial W/\partial c$ and $\partial W/\partial t$ most likely have opposite signs.

27. For a function $z = f(x, y)$, assume $f(1, 2) = 10$ and $f(1.1, 2) = 10.5$ and $f(1, 2.1) = 10.8$ and $f(1.1, 2.1) = 11.3$. Then $f_x(1, 2) \approx 5$.

28. For a function $z = f(x, y)$, assume $f(1, 2) = 10$ and $f(1.1, 2) = 10.5$ and $f(1, 2.1) = 10.8$ and $f(1.1, 2.1) = 11.3$. Then $f_y(1, 2) \approx 13$.

29. For a function $z = f(x, y)$, assume $f(5, 3) = 7$ and $f(5.1, 3) = 7.9$ and $f(5, 3.1) = 6.4$ and $f(5.1, 3.1) = 7.3$. Then $f_x(5, 3) \approx 9$.

30. For a function $z = f(x, y)$, assume $f(5, 3) = 7$ and $f(5.1, 3) = 7.9$ and $f(5, 3.1) = 6.4$ and $f(5.1, 3.1) = 7.3$. Then $f_y(5, 3) \approx 6$.

31. If $f(x, y) = x^2 y + 3x$ then $f_x(x, y) = 2xy + 3$.

32. If $f(x, y) = x^2 y + 3x$ then $f_y(1, 2) = 4$.

33. If $g(u, v) = ue^v$ then $g_u(0, 0) = 1$.

34. There exists a function f with $f_x(x, y) = f_y(x, y)$.

35. The function $Q(x, y) = x^3 y^2$ is decreasing in the y direction near $(1, 2)$.

36. If $P = 2N^{0.6} V^{0.4}$ then $\dfrac{\partial P}{\partial V} = 0.8 N^{0.6} V^{-0.6}$.

37. For all functions f it is always true that $f_{xx} = f_{yy}$ if f_{xx} and f_{yy} are continuous.

38. If $f(x, y) = 3x^2 e^{2y}$ then $f_y(1, 0) < f_x(1, 0)$.

39. If $z = p(A, B) = (1/2)(A^2 - B)$ then $\dfrac{\partial z}{\partial A} = A$.

40. If $z = V(r, h) = \pi r^2 h$ then $\dfrac{\partial^2 z}{\partial r \partial h} = 2\pi$.

41. If f is a function with $f_x(1, 2) = 0$ then $(1, 2)$ is a critical point of f.

42. The point $(1, 1)$ is a critical point of $f(x, y) = x^2 + y^2$.

43. The point $(3, 2)$ is a critical point of $g(u, v) = (u-3)^2 + (v-2)^2$.

44. The function $f(x, y) = xe^y$ has no critical points.

45. If $(0, 0)$ is a critical point of f then f has either a local maximum or local minimum at $(0, 0)$.

46. If $D > 0$ and $f_{xx} > 0$ in the second derivative test at the point (a, b) then f has a local minimum at (a, b).

47. To use the second derivative test on a function f at a point (a, b), we must have $f_x(a, b) = 0$ and $f_y(a, b) = 0$.

48. The function $f(x, y) = x^2 - y^2$ has a local minimum at $(0, 0)$.

49. The function $f(x, y) = 3xy$ has a neither a local maximum or minimum at $(0, 0)$.

50. A local minimum of f can also be a global minimum of f.

51. If f has a local maximum at P_0 subject to the constraint $g(x, y) = c$ then P_0 is a critical point of f.

52. If f has a local maximum at P_0 subject to the constraint $g(x, y) = c$ then P_0 satisfies the equation $g(x, y) = c$.

53. If we wish to maximize production given a fixed budget, then the production function is the constraint equation.

54. If we wish to minimize costs given a fixed production level, then the production equation is the constraint equation.

55. The minimum value of $f(x, y) = x^2 + y^2$ subject to the constraint $x - y = 0$ is $f(0, 0) = 0$.

56. The minimum value of $f(x, y) = x^2 + y^2$ subject to the constraint $2x + 3y = 12$ is $f(0, 0) = 0$.

57. If the minimum cost is \$50,000 given a fixed production level of 8,000 tons with $\lambda = 120$, then we expect production of 8,001 tons to cost about \$50,120.

58. If the maximum production level is 50,000 tons given a fixed budget of \$80,000 with $\lambda = 120$, then we expect production of \$50,001 tons to cost about \$80,120.

59. The second derivative test can be used to classify a point found by the Lagrange multiplier method as constrained maximum or minimum.

60. If the quantity Q of a good produced depends only on the quantities x and y of two raw materials used to manufacture the good, and there is a total of \$50,000 to spend on the two raw materials, then a budget constraint for Q is $x + y = 50000$.

PROJECTS FOR CHAPTER NINE

1. A Heater in a Room

Figure 9.59 shows the contours of the temperature along one wall of a heated room through one winter day, with time indicated as on a 24-hour clock. The room has a heater located at the left-most corner of the wall and one window in the wall. The heater is controlled by a thermostat about 2 feet from the window.

(a) Where is the window?   (b) When is the window open?

(c) When is the heat on?

(d) Draw graphs of the temperature along the wall of the room at 6 am, at 11 am, at 3 pm (15 hours) and at 5 pm (17 hours).

(e) Draw a graph of the temperature as a function of time at the heater, at the window and midway between them.

(f) The temperature at the window at 5 pm (17 hours) is less than at 11 am. Why do you think this might be?

(g) To what temperature do you think the thermostat is set? How do you know?

(h) Where is the thermostat?

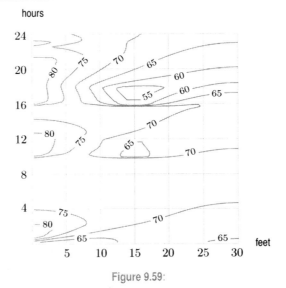

Figure 9.59:

2. Optimizing Relative Prices for Adults and Children

Some items are sold at a discount to senior citizens or children. The reason is that these groups are more sensitive to price, so a discount has greater impact on their purchasing decisions. The seller faces an optimization problem: How large a discount to offer in order to maximize profits? Suppose a theater can sell q_c child tickets and q_a adult tickets at prices p_c and p_a, according to the demand functions:

$$q_c = rp_c^{-4} \quad \text{and} \quad q_a = sp_a^{-2},$$

and has operating costs proportional to the total number of tickets sold. What should be the relative price of children's and adults' tickets?

3. Maximizing Production and Minimizing Cost: "Duality"

A company's production function is $P = 270x_1^{1/3}x_2^{2/3}$ for quantities x_1 and x_2 of two raw materials, costing $4 per unit and $27 per unit, respectively.

(a) How much of each raw material should be used to maximize production if the budget for raw materials is $324? What is the maximum production achieved, P_0?

(b) What is the minimum cost at which a production level of P_0 can be achieved? How much of each raw material is used at this minimum?

(c) Comment on the relationship between your answers to parts (a) and (b).

346

FOCUS ON THEORY

DERIVING THE FORMULA FOR A REGRESSION LINE

Suppose we want to find the "best fitting" line for some experimental data. In Appendix A, we use a computer or calculator to find the formula for this line. In this section, we derive this formula.

We decide which line fits the data best by using the following criterion. The data is plotted in the plane. The distance from a line to the data points is measured by adding the squares of the vertical distances from each point to the line. The smaller this sum of squares is, the better the line fits the data. The line with the minimum sum of square distances is called the *least-squares line*, or the *regression line*. If the data is nearly linear, the least-squares line will be a good fit; otherwise it may not be. (See Figure 9.60.)

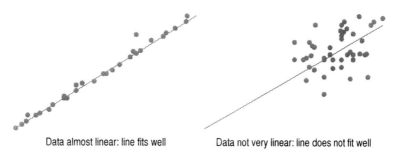

Data almost linear: line fits well   Data not very linear: line does not fit well

Figure 9.60: Fitting lines to data points

Example 1   Find a least-squares line for the following data points: $(1, 1)$, $(2, 1)$, and $(3, 3)$.

Solution   Suppose the line has equation $y = b + mx$. If we find b and m then we have found the line. So, for this problem, b and m are the two variables. We want to minimize the function $f(b, m)$ that gives the sum of the three squared vertical distances from the points to the line in Figure 9.61.

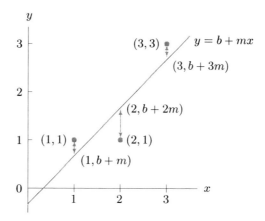

Figure 9.61: The least-squares line minimizes the sum of the squares of these vertical distances

The vertical distance from the point $(1, 1)$ to the line is the difference in the y-coordinates $1 - (b + m)$; similarly for the other points. Thus, the sum of squares is

$$f(b, m) = (1 - (b + m))^2 + (1 - (b + 2m))^2 + (3 - (b + 3m))^2.$$

347

To minimize f we look for critical points. First we differentiate f with respect to b:

$$f_b(b, m) = -2(1 - (b + m)) - 2(1 - (b + 2m)) - 2(3 - (b + 3m))$$
$$= -2 + 2b + 2m - 2 + 2b + 4m - 6 + 2b + 6m$$
$$= -10 + 6b + 12m.$$

Now we differentiate with respect to m:

$$f_m(b, m) = 2(1 - (b + m))(-1) + 2(1 - (b + 2m))(-2) + 2(3 - (b + 3m))(-3)$$
$$= -2 + 2b + 2m - 4 + 4b + 8m - 18 + 6b + 18m$$
$$= -24 + 12b + 28m.$$

The equations $f_b = 0$ and $f_m = 0$ give a system of two linear equations in two unknowns:

$$-10 + 6b + 12m = 0,$$
$$-24 + 12b + 28m = 0.$$

The solution to this pair of equations is the critical point $b = -1/3$ and $m = 1$. Since

$$D = f_{bb}f_{mm} - (f_{mb})^2 = (6)(28) - 12^2 = 24 \quad \text{and} \quad f_{bb} = 6 > 0,$$

we have found a local minimum. This local minimum is also the global minimum of f. Thus, the least squares line is

$$y = x - \frac{1}{3}.$$

As a check, notice that the line $y = x$ passes through the points $(1, 1)$ and $(3, 3)$. It is reasonable that introducing the point $(2, 1)$ moves the y-intercept down from 0 to $-1/3$.

Derivation of the Formulas for the Regression Line

We use the method of Example 1 to derive the formulas for the least-squares line $y = b + mx$ generated by data points $(x_1, y_1), (x_2, y_2), \ldots, (x_n, y_n)$. Notice that we are looking for the slope and y-intercept, so we think of m and b as the variables.

For each data point (x_i, y_i), the corresponding point directly above the line or below it on the line has the y-coordinate $b + mx_i$. Thus, the squares of the vertical distances from the point to the line is $(y_i - (b + mx_i))^2$. (See Figure 9.62.)

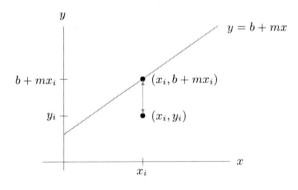

Figure 9.62: The vertical distance from a point to the line

We find the sum of the n squared distances from points to the line, and think of the sum as a function of m and b:

$$f(b, m) = \sum_{i=1}^{n} (y_i - (b + mx_i))^2.$$

To minimize this function, we first find the two partial derivatives, f_b and f_m. We use the chain rule and the properties of sums.

$$
\begin{aligned}
f_b(b, m) &= \frac{\partial}{\partial b} \left(\sum_{i=1}^{n} (y_i - (b + mx_i))^2 \right) = \sum_{i=1}^{n} \frac{\partial}{\partial b} (y_i - (b + mx_i))^2 \\
&= \sum_{i=1}^{n} 2(y_i - (b + mx_i)) \cdot \frac{\partial}{\partial b} (y_i - (b + mx_i)) \\
&= \sum_{i=1}^{n} 2(y_i - (b + mx_i)) \cdot (-1) \\
&= -2 \sum_{i=1}^{n} (y_i - (b + mx_i))
\end{aligned}
$$

$$
\begin{aligned}
f_m(b, m) &= \frac{\partial}{\partial m} \left(\sum_{i=1}^{n} (y_i - (b + mx_i))^2 \right) = \sum_{i=1}^{n} \frac{\partial}{\partial m} (y_i - (b + mx_i))^2 \\
&= \sum_{i=1}^{n} 2(y_i - (b + mx_i)) \cdot \frac{\partial}{\partial m} (y_i - (b + mx_i)) \\
&= \sum_{i=1}^{n} 2(y_i - (b + mx_i)) \cdot (-x_i) \\
&= -2 \sum_{i=1}^{n} (y_i - (b + mx_i)) \cdot x_i
\end{aligned}
$$

We now set the partial derivatives equal to zero and solve for m and b. This is easier than it looks: we simplify the appearance of the equations by temporarily substituting other symbols for the sums: write SY for $\sum y_i$, SX for $\sum x_i$, SXY for $\sum y_i x_i$ and SXX for $\sum x_i^2$. Remember, the x_i and y_i are all constants. We get a pair of simultaneous linear equations in m and b; solving for m and b gives us formulas in terms of SX, SY, SXY, and SXX. We separate $f_b(b, m)$ into three sums as shown:

$$f_b(b, m) = -2 \left(\sum_{i=1}^{n} y_i - b \sum_{i=1}^{n} 1 - m \sum_{i=1}^{n} x_i \right).$$

Similarly, we can separate $f_m(b, m)$ after multiplying through by x_i:

$$f_m(b, m) = -2 \left(\sum_{i=1}^{n} y_i x_i - b \sum_{i=1}^{n} x_i - m \sum_{i=1}^{n} x_i^2 \right).$$

Rewriting the sums as suggested and setting $\dfrac{\partial f}{\partial b}$ and $\dfrac{\partial f}{\partial m}$ equal to zero, we have:

$$0 = SY - bn - mSX$$
$$0 = SYX - bSX - mSXX$$

Solving this pair of simultaneous equations, we get the result:

$$b = ((SXX) \cdot (SY) - (SX) \cdot (SYX))/(n(SXX) - (SX)^2)$$
$$m = (n(SYX) - (SX) \cdot (SY))/(n(SXX) - (SX)^2)$$

Writing these expressions with summation notation, we arrive at the following result:

> The least-squares line for data points $(x_1, y_1), (x_2, y_2), \cdots, (x_n, y_n)$ is the line $y = b + mx$ where
>
> $$b = \left(\sum_{i=1}^{n} x_i^2 \sum_{i=1}^{n} y_i - \sum_{i=1}^{n} x_i \sum_{i=1}^{n} y_i x_i \right) / \left(n \sum_{i=1}^{n} x_i^2 - \left(\sum_{i=1}^{n} x_i \right)^2 \right)$$
>
> $$m = \left(n \sum_{i=1}^{n} y_i x_i - \sum_{i=1}^{n} x_i \sum_{i=1}^{n} y_i \right) / \left(n \sum_{i=1}^{n} x_i^2 - \left(\sum_{i=1}^{n} x_i \right)^2 \right).$$

Example 2   Use these formulas to find the best fitting line for the data point $(1, 5), (2, 4), (4, 3)$.

Solution   We compute the sums needed in the formulas:

$$\sum_{i=1}^{3} x_i = 1 + 2 + 4 = 7$$

$$\sum_{i=1}^{3} y_i = 5 + 4 + 3 = 12$$

$$\sum_{i=1}^{3} x_i^2 = 1^2 + 2^2 + 4^2 = 1 + 4 + 16 = 21$$

$$\sum_{i=1}^{3} y_i x_i = (5)(1) + (4)(2) + (3)(4) = 5 + 8 + 12 = 25.$$

Since $n = 3$, we have:

$$b = \left(\sum_{i=1}^{3} x_i^2 \sum_{i=1}^{3} y_i - \sum_{i=1}^{3} x_i \sum_{i=1}^{3} y_i x_i \right) / \left(3 \sum_{i=1}^{3} x_i^2 - \left(\sum_{i=1}^{3} x_i \right)^2 \right)$$
$$= ((21)(12) - (7)(25)) / (3(21) - (7^2))$$
$$= 77/14 = 5.5$$

and

$$m = \left(3 \sum_{i=1}^{3} y_i x_i - \sum_{i=1}^{3} x_i \sum_{i=1}^{3} y_i \right) / \left(3 \sum_{i=1}^{3} x_i^2 - \left(\sum_{i=1}^{3} x_i \right)^2 \right)$$
$$= (3(25) - (7)(12)) / (3(21) - (7^2))$$
$$= -9/14 = -0.64.$$

The least-squares line for these three points is

$$y = 5.5 - 0.64x.$$

To check this equation, plot the line and the three points together.

Many calculators have the formulas for the least-squares line built in, so that when you enter the data, out come the values of b and m. At the same time, you get the *correlation coefficient*, which measures how close the data points actually come to fitting the least-squares line.

Problems on Deriving the Formula for Regression Lines

In Problems 1–2, use the method of Example 1 to find the least-squares line. Check by graphing the points with the line.

1. $(-1, 2), (0, -1), (1, 1)$ **2.** $(0, 2), (1, 4), (2, 5)$

In Problems 3–5, use the formulas for b and m to check that you get the same result as in the problem or example specified.

3. $(-1, 2), (0, -1), (1, 1)$. See Problem 1.

4. $(0, 2), (1, 4), (2, 5)$. See Problem 2.

5. $(1, 1), (2, 1), (3, 3)$. See Example 1.

In Problems 6–7, we transform nonlinear data so that it looks more linear. For example, suppose the data points (x, y) fit the exponential equation,

$$y = Ce^{ax},$$

where a and C are constants. Taking the natural log of both sides, we get

$$\ln y = ax + \ln C.$$

Thus, $\ln y$ is a linear function of x. To find a and C, we can use least squares for the graph of $\ln y$ against x.

6. The population of the US was about 180 million in 1960, grew to 206 million in 1970, and 226 million in 1980.

   (a) Assuming that the population was growing exponentially, use logarithms and the method of least squares to estimate the population in 1990.

(b) According to the national census, the 1990 population was 249 million. What does this say about the assumption of exponential growth?

(c) Predict the population in the year 2010.

7. A biological rule of thumb states that as the area A of an island increases tenfold, the number of animal species, N, living on it doubles. The table contains data for islands in the West Indies. Assume that N is a power function of A.

   (a) Use the biological rule of thumb to find

      (i) N as a function of A

      (ii) $\ln N$ as a function of $\ln A$

   (b) Using the data given, tabulate $\ln N$ against $\ln A$ and find the line of best fit. Does your answer agree with the biological rule of thumb?

Island	Area (sq km)	Number of species
Redonda	3	5
Saba	20	9
Montserrat	192	15
Puerto Rico	8858	75
Jamaica	10854	70
Hispaniola	75571	130
Cuba	113715	125

APPENDICE

A FITTING FORMULAS TO DATA

In this section we see how the formulas that are used in a mathematical model can be developed. Some of the formulas we use are exact. However, many formulas we use are approximations, often constructed from data.

Fitting a Linear Function to Data

A company wants to understand the relationship between the amount spent on advertising, a, and total sales, S. The data they collect might look like that found in Table A.1.

Table A.1 *Advertising and sales: Linear relationship*

a (advertising in \$1000s)	3	4	5	6
S (sales in \$1000s)	100	120	140	160

The data in Table A.1 are linear, so a formula fits it exactly. The slope of the line is 20, and we can determine that the vertical intercept is 40, so the line is

$$S = 40 + 20a.$$

Now suppose that the company collected the data in Table A.2. This time the data are not linear. In general, it is difficult to find a formula to fit data exactly. We must be satisfied with a formula that is a good approximation to the data.

Table A.2 *Advertising and sales: Nonlinear relationship*

a (advertising in \$1000s)	3	4	5	6
S (sales in \$1000s)	105	117	141	152

Figure A.1 shows the data in Table A.2. Since the relationship is nearly, though not exactly, linear, it is well approximated by a line. Figure A.2 shows the line $S = 40 + 20a$ and the data.

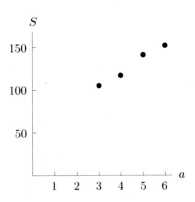

Figure A.1: The sales data from Table A.2

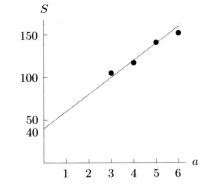

Figure A.2: The line $S = 40 + 20a$ and the data from Table A.2

The Regression Line

Is there a line that fits the data better than the one in Figure A.2? If so, how do we find it? The process of fitting a line to a set of data is called *linear regression* and the line of best fit is called the *regression line*. (Later in the section, we discuss what "best fit" means.) Many calculators and computer programs calculate the regression line from the data points. Alternatively, the regression line can be estimated by plotting the points on paper and fitting a line "by eye." In Chapter 9, we derive the formulas for the regression line. For the data in Table A.2, the regression line is

$$S = 54.5 + 16.5a.$$

This line is graphed with the data in Figure A.3.

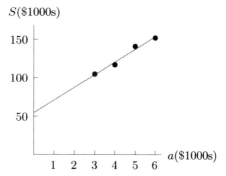

Figure A.3: The regression line $S = 54.5 + 16.5a$
and the data from Table A.2

Using the Regression Line to Make Predictions

We can use the formula for sales as a function of advertising to make predictions. For example, to predict total sales if $3500 is spent on advertising, substitute $a = 3.5$ into the regression line:

$$S = 54.5 + 16.5(3.5) = 112.25.$$

The regression line predicts sales of $112,250. To see that this is reasonable, compare it to the entries in Table A.2. When $a = 3$, we have $S = 105$, and when $a = 4$, we have $S = 117$. Predicted sales of $S = 112.25$ when $a = 3.5$ makes sense because it falls between 105 and 117. See Figure A.4. Of course, if we spent $3500 on advertising, sales would probably not be exactly $112,250. The regression equation allows us to make predictions, but does not provide exact results.

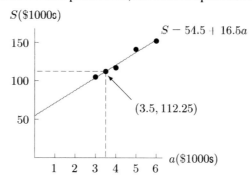

Figure A.4: Predicting sales when spending $3,500 on advertising

Example 1   Predict total sales given advertising expenditures of $4800 and $10,000.

Solution   When $4800 is spent on advertising, $a = 4.8$, so

$$S = 54.5 + 16.5(4.8) = 133.7.$$

Sales are predicted to be $133,700. When $10,000 is spent on advertising, $a = 10$, so

$$S = 54.5 + 16.5(10) = 219.5.$$

Sales are predicted to be $219,500.

Consider the two predictions made in Example 1 at $a = 4.8$ and $a = 10$. We have more confidence in the accuracy of the prediction when $a = 4.8$, because we are *interpolating* within an interval we already know something about. The prediction for $a = 10$ is less reliable, because we are *extrapolating* outside the interval defined by the data values in Table A.2. In general, interpolation is safer than extrapolation.

Interpreting the Slope of the Regression Line

The slope of a linear function is the change in the dependent variable divided by the change in the independent variable. For the sales and advertising regression line, the slope is 16.5. This tells us that S increases by about 16.5 whenever a increases by 1. If advertising expenses increase by \$1000, sales increase by about \$16,500. In general, the slope tells us the expected change in the dependent variable for a unit change in the independent variable.

How Regression Works: What "Best Fit" Means

Figure A.5 illustrates how a line is fitted to a set of data. We assume that the value of y is in some way related to the value of x, although other factors could influence y as well. Thus, we assume that we can pick the value of x exactly but that the value of y may be only partially determined by this x-value.

A calculator or computer finds the line that minimizes the sum of the squares of the vertical distances between the data points and the line. See Figure A.5. The regression line is also called a *least-squares line*, or the *line of best fit*.

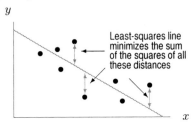

Figure A.5: Data and the corresponding least-squares regression line

Correlation

When a computer or calculator calculates a regression line, it also gives a *correlation coefficient*, r. This number lies between -1 and $+1$ and measures how well the regression line fits the data. If $r = 1$, the data lie exactly on a line of positive slope. If $r = -1$, the data lie exactly on a line of negative slope. If r is close to 0, the data may be completely scattered, or there may be a nonlinear relationship between the variables. (See Figure A.6.)

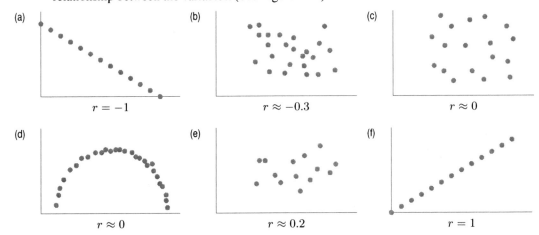

Figure A.6: Various data sets and correlation coefficients

Example 2　The correlation coefficient for the sales data in Table A.2 is $r \approx 0.99$. The fact that r is positive tells us that the regression line has positive slope. The fact that r is close to 1 tells us that the regression line fits the data well.

The Difference Between Relation, Correlation, and Causation

It is important to understand that a high correlation (either positive or negative) between two quantities does *not* imply causation. For example, there is a high correlation between children's reading level and shoe size.[1] However, large feet do not cause a child to read better (or vice versa). Larger feet and improved reading ability are both a consequence of growing older.

Notice also that a correlation of 0 does not imply that there is no relationship between x and y. For example, in Figure A.6(d) there is a relationship between x and y-values, while Figure A.6(c) exhibits no apparent relationship. Both data sets have a correlation coefficient of $r \approx 0$. Thus, a correlation of $r = 0$ usually implies there is no linear relationship between x and y, but this does not mean there is no relationship at all.

Regression When the Relationship Is Not Linear

Table A.3 shows the population of the US (in millions) from 1790 to 1860. These points are plotted in Figure A.7. Do the data look linear? Not really. It appears to make more sense to fit an exponential function than a linear function to this data. Finding the exponential function of best fit is called *exponential regression*. One algorithm used by a calculator or computer gives the exponential function that fits the data as

$$P = 3.9(1.03)^t,$$

where P is the US population in millions and t is years since 1790. Other algorithms may give different answers. See Figure A.8.

Since the base of this exponential function is 1.03, the US population was increasing at the rate of about 3% per year between 1790 and 1860. Is it reasonable to expect the population to continue to increase at this rate? It turns out that this exponential model does not fit the population of the US well beyond 1860. In Section 4.7, we see another function that is used to model the US population.

Table A.3   *US Population in millions, 1790–1860*

Year	1790	1800	1810	1820	1830	1840	1850	1860
Population	3.9	5.3	7.2	9.6	12.9	17.1	23.1	31.4

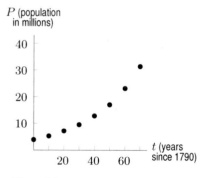

Figure A.7: US Population 1790–1860

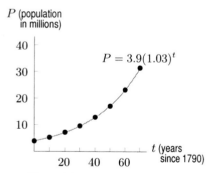

Figure A.8: US population and an exponential regression function

Calculators and computers can do linear regression, exponential regression, logarithmic regression, quadratic regression, and more. To fit a formula to a set of data, the first step is to graph the data and identify the appropriate family of functions.

---

[1] From *Statistics*, 2nd ed., by David Freedman, Robert Pisani, Roger Purves, Ani Adhikari, p. 142 (New York: W.W.Norton, 1991).

Example 3   The average fuel efficiency (miles per gallon of gasoline) of US automobiles declined until the 1960s and then started to rise as manufacturers made cars more fuel efficient.[2] See Table A.4.

(a) Plot the data. What family of functions should be used to model the data: linear, exponential, logarithmic, power function, or a polynomial? If a polynomial, state the degree and whether the leading coefficient is positive or negative.

(b) Use quadratic regression to fit a quadratic polynomial to the data; graph it with the data.

Table A.4   *What function fits these data?*

Year	1940	1950	1960	1970	1980	1986
Average miles per gallon	14.8	13.9	13.4	13.5	15.5	18.3

Solution

(a) The data are shown in Figure A.9, with time t in years since 1940. Miles per gallon decreases and then increases, so a good function to model the data is a quadratic (degree 2) polynomial. Since the parabola opens up, the leading coefficient is positive.

(b) If $f(t)$ is average miles per gallon, one algorithm for quadratic regression tells us that the quadratic polynomial that fits the data is

$$f(t) = 0.00617t^2 - 0.225t + 15.10.$$

In Figure A.10, we see that this quadratic does fit the data reasonably well.

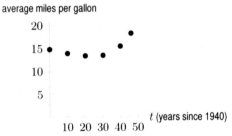

Figure A.9: Data showing fuel efficiency of US automobiles over time

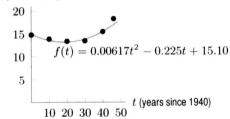

Figure A.10: Data and best quadratic polynomial, found using regression

Problems for Appendix A

1. Table A.5 gives the gross world product, G, which measures global output of goods and services.[3] If t is in years since 1950, the regression line for these data is

$$G = 3.543 + 0.734t.$$

(a) Plot the data and the regression line on the same axes. Does the line fit the data well?

(b) Interpret the slope of the line in terms of gross world product.

(c) Use the regression line to estimate gross world product in 2005 and in 2020. Comment on your confidence in the two predictions.

Table A.5   *G, in trillions of 1999 dollars*

Year	1950	1960	1970	1980	1990	2000
G	6.4	10.0	16.3	23.6	31.9	43.2

2. Table A.6 shows worldwide cigarette production as a function of t, the number of years since 1950.[4]

(a) Find the regression line for this data.

(b) Use the regression line to estimate world cigarette production in the year 2010.

(c) Interpret the slope of the line in terms of cigarette production.

(d) Plot the data and the regression line on the same axes. Does the line fit the data well?

Table A.6   *Cigarette production, P, in billions*

t	0	10	20	30	40	50
P	1686	2150	3112	4388	5419	5564

[2]C. Schaufele and N. Zumoff, *Earth Algebra, Preliminary Version*, p. 91 (New York: Harper Collins, 1993).
[3]The Worldwatch Institute, *Vital Signs* 2001, p. 57 (New York: W.W. Norton, 2001).
[4]The Worldwatch Institute, *Vital Signs* 2001, p. 77 (New York: W.W. Norton, 2001).

3. Table A.7 shows the US Gross National Product (GNP).[5]

  (a) Plot GNP against years since 1970. Does a line fit the data well?

  (b) Find the regression line and graph it with the data.

  (c) Use the regression line to estimate the GNP in 1985 and in 2020. Which estimate do you have more confidence in? Why?

Table A.7   *GNP in 2003 dollars*

Year	1970	1980	1990	2000
GNP (billions)	1045	2824	5838	9856

4. The acidity of a solution is measured by its pH, with lower pH values indicating more acidity. A study of acid rain was undertaken in Colorado between 1975 and 1978, in which the acidity of rain was measured for 150 consecutive weeks. The data followed a generally linear pattern and the regression line was determined to be

$$P = 5.43 - 0.0053t,$$

where P is the pH of the rain and t is the number of weeks into the study.[6]

  (a) Is the pH level increasing or decreasing over the period of the study? What does this tell you about the level of acidity in the rain?

  (b) According to the line, what was the pH at the beginning of the study? At the end of the study ($t = 150$)?

  (c) What is the slope of the regression line? Explain what this slope is telling you about the pH.

5. In a 1977 study[7] of 21 of the best American female runners, researchers measured the average stride rate, S, at different speeds, v. The data are given in Table A.8.

  (a) Find the regression line for these data, using stride rate as the dependent variable.

  (b) Plot the regression line and the data on the same axes. Does the line fit the data well?

  (c) Use the regression line to predict the stride rate when the speed is 18 ft/sec and when the speed is 10 ft/sec. Which prediction do you have more confidence in? Why?

6. Table A.9 shows the atmospheric concentration of carbon dioxide, CO_2 (in parts per million, ppm), at the Mauna Loa Observatory in Hawaii.[8]

  (a) Find the average rate of change of the concentration of carbon dioxide between 1980 and 2000. Give units and interpret your answer in terms of carbon dioxide.

  (b) Plot the data, and find the regression line for carbon dioxide concentration against years since 1980. Use the regression line to predict the concentration of carbon dioxide in the atmosphere in the year 2020.

Table A.9

Year	1980	1985	1990	1995	2000
CO_2	349.6	346.7	354.4	360.8	368.9

7. In Problem 6, carbon dioxide concentration was modeled as a linear function of time. However, if we include data for carbon dioxide concentration from as far back as 1900, the data appear to be more exponential than linear. (They looked linear in Problem 6 because we were only looking at a small piece of the graph.) If C is the CO_2 concentration in ppm and t is in years since 1900, an exponential regression function to fit the data is

$$C = 272.27(1.0026)^t.$$

  (a) What is the annual percent growth rate during this period? Interpret this rate in terms of CO_2 concentration.

  (b) What CO_2 concentration is given by the model for 1900? For 1980? Compare the 1980 estimate to the actual value in Table A.9.

8. (a) Fit an exponential function to the population data in Table A.10. Plot the data and the exponential function on the same axes.

  (b) At approximately what percentage rate was the population growing between 1960 and 2000?

  (c) If the population continues to grow at the same percentage rate, what population is projected for 2020?

Table A.8   *Stride rate, S, in steps/sec, and speed, v, in ft/sec*

v	15.86	16.88	17.50	18.62	19.97	21.06	22.11
S	3.05	3.12	3.17	3.25	3.36	3.46	3.55

Table A.10   *US Population 1960-2000*

t, years since 1960	0	10	20	30	40
population (m)	179.3	203.3	226.5	248.7	281.4

[5] *The World Almanac and Book of Facts 2005*, p. 111 (New York).

[6] William M. Lewis and Michael C. Grant, "Acid Precipitation in the Western United States," *Science* 207 (1980), pp. 176-177.

[7] R.C. Nelson, C.M. Brooks, and N.L. Pike, "Biomechanical Comparison of Male and Female Distance Runners." *The Marathon: Physiological, Medical, Epidemiological, and Psychological Studies*, ed. P. Milvy, pp. 793–807 (New York: New York Academy of Sciences, 1977).

[8] www.cmdl.noaa.gov/ccgg/iadv, accessed on February 20, 2005.

9. Table A.11 shows the public debt, D, of the US[9] in billions of dollars, t years after 1998.

   (a) Plot the public debt against the number of years since 1998.

   (b) Does the data look more linear or more exponential?

   (c) Fit an exponential function to the data and graph it with the data.

   (d) What annual percentage growth rate does the exponential model show?

   (e) Do you expect this model to give accurate predictions beyond 2004? Explain.

Table A.11

t	0	1	2	3	4	5	6
D	5526	5656	5674	5808	6228	6783	7379

10. A company collects the data in Table A.12. Find the regression line and interpret its slope. Sketch the data and the line. What is the correlation coefficient? Why is the value you get reasonable?

Table A.12   *Cost to produce various quantities of a product*

q (quantity in units)	25	50	75	100	125
C (cost in dollars)	500	625	689	742	893

11. Match the r values with scatter plots in Figure A.11.

$$r = -0.98, \quad r = -0.5, \quad r = -0.25,$$

$$r = 0, \quad r = 0.7, \quad r = 1.$$

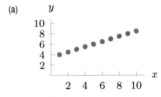
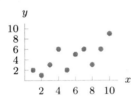
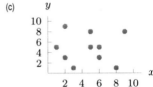
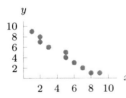
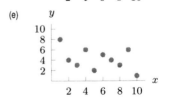

Figure A.11

12. Table A.13 shows the number of cars, N, in millions in the US[10] t years after 1940.

   (a) Plot the data, with number of passenger cars as the dependent variable.

   (b) Does a linear or exponential model appear to fit the data better?

   (c) Use a linear model first: Find the regression line for these data. Graph it with the data. Use the regression line to predict the number of passenger cars in the year 2010 ($t = 70$).

   (d) Interpret the slope of the regression line found in part (c) in terms of passenger cars.

   (e) Now use an exponential model: Find the exponential regression function for these data. Graph it with the data. Use the exponential function to predict the number of passenger cars in the year 2010 ($t = 70$). Compare your prediction with the prediction obtained from the linear model.

   (f) What annual percent growth rate in number of US passenger cars does your exponential model show?

Table A.13   *Number of passenger cars, in millions*

t	0	10	20	30	40	50	60
N	27.5	40.3	61.7	89.2	121.6	133.7	133.6

13. Table A.14 gives the population of the world in billions.

   (a) Plot these data. Does a linear or exponential model seem to fit the data best?

   (b) Find an exponential regression function.

   (c) What annual percent growth rate does the exponential function show?

   (d) Predict the population of the world in the year 2020 and in the year 2050. Comment on the relative confidence you have in these two estimates.

Table A.14   *World population in billions*

Year (since 1950)	0	10	20	30	40	50	58
Population (bn)	2.6	3.0	3.7	4.5	5.3	6.1	6.7

14. In 1969, all field goal attempts in the National Football League and American Football League were analyzed. See Table A.15. (The data has been summarized: all attempts between 10 and 19 yards from the goal post are listed as 14.5 yards out, etc.)

   (a) Graph the data, with success rate as the dependent variable. Discuss whether a linear or an exponential model fits best.

---

[9]*The World Almanac and Book of Facts 2005*, p. 119 (New York).
[10]*The World Almanac and Book of Facts 2005*, p. 237 (New York).

(b) Find the linear regression function; graph it with the data. Interpret the slope of the regression line in terms of football.

(c) Find the exponential regression function; graph it with the data. What success rate does this function predict from a distance of 50 yards?

(d) Using the graphs in parts (b) and (c), decide which model seems to fit the data best.

Table A.15   *Successful fraction of field goal attempts*

Distance from goal, x yards	14.5	24.5	34.5	44.5	52.0
Fraction successful, Y	0.90	0.75	0.54	0.29	0.15

15. Table A.16 shows the number of Japanese cars imported into the US.[11]

(a) Plot the number of Japanese cars imported against the number of years since 1964.

(b) Does the data look more linear or more exponential?

(c) Fit an exponential function to the data and graph it with the data.

(d) What annual percentage growth rate does the exponential model show?

(e) Do you expect this model to give accurate predictions beyond 1971? Explain.

Table A.16   *Imported Japanese cars, 1964–1971*

Year since 1964	0	1	2	3	4	5	6	7
Cars (thousands)	16	24	56	70	170	260	381	704

16. Figure A.12 shows oil production in the Middle East.[12] If you were to model this function with a polynomial, what degree would you choose? Would the leading coefficient be positive or negative?

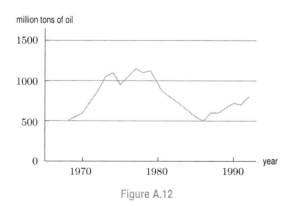

million tons of oil

Figure A.12

17. After the oil crisis in 1973, the average fuel efficiency, E, of cars increased until the early 1990s, when it started to decrease again.

(a) Plot the data [13] in Table A.17, using t in years since 1975. If you were to fit a quadratic polynomial to the data, what would be the sign of the leading coefficient?

(b) Fit a quadratic polynomial and plot it with the data.

Table A.17

Year	1975	1980	1985	1990	1995	2000
E, mpg	13.1	19.2	21.3	21.5	21.1	20.7

18. Table A.18 gives the area of rain forest destroyed for agriculture and development.[14]

(a) Plot these data.

(b) Are the data increasing or decreasing? Concave up or concave down? In each case, interpret your answer in terms of rain forest.

(c) Use a calculator or computer to fit a logarithmic function to this data. Plot this function on the axes in part (a).

(d) Use the curve you found in part (c) to predict the area of rain forest destroyed in 2010.

Table A.18   *Destruction of rain forest*

x (year)	1960	1970	1980	1988
y (million hectares)	2.21	3.79	4.92	5.77

In Problems 19–21, tables of data are given.[15]

(a) Use a plot of the data to decide whether a linear, exponential, logarithmic, or quadratic function fits the data best.

(b) Use regression to find a formula for the function you chose in part (a). If the function is linear or exponential, interpret the rate of change or percent rate of change.

(c) Use your function to predict the value of the function in the year 2015.

(d) Plot your function on the same axes as the data, and comment on the fit.

---

[11] *The World Almanac 1995.*
[12] Lester R. Brown, et al., *Vital Signs*, p. 49 (New York: W. W. Norton and Co., 1994).
[13] *The World Almanac and Book of Facts* (New York, 2005).
[14] C. Schaufele and N. Zumoff, *Earth Algebra, Preliminary Version*, p. 131 (New York: Harper Collins, 1993).
[15] The Worldwatch Institute, *Vital Signs 2007–2008* (New York: W.W. Norton & Company, 2007).

19.

World solar power, S, in megawatts; t in years since 1990

t	0	1	2	3	4	5	6	7
S	47	55	58	60	69	78	89	126
t	8	9	10	11	12	13	14	15
S	153	201	277	386	547	748	1194	1782

20.

Nuclear warheads, N, in thousands; t in years since 1960

t	0	5	10	15	20	25	30	35	40	45
N	22	38	39	48	55	65	59	40	33	28

21.

Carbon dioxide, C, in ppm; t in years since 1970

t	0	5	10	15	20	25	30	35
C	326	331	339	346	354	361	369	380

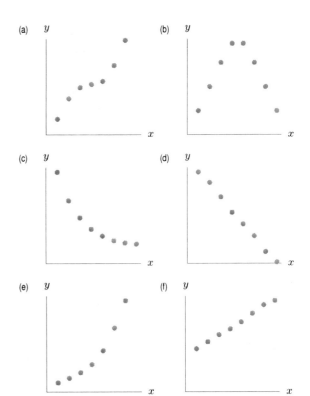

Figure A.13

22. For each graph in Figure A.13, decide whether the best fit for the data appears to be a linear function, an exponential function, or a polynomial.

B   COMPOUND INTEREST AND THE NUMBER e

If you have some money, you may decide to invest it to earn interest. The interest can be paid in many different ways—for example, once a year or many times a year. If the interest is paid more frequently than once per year and the interest is not withdrawn, there is a benefit to the investor since the interest earns interest. This effect is called *compounding*. You may have noticed banks offering accounts that differ both in interest rates and in compounding methods. Some offer interest compounded annually, some quarterly, and others daily. Some even offer continuous compounding.

What is the difference between a bank account advertising 8% compounded annually (once per year) and one offering 8% compounded quarterly (four times per year)? In both cases 8% is an annual rate of interest. The expression 8% *compounded annually* means that at the end of each year, 8% of the current balance is added. This is equivalent to multiplying the current balance by 1.08. Thus, if $100 is deposited, the balance, B, in dollars, will be

$$B = 100(1.08) \quad \text{after one year,}$$
$$B = 100(1.08)^2 \quad \text{after two years,}$$
$$B = 100(1.08)^t \quad \text{after } t \text{ years.}$$

The expression 8% *compounded quarterly* means that interest is added four times per year (every three months) and that $\frac{8}{4} = 2\%$ of the current balance is added each time. Thus, if $100 is deposited, at the end of one year, four compoundings have taken place and the account will contain $100(1.02)^4$. Thus, the balance will be

$$B = 100(1.02)^4 \quad \text{after one year,}$$
$$B = 100(1.02)^8 \quad \text{after two years,}$$
$$B = 100(1.02)^{4t} \quad \text{after } t \text{ years.}$$

Note that 8% is *not* the rate used for each three-month period; the annual rate is divided into four 2% payments. Calculating the total balance after one year under each method shows that

$$\text{Annual compounding:} \quad B = 100(1.08) = 108.00,$$
$$\text{Quarterly compounding:} \quad B = 100(1.02)^4 = 108.24.$$

Thus, more money is earned from quarterly compounding, because the interest earns interest as the year goes by. In general, the more often interest is compounded, the more money will be earned (although the increase may not be very large).

We can measure the effect of compounding by introducing the notion of *effective annual rate*. Since $100 invested at 8% compounded quarterly grows to $108.24 by the end of one year, we say that the effective annual rate in this case is 8.24%. We now have two interest rates that describe the same investment: the 8% compounded quarterly and the 8.24% effective annual rate. We call the 8% the *nominal rate* (nominal means "in name only"). However, it is the effective rate that tells you exactly how much interest the investment really pays. Thus, to compare two bank accounts, simply compare the effective annual rate. The next time you walk by a bank, look at the advertisements, which should (by law) include both the nominal rate and the effective annual rate.

Using the Effective Annual Yield

Example 1   Which is better: Bank X paying a 7% annual rate compounded monthly or Bank Y offering a 6.9% annual rate compounded daily?

Solution   We find the effective annual rate for each bank.

Bank X: There are 12 interest payments in a year, each payment being $0.07/12 = 0.005833$ times the current balance. If the initial deposit were $100, then the balance B would be

$$B = 100(1.005833) \quad \text{after one month,}$$
$$B = 100(1.005833)^2 \quad \text{after two months,}$$
$$B = 100(1.005833)^t \quad \text{after } t \text{ months.}$$

To find the effective annual rate, we look at one year, or 12 months, giving $B = 100(1.005833)^{12} = 100(1.072286)$, so the effective annual rate $\approx 7.23\%$.

Bank Y: There are 365 interest payments in a year (assuming it is not a leap year), each being $0.069/365 = 0.000189$ times the current balance. Then the balance is

$$B = 100(1.000189) \quad \text{after one day,}$$
$$B = 100(1.000189)^2 \quad \text{after two days,}$$
$$B = 100(1.000189)^t \quad \text{after } t \text{ days.}$$

so at the end of one year we have multiplied the initial deposit by

$$(1.000189)^{365} = 1.071413$$

so the effective annual rate for Bank Y $\approx 7.14\%$.

Comparing effective annual rates for the banks, we see that Bank X is offering a better investment, by a small margin.

Example 2   If $1000 is invested in each bank in Example 1, write an expression for the balance in each bank after t years.

Solution   For Bank X, the effective annual rate $\approx 7.23\%$, so after t years the balance, in dollars, will be

$$B = 100\left(1 + \frac{0.07}{12}\right)^{12t} = 1000(1.005833)^{12t} = 1000(1.0723)^t.$$

For Bank Y, the effective annual rate $\approx 7.14\%$, so after t years the balance, in dollars, will be

$$B = 1000\left(1 + \frac{0.069}{365}\right)^{365t} = 1000(1.0714)^t.$$

(Again, we are ignoring leap years.)

> If interest at an annual rate of r is compounded n times a year, then r/n times the current balance is added n times a year. Therefore, with an initial deposit of P, the balance t years later is
> $$B = P\left(1 + \frac{r}{n}\right)^{nt}.$$
> Note that r is the nominal rate; for example, $r = 0.05$ when the annual rate is 5%.

Increasing the Frequency of Compounding: Continuous Compounding

Let us look at the effect of increasing the frequency of compounding. How much effect does it have?

Example 3   Find the effective annual rate for a 7% annual rate compounded
(a)   1000 times a year.   (b)   10,000 times a year.

Solution   (a) In one year, a deposit is multiplied by

$$\left(1 + \frac{0.07}{1000}\right)^{1000} \approx 1.0725056,$$

giving an effective annual rate of about 7.25056%.
(b) In one year, a deposit is multiplied by

$$\left(1 + \frac{0.07}{10,000}\right)^{10,000} \approx 1.0725079,$$

giving an effective annual yield of about 7.25079%.

Notice that there's not a great deal of difference—7.25056% versus 7.25079%—between compounding 1000 times each year (about three times per day) and 10,000 times each year (about 30 times per day). What happens if we compound more often still? Every minute? Every second? Surprisingly, the effective annual rate does not increase indefinitely, but tends to a finite value. The benefit of increasing the frequency of compounding becomes negligible beyond a certain point.

For example, computing the effective annual rate on a 7% investment compounded n times per year for values of n larger than 100,000 gives

$$\left(1 + \frac{0.07}{n}\right)^n \approx 1.0725082.$$

So the effective annual rate is about 7.25082%. Even if you take $n = 1,000,000$ or $n = 10^{10}$, the effective annual rate does not change appreciably. The value 7.25082% is an upper bound that is approached as the frequency of compounding increases.

When the effective annual rate is at this upper bound, we say that the interest is being *compounded continuously*. (The word *continuously* is used because the upper bound is approached by compounding more and more frequently.) Thus, when a 7% nominal annual rate is compounded so frequently that the effective annual rate is 7.25082%, we say that the 7% is compounded *continuously*. This represents the most one can get from a 7% nominal rate.

Where Does the Number e Fit In?

It turns out that e is intimately connected to continuous compounding. To see this, use a calculator to check that $e^{0.07} \approx 1.0725082$, which is the same number we obtained by compounding 7% a large number of times. So you have discovered that for very large n

$$\left(1 + \frac{0.07}{n}\right)^n \approx e^{0.07}.$$

As n gets larger, the approximation gets better and better, and we write

$$\lim_{n \to \infty} \left(1 + \frac{0.07}{n}\right)^n = e^{0.07},$$

meaning that as n increases, the value of $(1 + 0.07/n)^n$ approaches $e^{0.07}$.

If P is deposited at an annual rate of 7% compounded continuously, the balance, B, after t years, is given by

$$B = P(e^{0.07})^t = Pe^{0.07t}.$$

If interest on an initial deposit of P is *compounded continuously* at an annual rate r, the balance t years later can be calculated using the formula

$$B = Pe^{rt}.$$

In working with compound interest, it is important to be clear whether interest rates are nominal rates or effective rates, as well as whether compounding is continuous or not.

Example 4   Find the effective annual rate of a 6% annual rate, compounded continuously.

Solution   In one year, an investment of P becomes $Pe^{0.06}$. Using a calculator, we see that

$$Pe^{0.06} = P(1.0618365).$$

So the effective annual rate is about 6.18%.

Example 5   You invest money in a certificate of deposit (CD) for your child's education, and you want it to be worth \$120,000 in 10 years. How much should you invest if the CD pays interest at a 9% annual rate compounded quarterly? Continuously?

Solution   Suppose you invest P initially. A 9% annual rate compounded quarterly has an effective annual rate given by $(1 + 0.09/4)^4 = 1.0930833$, or 9.30833%. So after 10 years you have

$$P(1.0930833)^{10} = 120,000.$$

Therefore, you should invest

$$P = \frac{120,000}{(1.0930833)^{10}} = \frac{120,000}{2.4351885} = 49,277.50.$$

On the other hand, if the CD pays 9% per year, compounded continuously, after 10 years you have

$$Pe^{(0.09)10} = 120,000.$$

So you would need to invest

$$P = \frac{120,000}{e^{(0.09)10}} = \frac{120,000}{2.4596031} = 48,788.36.$$

Notice that to achieve the same result, continuous compounding requires a smaller initial investment than quarterly compounding. This is to be expected since the effective annual rate is higher for continuous than for quarterly compounding.

Problems for Appendix B

1. A department store issues its own credit card, with an interest rate of 2% per month. Explain why this is not the same as an annual rate of 24%. What is the effective annual rate?

2. A deposit of $10,000 is made into an account paying a nominal yearly interest rate of 8%. Determine the amount in the account in 10 years if the interest is compounded:

   (a) Annually   **(b)** Monthly   **(c)** Weekly
   (d) Daily   **(e)** Continuously

3. A deposit of $50,000 is made into an account paying a nominal yearly interest rate of 6%. Determine the amount in the account in 20 years if the interest is compounded:

   (a) Annually   **(b)** Monthly   **(c)** Weekly
   (d) Daily   **(e)** Continuously

4. Use a graph of $y = (1 + 0.07/x)^x$ to estimate the number that $(1 + 0.07/x)^x$ approaches as $x \to \infty$. Confirm that the value you get is approximately $e^{0.07}$.

5. **(a)** Find $(1 + 0.04/n)^n$ for $n = 10{,}000$, and $100{,}000$, and $1{,}000{,}000$. Use the results to predict the effective annual rate of a 4% annual rate compounded continuously.
   (b) Confirm your answer by computing $e^{0.04}$.

6. Find the effective annual rate of a 6% annual rate, compounded continuously.

7. What nominal annual interest rate has an effective annual rate of 5% under continuous compounding?

8. What is the effective annual rate, under continuous compounding, for a nominal annual interest rate of 8%?

9. **(a)** Find the effective annual rate for a 5% annual interest rate compounded n times/year if

   (i) $n = 1000$ (ii) $n = 10{,}000$
   (iii) $n = 100{,}000$

   (b) Look at the sequence of answers in part (a), and predict the effective annual rate for a 5% annual rate compounded continuously.
   (c) Compute $e^{0.05}$. How does this confirm your answer to part (b)?

10. A bank account is earning interest at 6% per year compounded continuously.

    (a) By what percentage has the bank balance in the account increased over one year? (This is the effective annual rate.)
    (b) How long does it take the balance to double?
    (c) For a continuous interest rate r, find a formula for the doubling time in terms of r.

11. Explain how you can match the interest rates (a)–(e) with the effective annual rates I–V without calculation.

    (a) 5.5% annual rate, compounded continuously.
    (b) 5.5% annual rate, compounded quarterly.
    (c) 5.5% annual rate, compounded weekly.
    (d) 5% annual rate, compounded yearly.
    (e) 5% annual rate, compounded twice a year.

    I.   5%   II.   5.06%   III.   5.61%
    IV.   5.651%   V.   5.654%

Countries with very high inflation rates often publish monthly rather than yearly inflation figures, because monthly figures are less alarming. Problems 12–13 involve such high rates, which are called *hyperinflation*.

12. In 1989, US inflation was 4.6% a year. In 1989 Argentina had an inflation rate of about 33% a month.

    (a) What is the yearly equivalent of Argentina's 33% monthly rate?
    (b) What is the monthly equivalent of the US 4.6% yearly rate?

13. Between December 1988 and December 1989, Brazil's inflation rate was 1290% a year. (This means that between 1988 and 1989, prices increased by a factor of $1 + 12.90 = 13.90$.)

    (a) What would an article which cost 1000 cruzados (the Brazilian currency unit) in 1988 cost in 1989?
    (b) What was Brazil's monthly inflation rate during this period?

C SPREADSHEET PROJECTS

The following projects require the use of a spreadsheet. They can all be done using only the ideas in Chapter 1. In addition, Project 9 (Verhulst: The Logistic Model) and Project 10 (The Spread of Information) give another perspective on material in Chapters 4 and 10. Project 4 (Comparing Home Mortgages) uses a geometric series but can be done before Chapter 11.

1. MALTHUS: POPULATION OUTSTRIPS FOOD SUPPLY

In this project, we compare exponential and linear growth. We see the eventual dominance of exponential functions over linear functions.

One of the most famous models of population growth was made by Thomas Malthus in the early 19th century. Malthus believed that while human population increased exponentially, its means of subsistence increased linearly. The gloomy conclusion that Malthus drew from this observation was that the population of the earth would inevitably outstrip its means of subsistence, resulting in an inadequate supply of food. (Malthus went on to note that this state of affairs could only be averted by war, famine, epidemic disease, wide-scale sexual restraint, or other such drastic checks on population growth.)

The following table shows part of a spreadsheet showing such as a scenario.[16] The starting population is 1 million, while the available food feeds 2 million people. The population grows at an annual rate of 3% per year and the food production increases by 100,000 per year. These population and food growth rates are in cells on the right of the spreadsheet. The fourth column contains the ratio of available food per person in the population. The spreadsheet includes a safety ratio—so long as the food-to-population ratio is above this figure of 1.5, the fifth column displays "Yes"; whenever the ratio drops below this figure the fifth column displays "No" (as it does by the end of the 21st century)

We see that at first there is plenty of food—the ratio of food to population is 2, which means that there is twice as much food as is necessary to feed the population. For the first few years the ratio increases, but at a certain point it starts to decrease, and eventually the ratio drops below one.

Year	Population	Food supply	Ratio	Above safety ratio?	
1999	1000000	2000000	2.00	Yes	Annual pop growth rate 3.00%
2000	1030000	2100000	2.04	Yes	
2001	1060900	2200000	2.07	Yes	
2002	1092727	2300000	2.10	Yes	Annual food growth rate 100,000
2003	1125509	2400000	2.13	Yes	
2004	1159274	2500000	2.16	Yes	
2005	1194052	2600000	2.18	Yes	Safety ratio 1.5
2006	1229874	2700000	2.20	Yes	
2007	1266770	2800000	2.21	Yes	
2008	1304773	2900000	2.22	Yes	
⋮	⋮	⋮	⋮	⋮	
2098	18658866	11900000	0.64	No	
2099	19218632	12000000	0.62	No	
2100	19795191	12100000	0.61	No	

---

[16]From Graeme Bird.

1. Set up your spreadsheet to look like the one shown in the table, extending it to the year 2100. Virtually every cell must contain a formula — the exceptions being the six cells containing "1999," "1,000,000," "2,000,000," "3.00%," "100,000," and "1.5."

2. (a) About what year is the food-to-population ratio the highest?
   (b) In which year does this ratio reach 1?

3. There are at least two ways to improve upon the current situation: we can lower the population growth rate, or we can increase the food supply.

   (a) What would the population growth rate have to be lowered to, in order to have the food-to-population ratio not reach 1 until the year 2100? (Keep the food supply increasing at 100,000 per year.)

   (b) What would the food supply rate have to be increased to, in order to achieve this same goal, of the ratio not reaching 1 until the year 2100? (Keep the population growth rate at its original 3%.)

4. Using the original scenario, create each of the following charts (both line and column). Extend the charts to the year 2100, so that the point where the population outstrips the food supply is clearly evident.

   (a) Showing population and food, with the years on the horizontal axis.
   (b) Showing only the ratio, with the years on the horizontal axis.

2. CREDIT CARD DEBT

You have a credit card on which you owe $2000. Your credit card company charges a monthly interest rate of 1.5% and requires a minimum monthly payment of 2.5% of your current balance. (This payment scheme is similar to ones used by many credit card companies, but see Question 8.)
    [Note: For Questions 1–2, you will not need a spreadsheet, although you will need a calculator.]

1. If the monthly interest rate is 1.5%, what is the effective annual interest rate?

2. As a rule, the minimum monthly payment required exceeds the interest accrued in a month. (For example, here the minimum monthly payment of 2.5% exceeds the monthly interest charges of 1.5%.) Explain why this should be the case. What would happen to the card balance if the minimum required payment was less than the accrued interest?

Suppose that you decide to pay off your $2000 credit card debt by making only the minimum required payment every month. Assume that you make no further charges to the card, since you're trying to pay it off.

3. Since 2.5% of $2000 is $50, your first monthly payment is $50. Before you do any spreadsheet calculations, guess how long it will take to bring your total balance down from $2000 to less than $50, assuming that you make only the minimum required payment every month. A rough guess is fine; use common sense and explain your reasoning.

4. Over time, your monthly payments, which start at $50, will decrease. Explain why this happens. Does the fact that your monthly payments decrease affect the answer you gave to Question 3?

5. Although you only owe the credit card company $2000, you will end up paying quite a bit more than $2000, due to interest charges. Make your best guess (before making any specific calculations) as to how much, roughly, you will end up having paid the credit card company for your initial $2000 debt.

Set up a spreadsheet showing the number of months since you began paying off your debt, your current balance, the interest due that month, and the payment you make, each in a separate column. Each quantity should be calculated by a formula.

Example: To figure out what formulas you need, recall that your initial balance is $2000, the monthly interest charged is 1.5%, and the minimum required payment is 2.5%. Thus, at the beginning of Month 1, your balance is $2000, since you have paid off nothing yet. Therefore, at the end of Month 1, the interest that you owe is 1.5% of the $2000 balance, or $30. Your minimum payment is 2.5% of the $2000 balance, or $50. Thus, at the beginning of Month 2, your new balance is the old balance of $2000 plus the $30 in interest minus the $50 payment, or $1980. Notice that the figures for Month 2 depend on the figures for Month 1; similarly, the figures for Month 3 depend on Month 2 figures, and so on. Follow this procedure to figure out what formulas you need in each column of your spreadsheet.

Once you have set up a working spreadsheet, answer the following questions.

6. How good was your guess in Question 3? Using your spreadsheet, find out how many months it takes to bring your balance down to less than $50. Was your guess close, or were you surprised by how long it really takes?

7. How good was your guess in Question 5? Using your spreadsheet, figure out exactly how much you pay the credit card company to bring your balance down to less than $50. How does this figure compare to the original debt of $2000?

8. Use your spreadsheet to find out how long it takes to bring your balance down to $0. Or can't you tell? Does there ever come a point when you have exactly paid off your debt? [Hint: Eventually, the minimum monthly payments and the interest charges become unrealistic. In what way are they unrealistic? How do real credit card companies avoid this problem?]

9. Now let's try experimenting with the numbers and see what happens. In each of the following cases, make the appropriate changes to your spreadsheet. Assume that as soon as your balance is under $50, you pay it off in a lump sum.

   (a) If every month you pay $1 more than the minimum required payment, how long does it take to bring your debt down to less than $50? How much do you end up paying to your creditors in total? How much money do you save by using this payment scheme instead of the one in Question 5?

   (b) Your first monthly payment is $50. If you paid $50 every month, instead of the minimum required payment, how long does it take to bring your debt down to less than $50? How much money do you save by using this payment scheme instead of the one in Question 5?

   (c) Recently, many credit card companies have made offers similar to the following: if you transfer your debt from a competitor's card to their card, they will charge you a lower interest rate. Suppose you find a credit card company willing to make this transaction, and that their monthly interest rate is 1%, not 1.5%. Leaving all the other original assumptions unchanged, how long does it take to bring your debt down to less than $50, and how much do you pay your creditors in total? How much money do you save compared to what you would have paid your original card company?

10. Comparing the scheme used in Questions 3–5 with each of the schemes in Question 9, what conclusions can you reach about paying off a credit card debt?

3. CHOOSING A BANK LOAN

A local bank offers the following loan packages. Use a spreadsheet to decide which option is the best. The packages are as follows:

- A loan of $2000, at an annual rate of 9%, payable in 24 monthly installments.
- A loan of $2000, at an annual rate of 10%, payable in 36 monthly installments.
- A loan of $2000, at an annual rate of 9.25%, payable in 52 biweekly installments.

Interest is compounded with the same frequency as payments are made. Notice that the first and last loans have two-year payoff periods; the middle loan is for three years.

1. Use a spreadsheet to decide which loan is cheapest in terms of total payoff to the bank. (See the following hint.)

2. Use a spreadsheet to decide which loan is easiest to afford in terms of lowest monthly payment. (See the following hint.)

Hint: The difficult part of Questions 1 and 2 is figuring out your monthly (or biweekly) payments. There are formulas that give the payment based on the period and amount of the loan and the interest charged, but instead of using them, we will use a spreadsheet. The idea is that you can make an educated guess as to what the payment ought to be, and then use a spreadsheet to check your answer. By looking at the spreadsheet, you can decide whether your guess was too high or too low, and thus improve upon your original guess. It's surprising how quickly you can zero in on the required monthly payment, down to the nearest penny, by this guess-and-check method.

For example, consider the first loan, the two-year $2000 loan at 9%. Set up a spreadsheet with the initial balance of $2000, the interest for the first month, which is $(9\%/12) \cdot \$2000 = \$15$, and a guess at the monthly payment. There are lots of ways to make a guess at the monthly payment. One way is to say that if you were to borrow $2000 for 2 years at 9%, you would owe about $\$2000(1.09)^2 = \$2376$. (Never mind the monthly compounding—this is just a rough approximation.) To pay off this amount in 24 equal monthly payments would require $\$2376/24 = \$99$. So, we guess a monthly payment of $100. Using this guess, the second month's balance will be

$$(\$2000) + (\$15 \text{ in interest for month } 1) - (\text{Payment of } \$100) = \$1915.$$

Thus, the next month's interest will be $(9\%/12) \cdot \$1915$, and the next month's payment should be the same as the first month's payment, or $100. Continue this process until 24 months' (two years') worth of payments have been made. You will see that the final balance is negative, meaning you paid the bank more than you really owed. This means that $100 is too high a monthly payment to pay off your $2000 loan. (We could have predicted that this was the case when we made our estimate above. Do you see why?) So, since $100 is too high, you might guess that an $80 monthly payment would be right. If you do, you'll see that you'd still owe the bank some money after 24 months had passed. This tells you that $80 is too low a monthly payment, and that the actual payment is somewhere between $80 and $100. This procedure can be repeated until the exact monthly payment is reached.

4. COMPARING HOME MORTGAGES

To do this project, first go to any bank and ask for a fact sheet of their most recent *mortgage loan rates*. Banks are happy to provide them.

Obtain rates for a thirty-year loan, a fifteen-year loan, a thirty-year biweekly loan, and a twenty-year loan (if available) for $100,000 with zero points. (Note: Some loans include points. A point is an additional fee paid to the lender at the time of the loan equal to 1% of the amount borrowed. Typically, you get lower interest rates by paying a point or two. We will only consider loans with zero points.)

The following formula can be used to determine your payment, x:

$$x = \frac{Pr^n(r - 1)}{r^n - 1},$$

where P is the amount of the loan—in this case $100,000—and n is the number of payments. For a thirty-year loan with monthly payments, $n = 360$; for a thirty-year biweekly loan, $n = 780$ (there are 26 payments every year). Finally, r is the interest rate per period plus 1. (For example, if the interest rate is 2%, then $r = 1.02$.) In Question 4, you will derive this formula for x using the following formula for the sum of a geometric series:[17]

$$1 + r + r^2 + \cdots + r^{n-2} + r^{n-1} = \frac{1 - r^n}{1 - r}.$$

---

[17]Geometric series are discussed in detail in Chapter 11.

1. Using the information given, as well as the fact sheet you got from the bank, determine which loan (30-year, 30-year biweekly, 20-year or 15-year) is best if you intend to live in your house for the full term of the loan. Assume the best mortgage is the one that ends up costing you the least overall. (The situation in real life can be more complicated when points and taxes are considered.) Although it's possible to work this problem without a spreadsheet, you might want to set one up anyway.

2. Banks usually require that the monthly payments not exceed some stated fraction of the applicant's monthly income. For this reason, it is generally easier to qualify for loans with smaller monthly payments. Thus, it may be that the "best loan"—the one you found in Question 1— is not the "easiest" loan to qualify for. Which of the loans on your fact sheet has the lowest monthly payment? The highest?

3. Suppose that you expect to sell your house for $145,000 in five years. In this case, which loan should you take? [Hint: The goal here is to maximize profit. Figure out how much money you have paid to the bank after five years, and your remaining debt at that time. When you sell your house, the remaining debt is paid to the bank immediately, so your total profit is (Selling price of house)−(Loan payoff to bank)−(Amount paid to bank during first five years).]

4. Derive the formula for the monthly payment, x. [Hint: Set up a geometric series in terms of x and r to give your loan balance after n months. The loan balance equals 0 when you have paid off your loan; use this fact to solve for x. Simplify the resulting expression (by summing a geometric series) to get the formula given for x.]

5. PRESENT VALUE OF LOTTERY WINNINGS

On Thursday, February 24th, 1993, Bruce Hegarty of Dennis Port, MA, received the first installment of the $26,680,940 prize he won in the Mass Millions state lottery. Mr. Hegarty was scheduled to receive 19 more such installments on a yearly basis. Each check written by the Lottery Commission is for one twentieth of the total prize, or $1,334,047. Why doesn't the Lottery Commission pay all of Mr. Hegarty's prize up front, instead of making him wait for twenty years?

1. Compute the present value of the money paid out by the Lottery Commission, assuming annual discount rates (interest rates) of 5%, 10% and 15%. In each case, what percent does the present value represent of the face value of the prize, $26,680,940?

2. What discount rate would result in a present value of the payments worth only half the face value of the prize?

3. Graph the present value of the payments against the discount rate, ranging from a rate of 0% up to 15%. Describe the graph. What does it tell you about why the Lottery Commission does not pay the prize money up front?

6. COMPARING INVESTMENTS

Consider two investment projects. Project A is built in one year at an initial cost of $10,000. It then yields the following decreasing stream of benefits over a five-year period: $5000, $4000, $3000, $2000, $1000. Project B is built in two years. Initial costs are $10,000 in the first year and $5000 in the second year. It then yields yearly profits of $6000 for the next four years. Which of these investment projects is preferable?

1. Compute the present values of both projects assuming an annual discount rate (interest rate) of 4%. Which project seems preferable? [Hint: Treat expenditures as negative and income as positive.]

2. Compute the present values of both projects assuming an annual discount rate of 16%. Which project seems preferable now?

3. Describe in complete sentences why one of the investment projects is favored by a low discount rate, whereas the other is favored by a high discount rate.

4. The discount rate at which the present value of a project becomes zero is known as the *internal rate of return*. What is the internal rate of return of project A? Of project B? [Hint: Guess different discount rates until you find the one that brings the present value down to $0.]

5. Make a chart of the present value of the two investments against discount rates ranging from 0% to 30%. What features of this chart correspond to the internal rates of return of the two projects?

7. INVESTING FOR THE FUTURE: TUITION PAYMENTS

Parents of two teenagers, ages 13 and 17, deposit a sum of money into an account earning interest at the rate of 7% per year compounded annually. The deposit will be used for a series of eight annual college tuition payments of $10,000 each. Payments out of the account will begin one year after the initial deposit.

1. Use a spreadsheet to model the savings account that the parents opened. At the end of every year, the account earns 7% interest, and then there is a $10,000 withdrawal. Determine what initial deposit provides just enough money to make the eight yearly payments of $10,000. Do this by guessing different values, and seeing which value leaves you with nothing exactly nine years later.

2. Having answered Question 1, use a spreadsheet to compute the present value of eight yearly payments of $10,000 each, beginning one year in the future, at a discount rate of 7%.

3. Compare your answer to Question 1 with your answer to Question 2. Is this a coincidence? Discuss.

4. Suppose the parents have only $50,000 to deposit into the savings account. What annual interest rate must the account earn if the eight payments of $10,000 are to be made? [Hint: Compute the present value of the payments for various discount rates.]

8. NEW OR USED?

You are deciding whether to buy a new or used car (of the same make) and how many years to keep the car. You want to minimize your total costs, which consist of two parts: the loss in value of the car and the repairs. A new car costs $20,000 and loses 20% of its value each year. Repairs are $400 the first year and increase by 25% each subsequent year.

1. Set up a spreadsheet that gives, for each year, the value of the car, its loss in value, its repair costs, and the total cost for that year. The first two lines look like this, rounded to the nearest dollar:

Year	Value	Loss	Repair	Cost
1	20,000	4000	400	4400
2	16,000	3200	500	3700
⋮	⋮	⋮	⋮	⋮

2. Which year has the lowest cost?

3. You intend to keep your car 5 years. Compare your total costs for a new car and for a two-year-old car.

4. How old a car should you buy if you plan to keep it for 5 years?

5. Add two columns to your spreadsheet showing the average yearly cost for second-hand cars of different ages kept for 4 years and 5 years. Which is the best buy?

6. A new car costs $30,000 and loses 25% of its value each year; repairs start at $500 and increase at 10% per year. If you buy a seven-year-old car, should you keep it for 4 or 5 years? What is the average yearly cost in each case?

7. A car costs $30,000 when new. You buy a four-year-old car and keep it for 5 years; repairs are as in Problem 6. Find the rate at which it loses value if the average yearly cost is $2300.

9. VERHULST: THE LOGISTIC MODEL

The relative growth rate of a population, P, over a time interval, Δt, is given by

$$\text{Relative growth rate} = \frac{1}{P} \cdot \frac{\Delta P}{\Delta t}.$$

In exponential growth, the relative growth rate is a constant. Although exponential growth is often used to model populations, this model predicts that a population will increase without limit, which is unrealistic. In the 1830s, a Belgian mathematician, P. F. Verhulst, suggested the *logistic model*, in which the relative growth rate of a population decreases to 0 linearly as the population increases. Verhulst's model predicts that the population size eventually levels off to a value known as the *carrying capacity*.

To see how Verhulst's logistic model works, assume that a pair of breeding rabbits is introduced onto a small island with no rabbits. At the outset the rabbit population doubles every month. This means that initially the relative growth rate is 100% per month. Eventually, though, as the population grows, the relative growth rate drops down to 0% per month. Suppose that the growth rate reaches 0 when the population reaches 10,000 rabbits. (Thus, 10,000 is the carrying capacity of the island.) Using spreadsheets, we will model the rabbit population over time.

1. Let P be the population and r be the relative growth rate per month. Verhulst assumed that the relative growth rate decreases linearly as population increases. This means that r goes from 100% to 0% as P goes from 0 to 10,000 rabbits. Explain why the following formula for r corresponds to Verhulst's assumptions: $r = 0.0001(10{,}000 - P)$.

2. Use a spreadsheet to model the monthly rabbit population on the island for the first two years (24 months). Graph the rabbit population over time. Describe the behavior of the rabbit population. [Hint: Start with two rabbits, and compute the growth rate using the formula in Question 1. Then, for each month, update the rabbit population as well as the relative growth rate.]

3. Draw a graph comparing your logistic model of the rabbit population to a population growing exponentially at a constant relative growth rate of 100% per month. Both models should start out with two rabbits. Describe the similarities and the differences between the two charts. What advantages does the logistic model have over the exponential model? (You'll have to be careful when setting the chart parameters; otherwise, all you'll be able to see is the exponential population, which climbs so quickly that the logistic one will not be visible at all.)

4. The key to the logistic model is that the relative growth rate is decreasing linearly as the population increases. However, this does not mean that the relative growth rate is decreasing linearly over time. Make a chart of the relative growth rate against time for the first two years. Describe the behavior of the relative growth rate over time.

5. (a) Different assumptions about the growing rabbit population lead to different logistic curves. In Question 1, we assumed that the relative growth rate was 100% initially, dropping to 0% when the population reached 10,000. This led to the formula $r = 0.0001(10000 - P)$. Now assume that the initial relative growth rate is 10% (instead of 100%). What is the new formula relating r and P? (The carrying capacity is still 10,000, so $r = 10\%$ when $P = 0$, and r decreases to 0% as P increases to 10,000.)

(b) Using your new formula for the relative growth rate, r, let's see how different initial populations of rabbits lead to different logistic curves. Model the following scenarios, over a 5-year (60-month) period: a population starting at 100 rabbits, a population starting at 5,000 rabbits, a population starting at 12,500 rabbits, and a population starting at 17,500 rabbits. Place all of your data on the same chart. What happens to the rabbit population when it starts out above the island's rabbit carrying capacity? Why does this make sense?

10. THE SPREAD OF INFORMATION: A COMPARISON OF TWO MODELS

The spread of information through a population is important to policy makers. For example, agricultural ministries use mathematical models to understand the spread of technical innovations or new seed types through their countries.

In this project, you will compare two different models—one of them logistic —for the spread of information. In both cases, assume that the population is 10,000 and that initially only 100 people have the information. Let N be the number of people who have the information at time t.

Model 1: If the information is spread by mass media (TV, radio, newspapers), the absolute rate, $\Delta N/\Delta t$, at which the information is spread is assumed to be proportional to the number of people *not* having the information at that time. If t is in days, the constant of proportionality is 10%. For example, on the first day, the number of people not having the information is $10{,}000 - 100 = 9900$. Since 10% of 9900 is 990, the rate of spread of information is 990 people per day on the first day. This means that on the second day, the number of people not having the information is 8910 and that the rate of spread is 10% of 8910, or 891 people per day, and so on.

Model 2: If instead the information is spread by word of mouth, the absolute rate at which information is spread is assumed to be proportional to the product of the number of people who know and the number of people who do not know. If t is in days, the constant of proportionality is 0.002%. For example, on the first day, the product of the number of people who know and the number of people who do not know is $100 \cdot 9900 = 990{,}000$. Since 0.002% of 990,000 is about 20, the rate of spread of information is 20 people per day on the first day. This means that on the second day, the product of the number of people who know and the number who do not know is $120 \cdot 9880 = 1{,}188{,}600$, giving a rate of spread of 0.002% of 1,188,600 or about 24 people per day, and so on.

1. Using spreadsheets, compare the spread of information throughout the population using both models. Make a chart comparing both models' predictions of the number of people over time who have the information. Describe the similarities and differences between the two models. Why is Model 1 used when mass media are present? Why is Model 2 used when mass media are absent?

2. Which of the two models is logistic? How can you tell? What type of growth does the other model exhibit? How can you tell?

3. By definition, a population exhibits logistic growth if its relative rate of change is a decreasing linear function of the current population. Explain why Model 2 leads to logistic growth although it was defined in terms of absolute growth rates.

4. The solutions from our spreadsheets are only approximations. Discuss why this is the case. [Hint: There's more going on here than rounding error.]

11. THE FLU IN WORLD WAR I

During World War I, a particularly lethal form of flu killed about 40 million people around the world.[18] The epidemic started in an army camp of 45,000 soldiers outside of Boston, where the first soldier fell sick on September 7, 1918. In this problem you will make a spreadsheet for the SIR

---

[18]"Capturing a Killer Flu Virus," J. Taukenberger, A. Reid, T. Fanning, in *Scientific American*, Vol. 292, No. 1, Jan. 2005.

model of the 1918 flu outbreak. Starting from the initial values S_0 and I_0, for any increment in time Δt, the changes in the number of susceptibles and infecteds are approximated by

$$\Delta S \approx -aSI\Delta t$$
$$\Delta I \approx aSI\Delta t - bI\Delta t.$$

1. Choosing $\Delta t = 0.1$, $a = 0.0003$, $b = 10$, make a spreadsheet whose first few lines look like this:

t	S	I	ΔS	ΔI
0	44999	1	-1.3500	0.3500
0.1	44997.65	1.3500	-1.82240	0.4724
0.2	44995.828	1.8224	...	...

2. How many soldiers got sick on the fifth day? How many were susceptible on this day?

3. Alter the spreadsheet so that it accepts any values of $\Delta t, a, b$ input by the user.

4. Using the values $a = 0.000267$, $b = 9.865$ for the 1918 epidemic, decrease the value of Δt until a stable estimate is reached for the number of soldiers sick on September 16th. How many soldiers had been infected by this date?

5. Approximately how long did it take for the 1918 epidemic to run its course?

ANSWERS TO ODD-NUMBERED PROBLEMS

Section 1.1

1 (a) (IV)
  (b) (II)
  (c) (III)

3 Argentina produced 9 million metric tons of wheat in 2002

5 population

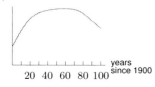

years since 1900

7 $f(5) = 13$

9 $f(5) = 3$

11 $f(5) = 4.1$

13 (b) CFC consumption in 1987
   (c) Year CFC consumption is zero

17

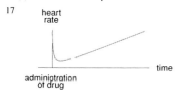

19

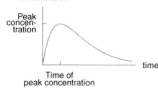

21

23 (a) (III)
   (b) Potato's temperature before put in oven

25 miles per gallon

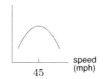

Section 1.2

1 Slope: $-12/7$
  Vertical intercept: $2/7$

3 Slope: 2
  Vertical intercept: $-2/3$

5 $y = (1/2)x + 2$

7 $y = (1/2)x + 2$

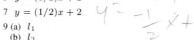

9 (a) l_1
  (b) l_3
  (c) l_2
  (d) l_4

11 (a) $P = 30,700 + 850t$
   (b) 39,200 people
   (c) In 2016

13 (a) $C_1 = 40 + 0.15m$
      $C_2 = 50 + 0.10m$

  (b) C (cost in dollars)

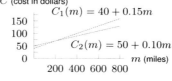

  (c) For distances less than 200 miles, C_1 is cheaper.
      For distances more than 200 miles, C_2 is cheaper.

15 (a) 1.8 billion dollars/year
   (b) 19.1 billion dollars
   (c) 28.1 billion dollars
   (d) 2011

17 (a) Linear
   (b) Linear
   (c) Not linear

19 (a) $q = -(1/3)p + 8$
   (b) $p = -3q + 24$

21 (a) $P = 11.3 + 0.4t$
   (b) 13.7%
   (c) 1.4%

23 (b) $P = 100 - 0.5d$
   (c) -0.5%/ft
   (d) 100%; 200 ft

25 (a) $C = 3.68 + 0.12w$
   (b) 0.12 $/gal
   (c) $3.68

27 (a) $\Delta w/\Delta h$ constant
   (b) $w = 5h - 174$; 5 lbs/in
   (c) $h = 0.2w + 34.8$; 0.2 in/lb

29 (c)

31 (a) 60, 40 years
   (b) (ii)
   (c) 6.375 beats/minute more under new formula

33 No

Section 1.3

1 Concave down

3 Concave up

5 Increases by 12.5%

7 Decreases by 6%

9 Decreasing
  Concave up

11

13 -3

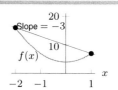

15 (a) 90 million bicycles
   (b) 1.8 million bicycles per year

17 (a) 115,000 people/year
   (b) 0.07, 0.08, 0.41, 0.06
   (c) 115,000 people/year

19 (a) $18,280 million
   (b) $3656 million per year

21 1 meter/sec

23 $72/7 = 10.286$ cm/sec

25 (a) Negative
   (b) Positive
   (c) Negative
   (d) Negative
   (e) Positive

27 1490 thousand people/year
   912.9 thousand people/year
   1879 thousand people/year

29 (a) $-$35 billion dollars
   (b) $-$7 billion dollars per year
   (c) Yes; 2006–2007, 2007–2008

31 (a) Negative
   (b) -0.087 mg/hour

33 15.468, 57.654, 135.899, 146.353, 158.549 people/min

35 (a) -11 cm/sec
   (b) -5.5 (cm/sec)/kg

37 (a) Concave up; no
   (b) 2.6 m/sec

39 Decreasing, concave down

41 (a)

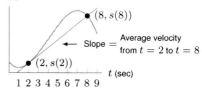

  (b) Between $t = 3$ and $t = 6$
  (c) Negative

43 The change in 1800-1810

45 The increase from $100,000 to $500,000

47 Decrease 12.8%

49

Year	2005	2006	2007	2008
Inflation	4.0%	2.1%	4.3%	0.0%

Section 1.4

1 (a) When more than roughly 335 items are produced and sold
   (b) About $650

3 (a) $75; $7.50 per unit
   (b) $150

5 (a) Price $12, sell 60
   (b) Decreasing

7 Vertical intercept: $p = 4$ dollars
   Horizontal intercept: $q = 6$ units

9 (a) $C(q) = 5000 + 30q$
      $R(q) = 50q$
   (b) $30/unit, $50/unit
   (c)

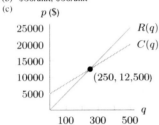

   (d) 250 chairs and $12,500

11 (a) $4000
   (b) $2
   (c) $10
   (d)

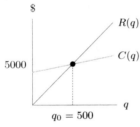

   (e) 500

13 (a) When there are more than 1000 customers
   (b)

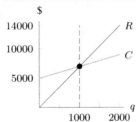

15 (a) $C(q) = 650,000 + 20q$
      $R(q) = 70q$
      $\pi(q) = 50q - 650,000$
   (b) $20/pair, $70/pair, $50/pair
   (c) More than 13,000 pairs

17 (a) Between 20 and 60 units
   (b) About 40 units

19 (a) $V(t) = -1500t + 15,000$
   (b) $V(3) = $10,500

21 (a) $V(t) = -2000t + 50,000$

(b)

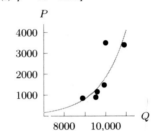

   (c) (0 years, $50,000) and (25 years, $0)

23 (a) $p = $10; q = 3000$
   (b) Suppliers produce 3500 units;
       Consumers buy 2500
   (c) Suppliers produce 2500 units;
       Consumers buy 3500

25 (a) $C = 5q + 7000$
      $R = 12q$
   (b) $q = 1520, \pi(12) = 3640
   (c) $C = 17,000 - 200p$
      $R = 2000p - 40p^2$
      $\pi(p) = -40p^2 + 2200p - 17,000$
   (d) At $27.50 per shirt the profit is $13,250

27 (a) $q = 820 - 20p$
   (b) $p = 41 - 0.05q$

29

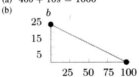

31 (a) $40b + 10s = 1000$
   (b)

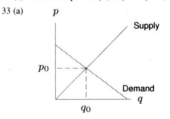

   (c) The intercepts are $(0, 25)$ and $(100, 0)$

33 (a)

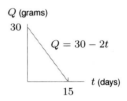

   (b) Equilibrium price will increase;
       equilibrium quantity will decrease
   (c) Equilibrium price and quantity will decrease

35 $q = 4p - 28$

37 (a) $p = 100, q = 500$
   (b) $p = 102, q = 460$
   (c) Consumer pays $2
       Producer pays $4
   (d) $2760

39 (a) Demand: $q = 100 - 2p$
       Supply: $q = 2.85p - 50$

(b) New equilibrium price $p \approx 30.93
    New equilibrium quantity $q \approx 38.14$ units
(c) Consumer pays $0.93
    Producer pays $0.62
    Total $1.55
(d) $59.12

Section 1.5

1 (a) (i), 12%
   (b) (ii), 1000
   (c) Yes, (iv)

3 (a) II
   (b) I
   (c) III
   (d) V

5 (a) $G = 310(1.03)^t$
   (b) $G = 310 + 8t$

7 (a) $Q = 30 - 2t$

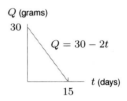

   (b) $Q = 30(0.88)^t$

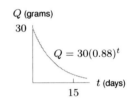

9 (a) $A = 50(0.94)^t$
   (b) 11.33 mg
   (c)

   (d) About 37 hours

11 CPI $= 211(1.028)^t$

13 (a)

x	0	1	2	3
e^x	1	2.72	7.39	20.09

(b)

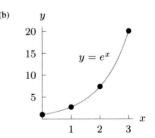

(c)

x	0	1	2	3
e^{-x}	1	0.37	0.14	0.05

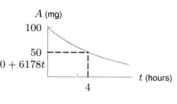

(d)

$y = e^{-x}$

15 $f(x) = 4.30(1.4)^x$

17 $y = 500(1.59)^t$

19 About 6.97 billion; close

21 22.6% per year

23 (a) $W = 40,300 + 16,178t$
 (b) $W = 40,300(1.25)^t = 40,300e^{0.22t}$
 (c)

W (Mw) $W = 40,300(1.25)^t$

200,000

100,000   (5, 121,188)

40,300

$W = 40,3000 + 6178t$

t (years since 2003)

2 4 6 8 10

25 (a) Neither
 (b) Exponential
  $s(t) = 30.12(0.6)^t$
 (c) Linear:
  $g(u) = -1.5u + 27$

27 Min. wage grew 4.69% per year

29 (a) 125%
 (b) 9 times

31 $d = 670(1.096)^{h/1000}$

Section 1.6

1 $t = (\ln 7)/(\ln 5) \approx 1.209$

3 $t = (\ln 2)/(\ln 1.02) \approx 35.003$

5 $t = (\ln 5)/(\ln 3) \approx 1.465$

7 $t = \ln 10 \approx 2.3026$

9 $t = (\ln 100)/3 \approx 1.535$

11 $t = 30.54$

13 $t = (\ln B - \ln P)/r$

15 $t = \ln 8 - \ln 5 \approx 0.47$

17 5; 7%

19 15; -6% (continuous)

21 (a) D
 (b) C
 (c) B

23 $P = 15(1.2840)^t$; growth

25 $P = P_0(1.2214)^t$; growth

27 $P = 15e^{0.4055t}$

29 $P = 174e^{-0.1054t}$

31 $P = 6.4e^{0.01252t}$

33 (a) $5 million; $3.704 million dollars
 (b) 4.108 years

35 (a) 12%
 (b) $P = 25e^{-0.128t}$, 12.8%

37 (a) $P = 5.4(1.034)^t$
 (b) $P = 5.4e^{0.0334t}$
 (c) Annual = 3.4%
  Continuous = 3.3%

39 9.53%

41 2009

Section 1.7

1 (a) $W = 18000e^{0.27t}$
 (b) About 2010

3 $14,918.25

5 About 11.6 years

7 About 10.24 years

9 (a) $5068.93
 (b) $4878.84

11 (a) $A = 100e^{-0.17t}$
 (b) $t \approx 4$ hours

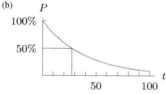

A (mg)

100

50

t (hours)

4

(c) $t = 4.077$ hours

13 A: continuous
 B: annual
 $20

15 (a) 47.6%
 (b) 23.7%

17 8.45%

19 (a) $P(t) = (0.975)^t$
 (b)

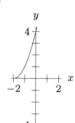

P

100%

50%

t

50   100

(c) About 27 years
 (d) About 8%

21 (a) 4 years
 (b) 4 years

23 About 173 hours

25 96.34 years

27 0.0345; 191 million tons

29 (a) (i) About 4.07 billion dollars
   (ii) About 7.98 trillion dollars
 (b) 1803

31 It is a fake

33 $35,365.34

35 $14,522.98

37 (a) Option 1
 (b) $2102.54, $2051.27, $2000
 (c) $2000, $1951.23, $1902.46

39 (a) Option 1
 (b) Option 1: $10.929 million;
  Option 2: $10.530 million

41 Loan

43 No

Section 1.8

1 (a) $h^2 + 6h + 11$
 (b) 11
 (c) $h^2 + 6h$

3 (a) 4

(b) 2
(c) $(x + 1)^2$
(d) $x^2 + 1$
(e) $t^2(t + 1)$

5 (a) e
 (b) e^2
 (c) e^{x^2}
 (d) e^{2x}
 (e) $e^t t^2$

7 (a) 9
 (b) 20
 (c) 25
 (d) 11

9 (a) $10x^2 + 3$
 (b) $20x^2 + 60x + 45$
 (c) $4x + 9$

11 (a) 3
 (b) 4
 (c) 11
 (d) 8
 (e) 12

13 (a) $y = 2^u$, $u = 3x - 1$
 (b) $P = \sqrt{u}$, $u = 5t^2 + 10$
 (c) $w = 2\ln u$, $u = 3r + 4$

15 $2z + 1$

17 $2zh - h^2$

19 0.4

21 -0.9

23 $3(y - 1)^3$ $(y - 1)^2$

25 18

27 Can't be done

29   (a)

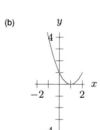

y

4

x

-2 2

-4

(b)

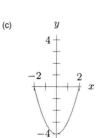

y

4

x

-2 2

-4

(c)

y

4

x

-2 2

-4

498

(d)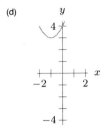

(e)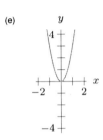

(f)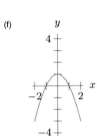

31 (a)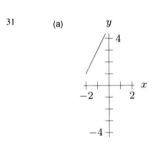

(b)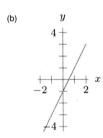

(c)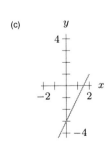

(d)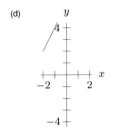

(e)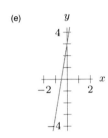

(f)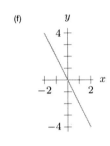

33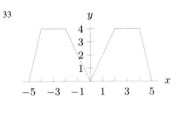

35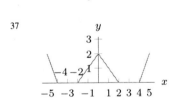

37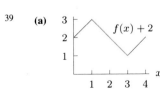

39 (a)

(b)

(c)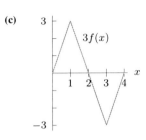

(d)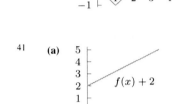

41 (a)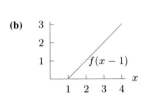

(b)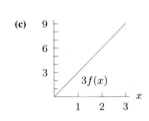

(c)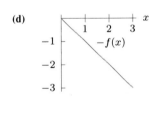

(d)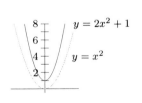

43 (a) $y = 2x^2 + 1$

(b) $y = 2(x^2 + 1)$
(c) No

380

Section 1.9

1 $y = 5x^{1/2}$

3 Not a power function.

5 $y = 9x^{10}$

7 Not a power function

9 $y = 8x^{-1}$

11 Not a power function

13 $S = kh^2$

15 $v = d/t$

17 $N = kA^{1/4}$, with $k > 0$,
Increasing, concave down

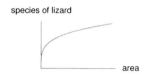

species of lizard

area

19 (a) $y = (x - 2)^3 + 1$
   (b) $y = -(x + 3)^2 - 2$

21 Yes; $k \approx 0.0087$

23 $N = k/L^2$; small

25 (a) $T = kB^{1/4}$
   (b) $k = 17.4$
   (c) 50.3 seconds

27 (a) $N = kP^{0.77}$
   (b) A has 5.888 times more than B
   (c) Town

29 (a) 0.5125
   (b) 0.3162
   (c) 201,583 dynes/cm^2

31 (a) $C = 115,000 - 700p$
      $R = 3000p - 20p^2$
   (b)

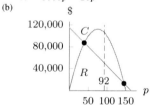

   (d) When it charges between \$40 and \$145
   (e) About \$92

Section 1.10

1 (b) Max at 7 pm; min at 9 am
   (c) Period = one day; amplitude $\approx 2°C$

3 sunscreen sales

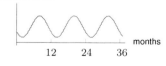

months

5 Amplitude = 3; Period = π

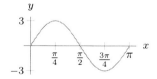

7 Amplitude = 4; Period = π

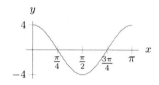

9 Amplitude = 1; Period = π

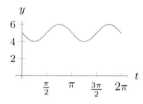

11 (a) 5
   (b) 8
   (c) $f(x) = 5\cos((\pi/4)x)$

13 $H = -16\cos((\pi/6)t) + 10$

15 $0.35\cos(2\pi t/5.4) + 4$

17 $y = 7\sin(\pi t/5)$

19 $f(x) = 5\cos(x/3)$

21 $f(x) = -4\sin(2x)$

23 $f(x) = -8\cos(x/10)$

25 $f(x) = 5\sin((\pi/3)x)$

27 $f(x) = 3 + 3\sin((\pi/4)x)$

29 Depth = $7 + 1.5\sin(\pi t/3)$

31 (a) Period is 12; amplitude is 4
   (b) $g(34) = 11$; $g(60) = 14$

Chapter 1 Review

1 5 kilometers, 23 minutes

3 temperature

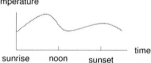

time

5 (b) 200 bushels
   (c) 80 lbs
   (d) $0 \le Y \le 550$
   (e) Decreasing
   (f) Concave down

7 Distance from Kalamazoo

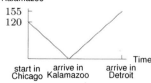

Time

9 $y = 8/3 - x/3$

11 $x = -1$

13 $y = 14x - 45$

15 (a) \$0.025/cubic foot

   (b) $c = 15 + 0.025w$
      c = cost of water
      w = cubic feet of water
   (c) 3400 cubic feet

17 $(\ln 3)/2 = 0.549$ mm/sec

19 0 mm/sec

21 0 mm/sec

23 (a) revenue

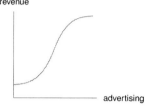

advertising

   (b) temperature

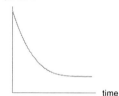

time

25 (a) customers

time

   (b) Concave down

27 (a) (i) Positive
       (ii) Positive
       (iii) Negative
       (iv) Positive
   (b) (i) $0 \le t \le 5$
       (ii) $0 \le t \le 20$
   (c) 25 m^3/week

29 (a) \$1000; \$15
   (b) It costs the company \$2500 to produce 100 items

31 $y = -(950/7)x + 950$

33 (a) $p = \$250$; $q = 750$
   (b) Suppliers produce 875 units;
      Consumers buy 625
   (c) Suppliers produce 625 units;
      Consumers would buy 875

35 $y = (-3/7)x + 3$

37 $y = 3e^{0.2197t}$
   or $y = 3(1.2457)^t$

39 $z = 1 - \cos\theta$

41 $f(x)$ is neither,
   $g(x) = 30.8 - 3.2x$ is linear,
   $h(x) = 15,000(0.6)^x$ is exponential

43 $(\log(2/5))/(\log 1.04) = -23.4$

45 $\ln(0.4)/3 = -0.305$

47 (a) 15%
   (b) $P = 10(1.162)^t$
   (c) 16.2%

(d) Graphs are the same since functions are equal

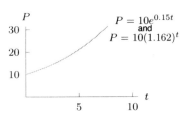

$P = 10e^{0.15t}$
and
$P = 10(1.162)^t$

49 (a) 15,678.7 years
(b) 5728.489, or about 5730 years

51 7.925 hours

53 Yes

55 (a) $\ln(2x + 3)$
(b) $2 \ln x + 3$
(c) $4x + 9$

57
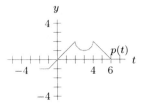

59
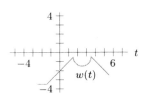

61

63 (a) A, B positive; C negative
(b) $A + B$
(c) A

65 $y = (x - 2)^2 - 5$

67 (a) $P = k/R$
(b) 300,000; 150,000; 100,000
(c) 6 million; 3 million, 2 million
(d) k is population of largest city

69 8π; 3

71 (a) Period = 12 months;
Amplitude = 4,500 cases
(b) About 2000 cases and 2000 cases

73 $f(x) = 2\sin(x/4) + 2$

Ch. 1 Understanding

1 False

3 True

5 True

7 True

9 False

11 False

13 True

15 False

17 False

19 False

21 False

23 False

25 True

27 True

29 True

31 True

33 False

35 Possible answer:
$$f(x) = \begin{cases} 1 & x \le 2 \\ x & x > 2 \end{cases}$$

37 Possible answer $f(x) = 1/(x + 7\pi)$

39 Possible answer:
$f(x) = (x - 1)/(x - 2)$

41 $f(x) = x, g(x) = -2x$

43 $f(x) = e^x, g(x) = e^{-2x}$

45 False; $f(x) = x/(x^2 + 1)$

47 False; $y = x + 1$ at points
$(1, 2)$ and $(2, 3)$

49 False; $y = 4x + 1$ starting at $x = 1$

51 False

53 True; $f(x) = 0$

55 True

57 True

59 False

61 True

63 False

65 False

67 (a) Follows
(b) Does not follow (although true)
(c) Follows
(d) Does not follow

Section 2.1

1 (a) Positive
(b) Negative

3 (a) (i) 6.3 m/sec
(ii) 6.03 m/sec
(iii) 6.003 m/sec
(b) 6 m/sec

5 (a) 8 ft/sec
(b) 6 ft/sec

7 (a) 63 cubic millimeters
(b) 10.5 cubic millimeters/month
(c) 44.4 cubic millimeters/month

9 $f'(2) \approx 40.268$

11 (a) Positive at C and G
Negative at A and E. Zero at B, D, and F
(b) Largest at G
Most negative at A

13 $f'(d) = 0$, $f'(b) = 0.5$, $f'(c) = 2$,
$f'(a) = -0.5$, $f'(e) = -2$

15 (a) The slopes of the two tangent lines at $x = a$
are equal for all a
(b) A vertical shift does not change the slope

17 $P'(0) = 10$

19 (a) $g(2) = 5$
(b) $g'(2) = -0.4$

21 (a) $f(4)$
(b) $f(2) - f(1)$

(c) $(f(2) - f(1))/(2 - 1)$
(d) $f'(1)$

23 300 million people;
2.867 million people/yr

Section 2.2

1 1.0, 0.3, -0.5, -1

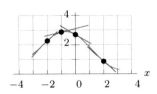

3
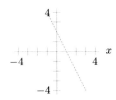

5
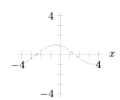

7
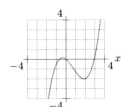

9 $-6, -3, -1.8, -1.2, -1.2$

11

13

15

17
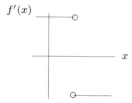

19

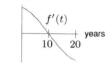

or

Other answers possible

21 VIII

23 II

25
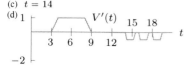

27 (a) Graph II
 (b) Graph I
 (c) Graph III

29 $f'(2) = 4$
 $f'(3) = 9$
 $f'(4) = 16$
 The pattern seems to be:
 $f'(x) = x^2$.

31 (a) $t = 3$
 (b) $t = 9$
 (c) $t = 14$
 (d)
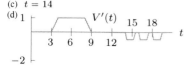

Section 2.3

1 dC/dW; dollars per pound

3 dP/dH; dollars per hour

5 (a) 12 pounds, 5 dollars
 (b) Positive
 (c) 12 pounds, 0.4 dollars/pound, extra pound costs about 40 cents

9 (a) Positive
 (b) °F/min

11 (a) Liters per centimeter
 (b) About 0.042 liters per centimeter
 (c) Cannot expand much more

13 (a) Investing the $1000 at 5% would yield about $1649 after 10 years
 (b) Extra percentage point would yield an increase of about $165; dollars/%

15 (a) Consuming 1800 Calories per day results in a weight of 155 pounds; Consuming 2000 Calories per day causes neither weight gain nor loss
 (b) Pounds/(Calories/day)

17 (a) Positive
 (b) Child weighs 45 pounds at 8 years
 (c) lbs/year
 (d) The child is growing at a rate of 4 lbs/year at 8 years of age
 (e) Decrease

19 $f'(t) \approx 6$ where t is retirement age in years, $f(t)$ is age of onset in weeks

21 (a) kg/week
 (b) Growing at 96 gm/wk at week 24

23 (a) Less
 (b) Greater

25 About 3.4, about 2.6

27 (a) In 2008: Net sales 5.1 bn dollars; rate increase 0.22 bn dollars/yr
 (b) About 5.98 billion dollars

29 (a) Dose for 140 lbs is 120 mg
 Dose increases by 3mg/lb
 (b) About 135 mg

31 (a) $f(0) = 80$; $f'(0) = 0.50$
 (b) $f(10) \approx 85$

33 (a) 1.7 (liters/minute)/hour
 (b) About 0.028 liter/minute
 (c) $g'(2) = 1.7$

35 (a) $f'(a)$ is always positive
 (c) $f'(100) = 2$: more
 $f'(100) = 0.5$: less

37 (a) Fat
 (b) Protein

39 (a) 2.0 kg/week
 (b) 0.6 kg/week
 (c) 0.3 kg/week

41 I-fat, II-protein

43 $f(4) = 200$ million users;
 $f'(4) \approx 12.5$ million users/month;
 Increasing at about 6.25%/month

45 0.50

Section 2.4

1 (a) Negative
 (b) Negative
 (c) Positive

3 $f'(x) > 0$
 $f''(x) > 0$

5 $f'(x) < 0$
 $f''(x) = 0$

7 $f'(x) > 0$
 $f''(x) < 0$

9 (a)
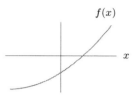

 (b)
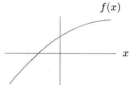

 (c)
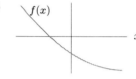

 (d)
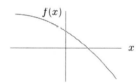

11 $s'(t)$: positive
 $s''(t)$: positive or zero

13 A positive second derivative indicates a successful campaign
 A negative second derivative indicates an unsuccessful campaign

15 (a) Positive; positive
 (b) Number of cars increasing at 3.24 million cars per year in 1975

17 Derivative:
 Pos. $-2.3 < t < -0.5$
 Neg. $-0.5 < t < 4$
 Second derivative:
 Pos. $0.5 < t < 4$
 Neg. $-2.3 < t < 0.5$

19

Point	f	f'	f''
A	−	0	+
B	+	0	−
C	+	−	−
D	−	+	+

21 (e)

23 22 only possible value

25 (a)
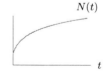

 (b) dN/dt is positive.
 d^2N/dt^2 is negative.

502

27 (a) utility

quantity

   (b) Derivative of utility is positive
      2nd derivative of utility is negative

Section 2.5

1 (a) Dollars/barrel
   (b) 101 barrels cost about \$3 more than 100 barrels

3 At $q = 5$;
   At $q = 40$

5 About \$16.67 (answers may vary)

7 $C'(2000) \approx \$0.37$/ton
   The marginal cost is smallest on the interval $2500 \leq q \leq 3000$.

9 (a) About \$2408
   (b) About \$2192

11 (a) About \$4348
   (b) \$11 profit
   (c) No, company will lose money

13 (a) \$1850 profit
   (b) About \$0.40 increase; increase production
   (c) About \$0.45 decrease; decrease production

15 (a) Fixed costs
   (b) Decreases slowly, then increases

Chapter 2 Review

1 (a) (i) 8.4 m/sec
     (ii) 8.04 m/sec
     (iii) 8.004 m/sec
   (b) 8 m/sec

3 (a) Negative
   (b) $f'(1) = -3$

5 Positive: A and D
   Negative: C and F
   Most positive: A
   Most negative: F

7 (a) (i) -1.00801 m/sec
     (ii) -0.8504 m/sec
     (iii) -0.834 m/sec
   (b) -0.83 m/sec

9

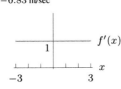

11

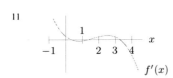

13

15 (a) $x_1 < x < x_3$
   (b) $0 < x < x_1; x_3 < x < x_5$

21 $f(21) \approx 65$
   $f(19) \approx 71$
   $f(25) \approx 53$

23 Wind stronger at 15.1 km than at 15 km

25 About 1.338 billion people in 2009; growing at 8 million people per year

27 (a) $f'(t) > 0$: depth increasing
     $f'(t) < 0$: depth decreasing
   (b) Depth increasing 20 cm/min
   (c) 12 meters/hr

31 (a) Negative
   (b) Degrees/min

33 Dollars/year; negative

35 (a)

   (b)

   (c)

   (d)

37 (a) Minutes/kilometer
   (b) Minutes/kilometer^2

39 (a) x_4, x_5
   (b) x_3, x_4
   (c) x_3, x_4
   (d) x_2, x_3
   (e) x_1, x_2, x_5
   (f) x_1, x_4, x_5

41 (a) t_3, t_4, t_5
   (b) t_2, t_3
   (c) t_1, t_2, t_5
   (d) t_1, t_2, t_4
   (e) t_3, t_4

43 (a)

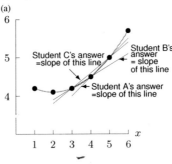

   (b) Student C's
   (c) $f'(x) = \dfrac{f(x+h) - f(x-h)}{2h}$

Ch. 2 Understanding

1 False

3 True

5 True

7 True

9 False

11 False

13 True

15 False

17 False

19 False

21 False

23 False

25 True

27 True

29 True

31 True

33 False

35 Possible answer:
$$f(x) = \begin{cases} 1 & x \leq 2 \\ x & x > 2 \end{cases}$$

37 Possible answer $f(x) = 1/(x + 7\pi)$

39 Possible answer:
   $f(x) = (x - 1)/(x - 2)$

41 $f(x) = x, g(x) = -2x$

43 $f(x) = e^x, g(x) = e^{-2x}$

45 False; $f(x) = x/(x^2 + 1)$

47 False; $y = x + 1$ at points $(1, 2)$ and $(2, 3)$

49 False; $y = 4x + 1$ starting at $x = 1$

51 False

53 True; $f(x) = 0$

55 True

57 True

59 False

61 True

63 False

65 False

67 (a) Follows
   (b) Does not follow (although true)
   (c) Follows
   (d) Does not follow

Theory: Limits, Derivatives

1

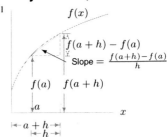

3 1

5 27

7 2.7

384

9 Yes

11 No; yes

13 Yes

15 Yes

17 No

19 Not continuous

21 Not continuous

23 Not continuous

Section 3.1

1 0

3 $12x^{11}$

5 $\frac{4}{3}x^{1/3}$

7 $12t^3 - 4t$

9 $-4x^{-5}$

11 $2x + 5$

13 $6x + 7$

15 $8.4q - 0.5$

17 $-5t^{-6}$

19 $-(7/2)r^{-9/2}$

21 $-(1/3)\theta^{-4/3}$

23 $15t^4 - \frac{5}{2}t^{-1/2} - 7t^{-2}$

25 $6t - 6/t^{3/2} + 2/t^3$

27 $(3/2)x^{1/2} + (1/2)x^{-1/2}$

29 $2kx$

31 $2aP + 3bP^2$

33 $b/(2\sqrt{t})$

35 $3ab^2$

37 (a) $P'(1)$: Positive;
$P'(3)$: Zero;
$P'(4)$: Negative

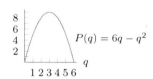

(b) $P'(1) = 4, P'(3) = 0, P'(4) = -2$

39 (a) $2t - 4$
(b) $f'(1) = -2, f'(2) = 0$

41 Height $= 625$ cm,
Changing (eroding) at -30 cm/year

43 $f(10) = 400$ tons;
$f'(10) = 60$ tons per year;
Relative rate $= 15\%$ per year

45 $f'(t) = 6t^2 - 8t + 3$
$f''(t) = 12t - 8$

47 (a) $f(100) = 11.11$ seconds
(b) $f'(100) = 0.05555$ seconds per foot

49 $y = 2x - 1$

51 $y = -2t + 16$

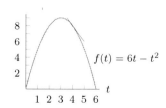

---

53 (a) $dA/dr = 2\pi r$
(b) Circumference of a circle

55 $100

57 (a) 770 bushels per acre
(b) 40 bushels per acre per pound of fertilizer
(c) Use more fertilizer

59 (a) $dC/dq = 0.24q^2 + 75$
(b) $C(50) = \$14,750; C'(50) = \$675$ per item

61 $f'(x) = 3x^2 - 12x - 15,$
$x = -1$ and $x = 5$

63 (a) $R(q) = bq + mq^2$
(b) $R'(q) = b + 2mq$

Section 3.2

1 $2e^x + 2x$

3 $10t + 4e^t$

5 $(\ln 2)2^x - 6x^{-4}$

7 $(\ln 2)2^x + 2(\ln 3)3^x$

9 $3 - 2(\ln 4)4^x$

11 $3e^{3t}$

13 $-4e^{-4t}$

15 $-30e^{-0.6t}$

17 $3000(\ln 1.02)(1.02)^t$

19 Ce^t

21 $Ae^x - 2Bx$

23 $3/q$

25 $2t + 5/t$

27 $2x + 4 - 3/x$

29 $f'(-1) \approx -0.736$
$f'(0) = -2$
$f'(1) \approx -5.437$

31 $y = -2t + 1$

33 (a) $13,394$ fish
(b) 8037 fish/month

35 $f(2) = 6065, f'(2) = -1516$

37 $f(5) = \$563.30;$
$f'(5) = \$70$ per week;
Relative rate $= 12.4\%$ per week

39 -444.3 people/year

41 $c = -1/\ln 2$

43 $C(50) \approx 1365, C'(50) \approx 18.27$

45 (a) 0.021 micrograms/year
(b) 779.4 years old in 1998

47 (a) $P = 1.166(1.015)^t$
(b) $\frac{dP}{dt} = 1.166(1.015)^t(\ln 1.015)$
$\frac{dP}{dt}|_{t=0} = 0.017$ billion people per year
$\frac{dP}{dt}|_{t=25} = 0.025$ billion people per year

49 (a) $y = x - 1$
(b) $0.1; 1$
(c) Yes

Section 3.3

1 $56x(4x^2 + 1)^6$

3 $8q(q^2 + 1)^3$

5 $300t^2(t^3 + 1)^{99}$

7 $3s^2/(2\sqrt{s^3 + 1})$

9 $216q(3q^2 - 5)^2$

11 $25e^{5t+1}$

13 $(e^{\sqrt{s}})/(2\sqrt{s})$

15 $1/(x - 1)$

17 $e^{-x}/(1 - e^{-x})$

19 $25/(5t + 1)$

21 $3/(3t + 2)$

23 $5 + 1/(x + 2)$

25 $0.5/(x(1 + \ln x)^{0.5})$

27 $-x/\sqrt{1 - x^2}$

29 $10/(10t + 5)$

31 5

33 $2/t$

35 1.5

37 $2/t$

39 $y = 3x - 5$

41 $v(t) = 10e^{\frac{t}{5}}$

43 $f(10) = 31.640$ feet;
$f'(10) = 4.741$ ft/sec,
Relative rate $= 15\%$ per second

45 0

47 $1/2$

49 Approx 1

51 Approx 1.9

Section 3.4

1 $f'(x) = 12x - 1$

3 $e^x(x + 1)$

5 $5e^{x^2} + 10x^2e^{x^2}$

7 $\ln x + 1$

9 $30t + 11$

11 $t + 2t\ln t$

13 $-(5/t^2) - (12/t^3)$

15 $e^{-t^2}(1 - 2t^2)$

17 $2p/(2p + 1) + \ln(2p + 1)$

19 $2we^{w^2}(5w^2 + 8)$

21 $(3t^2 + 5)(t^2 - 7t + 2) + (t^3 + 5t)(2t - 7)$

23 $(1 - x)/e^x$

25 $-2/(1 + t)^2$

27 $(15 + 10y + y^2)/(5 + y)^2$

29 $(ak - bc)/(cx + k)^2$

31 $ae^{-bx} - abxe^{-bx}$

33 $(1 - 2\alpha)e^{-2\alpha}e^{\alpha}e^{-2\alpha}$

35 $y = 0$

37 60.65 mg, 30.33 mg/hr,
41.04 mg, -12.31 mg/hr

39 $(kc)/(k + r)^2;$
Approx change in P per unit increase in r

41 Revenue $R(10) \approx 22,466$.
$R'(10) \approx \$449$/dollar.

43 $1/t$

45 $(fg)'/(fg) = (f'/f) + (g'/g)$

Section 3.5

1 $5\cos x$

3 $2t - 5\sin t$

5 $2q + 2\sin q$

7 $3\cos(3x)$

9 $-8t\sin(t^2)$

11 $2x\cos(x^2)$

13 $-4\sin(4\theta)$

15 $2x\cos x - x^2\sin x$

17 $3\theta^2\cos\theta - \theta^3\sin\theta$

19 $(2t\cos t + t^2\sin t)/(\cos t)^2$

21 $y = -x + \pi$

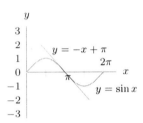

23 Decreasing, concave up

25 (a) max \$2600; min \$1400; April 1

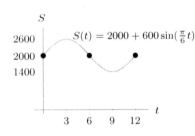

(b) $S(2) \approx 2519.62$; $S'(2) \approx 157.08$

27 (a) Falling, 0.38 m/hr
(b) Rising, 3.76 m/hr
(c) Rising, 0.75 m/hr
(d) Falling, 1.12 m/hr

29 (a) $H'(t) = (-10\pi/3)\sin((\pi/15)t)$
(b) $t = 0, 15, 30$ days; the percentage illuminated is not changing
(c) $H'(t)$ negative for $0 < t < 15$ and positive for $15 < t < 30$

Chapter 3 Review

1 $24t^3$

3 $2e^{2t}$

5 $0.08e^{0.08q}$

7 $e^{3x}(1 + 3x)$

9 $6(1 + 3t)e^{(1+3t)^2}$

11 $5 + 3.6e^{1.2t}$

13 $90(5x - 1)^2$

15 $10e^x(1 + e^x)^9$

17 $x(2\ln x + 1)$

19 $8t + 7\cos t$

21 $(-e^{-t} - 1)/(e^{-t} - t)$

23 $(50x - 25x^2)/(e^x)$

25 $2x\cos x - x^2\sin x$

27 e

29 $(e^x - 2 - e^{-x})/(1 - e^{-x})^2$

31 $xe^x + e^x$

33 $-3x^2\sin(x^3)$

35 $(t^2 + 6t + 13)/(t + 3)^2$

37 $((\ln 3)3^x)/3 - (33x^{-3/2})/2$

39 $6y^2e^{(y^3)}e^{2e^{(y^3)}}$

41 $f'(0) = 0$, $f'(1) = 2$, $f'(2) = 4$, $f'(-1) = -2$

43 $y = -4 - x$

45 30% per year

47 49.7% per year

49 73.7 m/yr

51 (a)
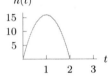

(b) $v(t) = 32 - 32t$

(c) $v(1) = 0$ ft/sec; $h(1) = 16$ ft

53 (a) $f'(0) = -1$
(b) $y = -x$

55 \$596.73/yr

57 (a) $\pi/\sqrt{9.8l}$
(b) Decr

59 (a) $dH/dt = -60e^{-2t}$
(b) $dH/dt < 0$
(c) At $t = 0$

61 (a) \$13,050
(b) $V'(t) = -4.063(0.85)^t$; thousands of dollars/year
(c) $V'(4) = -\$2121$ per year

63 (a) $P(d) = 4.07264 \cdot 10^{-10}d^{3/2}$
(b) Approximately 88 days
(c) $P'(d) = 6.10896 \cdot 10^{-10}d^{1/2}$ days/mile. Farther from the sun, time increases

65 $n = 4, a = 3/32$

67 (a) $H(4) = 1$
(b) $H'(4) = 30$
(c) $H(4) = 4$
(d) $H'(4) = 56$
(e) $H'(4) = -1$

69 $b = 1/40$ and $a = 169.36$

71 Approx 0.4

73 Approx 0.7

75 Approx -21.2

77 (a) $x < 1/2$
(b) $x > 1/4$
(c) $x < 0$ or $x > 0$

79 (a) $dy/dt = -(4.4\pi/6)\sin(\pi t/6)$ ft/hr
(b) Occurs at $t = 0, 6, 12, 18$, and 24 hrs

81 (c) $B'(10) \approx -3776.63$

83 If $a \neq e$, two solutions
If $a = e$, one solution

85 (a) 0.04 gal/mile; 0.06 gal/mile

(b) 25 mpg; 16.67 mpg
(d) 1.4 gallons
(e) 2.8 gal/hour; 1.8 gal/hour

87 $331.3 + 0.606T$ m/sec

Ch. 3 Understanding

1 True

3 True

5 True

7 False

9 True

11 False

13 False

15 False; $f(x) = |x|$

17 False; $\cos t + t^2$

19 False; $f(x) = 6, g(x) = 10$

21 False; $f(x) = 5x + 7, g(x) = x + 2$

23 False; $f(x) = x^2, g(x) = x^2 - 1$

25 False; $f(x) = e^{-x}, g(x) = x^2$

27 (a) Not a counterexample
(b) Not a counterexample
(c) Not a counterexample
(d) Counterexample

29 False

31 False

33 Possible answer
$$f(x) = \begin{cases} x & \text{if } 0 \leq x < 2 \\ 19 & \text{if } x = 2 \end{cases}$$

Practice: Differentiation

1 $2t + 4t^3$

3 $15x^2 + 14x - 3$

5 $-2/x^3 + 5/(2\sqrt{x})$

7 $10e^{2x} - 2 \cdot 3^x(\ln 3)$

9 $2pe^{p^2} + 10p$

11 $2x\sqrt{x^2 + 1} + x^3/\sqrt{x^2 + 1}$

13 $16/(2t + 1)$

15 $2^x(\ln 2) + 2x$

17 $6(2q + 1)^2$

19 bke^{kt}

21 $2x\ln(2x + 1) + 2x^2/(2x + 1)$

23 $10\cos(2x)$

25 $15\cos(5t)$

27 $2e^x + 3\cos x$

29 $17 + 12x^{-1/2}$

31 $20x^3 - 2/x^3$

33 $1, x \neq -1$

35 $-3\cos(2 - 3x)$

37 $6/(5r + 2)^2$

39 $ae^{ax}/(e^{ax} + b)$

41 $5(w^4 - 2w)^4(4w^3 - 2)$

43 $(\cos x - \sin x)/(\sin x + \cos x)$

45 $6/w^4 + 3/(2\sqrt{w})$

47 $(2t - ct^2)e^{-ct}$

49 $(\cos\theta)e^{\sin\theta}$

51 $3x^2/a + 2ax/b - c$

53 $2r(r + 1)/(2r + 1)^2$

55 $2e^t + 2te^t + 1/(2t^{3/2})$

57 $x^2 \ln x$

59 $6x\left(x^2 + 5\right)^2 \left(3x^3 - 2\right)\left(6x^3 + 15x - 2\right)$

61 $(2abr - ar^4)/(b + r^3)^2$

63 $20w/(a^2 - w^2)^3$

Section 4.1

1 One

3 Three

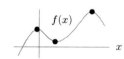

5 (a)

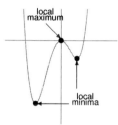

(b)

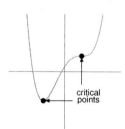

7 (a)

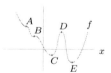

(b)

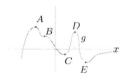

9 A: local max
B: local min
C: neither

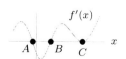

11 Increasing for all x, no critical point

13 Local min: $(2.3, -13.0)$

15 Alternately incr/decr

17

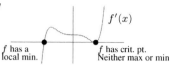

19 $t = 0.5 \ln(V/U)$

21 (a) $x \approx 2.5$ (or any $2 < x < 3$)
$x \approx 6.5$ (or any $6 < x < 7$)
$x \approx 9.5$ (or any $9 < x < 10$)
(b) $x \approx 2.5$: local max;
$x \approx 6.5$: local min;
$x \approx 9.5$: local max

23 $x = 0$: not max/min
$x = 3/7$: local max
$x = 1$: local min

25 (a)

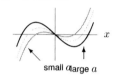

(b) 2 critical points move farther from origin
(c) $x = \pm\sqrt{a/3}$

27 $a = -6; b = 14$

29

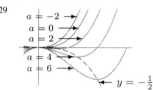

31 $a = -1/3$

33 (a) $x = 0, x = a^2/4$
(b) $a = \sqrt{20}$; Local minimum

35 Domain: All real numbers except $x = b$;
Critical points: $x = 0, x = 2b$

37 (a) $f(\theta) = 0$ at $\theta = 0$
(b) $f'(\theta) = 1 - \cos\theta > 0$
for $0 < \theta \le 1$.
Only zero is at origin

Section 4.2

1 Three

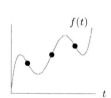

3 One

5 (a)

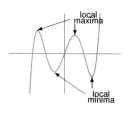

(b) 3

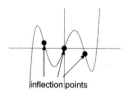

7

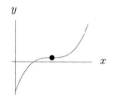

9 (a) 6 pm
(b) Noon; another between noon and 6 pm

11 Critical points: $x = \frac{5}{2}$, local minimum
Inflection points: None

13 Critical points:
$x = -3$ (local max) and $x = 2$ (local min)
Inflection point: $x = -1/2$

15 Critical points: $x = -1$, local min; $x = 0$ local
max; $x = 1$, local min
Inflection points: $x = -1/\sqrt{3}, x = 1/\sqrt{3}$

17 Critical points:
$x = 0$ (local max) and $x = \pm 2$ (local minima)
Inflection points: $x = \pm 2/\sqrt{3}$

19 Critical points: $x = 0$, local max; $x = 4$, local
min
Inflection point: $x = 3$

21 $x = -1, 1/2$

23

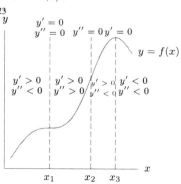

387

27 (a) Yes, at 2000 rabbits

population of rabbits

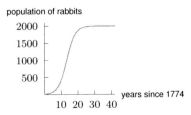

(b) 1787, 1000 rabbits
(c) 1787, 1000 rabbits

29 depth of water

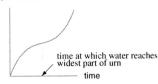

time at which water reaches widest part of urn

31

H (height)

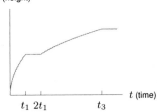

33 $y = x^3 - 3x^2 + 6$

35 (a)

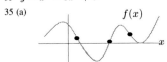

$f(x)$

(b)

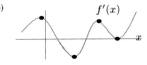

$f'(x)$

(c)

$f''(x)$

Section 4.3

1

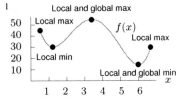

Local and global max
Local max
$f(x)$
Local max
Local min
Local and global min

---

3 (a) (IV)
(b) (I)
(c) (III)
(d) (II)

5

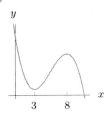

7

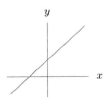

9

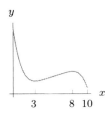

11

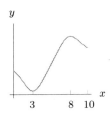

13 44.1 feet

15 (a) $f(0.91)$ is global minimum
$f(3.73)$ is global maximum
(b) $f(-3)$ is global minimum
$f(2)$ is global maximum

17 (a) $D = C$
(b) $D = C/2$

19 (a) $f'(x) = 6x^2 - 18x + 12$,
$f''(x) = 12x - 18$.
(b) $x = 1, 2$
(c) $x = 3/2$
(d) Local minimum: $x = 2$
Local maximum: $x = 1$
Global minimum: $x = -0.5$
Global maximum: $x = 3$
(e)

$f(x) = 2x^3 - 9x^2 + 12x + 1$

21 (a) $f'(x) = 1 + \cos x$,
$f''(x) = -\sin x$.

(b) $x = \pi$
(c) $x = 0, \pi, 2\pi$
(d) Global minimum: $x = 0$
Global maximum: $x = 2\pi$
(e)

$f(x) = x + \sin x$

23 Global max $= -1$ at $x = 2$
No global min

25 Global max $= 1/e$ at $t = 1$
No global min

27 Global max $= 1/2$ at $t = 1$
Global min $= -1/2$ at $t = -1$

29 $x = -b/2a$,
Max if $a < 0$, min if $a > 0$

31 (b) Yes, at $x = 0$
(c) Max: $x = -2$, Min: $x = 2$
(d) $5 > g(0) > g(2)$

33 1250 square feet

35 (a) $x + 27/x$ meters
(b) 5.2 meters

37 (a) $y = \begin{cases} 200x & x \le 200 \\ 600x - x^2 & x > 200 \end{cases}$

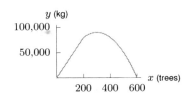

(b) 300 trees/km^2

39 (a) $H = 1/b$, $S = ae^{-1}/b$
(b) a: Increases
b: Decreases

41 (a) 10
(b) 9

43 (a) $k(\ln k - \ln S_0) - k + S_0 + I_0$
(b) Both

45 (a) 120 mm Hg, 80 mm Hg
(b) 0.8 sec
(c)

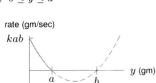

47 (a) $0 \le y \le a$

rate (gm/sec)

(b) $y = 0$

Section 4.4

1 $5.5 < q < 12.5$ positive;
   $0 < q < 5.5$ and $q > 12.5$ negative;
   Maximum at $q \approx 9.5$

3 (a) $q = 2500$
  (b) \$3 per unit
  (c) \$3000

5 (a) \$9
  (b) $-\$3$
  (c) $C'(78) = R'(78)$

7 (a) Increase production
  (b) $q = 8000$

9 \$0.20/item

11 $q = 0$ or $q = 3000$

13 One; between 40 and 50

15 (a) MR; increase production
   (b) MC; decrease production

17 Global maximum of \$6875 at $q = 75$

19 (a) Approximately \$1
   (b) No
   (c) About 400 items

21 (a) \$10; \$30,000; \$50,000
   (b) $R(q) = 70q - 0.02q^2$
   (c) 1750
   (d) \$35
   (e) \$61,250

23 \$14.

25 (a) $10,000 + 2q$
   (b) $q = 37,820 - 5544p$
   (c) $\pi = -0.00018q^2 + 4.822q - 10,000$
   (d) $13,394$ items, \$22,294

27 Maximum revenue = \$27,225
    Minimum = \$0

29 (a) q/r months
   (b) $(ra/q) + rb$ dollars
   (c) $C = (ra/q) + rb + kq/2$ dollars
   (d) $q = \sqrt{2ra/k}$

31 $L = [\beta pcK^\alpha/w]^{1/(1-\beta)}$

Section 4.5

1 (a) No
  (b) Yes

3 (a) $a(1000) \approx \$1.60$ per unit
  (b)

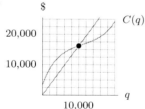

  (c) 18,000 units

5 $MC = \$20$; $a(q) = \$25$

7 (a) $q = 6$

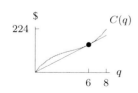

  (b) $q = 6$

9 (a) $C(q) = 0.01q^3 - 0.6q^2 + 13q$
  (b) \$1
  (c) $q = 30$, $a(30) = 4$
  (d) Marginal cost is 4

11 (a)

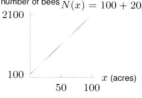

number of bees $N(x) = 100 + 20x$

  (b) (i) $N'(x) = 20$
       (ii) $N(x)/x = (100/x) + 20$

13

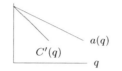

Section 4.6

1 (a) 1.5% decrease
  (b) 1.5% increase

5 Elastic

7 High

9 Low

11 (a) $E \approx 0.470$, inelastic
   (c) $P = 1.25$ and 1.50

13 (a) $q = 4960$
   (b) $E = 0.016$, so demand is inelastic

Section 4.7

1 (a) 40 billion
  (b)

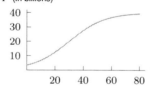

P (in billions)

  (c) 2020; 2095

3 (a)

total sales (thousands)

  (b) ·60,000

7 (a) 36 thousand; total number infected
  (b) $t \approx 16$, $n \approx 18$ thousand
  (c) Virus spreading fastest
  (d) Number infected half total

(a)& (b)

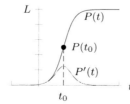

Chapter 4 Review

(c) P' maximum

11 (b) April 12, 2003
   (d) $t = 19$; 1600 cases
   (e) 1760 cases

15 (a)

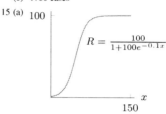

$R = \dfrac{100}{1+100e^{-0.1x}}$

  (b) About 46.05
  (c) Between 32.12 and 54.52 mg

17 (b) $f'(10) < 11$
       $f'(20) < 11$

19 10 mg to 18 mg

Section 4.8

1 (a)

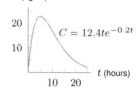

C (ng/ml)

$C = 12.4te^{-0.2t}$

  (b) 5 hrs, 22.8 ng/ml
  (c) 1 to 14.4 hrs
  (d) At least 20.8 hrs

9 $a = 49.3$; $b = 0.769$

13 (b) Min effective ≈ 0.2 Max safe ≈ 1.0
   (c) Min effective ≈ 1.2 Max safe ≈ 2.0

Chapter 4 Review

1

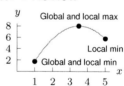

3 (a) $f(1)$ local minimum;
       $f(0)$, $f(2)$ local maxima
  (b) $f(1)$ global minimum
       $f(2)$ global maximum

5 Global min = 3 at $z = 1/2$
   No global max

7 Global max = $1/2$ at $x = 1$
   No global min

11 (a) Increasing for all x
   (b) No maxima or minima

13 (a) Incr: $-1 < x < 0$ and $x > 1$
       Decr: $x < -1$ and $0 < x < 1$

   (b) Local max: $f(0)$
        Local min: $f(-1)$ and $f(1)$

15 (a) cm/week

(b) Growing at 1.6 cm/wk in week 24

17 (a) Week 14
(b) Point of fastest growth

19 $a = 3e, b = -3$

21 (a) $5x^4 + 1$; positive
(b) One

25 (a) Polynomial; negative leading coefficient; degree 2
(b) Exponential.
(c) Logistic
(d) Logarithmic.
(e) Polynomial; positive leading coefficient; degree 2
(f) Exponential.
(g) Surge

27 (a) $0
(b) $96.56
(c) Raise the price by $5

29 (a) $\pi(q)$ max when
$R(q) > C(q)$ and R and Q are farthest apart

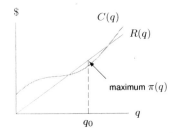

(b) $C'(q_0) = R'(q_0) = p$
(c)

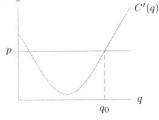

31

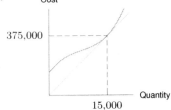

33 $C'(2)$

35 $(C(75) - C(50))/25$

37 $C(3)/3$

39 $E = 0.05$, demand is inelastic.

43 50 m by 50 m

45 (a) $1/5, 1$
(b) Local max: $p = 1/5$
Local min: $p = 1$
(c) Max: $256/3125$
Min: 0

47 (a) $E = 500e\left(\dfrac{2 - \cos\theta}{\sin\theta}\right) + 2000e$
$\left(\arctan\left(\dfrac{500}{2000}\right) \leq \theta \leq \pi/2\right)$

(b) $\theta = \pi/3$
(c) Independent of e, but dependent on $\overline{AB}/\overline{AL}$

49 (a) $k(1 - 2P/L) \cdot dP/dt$

51 Intercepts: $(0, 0)$, $(1.77, 0)$, $(2.51, 0)$
Critical Points: $(0, 0)$, $(1.25, 1)$, $(2.17, -1)$, $(2.80, 1)$
Inflection Points: $(0.81, 0.61)$, $(1.81, -0.13)$, $(2.52, 0.07)$

53 (a)

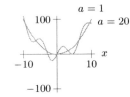

(b) $-2 < a < 2$

55 (a)

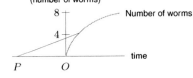

(b) 7 worms
(c) increases

57 Max slope $= 1/(3e)$ at $x = 1/3$

Ch. 4 Understanding

1 True
3 False
5 False
7 False
9 One possibility:
$f(x) = ax^2, a \neq 0$
11 (a) True
(b) False
(c) True
(d) False
(e) True
13 False
15 True
17 True
19 True
21 False
23 False
25 $f(x) = x^2 + 1$
27 $f(x) = -x^2 - 1$
29 Impossible

Section 5.1

1 (a)

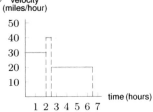

(b)

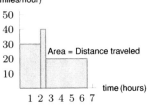

3 (a) Lower estimate $= 122$ ft
Upper estimate $= 298$ ft
(b)

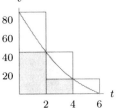

5 250 meters

7 ≈ 455 feet or 0.086 miles

9 (a) About 420 kg
(b) 336 and 504 kg

11 (a) 570 m³
(b) Every 2 minutes

13 (a)

(b) 125 feet
(c) 4 times as far

15 (a) Car A
(b) Car A
(c) Car B

17 60 m (Other answers possible.)

19 (b) $6151

Section 5.2

1 27
3 1692.5
5 About 543
7 (a) 224

(b) 96

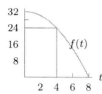

(c) 200

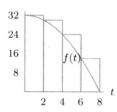

(d) 136

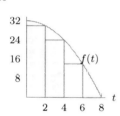

9 350

11 About 20

13 (a) 3.6

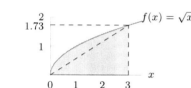

(b) 3.4641

15 (a) 2

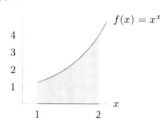

(b) 2.05045

17 (a) 78

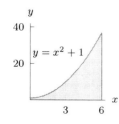

(b) 46; underestimate

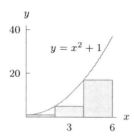

(c) 118; overestimate

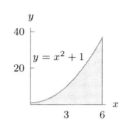

19 93.47

21 448.0

23 2.350

25 2.9

27 1.30

29 1.728, 1.816, 1.772, $n = 140$

31 (a) 4; 0, 4, 8, 12, 16; 25, 23, 22, 20, 17
   (b) 360; 328
   (c) 8; 0, 8, 16; 25, 22, 17
   (d) 376; 312

Section 5.3

1 8

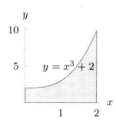

3 $\int_0^2 ((x + 5) - (2x + 1))\, dx = 6$

5 (a) 13
  (b) 1

7 Positive

9 Positive

11 −40

13 (a) −4
   (b) 0
   (c) 8

15 II

17 III

19 V < IV < II < III < I
   I, II, III positive
   IV, V negative

21 (a) −2
   (b) −$A/2$

23 About 3.34

25 0.80

27 14,052.5

29 14.688

31 13.457

33 2.762

Section 5.4

1 (a) Emissions 1970–2000, m. metric tons
  (b) 772.8 million metric tons

3 Change in position; meters

5 Change in world pop; bn people

7 Total amount = $\int_0^{60} f(t)\, dt.$

9 1417 antibodies

11 29,089.813 megawatts

13 3.4 ft

15 (a) Concave up
   (b) 3.1 kg

17 (a) 2023, at crossing of curves
   (b) 2023, at highest point of curve

19 2627 acres

21 Tree B is taller after 5 years.
   Tree A is taller after 10 years.

23 (a) Boys: black curve; girls: colored curve
   (b) About 43 cm
   (c) Boys: about 23 cm; girls: about 18 cm
   (d) About 13 cm taller

25 More fat

27 (a) About 750 liters
   (b) $\int_0^3 60g(t)\, dt$
   (c) About 150 liters

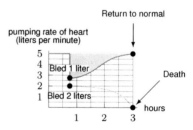

29 15 cm to the left

31 25 cm to the right

33 65 km from home
   3 hours
   90 km

35 Product B has a greater peak concentration
   Product A peaks sooner
   Product B has a greater overall bioavailability
   Product A should be used

37

39 About $13,800

41 (a) $\int_0^T 49(1 - (0.8187)^t)\, dt$
(meters)
(b) $T \approx 107$ seconds

Section 5.5

1 Dollars; cost of increasing production from 800 tons to 900 tons

3 (a)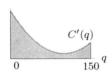

(b) \$22,775
(c) $C'(150) = 18.5$
(d) $C(151) \approx \$22,793.50$

5 7.65 million people

7 (a) \$18,650
(b) $C'(400) = 28$

9 (a)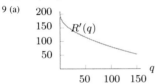

(b) \$12,000
(c) Marginal revenue is \$80/unit
Total revenue is \$12,080

11 (a) 7.54 inches after 8 hours
(b) 1.41 inches/hour

Chapter 5 Review

1 (a) 408
(b) 390

3 1096

5 Meters per second

7 Foot-pounds

9 1.44

11 1.15

13 -0.083

15 84

17 0.0833

19 About 0.1667

21 (a) 430 ft
(b) (ii)

23 (a)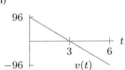

(b) 3 sec, 144 feet
(c) 80 feet

25 (a) Species B for both
(b) Species A

27 (a) Inflection point
(b) About 50 cm

29 (a) Negative
(b) Positive
(c) Negative
(d) Positive

33 (a) $\int_0^5 (10 + 8t - t^2)\, dt$
(b) 100
(c) 108.33

35 (a) Forward for $t < 3$, backward for $t > 3$
(b) Farthest forward: $t = 3$
Farthest backward:
no upper bound.

37 \$300,000

39 (a)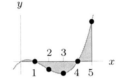

(b) I_1 is largest.
I_4 is smallest.
I_1 is the only positive value.

41 9 years

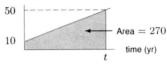

43 $F(x)$ is decreasing for
$x < -2$ and $x > 2$;
$F(x)$ is increasing for
$-2 < x < 2$

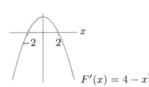

Ch. 5 Understanding

1 False
3 True
5 False
7 False
9 True
11 True
13 False
15 False
17 False
19 True
21 True
23 False
25 False

27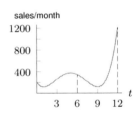

Theory: Second Fund. Thm.

1 x^3
3 xe^x
5 (a) 0
(b) F increases
(c) $F(1) \approx 1.4$, $F(2) \approx 4.3$, $F(3) \approx 10.1$
9 9
11 $8c$

Section 6.1

1 1.7
3 2
5 -3
7 About 8.5
9 (a) \$26,667 per year
(b) Less 25–65; more 65–85
11 \$6080
13 (a) 120 mm Hg
(b) 80 mm Hg
(c) 100 mm Hg
(d) Less
15 (a) 0.375 thousand/hour
(b) 1.75 thousands
17 (a) 9.9 hours
(b) 14.4 hours
(c) 12.0 hours
19 (a) Second half

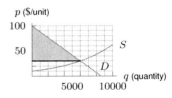

(b) \$1531.20, \$1963.20
(c) \$3494.40
(d) \$291.20/month

21 $(c) < (a) < (b)$

Section 6.2

1 (a) $p^* = \$30$, $q^* = 6000$
(b) Consumer surplus = \$210,000;
Producer surplus $\approx$ \$70,000

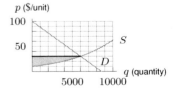

3 250

5 200

7 (a) Equilibrium price is $30
        Equilibrium quantity is 125 units
  (b) $3500
       $2000

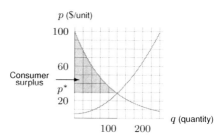

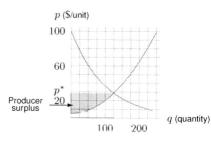

  (c) $5500

9 (a) Supplied: $11, demanded: $7
  (b) $p^* \approx \$9$
       $q^* \approx 400$

  (c) $1907
  (d) $1600

11 (a) Less
   (b) Can't tell
   (c) Less

13 (a) No, yes

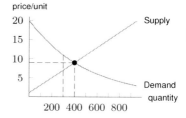

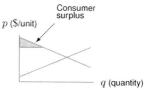

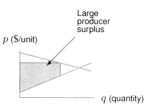

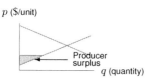

  (b) Yes, no

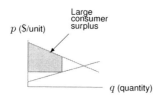

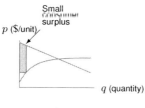

Section 6.3

1 Future value = $72,980.16
   Present value = $29,671.52

3 (a) $P = \$47,216.32$
       $F = \$77,846.55$
  (b) $60,000; $17,846.55

5 (a) (i) $18,846.59
       (ii) $16,484.00
  (b) (i) $21,249.47
       (ii) $24,591.24

7 (a) $65,022
  (b) $\approx$ 2.27 years

9 (a) $417,635.11
  (b) $228,174.64

11 $41,508

13 (a) $4.6 billion; $8.6 billion
   (b) $54.7 billion
   (c) $77.6 billion

15 About 1.75 years

17 In 10 years

Section 6.4

3 11%

5 Absolute:
   2003–2004: 3.1 m
   2006–2007: 2.0 m
   Relative:
   2003–2004: 10.3%
   2006–2007: 5.3%

7 (a) $P = 4000 - 100t$

  (b) $P = 4000(0.95)^t$
       Case (a)

9 22% increase

11 No change

13 Increasing $0 \le t \le 10$

15 Decreasing: $0 < t < 5$
    Increasing: $5 < t < 10$

17 (a) $P = 5.78(1.018)^t$
   (b) 0.10404 mil/yr
        0.10591 mil/yr
   (c) 1.8%, 1.8%

19 Decreases by about 9.52%

Chapter 6 Review

1 (a) 20
  (b) 10/3

3 (a) 0.79

5 (a) $Q(10) \approx 2.7$,
       $Q(20) \approx 1.8$
  (b) 2.21
  (c) 2.18

7 About 17

9 (a) $(1/5) \int_0^5 f(x)\,dx$
  (b) $(1/5)(\int_0^2 f(x)\,dx - \int_2^5 f(x)\,dx)$

11 (a)

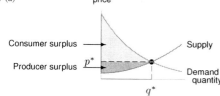

   (b)

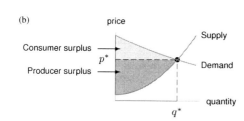

13 (a) Consumer: 87 units
        Suppliers: 100 units
        Down
   (b) $p^* = \$48$
        $q^* = 91$ units
   (c) Consumer surplus: $2096
        Producer surplus: $1143

15 (a) $5820 per year
   (b) $36,787.94

17 $635.37 per year

19 (a) $0 \le t \le 10$, 10.5%
   (b) $10 \le t \le 15$, 2.5%
   (c) 7.8% increase

21 In 45 years; 2035

Ch. 6 Understanding

1 False

3 True

5 False

7 False

9 True

11 True

13 False

15 False

17 False

19 True

21 True

23 False

25 False

27

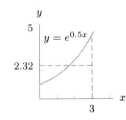

Section 7.1

1 $5x$

3 $t^3/3 + t^2/2$

5 $x^5/5$

7 $3x^4/2 + 4x$

9 $y^3 - y^4/4$

11 $x^3 + 5x$

13 $(x^3/3) - 3x^2 + 17x$

15 $5x^2/2 - 2x^{3/2}/3$

17 $\ln |z|$

19 $-1/2z^2$

21 $F(x) = x^7/7 + x^{-5}/35 + C$

23 $-e^{-3t}/3$

25 $G(t) = 5t + \sin t + C$

27 $F(x) = 3x$
(only possibility)

29 $F(x) = 2x + 2x^2 + (5/3)x^3$
(only possibility)

31 $F(x) = (2/3)x^{3/2}$
(only possibility)

33 $(5/2)x^2 + 7x + C$

35 $2x^3 + C$

37 $(x+1)^3/3 + C$

39 $t^3/3 + 5t^2/2 + t + C$

41 $t^4/4 + 2t^3 + C$

43 $2w^{3/2} + C$

45 $3\ln|t| + \dfrac{2}{t} + C$

47 $x^2/2 + 2x^{1/2} + C$

49 $e^t + 5x + C$

51 $e^{3t}/3 + C$

53 $-\cos t + C$

55 $25e^{4x} + C$

57 $3\sin x + 7\cos x + C$

59 $\frac{1}{2}\sin(x^2 + 4) + C$

61 $10x - 4\cos(2x) + C$

63 $F(x) = 3x^2 - 5x + 5$

65 $F(x) = -4\cos(2x) + 9$

67 $q^3 + 2q^2 + 6q + 200$

Section 7.2

1 $\frac{1}{5}(x^3 + 1)^5 + C$

3 $\frac{1}{4}(x + 10)^4 + C$

5 $e^{q^2+1} + C$

7 $(1/2)e^{t^2} + C.$

9 $(1/33)(t^3 - 3)^{11} + C$

11 $(1/9)(x^2 - 4)^{9/2} + C$

13 $-2\sqrt{4-x} + C$

15 $4\sin(x^3) + C$

17 $(1/5)x^5 + 2x^3 + 9x + C$

19 $\cos(3 - t) + C$

21 $-(2/9)(\cos 3t)^{3/2} + C$

23 $(1/7)\sin^7 \theta + C$

25 $(\sin x)^3/3 + C$

27 $-\frac{1}{8}\cos(4x^2) + C$

29 $\frac{1}{6}e^{3x^2} + C$

31 $\frac{1}{10}\ln(5q^2 + 8) + C$

33 $(1/2)\ln(y^2 + 4) + C$

35 $2e^{\sqrt{y}} + C$

37 $2\sqrt{x + e^x} + C$

39 $(1/2)\ln(x^2 + 2x + 19) + C$

41 (a) Yes; $-0.5\cos(x^2) + C$
(b) No
(c) No
(d) Yes; $-1/(2(1 + x^2)) + C$
(e) No
(f) Yes; $-\ln|2 + \cos x| + C$

43 (a) $x^4 + 2x^2 + C$
(b) $(x^2 + 1)^2 + C$
(c) Both correct but differ by a constant

Section 7.3

1 10

3 6

5 1/2

7 125

9 52

11 38/3

13 $(\ln 2)/2 \approx 0.35$

15 $5 - 5e^{-0.2} \approx 0.906$

17 $\sin 1 - \sin(-1) = 2\sin 1$

19 $20(e^{0.15} - 1)$

21 (a) $(\ln 2)/2$
(b) $(\ln 2)/2$

23 $\ln 10$

25 2

27 $\displaystyle\int_0^2 (6x^2 + 1)\,dx = 18$

29 2.32

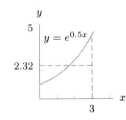

31 11

33 $(300)^{1/3} = 6.694$

35 (a) 6.9 billion, 7.8 billion
(b) 6.5 billion

37 $\displaystyle\int_1^\infty (1/x^2)\,dx$

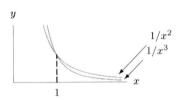

39 (a) 0.4988, 0.49998,
0.4999996,
0.499999998
(b) $(1 - e^{-2b})/2$
(c) Converges to 0.5

41 (a) $\displaystyle\int_0^\infty 1000te^{-0.5t}\,dt$
(b) r

43 1/30

45 (a) No
(b) Yes
(c) Yes

Section 7.4

1 $\frac{1}{5}te^{5t} - \frac{1}{25}e^{5t} + C$

3 $(1/4)(2z + 1)e^{2z} + C$

5 $(x^4/4)\ln x - (x^4/16) + C$

7 $(2/3)y(y + 3)^{3/2}$
$- (4/15)(y + 3)^{5/2} + C$

9 $-(z + 1)e^{-z} + C$

11 $-2y(5 - y)^{1/2}$
$- (4/3)(5 - y)^{3/2} + C$

13 $-t\cos t + \sin t + C$

15 $\frac{1}{5}t^2e^{5t} - \frac{2}{25}te^{5t} + \frac{2}{125}e^{5t} + C$

17 $5\ln 5 - 4 \approx 4.047$

19 $(9/2)\ln 3 - 2 \approx 2.944$

21 (a) Substitution
(b) Substitution
(c) Substitution
(d) Substitution
(e) Parts
(f) Parts

23 $1 - 3e^{-2}$

25 $4\ln 4 - 3\ln 3 - 1$

27 45.71 (ng/ml)-hours

29 Integrate by parts choosing $u = x^n, v' = e^x$

Section 7.5

1 3.5, 2, 1.5, 2, 2.5

3 $F(0) = 0$
$F(1) = 1$
$F(2) = 1.5$
$F(3) = 1$
$F(4) = 0$
$F(5) = -1$
$F(6) = -1.5$

5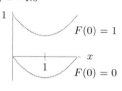

7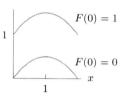

9 (a) Increasing for $x < -2$, $x > 2$
Decreasing for $-2 < x < 2$
Local maximum at $x = -2$
Local minimum at $x = 2$
(b)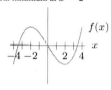

11 (a) $f(x)$ increasing when $2 < x < 5$
$f(x)$ decreasing when $x < 2$ or $x > 5$
f has a local minimum at $x = 2$
f has a local maximum at $x = 5$
(b)

13 size of leaf
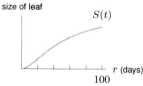

15 (a) (I) volume; (II) flow rate
(b) (I) is an antiderivative of (II)

17 Min: $(1.5, -20)$, max: $(4.67, 5)$

19 Critical points: $(0, 5), (2, 21), (4, 13), (5, 15)$
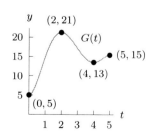

21 (a) $x = -1, x = 1, x = 3$
(b) Local min at $x = -1$, $x = 3$;
local max at $x = 1$
(c)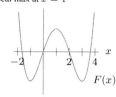

23 $f(3) - f(2)$,
$[f(4) - f(2)]/2$,
$f(4) - f(3)$

25 Slope$= \frac{f(b)-f(a)}{b-a}$
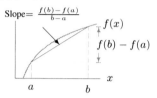

27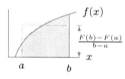

Chapter 7 Review

1 $2t^3/3 + 3t^4/4 + 4t^5/5$
3 $2x^3 - 4x^2 + 3x$
5 $P(r) = \pi r^2 + C$
7 $-1/t$
9 $G(x) = (x + 1)^4/4 + C$
11 $2t^2 + 7t + C$
13 $(x^4/4) - (x^2/2) + C$
15 $-5/t - 3/t^2 + C$
17 $4t^2 + 3t + C$
19 $2x^4 + \ln|x| + C$
21 $2\ln|x| - \pi \cos x + C$
23 34
25 $1 - \cos 1 \approx 0.460$
27 $1/2$
29 $F(x) = x^2$
(only possibility)
31 $2\sqrt{x^2 + 1} + C$
33 $-(1/2)e^{-x^2} + C$
35 $-1/(3(3x + 1)) + C$
37 $\frac{1}{55}(5x - 7)^{11} + C$
39 $\frac{1}{3}(x^2 + 1)^{3/2} + C$
41 $0.5 \sin(t^2) + C$
43 $\ln 3$
45 (a) $\int_0^5 462e^{0.019t}\, dt$
(b) About 2423 quadrillion BTUs
47
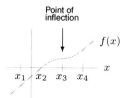
Point of inflection

49 $F(1) = -1, F(3) = 7, F(4) = 5$
51 (a)

$f(x) = xe^{-x}$

(b) $0.9596, 0.9995,$
0.99999996
(c) Converges to 1
53 250 mg
55 $-y \cos y + \sin y + C$

Ch. 7 Understanding

1 False
3 True
5 False
7 False
9 True
11 True
13 False
15 False
17 False
19 True
21 True
23 False
25 False
27

$f(x)$

0.4 10

Practice: Integration

1 $q^3/3 + 5q^2/2 + 2q + C$
3 $(x^3/3) + x + C$
5 $4x^{3/2} + C$
7 $(x^4/4) + 2x^2 + 8x + C$
9 $w^5/5 - 3w^4 + 2w^3 - 10w + C$
11 $2q^{1/2} + C$
13 $p^3/3 + 5\ln|p| + C$
15 $q^4/4 + 4q^2 + 15q + C$
17 $-5 \cos x + 3 \sin x + C$
19 $5\ln|w| + C$
21 $q^2/2 - 1/(2q^2) + C$
23 $3p^2q^5 + C$
25 $2.5e^{2q} + C$
27 $Ax^4/4 + Bx^2/2 + C$
29 $x^3/3 + 8x + e^x + C$
31 $t^3/3 - 3t^2 + 5t + C$
33 $Aq^2/2 + Bq + C$
35 $\frac{1}{2}e^{2t} + 5t + C$

37 $3\sin(4x) + C$

39 $(1/18)(y^2 + 5)^9 + C$

41 $-(A/B)\cos(Bt) + C$

43 $\ln(2 + e^x) + C$

45 $2\sqrt{1 + \sin x} + C$

47 $xe^x - e^x + C$

Section 8.1

1 (a) 0.25
 (b) 0.7
 (c) 0.15

3 (a) 0.4375
 (b) 0.49
 (c) 0.2475

7 (a) 0.19
 (b) Tenth; both same
 (c) 0.02, 0.38, 0.21

9 1/15

11 1/5

13

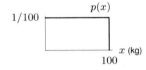

15

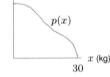

Section 8.2

1 (a)

t	0	1	2	3	4	$\cdots$
$P(t)$	0	0.60	0.975	1	1	$\cdots$

(b) fraction of patients

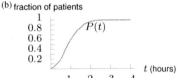

3 (a)

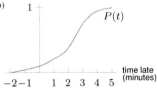

(b)

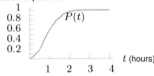

7 (a) 2/3

(b) 1/3
(c) Possibly many work just to pass
(d) fraction of students

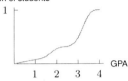

9 (a) First: January 30
 Last: August 28
 (b) 65%
 (c) 25%
 (d) 10%
 (e) April 10 – April 30

11 (a) 25%
 (b) 32.5%
 (c) 0

13 Fraction dead at time t

15 (a) $-e^{-2} + 1 \approx 0.865$
 (b) $-(\ln 0.05)/2 \approx 1.5$ km

Section 8.3

1 5.35 tons

3 (a) 0.2685
 (b) 0.067

5 2.48 weeks

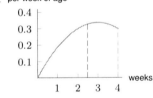

7 (a) 0.684 : 1
 (b) 1.6 hours
 (c) 1.682 hours

9 (a) 17.6%; 6%
 (b) About $45,200
 (c) False

Chapter 8 Review

1 (a)-(II), (b)-(I), (c)-(III)

3 fraction of population per foot

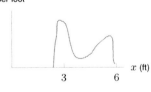

5 % of population per dollar of income

% of population having at least this income

7 0.04

9 0.008

11 (a) Twelfth
 (b) 1/4
 (c) 7/16

13 (b) About 3/4

15 (a) $f(r) = 0.2\,(0 < r < 5)$
 $f(r) = 0\,(5 \leq r)$
 (b) $F(r) = 0.2r$
 $(0 \leq r \leq 5)$
 $F(r) = 1\,(5 < r)$
 (c) $G(v) = 0.124r^{1/3}$
 $(0 \leq v \leq 523.6)$
 $G(v) = 1\,(523.6 < v)$
 (d) $g(v) = 0.0413v^{-2/3}\,(0 < v < 523.6)$
 $g(v) = 1\,(523.6 \leq v)$

17 (a) 22.1%
 (b) 33.0%
 (c) 30.1%
 (d) $C(h) = 1 - e^{-0.4h}$

19 14.6 days

21 False

23 True

25 True

Ch. 8 Understanding

1 True

3 False

5 False

7 False

9 True

11 True

13 False

15 False

17 False

19 True

21 False

23 False

25 False

27 True

29 False

Section 9.1

3 $f(20, p)$: 2.65, 2.59, 2.51, 2.43
 $f(100, p)$: 5.79, 5.77, 5.60, 5.53
 $f(I, 3.00)$: 2.65, 4.14, 5.11, 5.35, 5.79
 $f(I, 4.00)$ 2.51, 3.94, 4.97, 5.19, 5.60

7 $P = 0.052pf(I, p)/I$

9 Decreasing function of p
 Increasing function of a

11 (a) 81°F
 (b) 30%

13 Incr of A and r
 Decr of t

15

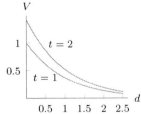

17 (a) 260 cal/m³

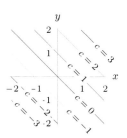

(b)

	w (gm/m³)			
T (°C)	0.1	0.2	0.3	0.4
0	150	290	425	590
10	110	240	330	450
20	100	180	260	350
30	70	150	220	300
40	65	140	200	270

Section 9.2

1 Decreasing function of x
Increasing function of y

3 Heat coming in window

5
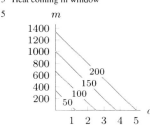

7 predicted high temperature
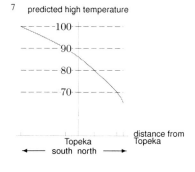

9 low temperature
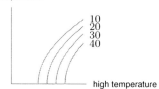

11 Contours evenly spaced

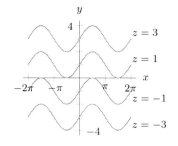

15 Contours evenly spaced

17 Contours evenly spaced

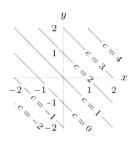

19 Contours evenly spaced

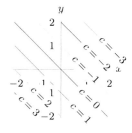

21 (a) False
(b) True
(c) False
(d) True

23 (a) 510.17 thousand pages/day
(b) 773.27 thousand pages/day
(c) 673.17 thousand pages/day
(d) 1020.34 thousand pages/day

25 (a) III
(b) II
(c) I
(d) IV

27 (a) 4 hours
(b) 40%
(c) Contours approx horizontal
(d) Increasing
(e) Increasing

Section 9.3

1 (a) Positive
(b) Negative
(c) Positive
(d) Zero

3 $\frac{\partial I}{\partial H}\big|_{(10,100)} \approx 0.4$
$\frac{\partial I}{\partial T}\big|_{(10,100)} \approx 1$

5 (a) f_c is negative
f_t is positive

7 (a) Concentration/distance
Rate of change of concentration with distance
$\partial c/\partial x < 0$

(b) Concentration/time
Rate of change of concentration with time
For small t, $\partial c/\partial t > 0$
For large t, $\partial c/\partial t < 0$

9 $z_x(1,0) \approx 2$
$z_x(0,1) \approx 0$
$z_y(0,1) \approx 10$

11 (a) Negative

13 (a) 3.3 % / in
(b) $-5\% / °$F

15 $\partial f/\partial P_1 < 0$
$\partial f/\partial P_2 > 0$

19 14 %

21 5.67

23 (a)
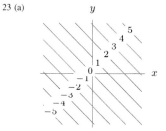

(b)
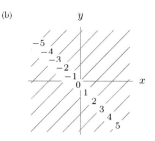

(c)
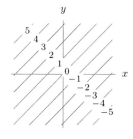

(d)
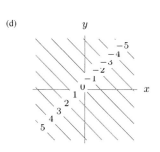

Section 9.4

1 $4x; 6y$

3 $f_x = 2x + 2y$
   $f_y = 2x + 3y^2$

5 $2u + 5v; 5u + 2v$

7 $5a^2 - 9ap^2$

9 $f_x = 10xy^3 + 8y^2 - 6x,$
   $f_y = 15x^2y^2 + 16xy$

11 $100te^{rt}$

13 $(1/2)v^2$

15 15; 5; 30

17 $Pre^{rt}; Pte^{rt}; e^{rt}$

19 80; 30; 313 tons
    2.9 tons per worker
    2.6 tons per $25,000

21 $f_{xx} = 2, f_{xy} = 2,$
    $f_{yy} = 2, f_{yx} = 2$

23 $f_{xx} = 0, f_{xy} = -2/y^2,$
    $f_{yy} = 4x/y^3, f_{yx} = -2/y^2$

25 $f_{xx} = y^2e^{xy},$
    $f_{xy} = (xy + 1)e^{xy},$
    $f_{yy} = x^2e^{xy},$
    $f_{yx} = (xy + 1)e^{xy}$

27 $V_{rr} = 2\pi h, V_{hh} = 0,$
    $V_{rh} = V_{hr} = 2\pi r$

29 $B_{xx} = 0, B_{tt} = 20xe^{-2t},$
    $B_{xt} = B_{tx} = -10e^{-2t}$

31 $f_{rr} = 100t^2e^{rt}, f_{tt} = 100r^2e^{rt},$
    $f_{tr} = f_{rt} = 100(rt + 1)e^{rt}$

35 (a) $-(c + 1)B/(c + r)^2$
   (b) Negative

Section 9.5

1 $f(0, 0) \approx 12.5$ is a local and global maximum
   $f(13, 30) \approx 4.5$ is a local minimum
   $f(37, 18) \approx 2.5$ is a local and global minimum
   $f(32, 34) \approx 10.5$ is a local maximum

3 Max: 11 at $(5.1, 4.9)$
   Min: -1 at $(1, 3.9)$

5 Max: 1 at $(\pi/2, 0); (\pi/2, 2\pi)$
   Min: -1 at $(\pi/2, \pi)$

7 $(-3, 6)$, neither

9 $(4, 2)$, neither

11 Saddle pts: $(1, -1), (-1, 1)$
    local max $(-1, -1)$
    local min $(1, 1)$

13 Local maximum at $(-1, 0)$,
    Saddle points at $(1, 0)$ and $(-1, 4)$,
    Local minimum at $(1, 4)$

15 Local max: $(1, 5)$

17 $A = 10, B = 4, C = -2$

19 (b) $p_1 = p_2 = 25$
    Max revenue is 4375

21 $q_1 = 300, q_2 = 225.$

Section 9.6

1 $f(66.7, 33.3) = 13,333$

3 $f(9.26, 2.32) = 201.9$

5 Min $= -\sqrt{2}$, max $= \sqrt{2}$

7 Min $= -22$, max $= 22$

9 Min $= -2$, max $= 2$

11 $x = 6; y = 6; f(6, 6) = 400$

13 (a) $C = 127x_1 + 92x_2$
    (b) $x_1^{0.6}x_2^{0.4} = 500$

---

15 (b) $L = 40, K = 30$

17 (a) Reduce K by $1/2$ unit,
    increase L by 1 unit.

19 (a) $P(x, y); C(x, y) = 50,000$

    (b) $C(x, y); P(x, y) = 2000$

21 (a) $P(K, L)$
    (b) $C(K, L) = 600,000$
    (c) Tons/dollar
    (d) Extra dollar produces approximately extra
    3.17 tons

23 (a) Quantity of fuel, $x_1 + x_2$
    (b) Terminal velocity (as function of x_1 and x_2)
    $= v_0$
    (c) Liters per meter/sec
    (d) 51 meters/sec requires about 8 more liters
    than 50 meters/sec

25 1820.04; about 209

Chapter 9 Review

1 Decreasing function of x
   Increasing function of y

3 Lines with slope $3/5$, evenly spaced

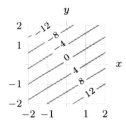

5 windspeed, v, (mph)

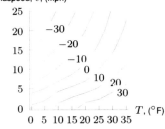

9 x-axis: price
   y-axis: advertising

11 Answers in $^\circ$C:

(a)

(b)

(c)

(d)

13 Table (a) matches (II)
    Table (b) matches (III)
    Table (c) matches (IV)
    Table (d) matches (I)

15 Positive, Negative, 10, 2, -4

17 $f_w(10, 25) \approx -0.4^\circ$F/mph

19 $f_w(5, 20) \approx -0.8$

21 (a) Both negative
    (b) Both negative

23 $(A) 0.06, -0.06$
    $(B) 0, -0.05$
    $(C) 0, 0$

25 $P_a = 2a - 2b^2, P_b = -4ab$

27 $\dfrac{\partial f}{\partial x} = 5e^{-2t}, \dfrac{\partial f}{\partial t} = -10xe^{-2t}$

29 $f_x = x/\sqrt{x^2 + y^2}$
    $f_y = y/\sqrt{x^2 + y^2}$

31 (a) $Q_K = 18.75K^{-0.25}L^{0.25},$
    $Q_L = 6.25K^{0.75}L^{-0.75}$
    (b) $Q = 1704.33$
    $Q_K = 21.3$
    $Q_L = 4.26$

33 (a)

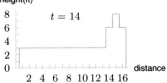

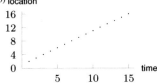

(b) location

(c) The "wave" at a sports arena

35 $(1, 0), (-1, 0); f(1, 0)$ is a local minimum

37 $p_1 = 110, p_2 = 115$

39 43

41 (a) $Q = x_1^{0.3} x_2^{0.7}$
   (b) $10x_1 + 25x_2 = 50,000$

43 (a) 2599

   (b) 129; $4,712,958

   (c) $8572.54 per car

45 (a) 475 units
   (b) 505 units

Ch. 9 Understanding

1 Could not be true

3 Might be true

5 True

7 True

9 False

11 False

13 True

15 True

17 False

19 False

21 True

23 True

25 False

27 True

29 True

31 True

33 False

35 False

37 True

39 False

41 False

43 False

45 True

47 False

49 True

51 True

53 True

55 False

57 True

59 True

61 True

63 False

65 False

Theory: Least Squares

1 $y = 2/3 - (1/2)x$

3 $y = 2/3 - (1/2)x$

5 $y = x - 1/3$

7 (a)   (i) Power function
       (ii) Linear function
   (b) $\ln N = 1.20 + 0.32 \ln A$
       Agrees with biological rule

Section 10.1

1 (a) (III)
   (b) (IV)
   (c) (I)
   (d) (II)

3 $dP/dt = kP, k > 0$

5 $dQ/dt = kQ, k < 0$

7 $dP/dt = -0.08P - 30$

9 $dA/dt = -1$

11 (a) $dA/dt = -0.17A$
   (b) -17 mg

13 (a) Increasing, decreasing
   (b) $W = 4$

15 $dN/dt = B + kN$

Section 10.2

1 $y = t^2 + C$

5 F

7 E

9 A

11 D

13 8, 8.5, 9.5, 11

15 4, 4, 4, 4

17 74, 78.8, 84.56 million

19 $k = -0.03$ and C is any number, or $C = 0$
   and k is any number

21 $k = 5$

Section 10.3

1 (a)

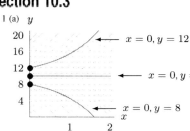

$x = 0, y = 12$

$x = 0, y = 10$

$x = 0, y = 8$

3 (a)

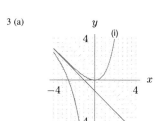

   (b) $y = -x - 1$

5 (a) Slopes = $2, 0, -1, -4$
   (b)

Slope = 0

Slope = -1

Slope = 2

Slope = -4

7 (a) II
   (b) VI
   (c) IV
   (d) I
   (e) III

(f) V

9 As x increases, $y \to \infty$.

11 As $x \to \infty$, y oscillates
   within a certain range

13 $y \to \infty$ as $x \to \infty$

Section 10.4

1 $w = 30e^{3r}$

3 $P = 20e^{0.02t}$

5 $Q = 50e^{(1/5)t}$

7 (a) $dM/dt = rM$
   (b) $M = 1000e^{rt}$
   (c)

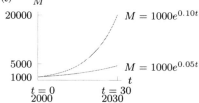

9 (a) $dB/dt = 0.10B + 1000$
   (b) $B = 10,000e^{0.1t} - 10,000$

11 Michigan: 72 years
   Ontario: 18 years

13 (a) $dQ/dt = -0.5365Q$
       $Q = Q_0 e^{-0.5365t}$
   (b) 4 mg

15 (a) $dy/dt = ky$
   (b) 0.2486 grams

17 (b) 2001

Section 10.5

1 $H = 75 - 75e^{3t}$

3 $P = 104e^t - 4$

5 $Q = 400 - 350e^{0.3t}$

7 $B = 25 + 75e^{2-2t}$

11 (a) $dV/dt = 0.02V - 80,000$
   (b) $V = \$4,000,000$
   (c) $V = 4,000,000 + Ce^{0.02t}$
   (d) $2,728,751$

13 (a) $dB/dt = 0.05B + 1200$
   (b) $B = 24,000(e^{0.05t} - 1)$
   (c) 6816.61

15 (a) $y = ce^{-t} + 100$
   (b)

$C = -50$

$C = -100$
$C = -150$

   (c) $y = 100 - 100e^{-t}$

17 (a) $dy/dt = -k(y - a)$
   (b) $y = (1 - a)e^{-kt} + a$
   (c) a: fraction remembered in the long run
       k: rate material is forgotten

19 (a) $y = 500$
   (b) $y = 500 + Ce^{0.5t}$

(c)

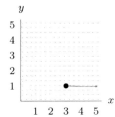

(d) Unstable

21 (a) $H = 200 - 180e^{-kt}$
  (b) $k \approx 0.027$ (if t is in minutes)

23 (a) $dT/dt = -k(T - 68)$
  (b) $T = 68 + 22.3e^{-0.06t}$;
    3:45 am.

Section 10.6

1 (a) $x \to \infty$ exponentially
    $y \to 0$ exponentially
  (b) Predator-prey

3 (a) $x \to \infty$ exponentially
    $y \to 0$ exponentially

  (b) y is helped by the presence of x

5 $dx/dt = x - xy$,
  $dy/dt = y - xy$

7 $dx/dt = -x - xy$,
  $dy/dt = -y - xy$

11 Symmetric about the line $r = w$;
    solutions closed curves

13 Robins:
    Max ≈ 2500
    Min ≈ 500
    When robins are at a max,
    the worm population is about 1 million

17 (a) $dw/dt = 0$
      $dr/dt = 1.2$
  (b) $w \approx 2.2, r \approx 1.1$
  (c) At $t = 0.2$:
      $w \approx 2.2, r \approx 1.3$
      At $t = 0.3$
      $w \approx 2.1, r \approx 1.4$

19 (a) r (predator)

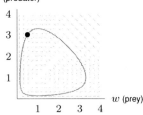

w (prey)

  (b) Down and left
  (c) $r = 3.3, w = 1$
  (d) $w = 3.3, r = 1$

21 (a)

(b)

(c)

(d)

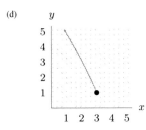

Section 10.7

5 (a) $I_0 = 1, S_0 = 349$
  (b) Increases; spreads

7 About 300 boys;
    $t \approx 6$ days

9 5

11 (a) b/a

Chapter 10 Review

1 (a) (III)
  (b) (V)
  (c) (I)
  (d) (II)
  (e) (IV)

3 Yes

5 (a) I is $y' = 1 + y$;
      II is $y' = 1 + x$
  (b) I

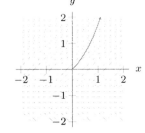

II

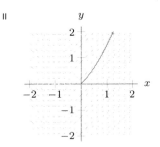

  (c) I: $y' = -1$, unstable;
      II: None

7 III: $y' = (1 + y)(2 - y)$

9 (a) $dB/dt = 0.07B$
  (b) $B = B_0 e^{0.07t}$
  (c) $B = 5000e^{0.07t}$
  (d) $B(10) \approx \$10,068.76$

11 $P = (1/2)t^2 + C$

13 $y = (5/2)t^2 + C$

15 $A = Ce^{-0.07t}$

17 $P = Ce^{-2t} + 5$

19 $y = Ce^{0.2x} + 40$

21 $dS/dt = -k(S - 65), k > 0$
    $S = 65 - 25e^{-kt}$

23 (a) $k \approx 0.000121$
  (b) 779.4 years

25 (a)

(b) $dQ/dt = -0.0187Q$
  (c) 3 days

27 (b) $dQ/dt = -0.347Q + 2.5$
  (c) $Q = 7.2$ mg

29 (a) $dW/dt = (1/3500)(I - 20W)$

  (b) $W = I/20 + (W_0 - I/20)\,e^{-(1/175)t}$
  (c)

31 (a) $y = 1, y = 8, y = 16$
  (d) Stable: $y = 1$
      Unstable: $y = 8, y = 16$

33 Initially $x = 0$; y decreases, x increases. Then x increases, y increases. Finally y increases, x decreases

35 Populations oscillate

Ch. 10 Understanding

1 True

3 True

5 True

7 False

9 True

11 False

13 False

15 False; $f(x) = |x|$

17 False; $\cos t + t^2$

19 False; $f(x) = 6, g(x) = 10$

21 False; $f(x) = 5x + 7, g(x) = x + 2$

23 False; $f(x) = x^2, g(x) = x^2 - 1$

25 False; $f(x) = e^{-x}, g(x) = x^2$

27 (a) Not a counterexample
   (b) Not a counterexample
   (c) Not a counterexample
   (d) Counterexample

29 False

31 False

33 Possible answer

$$f(x) = \begin{cases} x & \text{if } 0 \le x < 2 \\ 19 & \text{if } x = 2 \end{cases}$$

Theory: Separation of Vars

1 $P = e^{-2t}$

3 $P = \sqrt{2t + 1}$

5 $u = 1/(1 - (1/2)t)$

7 $R = 1 - 0.9e^{1-y}$

9 $z = -\ln(1 - t^2/2)$

11 $y = -2/(t^2 + 2t - 4)$

13 (a) Yes   (b) No   (c) Yes
   (d) No   (e) Yes   (f) Yes
   (g) No   (h) Yes   (i) No
   (j) Yes   (k) Yes   (l) No

15 $Q = b - Ae^{-t}$

17 $R = -(b/a) + Ae^{at}$

19 $y = -1/\left(k(t + t^3/3) + C\right)$

21 (a)

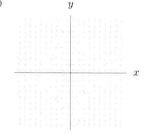

  (b)

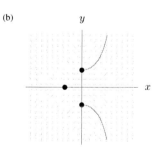

  (c) $y(x) = Ae^{x^2/2}$

Section 11.1

1 21

3 3,985,805

5 555.10

7 96.154

9 39,375

11 1039.482

13 400

15 30.51, 37.75,
   39.47, 39.87, Yes

17 (a) $3037.75, $2537.75
   (b) $6167.78, $5667.78

19 (a) 39 mg
   (b) 41.496 mg
   (c) 41.667 mg

21 (a) 40 mg
   (b) 0.57 mg/kg, Yes
   (c) (i) Greater than 100 kg
      (ii) Less than 13.3 kg

Section 11.2

1 Balance = $48,377.01,
   $20,000. from deposits,
   $28,377.01 from interest

3 $33,035.37

5 $42,567.82

7 17.54 payments

9 $44,407.33, $360,183.21

11 0.12%

13 $1081.11

15 (a) $1250
   (b) 12.50

17 (a) $400 billion
   (b) $900 billion

19 (a) $N + (1 - r)N$ dollars
   (b) $N + (1 - r)N + (1 - r)^2N$ dollars
   (c) N/r dollars

Section 11.3

1 230.159 billion barrels

3 (a) 98 mg
   (b) 121.5 mg
   (c) 125 mg

5 (a) 400 mg
   (b) $400(0.30) = 120$ mg

7 (a) 39.99 mg, 9.99 mg
   (b) 40 mg, 10 mg

9 (a) 4.10 mg
   (b) 7.54 mg
   (c) 14.46 mg
   (d) 35.24 mg
   (e) 69.87 mg

11 24.5 years

13 Until 2046

15 About 34 years

17 Lasts forever

Chapter 11 Review

1 2046

3 Does not exist

5 200

7 1.9961

9 (a) (i) $16.43 million
     (ii) $24.01 million
   (b) $16.87 million

11 $27,979.34

13 $400 million

15 (a) $N(k/(1 - k))$
   (b) $5.667N$

17 (a) $0.232323\ldots =$
     $0.23 + 0.23(0.01)$

     $+ 0.23(0.01)^2 + \cdots$
   (b) $0.23/(1 - 0.01) = (23)/(99)$

19 $25,503

Ch. 11 Understanding

1 False

3 True

5 False

7 False

9 False

11 False

13 False

15 False

17 True

19 False

21 False

23 False

25 False

27 False

29 True

31 False

33 False

35 False

37 False

39 True

41 True

43 True

45 True

Appendix A

1 (a)

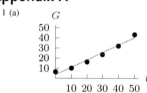

  (b) Increasing at a rate of $734 billion/year
  (c) $43.9 trillion, $54.9 trillion,
     More confidence in the 2005 prediction

3 (a)

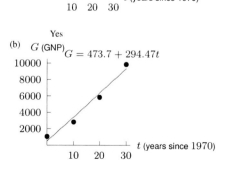

    Yes

  (b)

$$G = 294.47t + 473.7$$

(c) For 1985: 4891
For 2020: 15,197
More confidence in 1985

5 (a) $S = 0.08v + 1.77$
(b)
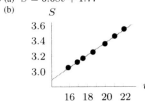

Yes
(c) At $v = 18$ ft/sec, $S = 3.21$
At $v = 10$ ft/sec, $S = 2.57$
$v = 18$ better

7 (a) $0.0026 = 0.26\%$
(b) For 1900, 272.27
For 1980, 335.1

9 (a)

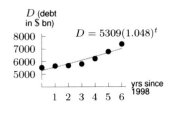
D (debt in $ bn)

(b) Exponential
(c) $D = 5309(1.048)^t$, answers may vary

D (debt in $ bn)
$D = 5309(1.048)^t$

(d) About 4.8%
(e) No

11 (a) $r = 1$
(b) $r = 0.7$
(c) $r = 0$
(d) $r = -0.98$
(e) $r = -0.25$
(f) $r = -0.5$

13 (a) Exponential

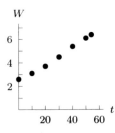

(b) $2.6(1.0165)^t$; answers may vary
(c) 1.65%
(d) At year 2020, 8.175 billion
At year 2050, 13.357 billion
2020 prediction more accurate

15 (a)
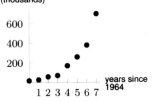
number of cars (thousands)

(b) Exponential
(c) $C = 15.9 \cdot (1.725)^t$, answers may vary

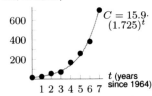
C (number of cars, thousands)
$C = 15.9 \cdot (1.725)^t$

(d) About 73%
(e) No

17 (a) Negative
(b) $f(t) = -0.03t^2 + 1.01t + 13.82$

E (mpg)

19 (a) Exponential
(b) $S = 29.96(1.30)^t$, answers may vary
Increasing 30%/yr
(c) 21,141 megawatts
(d)
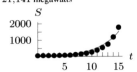
S

21 (a) Linear
(b) $C = 320 + 1.5t$
Increasing at 1.5 ppm/yr
(c) 387.5 ppm
(d)
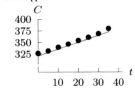
C

Appendix B

1 27%

3 (a) $160,356.77
(b) $165,510.22
(c) $165,891.05
(d) $165,989.48
(e) $166,005.85

5 (a) 1.0408107
1.0408108
1.0408108
4% compounded continuously
$\approx 4.08108\%$
(b) $e^{0.04} \approx 1.048108$

7 4.88%

9 (a) (i) $5.126978\ldots\%$
(ii) $5.127096\ldots\%$
(iii) $5.127108\ldots\%$
(b) 5.127%
(c) $e^{0.05} = 1.05127109\ldots$

11 (a) = V, (b) = III, (c) = IV,
(d) = I, (e) = II

13 (a) 13,900 cruzados
(b) 24.52%

Eighth Edition

Mathematics
An Applied
Approach

Michael Sullivan Chicago State University

Abe Mizrahi Indiana University Northwest

JOHN WILEY & SONS, INC. WILEY

ACQUISITIONS EDITOR	Michael Boezi
ASSOCIATE PUBLISHER	Laurie Rosatone
FREELANCE DEVELOPMENTAL EDITOR	Anne Scanlan-Rohrer
EXECUTIVE MARKETING MANAGER	Julie Lindstrom
SENIOR PRODUCTION EDITOR	Norine M. Pigliucci
SENIOR DESIGNER	Harry Nolan
COVER DESIGN	Howard Grossman
INTERIOR DESIGN	Jerry Wilke Design
ILLUSTRATION EDITOR	Sigmund Malinowski
ELECTRONIC ILLUSTRATIONS	Techsetters, Inc.
PHOTO EDITOR	Lisa Gee
ASSISTANT EDITOR	Jennifer Battista
PROGRAM ASSISTANT	Kelly Boyle
PRODUCTION MANAGEMENT SERVICES	Suzanne Ingrao/Ingrao Associates
COVER PHOTOS	(Peach) © Tim Turner/FoodPix/Getty Images
	(Apple) © Brian Hagiwara/FoodPix/Getty Images
	(Plum) © Richard Kolker/The Image Bank/Getty Images
INSET COVER PHOTOS	© Getty Images and Digital Vision

Excel is a trademark of Microsoft, Inc.

This book was set in Minion by Progressive Information Technologies and printed and bound by Von Hoffmann Corporation. The cover was printed by Von Hoffmann Corporation.

This book is printed on acid free paper. ∞

ISBN 0-471-32784-0

WIE ISBN 0-471-65664-X

Printed in the United States of America

10 9 8 7 6 5 4 3 2 1

Systems of Linear Equations; Matrices

Economists are always talking about the influence of consumer spending on the economy. What do they mean? Suppose you have $5000 in disposable income and could spend it on a new bathroom for your house or on a plasma TV or on a dream vacation in Fiji. Would your decision impact the economy? What if you decided to spend the $5000 on the construction of the new bathroom? This would help the construction industry but would not help the consumer electronics industry or the travel industry. What if 200 people were in the same position as you and each of them made the decision to spend $5000 on construction, resulting in increased demand for construction of $1,000,000? How would this impact other segments of the economy? Which ones does it help? Which ones does it hurt? A famous economic model, the Leontief Model, was constructed to answer such questions. We study this model in Section 2.7 and answer some of the questions listed here in the Chapter Project at the end of the chapter.

48

In Section 1.2 of Chapter 1, we discussed pairs of lines: coincident lines, parallel lines, and intersecting lines. Each line was given by a linear equation containing two variables. So a *pair* of lines is given by *two* linear equations containing two variables. We refer to this as a *system of two linear equations containing two variables.*

In this chapter we take up the problem of *solving* systems of linear equations containing two or more variables. As the section titles suggest, there are various ways to do this. The *method of substitution* for solving equations in several

unknowns goes back to ancient times. The *method of elimination,* though it had existed for centuries, was put into systematic order by Karl Friedrich Gauss (1777–1855) and by Camille Jordan (1838–1922). This method led to the *matrix method* that is now used for solving large systems by computer.

The theory of *matrices* was developed in 1857 by Arthur Cayley (1821–1895), though only later were matrices used as we use them in this chapter. Matrices have become a very flexible instrument, invaluable in almost all areas of mathematics.

2.1   Systems of Linear Equations: Substitution; Elimination*

PREPARING FOR THIS SECTION   *Before getting started, review the following:*

> Pairs of Lines (Section 1.2, pp 19–23)

OBJECTIVES   1   Solve systems of equations by substitution
                2   Solve systems of equations by elimination
                3   Identify inconsistent systems of equations containing two variables
                4   Express the solutions of a system of dependent equations containing two variables
                5   Solve systems of three equations containing three variables
                6   Identify inconsistent systems of equations containing three variables
                7   Express the solutions of a system of dependent equations containing three variables

We begin with an example.

EXAMPLE 1   Movie Theater Ticket Sales

A movie theater sells tickets for $8.00 each, with seniors receiving a discount of $2.00. One evening the theater took in $3580 in revenue. If x represents the number of tickets sold at $8.00 and y the number of tickets sold at the discounted price of $6.00, write an equation that relates these variables.

SOLUTION   Each nondiscounted ticket brings in $8.00, so x tickets will bring in $8x$ dollars. Similarly, y discounted tickets bring in $6y$ dollars. Since the total brought in is $3580, we must have

$$8x + 6y = 3580$$

        ◗

*Based on material from Precalculus, 6th ed., by Michael Sullivan. Used here with the permission of the author and Prentice-Hall, Inc.

The equation found in Example 1 is an example of a **linear equation containing two variables.** Some other examples of linear equations are

$$2x + 3y = 2 \qquad 5x - 2y + 3z = 10 \qquad 8x_1 + 8x_2 - 2x_3 + 5x_4 = 0$$

2 variables 3 variables 4 variables

In general, an equation containing n variables is said to be **linear** if it can be written in the form

$$a_1x_1 + a_2x_2 + \cdots + a_nx_n = b$$

where $x_1, x_2, \ldots, x_n$ are n distinct variables*, $a_1, a_2, \ldots, a_n, b$ are constants, and at least one of the a_i's is not 0.

In Example 1, suppose that we also know that 525 tickets were sold that evening. Then we have another equation relating the variables x and y.

$$x + y = 525$$

The two linear equations

$$8x + 6y = 3580$$
$$x + y = 525$$

form a *system* of linear equations.

In general, a **system of linear equations** is a collection of two or more linear equations, each containing one or more variables. Example 2 illustrates some systems of linear equations.

EXAMPLE 2   **Examples of Systems of Linear Equations**

(a) $\begin{cases} 2x + y = 5 & (1) \quad \text{Two equations containing} \\ -4x + 6y = -2 & (2) \quad \text{two variables, } x \text{ and } y \end{cases}$

(b) $\begin{cases} x + y + z = 6 & (1) \quad \text{Three equations containing} \\ 3x - 2y + 4z = 9 & (2) \quad \text{three variables, } x, y, \text{ and } z \\ x - y - z = 0 & (3) \end{cases}$

(c) $\begin{cases} x + y + z = 5 & (1) \quad \text{Two equations containing} \\ x - y = 2 & (2) \quad \text{three variables, } x, y, \text{ and } z \end{cases}$

(d) $\begin{cases} x + y + z = 6 & (1) \quad \text{Four equations containing} \\ 2x + 2z = 4 & (2) \quad \text{three variables, } x, y, \text{ and } z \\ y + z = 2 & (3) \\ x = 4 & (4) \end{cases}$

(e) $\begin{cases} x_1 - 2x_2 + x_3 - x_4 = 5 & (1) \quad \text{Two equations containing four} \\ 3x_1 + x_2 - x_3 - 5x_4 = 2 & (2) \quad \text{variables } x_1, x_2, x_3, \text{ and } x_4 \end{cases}$

We use a brace, as shown above, to remind us that we are dealing with a system of linear equations. We also will find it convenient to number each equation in the system.

A **solution** of a system of linear equations consists of values of the variables that are solutions of each equation of the system. To **solve** a system of linear equations means to find all solutions of the system.

---

*The notation x_n is read as "x sub n." The number n is called a **subscript** and should not be confused with an exponent. We use subscripts to distinguish one variable from another when a large or undetermined number of variables is required.

For example, $x = 2, y = 1$ is a solution of the system in Example 2(a) because

$$\begin{cases} 2x + y = 5 & (1) \\ -4x + 6y = -2 & (2) \end{cases} \qquad \begin{cases} 2(2) + 1 = 4 + 1 = 5 & (1) \\ -4(2) + 6(1) = -8 + 6 = -2 & (2) \end{cases}$$

A solution of the system in Example 2(b) is $x = 3, y = 2, z = 1$, because

$$\begin{cases} x + y + z = 6 & (1) \\ 3x - 2y + 4z = 9 & (2) \\ x - y - z = 0 & (3) \end{cases} \qquad \begin{cases} 3 + 2 + 1 = 6 & (1) \\ 3(3) - 2(2) + 4(1) = 9 - 4 + 4 = 9 & (2) \\ 3 - 2 - 1 = 0 & (3) \end{cases}$$

Note that $x = 3, y = 3, z = 0$ is not a solution of the system in Example 2(b).

$$\begin{cases} x + y + z = 6 & (1) \\ 3x - 2y + 4z = 9 & (2) \\ x - y - z = 0 & (3) \end{cases} \qquad \begin{cases} 3 + 3 + 0 = 6 & (1) \\ 3(3) - 2(3) + 4(0) = 3 \ne 9 & (2) \\ 3 - 3 - 0 = 0 & (3) \end{cases}$$

Although these values satisfy Equations (1) and (3), they do not satisfy Equation (2). Any solution of the system must satisfy *each* equation of the system.

When a system of equations has at least one solution, it is said to be **consistent;** otherwise, it is called **inconsistent.**

 NOW WORK PROBLEM 3.

Two Linear Equations Containing Two Variables

Based on the discussion in Section 1.2, we can view the problem of solving a system of two linear equations containing two variables as a geometry problem. Because the graph of each equation in such a system is a line, a system of two linear equations containing two variables represents a pair of lines. The lines either (1) are parallel or (2) are intersecting or (3) are coincident (that is, identical).

1. If the lines are parallel, then the system of equations has no solution, because the lines never intersect. The system is **inconsistent.**
2. If the lines intersect, then the system of equations has one solution, given by the point of intersection. The system is **consistent** and the equations are **independent.**
3. If the lines are coincident, then the system of equations has infinitely many solutions, represented by the totality of points on the line. The system is **consistent** and the equations are **dependent.**

Based on this, a system of equations is either

(I) Inconsistent; has no solution
   or
(II) Consistent; with
   (a) One solution (equations are independent)
      or
   (b) Infinitely many solutions (equations are dependent)

Figure 1 illustrates these conclusions.

FIGURE 1

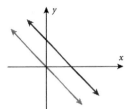 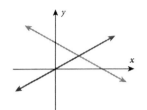 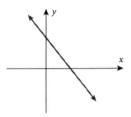

(a) Parallel lines; system has no solution and is inconsistent

(b) Intersecting lines; system has one solution and is consistent; the equations are independent

(c) Coincident lines; system has infinitely many solutions and is consistent; the equations are dependent

---

EXAMPLE 3   **Graphing a System of Linear Equations**

Graph the system: $\begin{cases} 2x + y = 5 & (1) \\ -4x + 6y = 12 & (2) \end{cases}$

SOLUTION   Equation (1) is a line with x-intercept $\left(\frac{5}{2}, 0\right)$ and y-intercept $(0, 5)$. Equation (2) is a line with x-intercept $(-3, 0)$ and y-intercept $(0, 2)$.

FIGURE 2

Figure 2 shows their graphs.

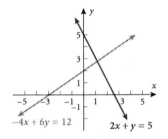

From the graph in Figure 2 we see that the lines intersect, so the system is consistent and the equations are independent. We can also use the graph as a means of approximating the solution. For this system the solution would appear to be close to the point $(1, 3)$. The actual solution, which you should verify, is $\left(\frac{9}{8}, \frac{11}{4}\right)$.

To obtain the exact solution, we use algebraic methods. The first algebraic method we take up is the *method of substitution*.

Method of Substitution

Solve systems of equations   **1**   We illustrate the method of substitution by solving the system of Example 3.
by substitution

EXAMPLE 4   **Solving a System of Equations Using Substitution**

Solve: $\begin{cases} 2x + y = 5 & (1) \\ -4x + 6y = 12 & (2) \end{cases}$

SOLUTION   We solve the first equation for y, obtaining

$$2x + y = 5 \qquad (1)$$

$$y = -2x + 5 \qquad \text{Subtract } 2x \text{ from each side}$$

We substitute this result for y in the second equation. This results in an equation containing one variable, which we can solve.

$$-4x + 6y = 12 \qquad (2)$$

$$-4x + 6(-2x + 5) = 12 \qquad \text{Substitute } y = -2x + 5 \text{ in (2)}$$

$$-4x - 12x + 30 = 12 \qquad \text{Remove parenthesis.}$$

$$-16x = -18 \qquad \text{Combine like terms; subtract 30 from each side.}$$

$$x = \frac{-18}{-16} = \frac{9}{8} \qquad \text{Divide each side by } -16.$$

Once we know that $x = \frac{9}{8}$, we can find the value of y by **back-substitution,** that is, by substituting $\frac{9}{8}$ for x in one of the original equations.

We will use the first equation.

$$2x + y = 5 \qquad\qquad (1)$$

$$2\left(\frac{9}{8}\right) + y = 5 \qquad\qquad \text{Substitute } x = \frac{9}{8} \text{ in (1).}$$

$$\frac{9}{4} + y = 5 \qquad\qquad \text{Simplify}$$

$$y = 5 - \frac{9}{4} \qquad\qquad \text{Subtract } \frac{9}{4} \text{ from each side.}$$

$$= \frac{20}{4} - \frac{9}{4} = \frac{11}{4}$$

The solution of the system is $x = \dfrac{9}{8} = 1.125, y = \dfrac{11}{4} = 2.75.$

✓ **CHECK:**

$$\begin{cases} 2x + y = 5. & 2\left(\frac{9}{8}\right) + \frac{11}{4} = \frac{9}{4} + \frac{11}{4} = \frac{20}{4} = 5 \\[2mm] -4x + 6y = 12: & -4\left(\frac{9}{8}\right) + 6\left(\frac{11}{4}\right) = -\frac{9}{2} + \frac{33}{2} = \frac{24}{2} = 12 \end{cases}$$ ▶

COMMENT: We can also verify our algebraic solution in Example 4 using a graphing utility.

First, we solve each equation for y. This is equivalent to writing each equation in slope–intercept form. Equation (1) in slope–intercept form is $Y_1 = -2x + 5$. Equation (2) in slope-intercept form is $Y_2 = \frac{2}{3}x + 2$. Figure 3 shows the graphs using a graphing utility. From the graph in Figure 3, we see that the lines intersect, so the system is consistent and the equations are independent. Using INTERSECT, we obtain the solution (1.125, 2.75), which is equivalent to $\left(\frac{9}{8}, \frac{11}{4}\right)$. ▶

The method used to solve the system in Example 4 is called **substitution.** The steps used are outlined in the box below.

FIGURE 3

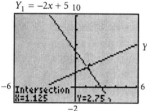

$Y_1 = -2x + 5$

$Y_2 = \frac{2}{3}x + 2$

Steps for Solving by Substitution

STEP 1 Pick one of the equations and solve for one of the variables in terms of the remaining variables.

STEP 2 Substitute the result in the remaining equations.

STEP 3 If one equation in one variable results, solve this equation. Otherwise, repeat Steps 1 and 2 until a single equation with one variable remains.

STEP 4 Find the values of the remaining variables by back-substitution.

STEP 5 Check the solution found.

EXAMPLE 5   **Solving a System of Equations Using Substitution**

Solve: $\begin{cases} 3x - 2y = 5 & (1) \\ 5x - y = 6 & (2) \end{cases}$

SOLUTION   **STEP 1** After looking at the two equations, we conclude that it is easiest to solve for the variable y in Equation (2):

$$5x - y = 6 \qquad (2)$$
$$y = 5x - 6 \qquad \text{Add } y \text{ and subtract 6 from each side.}$$

STEP 2 We substitute this result into Equation (1) and simplify:

$$3x - 2y = 5 \qquad (1)$$
$$3x - 2(5x - 6) = 5 \qquad y = 5x - 6$$

STEP 3
$$-7x + 12 = 5 \qquad \text{Simplify}$$
$$-7x = -7 \qquad \text{Simplify}$$
$$x = 1 \qquad \text{Solve for } x$$

STEP 4 Knowing $x = 1$, we can find y from the equation

$$y = 5x - 6 = 5(1) - 6 = -1 \qquad x = 1$$

STEP 5 *Check:* $\begin{cases} 3(1) - 2(-1) = 3 + 2 = 5 \\ 5(1) - (-1) = 5 + 1 = 6 \end{cases}$

The solution of the system is $x = 1, y = -1$.

NOW WORK PROBLEM 13 USING SUBSTITUTION.

Method of Elimination

Solve systems of equations **2** by elimination

A second method for solving a system of linear equations is the *method of elimination*. This method is usually preferred over substitution if substitution leads to fractions or if the system contains more than two variables. Elimination also provides the necessary motivation for solving systems using matrices (the subject of the next section).

The idea behind the method of elimination is to replace the original system of equations by an equivalent system so that adding two of the equations eliminates a variable. The rules for obtaining equivalent equations are the same as those studied earlier. However, we may also interchange any two equations of the system and/or replace any equation in the system by the sum (or difference) of that equation and any other equation in the system.

Rules for Obtaining an Equivalent System of Equations

1. Interchange any two equations in the system.
2. Multiply (or divide) each side of an equation by the same nonzero constant.
3. Replace any equation in the system by the sum (or difference) of that equation and a nonzero multiple of any other equation in the system.

An example will give you the idea. As you work through the example, pay particular attention to the pattern being followed.

EXAMPLE 6   **Solving a System of Linear Equations Using Elimination**

Solve: $\begin{cases} 2x + 3y = 1 & (1) \\ -x + y = -3 & (2) \end{cases}$

SOLUTION   We multiply each side of equation (2) by 2 so that the coefficients of x in the two equations are opposites of one another. The result is the equivalent system

$$\begin{cases} 2x + 3y = 1 & (1) \\ -2x + 2y = -6 & (2) \end{cases}$$

If we now replace Equation (2) of this system by the sum of the two equations, we obtain an equation containing just the variable y, which we can solve.

$$\begin{cases} 2x + 3y = 1 & (1) \\ -2x + 2y = -6 & (2) \end{cases}$$

$$5y = -5 \quad \text{Add (1) and (2).}$$

$$y = -1 \quad \text{Solve for } y.$$

We back-substitute this value for y in Equation (1) and simplify to get

$$2x + 3y = 1 \quad (1)$$

$$2x + 3(-1) = 1 \quad \text{Subsitute } y = -1 \text{ in (1).}$$

$$2x = 4 \quad \text{Simplify.}$$

$$x = 2 \quad \text{Solve for } x.$$

The solution of the original system is $x = 2$, $y = -1$. We leave it to you to check the solution.

The procedure used in Example 6 is called the **method of elimination.** Notice the pattern of the solution. First, we eliminated the variable x from the second equation. Then we back-substituted; that is, we substituted the value found for y back into the first equation to find x.

> **Steps for Solving by Elimination**
>
> **STEP 1** Select two equations from the system and eliminate a variable from them.
> **STEP 2** If there are additional equations in the system, pair off equations and eliminate the same variable from them.
> **STEP 3** Continue Steps 1 and 2 on successive systems until one equation containing one variable remains.
> **STEP 4** Solve for this variable and back-substitute in previous equations until all the variables have been found.

   NOW WORK PROBLEM 13 USING ELIMINATION.

Let's return to the movie theater example (Example 1).

EXAMPLE 7   Movie Theater Ticket Sales

A movie theater sells tickets for $8.00 each, with seniors receiving a discount of $2.00. One evening the theater sold 525 tickets and took in $3580 in revenue. How many of each type of ticket were sold?

SOLUTION   If x represents the number of tickets sold at $8.00 and y the number of tickets sold at the discounted price of $6.00, then the given information results in the system of equations

$$\begin{cases} 8x + 6y = 3580 & (1) \\ x + y = 525 & (2) \end{cases}$$

We use elimination and multiply equation (2) by -6 and then add the equations.

$$\begin{cases} 8x + 6y = 3580 & (1) \\ -6x - 6y = -3150 & (2) \end{cases}$$

$$2x = 430 \qquad \text{Add (1) and (2).}$$
$$x = 215 \qquad \text{Solve for } x.$$

Since $x + y = 525$, then $y = 525 - x = 525 - 215 = 310$. We conclude that 215 nondiscounted tickets and 310 senior discount tickets were sold.   ▌

Identify inconsistent ③
systems of equations
containing two variables

The previous examples dealt with consistent systems of equations that had one solution. The next two examples deal with two other possibilities that may occur, the first being a system that has no solution.

EXAMPLE 8   An Inconsistent System of Linear Equations

Solve: $\begin{cases} 2x + y = 5 & (1) \\ 4x + 2y = 8 & (2) \end{cases}$

SOLUTION   We choose to use the method of substitution and solve equation (1) for y.

$$2x + y = 5 \qquad (1)$$
$$y = -2x + 5 \qquad \text{Subtract } 2x \text{ from each side.}$$

Now substitute $y = -2x + 5$ for y in equation (2) and solve for x.

$$4x + 2y = 8 \qquad (2)$$
$$4x + 2(-2x + 5) = 8 \qquad \text{Substitute } y = -2x + 5 \text{ in (2).}$$
$$4x - 4x + 10 = 8 \qquad \text{Remove parentheses.}$$
$$0 \cdot x = -2 \qquad \text{Subtract 10 from each side.}$$

This equation has no solution. We conclude that the system itself has no solution and is therefore inconsistent.   ▶

Figure 4 illustrates the pair of lines whose equations form the system in Example 8. Notice that the graphs of the two equations are lines, each with slope -2; one line has y-intercept $(0, 5)$, the other has y-intercept $(0, 4)$. The lines are parallel and have no point of intersection. This geometric statement is equivalent to the algebraic statement that the system is inconsistent and has no solution.

FIGURE 4

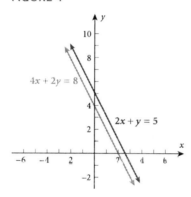

 NOW WORK PROBLEM 19.

The next example is an illustration of a system with infinitely many solutions.

4 **EXAMPLE 9**   **Solving a System of Linear Equations with Infinitely Many Solutions**

Solve: $\begin{cases} 2x + y = 4 & (1) \\ -6x - 3y = -12 & (2) \end{cases}$

SOLUTION   We choose to use the method of elimination:

$$\begin{cases} 2x + y = 4 & (1) \\ -6x - 3y = -12 & (2) \end{cases}$$

$$\begin{cases} 6x + 3y = 12 & (1) \quad \text{Multiply each side of equation (1) by 3.} \\ -6x - 3y = -12 & (2) \end{cases}$$

$$\begin{cases} 6x + 3y = 12 & (1) \quad \text{Replace equation (2) by the sum of} \\ 0 = 0 & (2) \quad \text{equations (1) and (2).} \end{cases}$$

The original system is equivalent to a system containing one equation, so the equations are dependent. This means that any values of x and y for which $6x + 3y = 12$ (or, equivalently, $2x + y = 4$) are solutions. For example, $x = 2$, $y = 0$; $x = 0$, $y = 4$; $x = -2$, $y = 8$; $x = 4$, $y = -4$; and so on, are solutions. There are, in fact, infinitely many values of x and y for which $2x + y = 4$, so the original system has infinitely many solutions. We will write the solutions of the original system either as

$$y = 4 - 2x$$

where x can be any real number, or as

$$x = 2 - \tfrac{1}{2}y$$

where y can be any real number. ▶

FIGURE 5

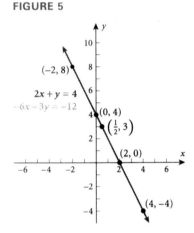

Figure 5 illustrates the situation presented in Example 9. Notice that the graphs of the two equations are lines, each with slope -2 and each with y-intercept $(0, 4)$. The lines are coincident. Notice also that Equation (2) in the original system is just -3 times Equation (1), indicating that the two equations are dependent.

For the system in Example 9 we can find some of the infinite number of solutions by assigning values to x and then finding $y = 4 - 2x$.

When we express the solution in this way, we call x a **parameter**. Thus:

If $x = 4$, then $y = -4$.   This is the point $(4, -4)$ on the graph.

If $x = 0$, then $y = 4$.   This is the point $(0, 4)$ on the graph.

If $x = \tfrac{1}{2}$, then $y = 3$.   This is the point $(\tfrac{1}{2}, 3)$ on the graph.

Alternatively, if we express the solution in the form $x = 2 - \tfrac{1}{2}y$, then y is the parameter and we can assign values to y in order to find x.

If $y = -4$, then $x = 2 - \tfrac{1}{2}(-4) = 4$

If $y = 0$, then $x = 2 - \tfrac{1}{2}(0) = 2$

If $y = 8$, then $x = 2 - \tfrac{1}{2}(8) = -2$

NOW WORK PROBLEM 23.

Three Equations Containing Three Variables

Just as with a system of two linear equations containing two variables, a system of three linear equations containing three variables also has either (1) exactly one solution (a consistent system with independent equations), or (2) no solution (an inconsistent system), or (3) infinitely many solutions (a consistent system with dependent equations).

We can view the problem of solving a system of three linear equations containing three variables as a geometry problem. The graph of each equation in such a system is a plane in space. A system of three linear equations containing three variables represents three planes in space. Figure 6 illustrates some of the possibilities.

Recall that a **solution** to a system of equations consists of values for the variables that are solutions of each equation of the system. For example, $x = 3$, $y = -1$, $z = -5$ is a solution to the system of equations

$$\begin{cases} x + y + z = -3 & (1) \quad 3 + (-1) + (-5) = -3 \\ 2x - 3y + 6z = -21 & (2) \quad 2(3) - 3(-1) + 6(-5) = 6 + 3 - 30 = -21 \\ -3x + 5y = -14 & (3) \quad -3(3) + 5(-1) = -9 - 5 = -14 \end{cases}$$

because these values of the variables are solutions of each equation.

FIGURE 6

| (a) Consistent system; one solution | (b) Consistent system; infinite number of solutions | (c) Inconsistent system; no solution |

Typically, when solving a system of three linear equations containing three variables, we use the method of elimination. Recall that the idea behind the method of elimination is to form equivalent equations so that adding two of the equations eliminates a variable.

Solve systems of three equations containing three variables **5**   Let's see how elimination works on a system of three equations containing three variables.

EXAMPLE 10   **Solving a System of Three Linear Equations with Three Variables**

Use the method of elimination to solve the system of equations.

$$\begin{cases} x + y - z = -1 & (1) \\ 4x - 3y + 2z = 16 & (2) \\ 2x - 2y - 3z = 5 & (3) \end{cases}$$

SOLUTION   For a system of three equations, we attempt to eliminate one variable at a time, using pairs of equations, until an equation with a single variable remains. Our plan of attack on this system will be to use Equation (1) to eliminate the variable x from Equations (2) and (3).

We begin by multiplying each side of Equation (1) by -4 and adding the result to Equation (2). (Do you see why? The coefficients of x are now opposites of each other.) We also multiply Equation (1) by -2 and add the result to Equation (3). Notice that these two procedures result in the removal of the x-variable from Equations (2) and (3).

$$\begin{aligned} x + y - z &= -1 \quad \text{(1) Multiply by } -4 \\ 4x - 3y + 2z &= 16 \quad \text{(2)} \end{aligned}$$

$$\begin{aligned} -4x - 4y + 4z &= 4 \quad \text{(1)} \\ 4x - 3y + 2z &= 16 \quad \text{(2)} \\ \hline -7y + 6z &= 20 \quad \text{Add} \end{aligned}$$

$$\begin{aligned} x + y - z &= -1 \quad \text{(1) Multiply by } -2 \\ 2x - 2y - 3z &= 5 \quad \text{(3)} \end{aligned}$$

$$\begin{aligned} -2x - 2y + 2z &= 2 \quad \text{(1)} \\ 2x - 2y - 3z &= 5 \quad \text{(3)} \\ \hline -4y - z &= 7 \quad \text{Add} \end{aligned}$$

$$\begin{cases} x + y - z = -1 & (1) \\ -7y + 6z = 20 & (2) \\ -4y - z = 7 & (3) \end{cases}$$

We now concentrate on Equations (2) and (3), treating them as a system of two equations containing two variables. It is easier to eliminate z. We multiply each side of Equation (3) by 6 and add Equations (2) and (3). The result is the new Equation (3).

423

$$\begin{array}{ll} -7y + 6z = 20 & (2) \\ -4y - z = 7 & (3)\ \text{Multiply by } 6 \end{array}$$

$$\begin{array}{ll} -7y + 6z = 20 & (2) \\ \underline{-24y - 6z = 42} & (3) \\ -31y = 62 & \text{Add} \end{array}$$

$$\begin{cases} x + y - z = -1 & (1) \\ -7y + 6z = 20 & (2) \\ -31y = 62 & (3) \end{cases}$$

We now solve Equation (3) for y by dividing both sides of the equation by -31.

$$\begin{cases} x + y - z = -1 & (1) \\ -7y + 6z = 20 & (2) \\ y = -2 & (3) \end{cases}$$

Back-substitute $y = -2$ in Equation (2) and solve for z.

$$\begin{array}{ll} -7y + 6z = 20 & (2) \\ -7(-2) + 6z = 20 & \text{Substitute } y = -2 \text{ in (2).} \\ 6z = 6 & \text{Subtract 14 from each side.} \\ z = 1 & \text{Divide each side by 6.} \end{array}$$

Finally, we back-substitute $y = -2$ and $z = 1$ in Equation (1) and solve for x.

$$\begin{array}{ll} x + y - z = -1 & (1) \\ x + (-2) - 1 = -1 & \text{Substitute } y = -2 \text{ and } z = 1 \text{ in (1).} \\ x - 3 = -1 & \text{Simplify.} \\ x = 2 & \text{Add 3 to each side.} \end{array}$$

The solution of the original system is $x = 2$, $y = -2$, $z = 1$. You should verify this solution.

Look back over the solution given in Example 10. Note the pattern of removing one of the variables from two of the equations, followed by solving this system of two equations and two unknowns. Although which variables to remove is your choice, the methodology remains the same for all systems.

 NOW WORK PROBLEM 35.

Identify inconsistent **6**   The previous example was a consistent system that had a unique solution. The next
systems of equations   two examples deal with the two other possibilities that may occur.
containing three variables

EXAMPLE 11   An Inconsistent System of Linear Equations

Solve: $\begin{cases} 2x + y - z = -2 & (1) \\ x + 2y - z = -9 & (2) \\ x - 4y + z = 1 & (3) \end{cases}$

SOLUTION   Our plan of attack is the same as in Example 10. However, in this system, it seems easiest to eliminate the variable z first. Do you see why?

Multiply each side of Equation (1) by -1 and add the result to Equation (2). Add
Equations (2) and (3).

$$-2x - y + z = 2 \quad \text{(1) Multiply by } -1.$$
$$\underline{x + 2y - z = -9} \quad \text{(2)}$$
$$-x + y \quad\quad = -7 \quad \text{Add}$$

$$x + 2y - z = -9 \quad \text{(2)}$$
$$\underline{x - 4y + z = 1} \quad \text{(3)}$$
$$2x - 2y \quad\quad = -8 \quad \text{Add}$$

$$\begin{cases} 2x + y - z = -2 & \text{(1)} \\ -x + y \quad\quad = -7 & \text{(2)} \\ 2x - 2y \quad\quad = -8 & \text{(3)} \end{cases}$$

We now concentrate on Equations (2) and (3), treating them as a system of two equations containing two variables. Multiply each side of Equation (2) by 2 and add the result to Equation (3).

$$-x + y = -7 \quad \text{(2) Multiply by 2.}$$
$$2x - 2y = -8 \quad \text{(3)}$$

$$-2x + 2y = -14 \quad \text{(2)}$$
$$\underline{2x - 2y = -8} \quad \text{(3)}$$
$$0 = -22 \quad \text{Add}$$

$$\begin{cases} 2x + y - z = -2 & \text{(1)} \\ -x + y \quad\quad = -7 & \text{(2)} \\ 0 = -22 & \text{(3)} \end{cases}$$

Equation (3) has no solution and the system is inconsistent.   ▸

NOW WORK PROBLEM 37.

Express the solutions of **7** **a system of dependent equations containing three variables**

Now let's look at a system of dependent equations.

EXAMPLE 12   Solving a System of Dependent Equations

Solve: $\begin{cases} x - 2y - z = 8 & \text{(1)} \\ 2x - 3y + z = 23 & \text{(2)} \\ 4x - 5y + 5z = 53 & \text{(3)} \end{cases}$

SOLUTION   Multiply each side of Equation (1) by -2 and add the result to Equation (2). Also, multiply each side of Equation (1) by -4 and add the result to Equation (3).

$$x - 2y - z = 8 \quad \text{(1) Multiply by } -2.$$
$$2x - 3y + z = 23 \quad \text{(2)}$$

$$-2x + 4y + 2z = -16 \quad \text{(1)}$$
$$\underline{2x - 3y + z = 23} \quad \text{(2)}$$
$$y + 3z = 7 \quad \text{Add}$$

$$x - 2y - z = 8 \quad \text{(1) Multiply by } -4.$$
$$4x - 5y + 5z = 53 \quad \text{(3)}$$

$$-4x + 8y + 4z = -32 \quad \text{(1)}$$
$$\underline{4x - 5y + 5z = 53} \quad \text{(2)}$$
$$3y + 9z = 21 \quad \text{Add}$$

$$\begin{cases} x - 2y - z = 8 & \text{(1)} \\ y + 3z = 7 & \text{(2)} \\ 3y + 9z = 21 & \text{(3)} \end{cases}$$

Treat Equations (2) and (3) as a system of two equations containing two variables, and eliminate the y-variable by multiplying each side of Equation (2) by -3 and adding the result to Equation (3).

$$y + 3z = 7 \quad \text{Multiply by } -3.$$
$$3y + 9z = 21$$

$$-3y - 9z = -21$$
$$\underline{3y + 9z = 21}$$
$$0 = 0 \quad \text{Add}$$

$$\begin{cases} x - 2y - z = 8 & \text{(1)} \\ y + 3z = 7 & \text{(2)} \\ 0 = 0 & \text{(3)} \end{cases}$$

The original system is equivalent to a system containing two equations, so the equations are dependent and the system has infinitely many solutions. If we let z represent any real number, then, solving Equation (2) for y, we determine that $y = -3z + 7$. Substitute this expression into Equation (1) to determine x in terms of z.

$$x - 2y - z = 8 \qquad (1)$$
$$x - 2(-3z + 7) - z = 8 \qquad \text{Substitute } y = -3z + 7 \text{ in (1).}$$
$$x + 6z - 14 - z = 8 \qquad \text{Remove parentheses.}$$
$$x + 5z = 22 \qquad \text{Combine like terms.}$$
$$x = -5z + 22 \qquad \text{Solve for } x.$$

We will write the solution to the system as

$$\begin{cases} x = -5z + 22 \\ y = -3z + 7 \end{cases}$$

where z, the parameter, can be any real number.

To find specific solutions to the system, choose any value of z and use the equations $x = -5z + 22$ and $y = -3z + 7$ to determine x and y. For example, if $z = 0$, then $x = 22$ and $y = 7$, and if $z = 1$, then $x = 17$ and $y = 4$. ▶

NOW WORK PROBLEM 39.

EXERCISE 2.1   **Answers to Odd-Numbered Problems Begin on Page AN-9.**

In Problems 1–10, decide whether the values of the variables listed are solutions of the system of equations.

1. $\begin{cases} 2x - y = 5 \\ 5x + 2y = 8 \end{cases}$
$x = 2, y = -1$

2. $\begin{cases} 3x + 2y = 2 \\ x - 7y = -30 \end{cases}$
$x = 2, y = 4$

3. $\begin{cases} 3x + 4y = 4 \\ \dfrac{1}{2}x - 3y = -\dfrac{1}{2} \end{cases}$
$x = 2, y = \dfrac{1}{2}$

4. $\begin{cases} 2x + \dfrac{1}{2}y = 0 \\ 3x - 4y = -\dfrac{19}{2} \end{cases}$
$x = -\dfrac{1}{2}, y = 2$

5. $\begin{cases} x - y = 3 \\ \dfrac{1}{2}x + y = 3 \end{cases}$
$x = 4, y = 1$

6. $\begin{cases} x - y = 3 \\ -3x + y = 1 \end{cases}$
$x = -2, y = -5$

7. $\begin{cases} 3x + 3y + 2z = 4 \\ x - y - z = 0 \\ 2y - 3z = -8 \end{cases}$
$x = 1, y = -1, z = 2$

8. $\begin{cases} 4x - z = 7 \\ 8x + 5y - z = 0 \\ -x - y + 5z = 6 \end{cases}$
$x = 2, y = -3, z = 1$

9. $\begin{cases} 3x + 3y + 2z = 4 \\ x - 3y + z = 10 \\ 5x - 2y - 3z = 8 \end{cases}$
$x = 2, y = -2, z = 2$

10. $\begin{cases} 4x - 5z = 6 \\ 5y - z = -17 \\ -x - 6y + 5z = 24 \end{cases}$
$x = 4, y = -3, z = 2$

In Problems 11–46, solve each system of equations. If the system has no solution, say that it is inconsistent.

11. $\begin{cases} x + y = 8 \\ x - y = 4 \end{cases}$

12. $\begin{cases} x + 2y = 5 \\ x + y = 3 \end{cases}$

13. $\begin{cases} 5x - y = 13 \\ 2x + 3y = 12 \end{cases}$

14. $\begin{cases} x + 3y = 5 \\ 2x - 3y = -8 \end{cases}$

15. $\begin{cases} 3x = 24 \\ x + 2y = 0 \end{cases}$

16. $\begin{cases} 4x + 5y = -2 \\ -2y = -4 \end{cases}$

17. $\begin{cases} 3x - 6y = 2 \\ 5x + 4y = 1 \end{cases}$

18. $\begin{cases} 2x + 4y = \dfrac{2}{3} \\ 3x - 5y = -10 \end{cases}$

19. $\begin{cases} 2x + y = 1 \\ 4x + 2y = 3 \end{cases}$

20. $\begin{cases} x - y = 5 \\ -3x + 3y = 2 \end{cases}$

21. $\begin{cases} 2x - y = 0 \\ 3x + 2y = 7 \end{cases}$

22. $\begin{cases} 3x + 3y = -1 \\ 4x + y = \dfrac{8}{3} \end{cases}$

23. $\begin{cases} x + 2y = 4 \\ 2x + 4y = 8 \end{cases}$

24. $\begin{cases} 3x - y = 7 \\ 9x - 3y = 21 \end{cases}$

25. $\begin{cases} 2x - 3y = -1 \\ 10x + y = 11 \end{cases}$

26. $\begin{cases} 3x - 2y = 0 \\ 5x + 10y = 4 \end{cases}$

27. $\begin{cases} 2x + 3y = 6 \\ x - y = \dfrac{1}{2} \end{cases}$

28. $\begin{cases} \dfrac{1}{2}x + y = -2 \\ x - 2y = 8 \end{cases}$

29. $\begin{cases} \dfrac{1}{2}x + \dfrac{1}{3}y = 3 \\ \dfrac{1}{4}x - \dfrac{2}{3}y = -1 \end{cases}$

30. $\begin{cases} \dfrac{1}{3}x - \dfrac{3}{2}y = -5 \\ \dfrac{3}{4}x + \dfrac{1}{3}y = 11 \end{cases}$

31. $\begin{cases} 3x - 5y = 3 \\ 15x + 5y = 21 \end{cases}$

32. $\begin{cases} 2x - y = -1 \\ x + \dfrac{1}{2}y = \dfrac{3}{2} \end{cases}$

33. $\begin{cases} x - y = 6 \\ 2x - 3z = 16 \\ 2y + z = 4 \end{cases}$

34. $\begin{cases} 2x + y = -4 \\ -2y + 4z = 0 \\ 3x - 2z = -11 \end{cases}$

35. $\begin{cases} x - 2y + 3z = 7 \\ 2x + y + z = 4 \\ -3x + 2y - 2z = -10 \end{cases}$

36. $\begin{cases} 2x + y - 3z = -2 \\ -2x + 2y + z = -9 \\ 3x - 4y - 3z = 15 \end{cases}$

37. $\begin{cases} x - y - z = 1 \\ 2x + 3y + z = 2 \\ 3x + 2y = 0 \end{cases}$

38. $\begin{cases} 2x - 3y - z = 0 \\ -x + 2y + z = 5 \\ 3x - 4y - z = 1 \end{cases}$

39. $\begin{cases} x - y - z = 1 \\ -x + 2y - 3z = -4 \\ 3x - 2y - 7z = 0 \end{cases}$

40. $\begin{cases} 2x - 3y - z = 0 \\ 3x + 2y + 2z = 2 \\ x + 5y + 3z = 2 \end{cases}$

41. $\begin{cases} 2x - 2y + 3z = 6 \\ 4x - 3y + 2z = 0 \\ -2x + 3y - 7z = 1 \end{cases}$

42. $\begin{cases} 3x - 2y + 2z = 6 \\ 7x - 3y + 2z = -1 \\ 2x - 3y + 4z = 0 \end{cases}$

43. $\begin{cases} x + y - z = 6 \\ 3x - 2y + z = -5 \\ x + 3y - 2z = 14 \end{cases}$

44. $\begin{cases} x - y + z = -4 \\ 2x - 3y + 4z = -15 \\ 5x + y - 2z = 12 \end{cases}$

45. $\begin{cases} x + 2y - z = -3 \\ 2x - 4y + z = -7 \\ -2x + 2y - 3z = 4 \end{cases}$

46. $\begin{cases} x + 4y - 3z = -8 \\ 3x - y + 3z = 12 \\ x + y + 6z = 1 \end{cases}$

---

47. **Dimensions of a Floor** The perimeter of a rectangular floor is 90 feet. Find the dimensions of the floor if the length is twice the width.

48. **Dimensions of a Field** The length of fence required to enclose a rectangular field is 3000 meters. What are the dimensions of the field if the difference between its length and width is 50 meters?

49. **Agriculture** According to the U.S. Department of Agriculture, in 1996–1997 the production cost for planting corn was $246 per acre and the cost for planting soybeans was $140 per acre. The average farm used 445 acres of land to raise corn and soybeans and budgeted $85,600

for planting these crops. If all the land and all the money budgeted is used, how many acres of each crop should they plant?

Source: USDA, National Agricultural Statistics Service.

50. **Movie Theater Tickets** A movie theater charges $9.00 for adults and $7.00 for senior citizens. On a day when 325 people paid an admission, the total receipts were $2495. How many who paid were adults? How many were seniors?

51. **Mixing Nuts** A store sells cashews for $5.00 per pound and peanuts for $1.50 per pound. The manager decides to mix 30 pounds of peanuts with some cashews and sell the mixture for $3.00 per pound. How many pounds of cashews should be mixed with the peanuts so that the mixture will produce the same revenue as would selling the nuts separately?

52. **Financial Planning** A recently retired couple need $12,000 per year to supplement their Social Security. They have $150,000 to invest to obtain this income. They have decided on two investment options: AA bonds yielding 10% per annum and a Bank Certificate yielding 5%.

(a) How much should be invested in each to realize exactly $12,000?

(b) If, after two years, the couple require $14,000 per year in income, how should they reallocate their investment to achieve the new amount?

53. **Cost of Food in Japan** In Kyotoshi, Japan, the cost of three bowls of noodles and two cartons of fresh milk is 2153 yen. Three cartons of fresh milk cost 89 yen more than one bowl of noodles. What is the cost of a bowl of noodles? A carton of fresh milk?

    Source: Statistics Bureau and Statistics Center, Ministry of Public Management, Home Affairs, Posts and Telecommunications, Japan, 2002.

54. **Cost of Fast Food** Four large cheeseburgers and two chocolate shakes cost a total of $7.90. Two shakes cost 15¢ more than one cheeseburger. What is the cost of a cheeseburger? A shake?

55. **Computing a Refund** The grocery store we use does not mark prices on its goods. My wife went to this store, bought three 1-pound packages of bacon and two cartons of eggs, and paid a total of $7.45. Not knowing that she went to the store, I also went to the same store, purchased two 1-pound packages of bacon and three cartons of eggs, and paid a total of $6.45. Now we want to return two 1-pound packages of bacon and two cartons of eggs. How much will be refunded?

56. **Blending Coffees** A coffee manufacturer wants to market a new blend of coffee that will cost $5 per pound by mixing $3.75-per-pound coffee and $8-per-pound coffee. What amounts of the $3.75-per-pound coffee and $8-per-pound coffee should be blended to obtain the desired mixture? [*Hint*: Assume the total weight of the desired blend is 100 pounds.]

57. **Pharmacy** A doctor's prescription calls for a daily intake of liquid containing 40 mg of vitamin C and 30 mg of vitamin D. Your pharmacy stocks two liquids that can be used: one contains 20% vitamin C and 30% vitamin D, the other 40% vitamin C and 20% vitamin D. How many milligrams of each liquid should be mixed to fill the prescription?

58. **Pharmacy** A doctor's prescription calls for the creation of pills that contain 12 units of vitamin B_{12} and 12 units of vitamin E. Your pharmacy stocks two powders that can be used to make these pills: one contains 20% vitamin B_{12} and 30% vitamin E, the other 40% vitamin B_{12} and 20% vitamin E. How many units of each powder should be mixed in each pill?

59. **Diet Preparation** A 600- to 700-pound yearling horse needs 33.0 grams of calcium and 21.0 grams of phosphorus per day for a healthy diet. A farmer provides a combination of rolled oats and molasses to provide those nutrients. Rolled oats provide 0.41 grams of calcium per pound and 1.95 grams of phosphorus per pound, while molasses provides 3.35 grams of calcium per pound and 0.36 grams of phosphorus per pound. How many pounds each of rolled oats and molasses should the farmer feed the yearling in order to meet the daily requirements?

    Source: Balancing Rations for Horses, R. D. Setzler, Washington State University.

60. **Restaurant Management** A restaurant manager wants to purchase 200 sets of dishes. One design costs $25 per set, while another costs $45 per set. If she only has $7400 to spend, how many of each design should be ordered?

61. **Theater Revenues** A Broadway theater has 500 seats, divided into orchestra, main, and balcony seating. Orchestra seats sell for $50, main seats for $35, and balcony seats for $25. If all the seats are sold, the gross revenue to the theater is $17,100. If all the main and balcony seats are sold, but only half the orchestra seats are sold, the gross revenue is $14,600. How many are there of each kind of seat?

62. **Theater Revenues** A movie theater charges $8.00 for adults, $4.50 for children, and $6.00 for senior citizens. One day the theater sold 405 tickets and collected $2320 in receipts. There were twice as many children's tickets sold as adult tickets. How many adults, children, and senior citizens went to the theater that day?

63. **Investments** Kelly has $20,000 to invest. As her financial planner, you recommend that she diversify into three investments: Treasury Bills that yield 5% simple interest, Treasury Bonds that yield 7% simple interest, and corporate bonds that yield 10% simple interest. Kelly wishes to earn $1390 per year in income. Also, Kelly wants her investment in Treasury Bills to be $3000 more than her investment in corporate bonds. How much money should Kelly place in each investment?

64. Make up a system of two linear equations containing two variables that has:

    (a) No solution
    (b) Exactly one solution
    (c) Infinitely many solutions

    Give the three systems to a friend to solve and critique.

65. Write a brief paragraph outlining your strategy for solving a system of two linear equations containing two variables.

66. Do you prefer the method of substitution or the method of elimination for solving a system of two linear equations containing two variables? Give reasons.

2.2 Systems of Linear Equations: Matrix Method

OBJECTIVES
1. Write the augmented matrix of a system of linear equations
2. Write the system from the augmented matrix
3. Perform row operations on a matrix
4. Solve systems of linear equations using matrices
5. Express the solution of a system with an infinite number of solutions

The systematic approach of the method of elimination for solving a system of linear equations provides another method of solution that involves a simplified notation using a *matrix*.

A **matrix** is defined as a rectangular array of numbers, enclosed by brackets. The numbers are referred to as the **entries** of the matrix. A matrix is further identified by naming its *rows* and *columns*. Some examples of matrices are

Column 1 Column 2

$$
\begin{array}{c}
\text{Row 1} \\
\text{Row 2} \\
\text{Row 3}
\end{array}
\begin{bmatrix}
8 & 0 \\
1 & 3 \\
2 & 4
\end{bmatrix}
$$

(a)

Column 1 Column 2 Column 3

$$
\begin{array}{c}
\text{Row 1} \\
\text{Row 2}
\end{array}
\begin{bmatrix}
4 & 1 & -3 \\
2 & 1 & 2
\end{bmatrix}
$$

(b)

Column 1 Column 2

$$
\text{Row 1} \begin{bmatrix} 4 & 3 \end{bmatrix}
$$

(c)

Matrix Representation of a System of Linear Equations

Consider the following two systems of two linear equations containing two variables

$$
\begin{cases} x + 4y = 14 \\ 3x - 2y = 0 \end{cases} \quad \text{and} \quad \begin{cases} u + 4v = 14 \\ 3u - 2v = 0 \end{cases}
$$

We observe that, except for the symbols used to represent the variables, these two systems are identical. As a result, we can dispense altogether with the letters used to symbolize the variables, provided we have some means of keeping track of them. A matrix serves us well in this regard.

When a matrix is used to represent a system of linear equations, it is called the **augmented matrix** of the system. For example, the system

$$
\begin{cases} x + 4y = 14 & (1) \\ 3x - 2y = 0 & (2) \end{cases}
$$

can be represented by the augmented matrix

	Column 1 x	Column 2 y	Column 3 right-hand side

$$
\begin{array}{c}
\text{Row 1 [Equation (1)]} \\
\text{Row 2 [Equation (2)]}
\end{array}
\left[
\begin{array}{cc|c}
1 & 4 & 14 \\
3 & -2 & 0
\end{array}
\right]
$$

Here it is understood that column 1 contains the coefficients of the variable x, column 2 contains the coefficients of the variable y, and column 3 contains the numbers to the right of the equal sign. Each row of the matrix represents an equation of the system. Although not required, it has become customary to place a vertical bar in the matrix as a reminder of the equal sign.

In this book we shall follow the practice of using x and y to denote the variables for systems containing two variables. We will use x, y, and z for systems containing three variables; we will use subscripted variables (x_1, x_2, x_3, x_4, etc.) for systems containing four or more variables.

In writing the augmented matrix of a system, the variables of each equation must be on the left side of the equal sign and the constants on the right side. A variable that does not appear in an equation has a coefficient of 0.

1   **EXAMPLE 1**   **Writing the Augmented Matrix of a System of Linear Equations**

Write the augmented matrix of each system of equations.

(a) $\begin{cases} 3x - 4y = -6 & (1) \\ 2x - 3y = -5 & (2) \end{cases}$ (b) $\begin{cases} 2x - y + z = 0 & (1) \\ x + z - 1 = 0 & (2) \\ x + 2y - 8 = 0 & (3) \end{cases}$ (c) $\begin{cases} 3x_1 - x_2 + x_3 + x_4 = 5 & (1) \\ 2x_1 + 6x_3 = 2 & (2) \end{cases}$

SOLUTION   (a) The augmented matrix is

$$\begin{bmatrix} 3 & -4 & | & -6 \\ 2 & -3 & | & -5 \end{bmatrix}$$

(b) Care must be taken that the system be written so that the coefficients of all variables are present (if any variable is missing, its coefficient is 0). Also, all constants must be to the right of the equal sign. We need to rearrange the given system as follows:

$$\begin{cases} 2x - y + z = 0 & (1) \\ x + z - 1 = 0 & (2) \\ x + 2y - 8 = 0 & (3) \end{cases}$$

$$\begin{cases} 2x - y + z = 0 & (1) \\ x + 0 \cdot y + z = 1 & (2) \\ x + 2y + 0 \cdot z = 8 & (3) \end{cases}$$

The augmented matrix is

$$\begin{bmatrix} 2 & -1 & 1 & | & 0 \\ 1 & 0 & 1 & | & 1 \\ 1 & 2 & 0 & | & 8 \end{bmatrix}$$

(c) The augmented matrix is

$$\begin{bmatrix} 3 & -1 & 1 & 1 & | & 5 \\ 2 & 0 & 6 & 0 & | & 2 \end{bmatrix}$$

 NOW WORK PROBLEM 1.

Given an augmented matrix, we can write the corresponding system of equations.

2   **EXAMPLE 2**   **Writing the System of Linear Equations from the Augmented Matrix**

Write the system of linear equations corresponding to each augmented matrix.

(a) $\begin{bmatrix} 5 & 2 & | & 13 \\ -3 & 1 & | & -10 \end{bmatrix}$ **(b)** $\begin{bmatrix} 3 & -1 & -1 & | & 7 \\ 4 & 0 & 2 & | & 8 \\ 0 & 1 & 1 & | & 0 \end{bmatrix}$

SOLUTION   **(a)** The matrix has two rows and so represents a system of two equations. The two columns to the left of the vertical bar indicate that the system has two variables. If x and y are used to denote these variables, the system of equations is

$$\begin{cases} 5x + 2y = 13 & (1) \\ -3x + y = -10 & (2) \end{cases}$$

(b) Since the augmented matrix has three rows, it represents a system of three equations. Since there are three columns to the left of the vertical bar, the system contains three variables. If x, y, and z are the three variables, the system of equations is

$$\begin{cases} 3x - y - z = 7 & (1) \\ 4x + 2z = 8 & (2) \\ y + z = 0 & (3) \end{cases}$$

▶

Row Operations on a Matrix

Perform row operations on **3**
a matrix

Row operations on a matrix are used to solve systems of equations when the system is written as an augmented matrix. There are three basic row operations.

> ### Row Operations
>
> 1. Interchange any two rows.
> 2. Replace a row by a nonzero multiple of that row.
> 3. Replace a row by the sum of that row and a constant nonzero multiple of some other row.

These three row operations correspond to the three rules given earlier for obtaining an equivalent system of equations. When a row operation is performed on a matrix, the resulting matrix represents a system of equations equivalent to the system represented by the original matrix.

For example, consider the augmented matrix

$$\begin{bmatrix} 1 & 2 & | & 3 \\ 4 & -1 & | & 2 \end{bmatrix}$$

Suppose that we want to apply a row operation to this matrix that results in a matrix whose entry in row 2, column 1 is a 0. The row operation to use is

Multiply each entry in row 1 by -4 and add the result   **(1)**
to the corresponding entries in row 2.

If we use R_2 to represent the new entries in row 2 and we use r_1 and r_2 to represent the original entries in rows 1 and 2, respectively, then we can represent the row operation in statement (1) by

$$R_2 = -4r_1 + r_2$$

Then

$$\begin{bmatrix} 1 & 2 & | & 3 \\ 4 & -1 & | & 2 \end{bmatrix} \xrightarrow{R_2 \,=\, -4r_1 + r_2} \begin{bmatrix} 1 & 2 & | & 3 \\ -4(1) + 4 & -4(2) + (-1) & | & -4(3) + 2 \end{bmatrix} = \begin{bmatrix} 1 & 2 & | & 3 \\ 0 & -9 & | & -10 \end{bmatrix}$$

As desired, we now have the entry 0 in row 2, column 1.

EXAMPLE 3   Applying a Row Operation to an Augmented Matrix

Apply the row operation $R_2 = -3r_1 + r_2$ to the augmented matrix

$$\begin{bmatrix} 1 & -2 & | & 2 \\ 3 & -5 & | & 9 \end{bmatrix}$$

SOLUTION   The row operation $R_2 = -3r_1 + r_2$ tells us that the entries in row 2 are to be replaced by the entries obtained after multiplying each entry in row 1 by -3 and adding the result to the corresponding entries in row 2. Thus,

$$\begin{bmatrix} 1 & -2 & | & 2 \\ 3 & -5 & | & 9 \end{bmatrix} \xrightarrow{R_2 \,=\, -3r_1 + r_2} \begin{bmatrix} 1 & -2 & | & 2 \\ -3(1) + 3 & (-3)(-2) + (-5) & | & -3(2) + 9 \end{bmatrix} = \begin{bmatrix} 1 & -2 & | & 2 \\ 0 & 1 & | & 3 \end{bmatrix}$$

NOW WORK PROBLEM 13.

EXAMPLE 4   Finding a Particular Row Operation

Using the augmented matrix

$$\begin{bmatrix} 1 & -2 & | & 2 \\ 0 & 1 & | & 3 \end{bmatrix}$$

find a row operation that will result in this augmented matrix having a 0 in row 1, column 2.

SOLUTION   We want a 0 in row 1, column 2. This result can be accomplished by multiplying row 2 by 2 and adding the result to row 1. That is, we apply the row operation $R_1 = 2r_2 + r_1$.

$$\begin{bmatrix} 1 & -2 & | & 2 \\ 0 & 1 & | & 3 \end{bmatrix} \xrightarrow{R_1 \,=\, 2r_2 + r_1} \begin{bmatrix} 2(0) + 1 & 2(1) + (-2) & | & 2(3) + 2 \\ 0 & 1 & | & 3 \end{bmatrix} = \begin{bmatrix} 1 & 0 & | & 8 \\ 0 & 1 & | & 3 \end{bmatrix}$$

A word about the notation that we have introduced. A row operation such as $R_1 = 2r_2 + r_1$ changes the entries in row 1. Note also that for this type of row operation we change the entries in a given row by multiplying the entries in some other row by an appropriate nonzero number and adding the results to the original entries of the row to be changed.

Solving a System of Linear Equations Using Matrices

To solve a system of linear equations using matrices, we use row operations on the augmented matrix of the system to obtain a matrix that is in *row echelon form*.

> A matrix is in **row echelon form** when
>
> 1. The entry in row 1, column 1 is a 1, and 0s appear below it.
> 2. The first nonzero entry in each row after the first row is a 1, 0s appear below it, and it appears to the right of the first nonzero entry in any row above.
> 3. Any rows that contain all 0s to the left of the vertical bar appear at the bottom.

For example, for a system of two linear equations containing two variables with a unique solution, the augmented matrix is in row echelon form if it is of the form

$$\begin{bmatrix} 1 & a & | & b \\ 0 & 1 & | & c \end{bmatrix}$$

where a, b, and c are real numbers. The second row tells us that $y = c$. We can then find the value of x by back-substituting $y = c$ into the equation given by the first row: $x + ay = b$ and solving for x.

For a system of three equations containing three variables with a unique solution, the augmented matrix is in row echelon form if it is of the form

$$\begin{bmatrix} 1 & a & b & | & d \\ 0 & 1 & c & | & e \\ 0 & 0 & 1 & | & f \end{bmatrix}$$

where a, b, c, d, e, and f are real numbers. The last row of the augmented matrix states that $z = f$. We can then determine the value of y using back-substitution with $z = f$, since row 2 represents the equation $y + cz = e$. Finally, x is determined using back-substitution again.

Two advantages of solving a system of equations by writing the augmented matrix in row echelon form are the following:

1. The process is algorithmic; that is, it consists of repetitive steps that can be programmed on a computer.
2. The process works on any system of linear equations, no matter how many equations or variables are present.

Let's see how row operations are used to solve a system of linear equations. To see what is happening, we'll write the corresponding system of equations next to the matrix obtained after a row operation is performed.

EXAMPLE 5   **Solving a System of Linear Equations Using Matrices (Row Echelon Form)**

Solve: $\begin{cases} 4x + 3y = 11 \\ x - 3y = -1 \end{cases}$

SOLUTION   We write the augmented matrix that represents this system:

$$\begin{bmatrix} 4 & 3 & | & 11 \\ 1 & -3 & | & -1 \end{bmatrix} \qquad \begin{cases} 4x + 3y = 11 \\ x - 3y = -1 \end{cases}$$

The next step is to place a 1 in row 1, column 1. An interchange of rows 1 and 2 is the easiest way to do this.

$$\begin{bmatrix} 4 & 3 & | & 11 \\ 1 & -3 & | & -1 \end{bmatrix} \xrightarrow[\substack{R_1 = r_2 \\ R_2 = r_1}]{} \begin{bmatrix} 1 & -3 & | & -1 \\ 4 & 3 & | & 11 \end{bmatrix} \qquad \begin{cases} x - 3y = -1 \\ 4x + 3y = 11 \end{cases}$$

Now we want a 0 under the entry 1 in column 1. (This eliminates the variable x from the second equation.) We use the row operation $R_2 = -4r_1 + r_2$.

$$\begin{bmatrix} 1 & -3 & | & -1 \\ 4 & 3 & | & 11 \end{bmatrix} \xrightarrow[R_2 = -4r_1 + r_2]{} \begin{bmatrix} 1 & -3 & | & -1 \\ 0 & 15 & | & 15 \end{bmatrix} \qquad \begin{cases} x - 3y = -1 \\ 15y = 15 \end{cases}$$

Now we want the entry 1 in row 2, column 2. (This makes it easy to solve for y.) We use $R_2 = \frac{1}{15}r_2$.

$$\begin{bmatrix} 1 & -3 & | & -1 \\ 0 & 15 & | & 15 \end{bmatrix} \xrightarrow[R_2 = \frac{1}{15}r_2]{} \begin{bmatrix} 1 & -3 & | & -1 \\ 0 & 1 & | & 1 \end{bmatrix} \qquad \begin{cases} x - 3y = -1 \\ y = 1 \end{cases}$$

This matrix is the row echelon form of the augmented matrix. The second row of the matrix on the right represents the equation $y = 1$. Using $y = 1$, we back-substitute into the equation $x - 3y = -1$ (from the first row) to get

$$x - 3y = -1$$
$$x - 3(1) = -1 \quad y = 1$$
$$x = 2$$

The solution of the system is $x = 2, y = 1$.   ▶

NOW WORK PROBLEM 33.

The steps we used to solve the system of linear equations in Example 5 can be summarized as follows:

Matrix Method for Solving a System of Linear Equations (Row Echelon Form)

STEP 1 Write the augmented matrix that represents the system.

STEP 2 Perform row operations that place the entry 1 in row 1, column 1.

STEP 3 Perform row operations that leave the entry 1 in row 1, column 1 unchanged, while causing 0s to appear below it in column 1.

STEP 4 Perform row operations that place the entry 1 in row 2, column 2, but leave the entries in columns to the left unchanged. If it is impossible to place a 1 in row 2, column 2, then proceed to place a 1 in row 2, column 3. Once a 1 is in place, perform row operations to place 0s below it.

(If any rows are obtained that contain only 0s on the left side of the vertical bar, place such rows at the bottom of the matrix.)

STEP 5 Now repeat Step 4, placing a 1 in the next row, but one column to the right. Continue until the bottom row or the vertical bar is reached.

STEP 6 The matrix that results is the row echelon form of the augmented matrix. Analyze the system of equations corresponding to it to solve the original system.

In the next example, we solve a system of three linear equations containing three variables using the steps of the matrix method.

EXAMPLE 6   **Solving a System of Linear Equations Using the Matrix Method (Row Echelon Form)**

$$\text{Solve: } \begin{cases} x - y + z = 8 & (1) \\ 2x + 3y \quad z - -2 & (2) \\ 3x - 2y - 9z = 9 & (3) \end{cases}$$

SOLUTION   **STEP 1** The augmented matrix of the system is

$$\begin{bmatrix} 1 & -1 & 1 & 8 \\ 2 & 3 & -1 & -2 \\ 3 & -2 & -9 & 9 \end{bmatrix}$$

STEP 2 Because the entry 1 is already present in row 1, column 1, we can go to Step 3.

STEP 3 Perform the row operations $R_2 = -2r_1 + r_2$ and $R_3 = -3r_1 + r_3$. Each of these leaves the entry 1 in row 1, column 1 unchanged, while causing 0's to appear under it.*

$$\begin{bmatrix} 1 & -1 & 1 & 8 \\ 2 & 3 & -1 & -2 \\ 3 & -2 & -9 & 9 \end{bmatrix} \xrightarrow[\substack{R_2 = -2r_1 + r_2 \\ R_3 = -3r_1 + r_3}]{} \begin{bmatrix} 1 & -1 & 1 & 8 \\ 0 & 5 & -3 & -18 \\ 0 & 1 & -12 & -15 \end{bmatrix}$$

STEP 4 The easiest way to obtain the entry 1 in row 2, column 2 without altering column 1 is to interchange rows 2 and 3 (another way would be to multiply row 2 by $\frac{1}{5}$, but this introduces fractions).

$$\begin{bmatrix} 1 & -1 & 1 & 8 \\ 0 & 1 & -12 & -15 \\ 0 & 5 & -3 & 18 \end{bmatrix}$$

To get a 0 under the 1 in row 2, column 2, perform the row operation $R_3 = -5r_2 + r_3$.

$$\begin{bmatrix} 1 & -1 & 1 & 8 \\ 0 & 1 & -12 & -15 \\ 0 & 5 & -3 & -18 \end{bmatrix} \xrightarrow[R_3 = -5r_2 + r_3]{} \begin{bmatrix} 1 & -1 & 1 & 8 \\ 0 & 1 & -12 & -15 \\ 0 & 0 & 57 & 57 \end{bmatrix}$$

You should convince yourself that doing both of these simultaneously is the same as doing the first followed by the second.

STEP 5 Continuing, we obtain a 1 in row 3, column 3 by using $R_3 = \frac{1}{57} r_3$.

$$\begin{bmatrix} 1 & -1 & 1 & | & 8 \\ 0 & 1 & -12 & | & -15 \\ 0 & 0 & 57 & | & 57 \end{bmatrix} \xrightarrow[R_3 = \frac{1}{57} r_3]{} \begin{bmatrix} 1 & -1 & 1 & | & 8 \\ 0 & 1 & -12 & | & -15 \\ 0 & 0 & 1 & | & 1 \end{bmatrix}$$

STEP 6 The matrix on the right is the row echelon form of the augmented matrix. The system of equations represented by the matrix in row echelon form is

$$\begin{cases} x - y + z = 8 & (1) \\ y - 12z = -15 & (2) \\ z = 1 & (3) \end{cases}$$

Using $z = 1$, we back-substitute to get

$$\begin{cases} x - y + 1 = 8 & (1) \\ y - 12(1) = -15 & (2) \end{cases} \xrightarrow{\text{Simplify.}} \begin{cases} x - y = 7 & (1) \\ y = -3 & (2) \end{cases}$$

We get $y = -3$, and back-substituting into $x - y = 7$, we find that $x = 4$. The solution of the system is $x = 4, y = -3, z = 1$. ▶

GRAPHING UTILITY SOLUTION   A graphing utility can be used to obtain the row echelon form of the augmented matrix. The augmented matrix of the system given in Example 6 is

$$\begin{bmatrix} 1 & -1 & 1 & | & 8 \\ 2 & 3 & -1 & | & -2 \\ 3 & -2 & -9 & | & 9 \end{bmatrix}$$

We enter this matrix into our graphing utility and name it A. See Figure 7(a). Using the REF (row echelon form) command on matrix A, we obtain the results shown in Figure 7(b). Since the entire matrix does not fit on the screen, we need to scroll right to see the rest of it. See Figure 7(c).

FIGURE 7

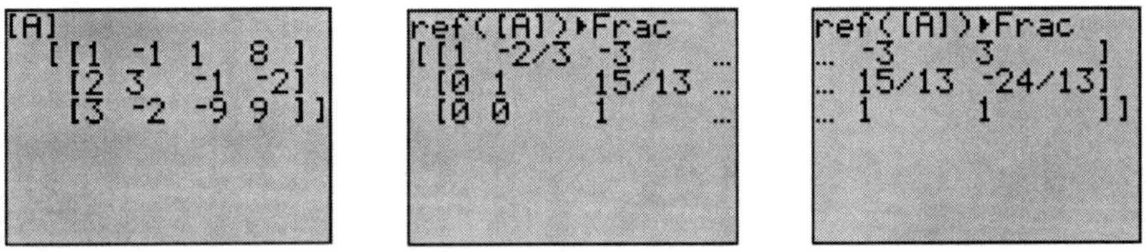

(a)   (b)   (c)

The system of equations represented by the matrix in row echelon form is

$$\begin{cases} x - \dfrac{2}{3}y - 3z = 3 & (1) \\ y + \dfrac{15}{13}z = -\dfrac{24}{13} & (2) \\ z = 1 & (3) \end{cases}$$

Using $z = 1$, we back-substitute to get

$$\begin{cases} x - \dfrac{2}{3}y - 3(1) = 3 & (1) \\ y + \dfrac{15}{13}(1) = -\dfrac{24}{13} & (2) \end{cases} \xrightarrow{\text{Simplify.}} \begin{cases} x - \dfrac{2}{3}y = 6 & (1) \\ y = -\dfrac{39}{13} = -3 & (2) \end{cases}$$

Solving the second equation for y, we find that $y = -3$. Back-substituting $y = -3$ into $x - \frac{2}{3}y = 6$, we find that $x = 4$. The solution of the system is $x = 4, y = -3, z = 1$. ▶

Notice that the row echelon form of the augmented matrix using the graphing utility differs from the row echelon form in our algebraic solution, yet both matrices provide the same solution! This is because the two solutions used different row operations to obtain the row echelon form. In all likelihood, the two solutions parted ways in Step 4 of the algebraic solution, where we avoided introducing fractions by interchanging rows 2 and 3.

Sometimes it is advantageous to write a matrix in **reduced row echelon form.** In this form, row operations are used to obtain entries that are 0 above (as well as below) the leading 1 in a row. For example, the row echelon form obtained in the algebraic solution to Example 6 is

$$\left[\begin{array}{ccc|c} 1 & -1 & 1 & 8 \\ 0 & 1 & -12 & -15 \\ 0 & 0 & 1 & 1 \end{array}\right]$$

To write this matrix in reduced row echelon form, we proceed as follows:

$$\left[\begin{array}{ccc|c} 1 & -1 & 1 & 8 \\ 0 & 1 & -12 & -15 \\ 0 & 0 & 1 & 1 \end{array}\right] \xrightarrow[R_1 = r_2 + r_1]{} \left[\begin{array}{ccc|c} 1 & 0 & -11 & -7 \\ 0 & 1 & -12 & -15 \\ 0 & 0 & 1 & 1 \end{array}\right] \xrightarrow[\substack{R_1 = 11r_3 + r_1 \\ R_2 = 12r_3 + r_2}]{} \left[\begin{array}{ccc|c} 1 & 0 & 0 & 4 \\ 0 & 1 & 0 & -3 \\ 0 & 0 & 1 & 1 \end{array}\right]$$

The matrix is now written in reduced row echelon form. The advantage of writing the matrix in this form is that the solution to the system, $x = 4, y = -3, z = 1$, is readily found, without the need to back-substitute. Another advantage will be seen in Section 2.6, where the inverse of a matrix is discussed.

FIGURE 8

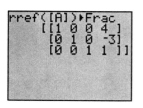

COMMENT: Most graphing utilities also have the ability to put a matrix in reduced row echelon form. Figure 8 shows the reduced row echelon form of the augmented matrix from Example 6 using the RREF command on a TI-83 Plus graphing calculator. ▶

NOW WORK PROBLEMS 49 AND 59.

EXAMPLE 7   **Calculating Production Output**

FoodPerfect Corporation manufactures three models of the Perfect Foodprocessor. Each Model X processor requires 30 minutes of electrical assembly, 40 minutes of mechanical assembly, and 30 minutes of testing; each Model Y requires 20 minutes of electrical assembly, 50 minutes of mechanical assembly, and 30 minutes of testing; and each Model Z requires 30 minutes of electrical assembly, 30 minutes of mechanical assembly, and 20 minutes of testing. If 2500 minutes of electrical assembly, 3500 minutes of mechanical assembly, and 2400 minutes of testing are used in one day, how many of each model will be produced?

SOLUTION   The table on page 74 summarizes the given information:

	Model			Time
	X	Y	Z	Used
Electrical Assembly	30	20	30	2500
Mechanical Assembly	40	50	30	3500
Testing	30	30	20	2400

We assign variables to represent the unknowns:

$$x = \text{Number of Model X produced}$$
$$y = \text{Number of Model Y produced}$$
$$z = \text{Number of Model Z produced}$$

Based on the table, we obtain the following system of equations:

$$\begin{cases} 30x + 20y + 30z = 2500 \\ 40x + 50y + 30z = 3500 \\ 30x + 30y + 20z = 2400 \end{cases} \xrightarrow[\substack{\text{Divide each} \\ \text{equation by 10.}}]{} \begin{cases} 3x + 2y + 3z = 250 & (1) \\ 4x + 5y + 3z = 350 & (2) \\ 3x + 3y + 2z = 240 & (3) \end{cases}$$

The augmented matrix of this system is

$$\begin{bmatrix} 3 & 2 & 3 & | & 250 \\ 4 & 5 & 3 & | & 350 \\ 3 & 3 & 2 & | & 240 \end{bmatrix}$$

We could obtain a 1 in row 1, column 1 by using the row operation $R_1 = \frac{1}{3}r_1$, but the introduction of fractions is best avoided. Instead, we use

$$R_2 = -1r_1 + r_2$$

to place a 1 in row 2, column 1. The result is

$$\begin{bmatrix} 3 & 2 & 3 & | & 250 \\ 4 & 5 & 3 & | & 350 \\ 3 & 3 & 2 & | & 240 \end{bmatrix} \xrightarrow{R_2 = -r_1 + r_2} \begin{bmatrix} 3 & 2 & 3 & | & 250 \\ 1 & 3 & 0 & | & 100 \\ 3 & 3 & 2 & | & 240 \end{bmatrix}$$

Next, interchange row 1 and row 2:

$$\begin{bmatrix} 3 & 2 & 3 & | & 250 \\ 1 & 3 & 0 & | & 100 \\ 3 & 3 & 2 & | & 240 \end{bmatrix} \xrightarrow[R_2 = r_1]{R_1 = r_2} \begin{bmatrix} 1 & 3 & 0 & | & 100 \\ 3 & 2 & 3 & | & 250 \\ 3 & 3 & 2 & | & 240 \end{bmatrix}$$

Use $R_2 = -3r_1 + r_2$ and $R_3 = -3r_1 + r_3$ to obtain

$$\begin{bmatrix} 1 & 3 & 0 & | & 100 \\ 3 & 2 & 3 & | & 250 \\ 3 & 3 & 2 & | & 240 \end{bmatrix} \xrightarrow[R_3 = -3r_1 + r_3]{R_2 = -3r_1 + r_2} \begin{bmatrix} 1 & 3 & 0 & | & 100 \\ 0 & -7 & 3 & | & -50 \\ 0 & -6 & 2 & | & -60 \end{bmatrix}$$

We use $R_2 = -1r_2$ followed by $R_2 = r_3 + r_2$:

$$\begin{bmatrix} 1 & 3 & 0 & | & 100 \\ 0 & -7 & 3 & | & -50 \\ 0 & -6 & 2 & | & -60 \end{bmatrix} \xrightarrow{R_2 = -1r_2} \begin{bmatrix} 1 & 3 & 0 & | & 100 \\ 0 & 7 & -3 & | & 50 \\ 0 & -6 & 2 & | & -60 \end{bmatrix} \xrightarrow{R_2 = r_3 + r_2} \begin{bmatrix} 1 & 3 & 0 & | & 100 \\ 0 & 1 & -1 & | & -10 \\ 0 & -6 & 2 & | & -60 \end{bmatrix}$$

Next, use $R_3 = 6r_2 + r_3$ to obtain

$$\begin{bmatrix} 1 & 3 & 0 & | & 100 \\ 0 & 1 & -1 & | & -10 \\ 0 & -6 & 2 & | & -60 \end{bmatrix} \xrightarrow{R_3 = 6r_2 + r_3} \begin{bmatrix} 1 & 3 & 0 & | & 100 \\ 0 & 1 & -1 & | & -10 \\ 0 & 0 & -4 & | & -120 \end{bmatrix}$$

Next, use $R_3 = -\frac{1}{4}r_3$. The result is

$$\begin{bmatrix} 1 & 3 & 0 & | & 100 \\ 0 & 1 & -1 & | & -10 \\ 0 & 0 & -4 & | & -120 \end{bmatrix} \xrightarrow{R_3 = -\frac{1}{4}r_3} \begin{bmatrix} 1 & 3 & 0 & | & 100 \\ 0 & 1 & -1 & | & -10 \\ 0 & 0 & 1 & | & 30 \end{bmatrix}$$

The matrix is now in row echelon form. We find $z = 30$. From row 2, we have $y - z = -10$ so that $y = z - 10 = 30 - 10 = 20$. Finally, from row 1, we have $x + 3y = 100$ so $x = -3y + 100 = -60 + 100 = 40$. The solution of the system is $x = 40$, $y = 20$, $z = 30$. In one day 40 Model X, 20 Model Y, and 30 Model Z processors were produced. ▶

NOW WORK PROBLEM 73.

The matrix method for solving a system of linear equations also identifies systems that have infinitely many solutions and systems that are inconsistent. Let's see how.

5 **EXAMPLE 8** **Solving a System of Linear Equations Using Matrices (Infinitely Many Solutions)**

Solve: $\begin{cases} 2x - 3y = 5 \\ 4x - 6y = 10 \end{cases}$

SOLUTION The augmented matrix representing this system is

$$\begin{bmatrix} 2 & -3 & | & 5 \\ 4 & -6 & | & 10 \end{bmatrix}$$

To place a 1 in row 1, column 1, we use $R_1 = \frac{1}{2}r_1$:

$$\begin{bmatrix} 2 & -3 & | & 5 \\ 4 & -6 & | & 10 \end{bmatrix} \xrightarrow{R_1 = \frac{1}{2}r_1} \begin{bmatrix} 1 & -\frac{3}{2} & | & \frac{5}{2} \\ 4 & -6 & | & 10 \end{bmatrix}$$

To place a 0 in column 1, row 2, we use $R_2 = -4r_1 + r_2$:

$$\begin{bmatrix} 1 & -\frac{3}{2} & | & \frac{5}{2} \\ 4 & -6 & | & 10 \end{bmatrix} \xrightarrow{R_2 = -4r_1 + r_2} \begin{bmatrix} 1 & -\frac{3}{2} & | & \frac{5}{2} \\ 0 & 0 & | & 0 \end{bmatrix}$$

The system of equations looks like

$$\begin{cases} x - \dfrac{3}{2}y = \dfrac{5}{2} \\ 0x + 0y = 0 \end{cases}$$

The second equation is true for any choice of x and y, so all numbers x and y that obey the first equation are solutions of the system. Since any point on the line $x - \frac{3}{2}y = \frac{5}{2}$ is a solution, there are an infinite number of solutions.

Using y as parameter, we can list some of these solutions by assigning values to y and then calculating x from the equation $x = \frac{3}{2}y + \frac{5}{2}$.

If $y = 0$, then $x = \frac{5}{2}$, so $x = \frac{5}{2}, y = 0$ is a solution.

If $y = 1$, then $x = 4$, so $x = 4, y = 1$ is a solution.

If $y = 5$, then $x = 10$, so $x = 10, y = 5$ is a solution.

If $y = -3$, then $x = -2$, so $x = -2, y = -3$ is a solution.

And so on.

We write the solution as

$$x = \frac{3}{2}y + \frac{5}{2}, \quad y \text{ is any real number}$$

NOW WORK PROBLEM 43.

Let's solve a system of three equations containing three variables.

EXAMPLE 9   **Solving a System of Linear Equations Using Matrices (Infinitely Many Solutions)**

Solve: $\begin{cases} 6x - y - z = 4 & (1) \\ -12x + 2y + 2z = -8 & (2) \\ 5x + y - z = 3 & (3) \end{cases}$

SOLUTION   We start with the augmented matrix of the system. We then use row operations to obtain a 1 in row 1, column 1 and 0's in the remainder of column 1.

$$\begin{bmatrix} 6 & -1 & -1 & | & 4 \\ -12 & 2 & 2 & | & -8 \\ 5 & 1 & -1 & | & 3 \end{bmatrix} \xrightarrow{R_1 = -r_3 + r_1} \begin{bmatrix} 1 & -2 & 0 & | & 1 \\ -12 & 2 & 2 & | & -8 \\ 5 & 1 & -1 & | & 3 \end{bmatrix} \xrightarrow[R_3 = -5r_1 + r_3]{R_2 = 12r_1 + r_2} \begin{bmatrix} 1 & -2 & 0 & | & 1 \\ 0 & -22 & 2 & | & 4 \\ 0 & 11 & -1 & | & -2 \end{bmatrix}$$

Obtaining a 1 in row 2, column 2 without altering column 1 can be accomplished by $R_2 = -\frac{1}{22}r_2$, or by $R_3 = \frac{1}{11}r_3$ and interchanging rows 2 and 3, or by $R_2 = \frac{23}{11}r_3 + r_2$. We shall use the first of these.

$$\begin{bmatrix} 1 & -2 & 0 & | & 1 \\ 0 & -22 & 2 & | & 4 \\ 0 & 11 & -1 & | & -2 \end{bmatrix} \xrightarrow{R_2 = -\frac{1}{22}r_2} \begin{bmatrix} 1 & -2 & 0 & | & 1 \\ 0 & 1 & -\frac{1}{11} & | & -\frac{2}{11} \\ 0 & 11 & -1 & | & -2 \end{bmatrix} \xrightarrow{R_3 = -11r_2 + r_3} \begin{bmatrix} 1 & -2 & 0 & | & 1 \\ 0 & 1 & -\frac{1}{11} & | & -\frac{2}{11} \\ 0 & 0 & 0 & | & 0 \end{bmatrix}$$

This matrix is in row echelon form. Because the bottom row consists entirely of 0s, the system actually consists of only two equations.

$$\begin{cases} x - 2y = 1 & (1) \\ y - \dfrac{1}{11}z = -\dfrac{2}{11} & (2) \end{cases}$$

From the second equation we get $y = \frac{1}{11}z - \frac{2}{11}$, and then we back-substitute the solution for y from the second equation into the first equation to get

$$x = 2y + 1 = 2\left(\frac{1}{11}z - \frac{2}{11}\right) + 1 = \frac{2}{11}z + \frac{7}{11}$$

The original system is equivalent to the system

$$\begin{cases} x = \dfrac{2}{11}z + \dfrac{7}{11} & (1) \\[2mm] y = \dfrac{1}{11}z - \dfrac{2}{11} & (2) \end{cases}$$

where z, the parameter, can be any real number.

Let's look at the situation. The original system of three equations is equivalent to a system containing two equations. This means that any values of x, y, z that satisfy both

$$x = \frac{2}{11}z + \frac{7}{11} \quad \text{and} \quad y = \frac{1}{11}z - \frac{2}{11}$$

will be solutions. For example, $z = 0$, $x = \frac{7}{11}$, $y = -\frac{2}{11}$; $z = 1$, $x = \frac{9}{11}$, $y = -\frac{1}{11}$; and $z = -1$, $x = \frac{5}{11}$, $y = -\frac{3}{11}$ are some of the solutions of the original system. There are, in fact, infinitely many values of x, y, and z for which the two equations are satisfied. That is, the original system has infinitely many solutions. We will write the solution of the original system as

$$\begin{cases} x = \dfrac{2}{11}z + \dfrac{7}{11} \\[2mm] y = \dfrac{1}{11}z - \dfrac{2}{11} \end{cases}$$

where z, the parameter, can be any real number.

We can also find the solution by writing the augmented matrix in reduced row echelon form. Starting with the row echelon form, we have

$$\begin{bmatrix} 1 & -2 & 0 & | & 1 \\ 0 & 1 & -\dfrac{1}{11} & | & -\dfrac{2}{11} \\ 0 & 0 & 0 & | & 0 \end{bmatrix} \xrightarrow{R_1 = 2r_2 + r_1} \begin{bmatrix} 1 & 0 & -\dfrac{2}{11} & | & \dfrac{7}{11} \\ 0 & 1 & -\dfrac{1}{11} & | & -\dfrac{2}{11} \\ 0 & 0 & 0 & | & 0 \end{bmatrix}$$

The matrix on the right is in reduced row echelon form. The corresponding system of equations is

$$\begin{cases} x - \dfrac{2}{11}z = \dfrac{7}{11} & (1) \\[2mm] y - \dfrac{1}{11}z = -\dfrac{2}{11} & (2) \end{cases}$$

or, equivalently,

$$\begin{cases} x = \dfrac{2}{11}z + \dfrac{7}{11} & (1) \\[2mm] y = \dfrac{1}{11}z - \dfrac{2}{11} & (2) \end{cases}$$

where z, the parameter, can be any real number.

NOW WORK PROBLEM 51.

441

EXAMPLE 10   Solving a System of Linear Equations Using Matrices

Solve: $\begin{cases} x + y + z = 6 \\ 2x - y - z = 3 \\ x + 2y + 2z = 0 \end{cases}$

SOLUTION   We proceed as follows, beginning with the augmented matrix.

$$\begin{bmatrix} 1 & 1 & 1 & | & 6 \\ 2 & -1 & -1 & | & 3 \\ 1 & 2 & 2 & | & 0 \end{bmatrix} \xrightarrow[R_3 = -r_1 + r_3]{R_2 = -2r_1 + r_2} \begin{bmatrix} 1 & 1 & 1 & | & 6 \\ 0 & -3 & -3 & | & -9 \\ 0 & 1 & 1 & | & -6 \end{bmatrix} \xrightarrow[\text{rows 2 and 3.}]{\text{Interchange}} \begin{bmatrix} 1 & 1 & 1 & | & 6 \\ 0 & 1 & 1 & | & -6 \\ 0 & -3 & -3 & | & -9 \end{bmatrix} \xrightarrow{R_3 = 3r_2 + r_3} \begin{bmatrix} 1 & 1 & 1 & | & 6 \\ 0 & 1 & 1 & | & -6 \\ 0 & 0 & 0 & | & -27 \end{bmatrix}$$

This matrix is in row echelon form. The bottom row is equivalent to the equation

$$0x + 0y + 0z = -27$$

which has no solution. The original system is inconsistent.

 NOW WORK PROBLEM 23.

SUMMARY   After finding the row echelon form of the augmented matrix of a system of two linear equations containing two variables, one of the following matrices will result.

$\begin{bmatrix} 1 & a & | & c \\ 0 & 1 & | & d \end{bmatrix}$ **Unique solution:** $y = d$
Back-substitute to find x.

$\begin{bmatrix} a & b & | & c \\ 0 & 0 & | & 0 \end{bmatrix}$ **Infinite number of solutions:** $ax + by = c$
Either x or y can be used as parameter.

$\begin{bmatrix} a & b & | & c \\ 0 & 0 & | & \text{nonzero number} \end{bmatrix}$ **No solution**

EXERCISE 2.2   Answers to Odd-Numbered Problems Begin on Page AN-10.

In Problems 1–12, write the augmented matrix of each system of equations.

1. $\begin{cases} 2x - 3y = 5 \\ x - y = 3 \end{cases}$

2. $\begin{cases} 4x + y = 5 \\ 2x + y = 5 \end{cases}$

3. $\begin{cases} 2x + y + 6 = 0 \\ 3x + y = -1 \end{cases}$

4. $\begin{cases} -3x - y = -3 \\ 4x - y + 2 = 0 \end{cases}$

5. $\begin{cases} 2x - y - z = 0 \\ x - y + z = 1 \\ 3x - y = 2 \end{cases}$

6. $\begin{cases} x + y + z = 3 \\ 2x + z = 0 \\ 3x - y - z = 1 \end{cases}$

7. $\begin{cases} 2x - 3y + z - 7 = 0 \\ x + y - z = 1 \\ 2x + 2y - 3z + 4 = 0 \end{cases}$

8. $\begin{cases} 5x - 3y + 6z + 1 = 0 \\ -x - y + z = 1 \\ 2x + 3y + 5 = 0 \end{cases}$

9. $\begin{cases} 4x_1 - x_2 + 2x_3 - x_4 = 4 \\ x_1 + x_2 + 6 = 0 \\ 2x_2 - x_3 + x_4 = 5 \end{cases}$

10. $\begin{cases} 3x_1 - 5x_2 + x_3 = 2 \\ x_1 - x_2 + x_3 = 6 \\ 2x_1 + x_3 + 4 = 0 \end{cases}$

11. $\begin{cases} x_1 - x_2 + x_3 - x_4 = 0 \\ 2x_1 + 3x_2 - x_3 + 4x_4 = 5 \end{cases}$

12. $\begin{cases} x_1 + x_2 + x_3 + x_4 = 4 \\ x_1 - 2x_2 + 3x_3 - 4x_4 = 5 \end{cases}$

In Problems 13–20, perform each row operation on the given augmented matrix.

$\begin{bmatrix} 1 & -3 & | & -2 \\ 2 & -5 & | & 5 \end{bmatrix}$ (a) $R_2 = -2r_1 + r_2$

14. $\begin{bmatrix} 1 & -3 & | & -3 \\ 2 & -5 & | & -4 \end{bmatrix}$ (a) $R_2 = -2r_1 + r_2$

15. $\begin{bmatrix} 1 & -3 & 4 & | & 3 \\ 2 & -5 & 6 & | & 6 \\ -3 & 3 & 4 & | & 6 \end{bmatrix}$ (a) $R_2 = -2r_1 + r_2$ (b) $R_3 = 3r_1 + r_3$

16. $\begin{bmatrix} 1 & -3 & 3 & | & -5 \\ 2 & -5 & -3 & | & -5 \\ -3 & -2 & 4 & | & 6 \end{bmatrix}$ (a) $R_2 = -2r_1 + r_2$ (b) $R_3 = 3r_1 + r_3$

17. $\begin{bmatrix} 1 & -3 & 2 & | & -6 \\ 2 & -5 & 3 & | & -4 \\ -3 & -6 & 2 & | & 6 \end{bmatrix}$ (a) $R_2 = -2r_1 + r_2$ (b) $R_3 = 3r_1 + r_3$

18. $\begin{bmatrix} 1 & -3 & -4 & | & -6 \\ 2 & -5 & 6 & | & -6 \\ -3 & 1 & 4 & | & 6 \end{bmatrix}$ (a) $R_2 = -2r_1 + r_2$ (b) $R_3 = 3r_1 + r_3$

19. $\begin{bmatrix} 1 & -3 & 1 & | & -2 \\ 2 & -5 & 6 & | & -2 \\ -3 & 1 & 4 & | & 6 \end{bmatrix}$ (a) $R_2 = -2r_1 + r_2$ (b) $R_3 = 3r_1 + r_3$

20. $\begin{bmatrix} 1 & -3 & -1 & | & 2 \\ 2 & -5 & 2 & | & 6 \\ -3 & -6 & 4 & | & 6 \end{bmatrix}$ (a) $R_2 = -2r_1 + r_2$ (b) $R_3 = 3r_1 + r_3$

In Problems 21–32, the row echelon form of a system of linear equations is given. Write the system of equations corresponding to the given matrix. Use x, y; or x, y, z; or x_1, x_2, x_3, x_4 as variables.
Determine whether the system is consistent or inconsistent. If it is consistent, give the solution.

21. $\begin{bmatrix} 1 & 2 & | & 5 \\ 0 & 1 & | & -1 \end{bmatrix}$

22. $\begin{bmatrix} 1 & -3 & | & -4 \\ 0 & 1 & | & 0 \end{bmatrix}$

23. $\begin{bmatrix} 1 & 2 & 3 & | & 1 \\ 0 & 1 & 4 & | & 2 \\ 0 & 0 & 0 & | & 3 \end{bmatrix}$

24. $\begin{bmatrix} 1 & 2 & -1 & | & 0 \\ 0 & 1 & -1 & | & 1 \\ 0 & 0 & 0 & | & 2 \end{bmatrix}$

25. $\begin{bmatrix} 1 & 0 & 2 & | & -1 \\ 0 & 1 & -4 & | & -2 \\ 0 & 0 & 0 & | & 0 \end{bmatrix}$

26. $\begin{bmatrix} 1 & 0 & 4 & | & 4 \\ 0 & 1 & 3 & | & 2 \\ 0 & 0 & 0 & | & 0 \end{bmatrix}$

27. $\begin{bmatrix} 1 & 2 & -1 & 1 & | & 1 \\ 0 & 1 & 4 & 1 & | & 2 \\ 0 & 0 & 1 & 2 & | & 3 \end{bmatrix}$

28. $\begin{bmatrix} 1 & 2 & 4 & 0 & | & 1 \\ 0 & 1 & -1 & 2 & | & 2 \\ 0 & 0 & 1 & 3 & | & 0 \end{bmatrix}$

29. $\begin{bmatrix} 1 & 2 & 0 & 4 & | & 2 \\ 0 & 1 & 1 & 3 & | & 3 \\ 0 & 0 & 0 & 0 & | & 0 \end{bmatrix}$

30. $\begin{bmatrix} 1 & 0 & 3 & 0 & | & 1 \\ 0 & 1 & 4 & 3 & | & 2 \\ 0 & 0 & 1 & 2 & | & 3 \end{bmatrix}$

31. $\begin{bmatrix} 1 & -2 & 0 & 1 & | & -2 \\ 0 & 1 & -3 & 2 & | & 2 \\ 0 & 0 & 1 & -1 & | & 0 \\ 0 & 0 & 0 & 0 & | & 0 \end{bmatrix}$

32. $\begin{bmatrix} 1 & 3 & 0 & 4 & | & 1 \\ 0 & 1 & 2 & -1 & | & 2 \\ 0 & 0 & 1 & 2 & | & 3 \\ 0 & 0 & 0 & 1 & | & 0 \end{bmatrix}$

In Problems 33–58, solve each system of equations using matrices. If the system has no solution, say it is inconsistent.

33. $\begin{cases} x + y = 6 \\ 2x - y = 0 \end{cases}$

34. $\begin{cases} x - y = 2 \\ 2x + y = 1 \end{cases}$

35. $\begin{cases} 2x + y = 5 \\ x - y = 1 \end{cases}$

36. $\begin{cases} 3x + 2y = 7 \\ x + y = 3 \end{cases}$

37. $\begin{cases} 2x + 3y = 7 \\ 3x - y = 5 \end{cases}$

38. $\begin{cases} 2x - 3y = 5 \\ 3x + y = 2 \end{cases}$

39. $\begin{cases} 2x - 3y = 6 \\ 6x - 9y = 10 \end{cases}$

40. $\begin{cases} 3x + 9y = 4 \\ 2x + 6y = 1 \end{cases}$

41. $\begin{cases} 2x - 3y = 0 \\ 4x + 9y = 5 \end{cases}$

42. $\begin{cases} 3x - 4y = 3 \\ 6x + 2y = 1 \end{cases}$

43. $\begin{cases} 2x + 6y = 4 \\ 5x + 15y = 10 \end{cases}$

44. $\begin{cases} 3x + 5y = 5 \\ 6x + 10y = 10 \end{cases}$

45. $\begin{cases} \frac{1}{2}x + \frac{1}{3}y = 2 \\ x + y = 5 \end{cases}$

46. $\begin{cases} x - \frac{1}{4}y = 0 \\ \frac{1}{2}x + \frac{1}{2}y = \frac{5}{2} \end{cases}$

47. $\begin{cases} x + y = 1 \\ 3x - 2y = \frac{4}{3} \end{cases}$

48. $\begin{cases} 4x - y = \frac{11}{4} \\ 3x + y = \frac{5}{2} \end{cases}$

49. $\begin{cases} 2x + y + z = 6 \\ x - y - z = -3 \\ 3x + y + 2z = 7 \end{cases}$

50. $\begin{cases} x + y + z = 5 \\ 2x - y + z = 2 \\ x + 2y - z = 3 \end{cases}$

51. $\begin{cases} 2x - 2y - z = 2 \\ 2x + 3y + z = 2 \\ 3x + 2y = 0 \end{cases}$

52. $\begin{cases} 2x - y - z = -5 \\ x + y + z = 2 \\ x + 2y + 2z = 5 \end{cases}$

53. $\begin{cases} 2x + y - z = 2 \\ x + 3y + 2z = 1 \\ x + y + z = 2 \end{cases}$

54. $\begin{cases} 2x + 2y + z = 6 \\ x - y - z = -2 \\ x - 2y - 2z = -5 \end{cases}$

55. $\begin{cases} x + y - z = 0 \\ 4x + 4y - 4z = -1 \\ 2x + y + z = 2 \end{cases}$

56. $\begin{cases} x + y - z = 0 \\ 4x + 2y - 4z = 0 \\ x + 2y + z = 0 \end{cases}$

57. $\begin{cases} 3x + y - z = \frac{2}{3} \\ 2x - y + z = 1 \\ 4x + 2y = \frac{8}{3} \end{cases}$

58. $\begin{cases} x + y = 1 \\ 2x - y + z = 1 \\ x + 2y + z = \frac{8}{3} \end{cases}$

In Problems 59–64, use a graphing utility to find the row echelon form (REF) and the reduced row echelon form (RREF) of the augmented matrix of each of the following systems. Solve each system. If the system has no solution, say it is inconsistent.

59. $\begin{cases} 2x - 2y + z = 2 \\ x - \frac{1}{2}y + 2z = 1 \\ 2x + \frac{1}{3}y - z = 0 \end{cases}$

60. $\begin{cases} x + y = -1 \\ x - z = 0 \\ y - z = 1 \end{cases}$

61. $\begin{cases} x + y + z = 4 \\ x - y - z = 0 \\ y - z = -4 \end{cases}$

62. $\begin{cases} 2x + y + z = 6 \\ x - y - z = -3 \\ 3x + y + 2z = 7 \end{cases}$

63. $\begin{cases} x_1 + x_2 + x_3 + x_4 = 20 \\ x_2 + x_3 + x_4 = 0 \\ x_3 + x_4 = 13 \\ x_2 - 2x_4 = -5 \end{cases}$

64. $\begin{cases} x_1 - 2x_2 + 3x_3 - 4x_4 = 40 \\ 4x_2 + 6x_4 = -10 \\ x_3 - x_4 = 12 \\ x_2 + 2x_4 = -10 \end{cases}$

65. Theater Seating The Fox Theatre in St. Louis offers three levels of seating. One group of patrons buys 4 mezzanine tickets and 6 lower balcony tickets for $444. Another group spends $614 for 2 mezzanine tickets, 7 lower balcony tickets, and 8 middle balcony tickets. A third group purchases 3 lower balcony tickets and 12 middle balcony tickets for $474. What is the individual price of a mezzanine ticket, a lower balcony ticket, and a middle balcony ticket?

Sources: Fox Theatre, St. Louis, Missouri, 2002; MetroTix, 2003.

66. Mixture A store sells almonds for $6 per pound, cashews for $5 per pound, and peanuts for $2 per pound. One week the manager decides to prepare 100 16-ounce packages of nuts by mixing 40 pounds of peanuts with some almonds and cashews. Each package will be sold for $4. How many pounds of almonds and cashews should be mixed with the peanuts so that the mixture will produce the same revenue as selling the nuts separately?

67. Laboratory Work Stations A chemistry laboratory can be used by 38 students at one time. The laboratory has 16 work stations, some set up for 2 students each and the others set up for 3 students each. How many are there of each kind of work station?

68. Cost of Fast Food One group of people purchased 10 hot dogs and 5 soft drinks at a cost of $12.50. A second group bought 7 hot dogs and 4 soft drinks at a cost of $9. What is the cost of a single hot dog? A single soft drink?

69. Financial Planning Carletta has $10,000 to invest. As her financial consultant, you recommend that she invest in Treasury Bills that yield 6%, Treasury Bonds that yield 7%, and corporate bonds that yield 8%. Carletta wants to have an annual income of $680, and the amount invested in corporate bonds must be half that invested in Treasury Bills. Find the amount in each investment.

70. Financial Planning John has $20,000 to invest. As his financial consultant, you recommend that he invest in Treasury Bills that yield 5%, Treasury bonds that yield 7%, and corporate Bonds that yield 9%. John wants to have an annual income of $1280, and the amount invested in Treasury Bills must be two times the amount invested in corporate bonds. Find the amount in each investment.

71. Diet Preparation A hospital dietician is planning a meal consisting of three foods whose ingredients are summarized as follows:

	Chicken (2-oz. Serving)	Potatoes (1/2-cup serving)	Spinach (1-cup serving)
Grams of Protein	14	1	6
Grams of Carbohydrates	0	18	8
Grams of Fat	4.5	0	1

Determine the number of servings of each food needed to create a meal containing 30 grams of protein, 38 grams of carbohydrates, and 7 grams of fat.

Source: Food and Nutrition Service, United States Department of Agriculture, 2002.

72. Mixture Sally's Girl Scout troop is selling cookies for the Christmas season. There are three different kinds of cookies in three different containers. *bags* that hold 1 dozen chocolate chip and 1 dozen oatmeal; *gift boxes* that hold 2 dozen chocolate chip, 1 dozen mint, and 1 dozen oatmeal; and *cookie tins* that hold 3 dozen mint and 2 dozen chocolate chip. Sally's mother is having a Christmas party and wants 6 dozen oatmeal; 10 dozen mint, and 14 dozen chocolate chip cookies. How can Sally fill her mother's order?

73. Production A citrus company completes the preparation of its products by cleaning, filling, and labeling bottles. Each case of orange juice requires 10 minutes in the cleaning machine, 4 minutes in the filling machine, and 2 minutes in the labeling machine. For each case of tomato juice, the times are 12 minutes of cleaning, 4 minutes of filling, and 1 minute of labeling. Pineapple juice requires 9 minutes of cleaning, 6 minutes of filling, and 1 minute of labeling per case. If the company runs the cleaning machine for 398 minutes, the filling machine for 164 minutes, and the labeling machine for 58 minutes, how many cases of each type of juice are prepared?

74. Production The manufacture of an automobile requires painting, drying, and polishing. The Rome Motor Company produces three types of cars: the Centurion, the Tribune, and the Senator. Each Centurion requires 8 hours for painting, 2 hours for drying, and 1 hour for polishing. A Tribune needs 10 hours for painting, 3 hours for drying, and 2 hours for polishing. It takes 16 hours of painting, 5 hours of drying, and 3 hours of polishing to prepare a Senator. If the company uses 240 hours for painting, 69 hours for drying, and 41 hours for polishing in a given month, how many of each type of car are produced?

75. Inventory Control An art teacher finds that colored paper can be bought in three different packages. The first package has 20 sheets of white paper, 15 sheets of blue paper, and 1 sheet of red paper. The second package has 3 sheets of blue paper and 1 sheet of red paper. The last package has 40 sheets of white paper and 30 sheets of blue paper. Suppose he needs 200 sheets of white paper, 180 sheets of blue paper, and 12 sheets of red paper. How many of each type of package should he order?

76. Inventory Control An interior decorator has ordered 12 cans of sunset paint, 35 cans of brown, and 18 cans of fuchsia. The paint store has special pair packs, containing 1 can each of sunset and fuchsia; darkening packs, containing 2 cans of sunset, 5 cans of brown, and 2 cans of fuchsia; and economy packs, containing 3 cans of sunset, 15 cans of brown, and 6 cans of fuchsia. How many of each type of pack should the paint store send to the interior decorator?

77. Packaging A recreation center wants to purchase compact discs (CDs) to be used in the center. There is no requirement as to the artists. The only requirement is that they purchase 40 rock CDs, 32 western CDs, and 14 blues CDs. There are three different shipping packages offered by the company. They are an *assorted* carton, containing 2 rock CDs, 4 western CDs, and 1 blues CD; a *mixed* carton containing 4 rock and 2 western CDs; and a *single* carton containing 2 blues CDs. What combination of these packages is needed to fill the center's order?

78. Production A luggage manufacturer produces three types of luggage: economy, standard, and deluxe. The company produces 1000 pieces of luggage at a cost of $20, $25, and $30 for the economy, standard, and deluxe luggage, respectively. The manufacturer has a budget of $20,700. Each economy luggage requires 6 hours of labor, each standard luggage requires 10 hours of labor, and each deluxe model requires 20 hours of labor. The manufacturer has a maximum of 6800 hours of labor available. If the manufacturer sells all the luggage, consumes the entire budget, and uses all the available labor, how many of each type of luggage should be produced?

79. Mixture Suppose that a store has three sizes of cans of nuts. The *large* size contains 2 pounds of peanuts and 1 pound of cashews. The *mammoth* size contains 1 pound of walnuts, 6 pounds of peanuts, and 2 pounds of cashews. The *giant* size contains 1 pound of walnuts, 4 pounds of peanuts, and 2 pounds of cashews. Suppose that the store receives an order for 5 pounds of walnuts, 26 pounds of peanuts, and 12 pounds of cashews. How can it fill this order with the given sizes of cans?

80. Mixture Suppose that the store in Problem 79 receives a new order for 6 pounds of walnuts, 34 pounds of peanuts, and 15 pounds of cashews. How can this order be filled with the given cans?

81. Write a brief paragraph or two that outlines your strategy for solving a system of linear equations using matrices.

82. When solving a system of linear equations using matrices, do you prefer to place the augmented matrix in row echelon form or in reduced row echelon form? Give reasons for your choice.

83. Make up a system of three linear equations containing three variables that has:

(a) No solution
(b) Exactly one solution
(c) Infinitely many solutions

Give the three systems to a friend to solve and critique.

2.3 Systems of *m* Linear Equations Containing *n* Variables

OBJECTIVES
1 Analyze the reduced row echelon form of an augmented matrix
2 Solve a system of *m* linear equations containing *n* variables
3 Express the solution of a system with an infinite number of solutions

We saw in the previous two sections that systems of two linear equations containing two variables and systems of three linear equations containing three variables each have either one solution, no solution, or infinitely many solutions. As it turns out, no matter how many equations are in a system of linear equations and no matter how many variables a system has, only these three possibilities can arise.

For example, the system of three linear equations containing four variables

$$\begin{cases} x_1 + 3x_2 + 5x_3 + x_4 = 2 \\ 2x_1 + 3x_2 + 4x_3 + 2x_4 = 1 \\ x_1 + 2x_2 + 3x_3 + x_4 = 1 \end{cases}$$

will have either no solution, one solution, or infinitely many solutions.

A general definition of a system of *m* linear equations containing *n* variables is given next.

System of *m* Linear Equations Containing *n* Variables

A system of *m* linear equations containing *n* variables $x_1, x_2, \ldots, x_n$ is of the form

$$\begin{cases} a_{11}x_1 + a_{12}x_2 + \cdots + a_{1n}x_n = b_1 \\ a_{21}x_1 + a_{22}x_2 + \cdots + a_{2n}x_n = b_2 \\ a_{31}x_1 + a_{32}x_2 + \cdots + a_{3n}x_n = b_3 \\ \vdots \qquad \vdots \qquad \qquad \vdots \quad \vdots \\ a_{i1}x_1 + a_{i2}x_2 + \cdots + a_{in}x_n = b_i \\ \vdots \qquad \vdots \qquad \qquad \vdots \quad \vdots \\ a_{m1}x_1 + a_{m2}x_2 + \cdots + a_{mn}x_n = b_m \end{cases}$$

where a_{ij} and b_i are real numbers, $i = 1, 2 \ldots, m, j = 1, 2, \ldots, n$.

A **solution** of a system of *m* linear equations containing *n* variables $x_1, x_2, \ldots, x_n$ is any ordered set $(x_1, x_2, \ldots, x_n)$ of real numbers for which *each* of the *m* linear equations of the system is satisfied.

Reduced Row Echelon Form

A system of m linear equations containing n variables will have either no solution, one solution, or infinitely many solutions. We can determine which of these possibilities occurs and, if solutions exist, find them, by performing row operations on the augmented matrix of the system until we arrive at the reduced row echelon form of the augmented matrix.

Let's review the conditions required for the reduced row echelon form:

Conditions for the Reduced Row Echelon Form of a Matrix

1. The first nonzero entry in each row is 1 and it has 0s above it and below it.
2. The leftmost 1 in any row is to the right of the leftmost 1 in the row above.
3. Any rows that contain all 0s to the left of the vertical bar appear at the bottom.

The next two examples will help you recognize when an augmented matrix is in reduced row echelon form.

EXAMPLE 1   Examples of Matrices That Are in Reduced Row Echelon Form

(a) $\left[\begin{array}{cc|c} 1 & 0 & 2 \\ 0 & 1 & 3 \end{array}\right]$

(b) $\left[\begin{array}{ccc|c} 1 & 0 & -3 & 4 \\ 0 & 1 & -2 & 2 \end{array}\right]$

(c) $\left[\begin{array}{ccc|c} 1 & -2 & 0 & 1 \\ 0 & 0 & 1 & 3 \\ 0 & 0 & 0 & 5 \end{array}\right]$

(d) $\left[\begin{array}{cc|c} 1 & 0 & 3 \\ 0 & 1 & 4 \\ 0 & 0 & 0 \end{array}\right]$

EXAMPLE 2   Examples of Matrices That Are Not in Reduced Row Echelon Form

(a) $\left[\begin{array}{cc|c} 1 & 0 & 0 \\ 0 & 0 & 0 \\ 0 & 1 & 0 \end{array}\right]$ The second row contains all 0s and the third does not—this violates the rule that states that any rows with all 0s are at the bottom.

(b) $\left[\begin{array}{ccc|c} 1 & 0 & 2 & 4 \\ 0 & 2 & 4 & 3 \\ 0 & 0 & 0 & 1 \end{array}\right]$ The first nonzero entry in row 2 is not a 1.

(c) $\left[\begin{array}{ccc|c} 1 & 0 & 0 & 0 \\ 0 & 0 & 1 & 1 \\ 0 & 1 & 0 & 0 \end{array}\right]$ The leftmost 1 in the third row is not to the right of the leftmost 1 in the row above it.

Analyze the reduced row echelon form of an augmented matrix 1

NOW WORK PROBLEM 1.

Now let's analyze the reduced row echelon form of an augmented matrix.

EXAMPLE 3   **Analyzing the Reduced Row Echelon Form of an Augmented Matrix**

The matrix

$$\begin{bmatrix} 1 & 0 & 0 & | & 3 \\ 0 & 1 & 0 & | & 8 \\ 0 & 0 & 1 & | & -4 \end{bmatrix}$$

is the reduced row echelon form of the augmented matrix of a system of three linear equations containing three variables. If the variables are x, y, z, this matrix represents the system of equations

$$\begin{cases} x = 3 \\ y = 8 \\ z = -4 \end{cases}$$

The system has one solution: $x = 3, y = 8, z = -4$.   ▶

EXAMPLE 4   **Analyzing the Reduced Row Echelon Form of an Augmented Matrix**

The matrix

$$\begin{bmatrix} 1 & 0 & 3 & | & 0 \\ 0 & 1 & 2 & | & 0 \\ 0 & 0 & 0 & | & 1 \\ 0 & 0 & 0 & | & 0 \end{bmatrix}$$

is the reduced row echelon form of the augmented matrix of a system of four equations containing three variables. If the variables are x, y, z, the equation represented by the third row is

$$0 \cdot x + 0 \cdot y + 0 \cdot z = 1 \qquad \text{or} \qquad 0 = 1$$

Since $0 = 1$ is a contradiction, we conclude the system is inconsistent.   ▶

EXAMPLE 5   **Analyzing the Reduced Row Echelon Form of an Augmented Matrix**

The matrix

$$\begin{bmatrix} 1 & 0 & 0 & 2 & | & 5 \\ 0 & 1 & 0 & 1 & | & 2 \\ 0 & 0 & 1 & 3 & | & 4 \end{bmatrix}$$

is the reduced row echelon form of a system of three equations containing four variables. If x_1, x_2, x_3, x_4 are the variables, the system of equations is

$$\begin{cases} x_1 + 2x_4 = 5 \\ x_2 + x_4 = 2 \\ x_3 + 3x_4 = 4 \end{cases} \qquad \text{or} \qquad \begin{cases} x_1 = -2x_4 + 5 \\ x_2 = -x_4 + 2 \\ x_3 = -3x_4 + 4 \end{cases}$$

The system has infinitely many solutions. In this form, the variable x_4 is the parameter. We assign values to the parameter x_4 from which the variables x_1, x_2, x_3 can be calculated. Some of the possibilities are

$$\text{If } x_4 = 0, \text{ then } x_1 = 5, x_2 = 2, x_3 = 4.$$
$$\text{If } x_4 = 1, \text{ then } x_1 = 3, x_2 = 1, x_3 = 1.$$
$$\text{If } x_4 = 2, \text{ then } x_1 = 1, x_2 = 0, x_3 = -2.$$

And so on.

 NOW WORK PROBLEM 17.

Let's review the procedure for obtaining the reduced row echelon form of a matrix. Then we will use this procedure to solve systems of linear equations.

EXAMPLE 6   **Finding the Reduced Row Echelon Form of a Matrix**

Find the reduced row echelon form of

$$A = \begin{bmatrix} 1 & -1 & | & 2 \\ 2 & 3 & | & 2 \\ 3 & -5 & | & 2 \end{bmatrix}$$

SOLUTION   The entry in row 1, column 1 is 1. We proceed to obtain a matrix in which all the remaining entries in column 1 are 0s. We can obtain such a matrix by performing the row operations

$$R_2 = -2r_1 + r_2$$
$$R_3 = -3r_1 + r_3$$

The new matrix is

$$\begin{bmatrix} 1 & -1 & | & 2 \\ 0 & -1 & | & -2 \\ 0 & -2 & | & -4 \end{bmatrix}$$

We want the entry in row 2, column 2 (now -1), to be 1. By multiplying row 2 by -1, $R_2 = (-1)r_2$, we obtain

$$\begin{bmatrix} 1 & -1 & | & 2 \\ 0 & 1 & | & 2 \\ 0 & -2 & | & -4 \end{bmatrix}$$

Now we want the entry in row 1, column 2 and in row 3, column 2 to be 0. This can be accomplished by applying the row operations

$$R_1 = r_2 + r_1$$
$$R_3 = 2r_2 + r_3$$

The new matrix is

$$\begin{bmatrix} 1 & 0 & | & 4 \\ 0 & 1 & | & 2 \\ 0 & 0 & | & 0 \end{bmatrix}$$

This is the reduced row echelon form of A.

2 **EXAMPLE 7** **Solving a System of Three Linear Equations Containing Two Variables**

Solve: $\begin{cases} x - y = 2 \\ 2x - 3y = 2 \\ 3x - 5y = 2 \end{cases}$

SOLUTION The augmented matrix of this system is

$$\left[\begin{array}{cc|c} 1 & -1 & 2 \\ 2 & -3 & 2 \\ 3 & -5 & 2 \end{array}\right]$$

Using the solution to Example 6, the reduced row echelon form of this augmented matrix is

$$\left[\begin{array}{cc|c} 1 & 0 & 4 \\ 0 & 1 & 2 \\ 0 & 0 & 0 \end{array}\right]$$

We conclude that the system has the solution $x = 4$, $y = 2$. ▶

  COMMENT: A graphing utility can be used to solve systems of m linear equations containing n variables. Check the solution to Example 7 using the RREF feature of your graphing utility. ▶

EXAMPLE 8 **Solving a System of Four Linear Equations Containing Three Variables**

Solve: $\begin{cases} x - y + 2z = 2 \\ 2x - 3y + 2z = 1 \\ 3x - 5y + 2z = -3 \\ -4x + 12y + 8z = 10 \end{cases}$

SOLUTION We need to find the reduced row echelon form of the augmented matrix of this system, namely,

$$\left[\begin{array}{ccc|c} 1 & -1 & 2 & 2 \\ 2 & -3 & 2 & 1 \\ 3 & -5 & 2 & -3 \\ -4 & 12 & 8 & 10 \end{array}\right]$$

The entry 1 is already present in row 1, column 1. To obtain 0s elsewhere in column 1, we use the row operations

$$R_2 = -2r_1 + r_2 \qquad R_3 = -3r_1 + r_3 \qquad R_4 = 4r_1 + r_4$$

The new matrix is

$$\left[\begin{array}{ccc|c} 1 & -1 & 2 & 2 \\ 0 & -1 & -2 & -3 \\ 0 & -2 & -4 & -9 \\ 0 & 8 & 16 & 18 \end{array}\right]$$

To obtain the entry 1 in row 2, column 2, we use $R_2 = -r_2$, obtaining

$$\begin{bmatrix} 1 & -1 & 2 & | & 2 \\ 0 & 1 & 2 & | & 3 \\ 0 & -2 & -4 & | & -9 \\ 0 & 8 & 16 & | & 18 \end{bmatrix}$$

To obtain 0s elsewhere in column 2, we use

$$R_1 = r_2 + r_1 \qquad R_3 = 2r_2 + r_3 \qquad R_4 = -8r_2 + r_4$$

The new matrix is

$$\begin{bmatrix} 1 & 0 & 4 & | & 5 \\ 0 & 1 & 2 & | & 3 \\ 0 & 0 & 0 & | & -3 \\ 0 & 0 & 0 & | & -6 \end{bmatrix}$$

We can stop here even though the matrix is not in reduced row echelon form. Because the third row yields the equation

$$0 \cdot x + 0 \cdot y + 0 \cdot z = -3$$

we conclude the system is inconsistent.

NOW WORK PROBLEM 31.

Express the solution of a system with an infinite number of solutions **3**

Infinite Number of Solutions

We have seen several examples of systems of linear equations that have an infinite number of solutions. Let's look at a few more examples.

EXAMPLE 9 **Solving a System of Two Linear Equations Containing Three Variables**

Solve: $\begin{cases} x + y + z = 7 \\ x - y - 3z = 1 \end{cases}$

SOLUTION The augmented matrix of the system is

$$\begin{bmatrix} 1 & 1 & 1 & | & 7 \\ 1 & -1 & -3 & | & 1 \end{bmatrix}$$

The reduced row echelon form (as you should verify) is

$$\begin{bmatrix} 1 & 0 & -1 & | & 4 \\ 0 & 1 & 2 & | & 3 \end{bmatrix}$$

The system of equations represented by this matrix is

$$\begin{cases} x - z = 4 \\ y + 2z = 3 \end{cases} \quad \text{or} \quad \begin{cases} x = z + 4 \\ y = -2z + 3 \end{cases} \tag{1}$$

The system has infinitely many solutions. In the form (1), the variable z is the parameter. We can assign any value to z and use it to compute values of x and y.

✓ CHECK: We check the solution to Example 9 as follows:

$$x + y + \ z = (z + 4) + (-2z + 3) + \ z = 7 + z - 2z + \ z = 7$$
$$x - y - 3z = (z + 4) - (-2z + 3) - 3z = 1 + z + 2z - 3z = 1$$

The solution is verified.   ▶

The next example illustrates a system having an infinite number of solutions with two parameters.

EXAMPLE 10 **Solving a System of Three Linear Equations Containing Four Variables**

Solve: $\begin{cases} x_1 + x_2 + 2x_3 + 2x_4 = 2 \\ \quad\ x_1 + x_3 + x_4 = 0 \\ \quad\ x_2 + x_3 + x_4 = 2 \end{cases}$

SOLUTION The augmented matrix of the system is

$$\begin{bmatrix} 1 & 1 & 2 & 2 & | & 2 \\ 1 & 0 & 1 & 1 & | & 0 \\ 0 & 1 & 1 & 1 & | & 2 \end{bmatrix}$$

The reduced row echelon form (as you should verify) is

$$\begin{bmatrix} 1 & 0 & 1 & 1 & | & 0 \\ 0 & 1 & 1 & 1 & | & 2 \\ 0 & 0 & 0 & 0 & | & 0 \end{bmatrix}$$

The equations represented by this system are

$$\begin{cases} x_1 + x_3 + x_4 = 0 \\ x_2 + x_3 + x_4 = 2 \end{cases}$$

We can rewrite this system in the form

$$\begin{cases} x_1 = -x_3 - x_4 \\ x_2 = -x_3 - x_4 + 2 \end{cases} \qquad (2)$$

The system has infinitely many solutions. In the form (2), the system has two parameters x_3 and x_4. Solutions are obtained by assigning the two parameters x_3 and x_4 arbitrary values. Some choices are shown in Table 1 below.

TABLE 1

x_3	x_4	x_1	x_2	Solution (x_1, x_2, x_3, x_4)
0	0	0	2	$(0, 2, 0, 0)$
1	0	-1	1	$(-1, 1, 1, 0)$
0	2	-2	0	$(-2, 0, 0, 2)$

▶

The variables used as parameters are not unique. We could have chosen x_1 and x_4 as parameters by rewriting Equations (2) in the following manner.

From the first equation

$$x_3 = -x_1 - x_4$$

We can replace the parameter x_3 in the second equation by the above to produce

$$x_2 = 2 - x_3 - x_4 = 2 + x_1 + x_4 - x_4 = 2 + x_1$$

We then obtain the system

$$\begin{cases} x_2 = x_1 + 2 \\ x_3 = -x_1 - x_4 \end{cases}$$

showing the solution with x_1 and x_4 as parameters.

 NOW WORK PROBLEM 27.

EXAMPLE 11 **Investment Goals**

A couple have $25,000 available and want to invest in U.S. Savings Bonds. As of January 2003, the rates were 3.2% for EE/E bonds, 4% for I bonds, and 1.5% for HH/H bonds. Their goal is to invest in all three types of bonds in such a way that they obtain $500 in interest per year. Prepare a table showing the various ways this couple can achieve their goal.

Source: US Department of the Treasury, 2003.

SOLUTION   Let x represent the amount invested in EE/E bonds, y represent the amount invested in I bonds, and z represent the amount invested in HH/H bonds. Since the couple have $25,000 to invest and want a $500 return on their investment, we need to solve the system of equations

$$\begin{cases} x + y + z = 25000 \quad (1) \\ .032x + .04y + .015z = 500 \quad (2) \end{cases}$$

The augmented matrix of this system and its reduced row echelon form (which you should verify) are given by

$$\begin{bmatrix} 1 & 1 & 1 & | & 25{,}000 \\ .032 & .04 & .015 & | & 500 \end{bmatrix} \longrightarrow \begin{bmatrix} 1 & 0 & 3.125 & | & 62{,}500 \\ 0 & 1 & -2.125 & | & -37{,}500 \end{bmatrix}$$

The augmented matrix represents the system of equations

$$\begin{cases} x + 3.125z = 62{,}500 \\ y - 2.125z = -37{,}500 \end{cases} \xrightarrow[\text{x and y.}]{\text{Solve for}} \begin{cases} x = 62{,}500 - 3.125z \\ y = -37{,}500 + 2.125z \end{cases} \quad (3)$$

where z is the parameter. Since we want to invest in each of the three bond types, the conditions $x > 0$, $y > 0$, and $z > 0$ must hold. So we can determine from the system (3), that

$$62{,}500 - 3.125z > 0 \ \text{ so } \ z < 20{,}000$$
$$-37{,}500 + 2.125z > 0 \ \text{ so } \ z > 17{,}647$$

Several of the possible solutions are listed in Table 2 on page 90. The couple's final decision on asset allocation will usually reflect their attitude toward risk.

TABLE 2	Amount in EE/E	Amount in I	Amount in HH/H
	$6250	750	18,000
	5469	1281	18,250
	4688	1812	18,500
	3906	2344	18,750
	3125	2875	19,000
	2344	3406	19,250
	1563	3937	19,500
	781	4469	19,750

NOW WORK PROBLEM 53.

SUMMARY

Steps for Solving a System of m Linear Equations Containing n Variables

STEP 1 Write the augmented matrix.

STEP 2 Find the reduced row echelon form of the augmented matrix.

STEP 3 Analyze this matrix to determine if the system has no solution, one solution, or infinitely many solutions.

EXERCISE 2.3   Answers to Odd-Numbered Problems Begin on Page AN-11.

In Problems 1–12, tell whether the given matrix is in reduced row echelon form. If it is not, tell why.

1. $\begin{bmatrix} 1 & 2 & | & 3 \\ 0 & 0 & | & 0 \\ 0 & 0 & | & 1 \end{bmatrix}$

2. $\begin{bmatrix} 1 & 2 & | & 3 \\ 0 & 0 & | & 0 \\ 0 & 0 & | & 0 \end{bmatrix}$

3. $\begin{bmatrix} 1 & 1 & | & 0 \\ 0 & 1 & | & 0 \end{bmatrix}$

4. $\begin{bmatrix} 1 & 0 & | & 3 \\ 0 & 1 & | & 0 \end{bmatrix}$

5. $\begin{bmatrix} 0 & | & 1 \\ 1 & | & 0 \end{bmatrix}$

6. $\begin{bmatrix} 0 & 1 & | & 0 \\ 0 & 0 & | & 1 \\ 0 & 0 & | & 0 \end{bmatrix}$

7. $\begin{bmatrix} 1 & 2 & | & 1 \\ 0 & 0 & | & 0 \end{bmatrix}$

8. $\begin{bmatrix} 1 & 0 & | & 8 \\ 0 & 2 & | & 9 \end{bmatrix}$

9. $\begin{bmatrix} 1 & 0 & 0 & 0 & | & 0 \\ 0 & 0 & 1 & 2 & | & 0 \\ 0 & 0 & 0 & 0 & | & 1 \\ 0 & 0 & 0 & 0 & | & 0 \end{bmatrix}$

10. $\begin{bmatrix} 1 & 1 & | & 0 \\ 0 & 0 & | & 2 \end{bmatrix}$

11. $\begin{bmatrix} 1 & 0 & | & 1 \\ 0 & 1 & | & 2 \\ 0 & 0 & | & 0 \end{bmatrix}$

12. $\begin{bmatrix} 1 & 0 & | & 2 \\ 0 & 1 & | & 2 \end{bmatrix}$

In Problems 13–28, the reduced row echelon form of the augmented matrix of a system of linear equations is given. Tell whether the system has one solution, no solution, or infinitely many solutions. Write the solutions or, if there is no solution, say the system is inconsistent.

13. $\begin{bmatrix} 1 & 1 & | & 1 \\ 0 & 0 & | & 0 \end{bmatrix}$

14. $\begin{bmatrix} 1 & 0 & | & 0 \\ 0 & 0 & | & 1 \end{bmatrix}$

15. $\begin{bmatrix} 1 & 0 & | & 4 \\ 0 & 1 & | & 5 \end{bmatrix}$

16. $\begin{bmatrix} 1 & 0 & 0 & | & 0 \\ 0 & 1 & 0 & | & 0 \\ 0 & 0 & 1 & | & 6 \end{bmatrix}$

17. $\begin{bmatrix} 1 & 0 & -2 & | & 6 \\ 0 & 1 & 3 & | & 1 \end{bmatrix}$

18. $\begin{bmatrix} 1 & 0 & 0 & | & 0 \\ 0 & 1 & 0 & | & 5 \\ 0 & 0 & 0 & | & 0 \end{bmatrix}$

19. $\begin{bmatrix} 1 & 2 & 0 & | & 1 \\ 0 & 0 & 1 & | & 2 \\ 0 & 0 & 0 & | & 0 \end{bmatrix}$

20. $\begin{bmatrix} 1 & 2 & 0 & | & 0 \\ 0 & 0 & 1 & | & 0 \\ 0 & 0 & 0 & | & 1 \end{bmatrix}$

21. $\begin{bmatrix} 1 & 0 & 0 & | & 1 \\ 0 & 1 & 0 & | & 2 \\ 0 & 0 & 0 & | & 0 \end{bmatrix}$

22. $\begin{bmatrix} 1 & 0 & 1 & -1 & | & 0 \\ 0 & 1 & 2 & 1 & | & 1 \\ 0 & 0 & 0 & 0 & | & 0 \end{bmatrix}$

23. $\begin{bmatrix} 1 & 0 & 0 & | & -1 \\ 0 & 1 & 0 & | & 3 \\ 0 & 0 & 1 & | & 4 \\ 0 & 0 & 0 & | & 0 \end{bmatrix}$

24. $\begin{bmatrix} 1 & 2 & 0 & 0 & | & -4 \\ 0 & 0 & 1 & 0 & | & -3 \\ 0 & 0 & 0 & 1 & | & 2 \\ 0 & 0 & 0 & 0 & | & 0 \end{bmatrix}$

25. $\begin{bmatrix} 1 & 0 & -1 & | & 1 \\ 0 & 1 & 2 & | & 1 \end{bmatrix}$

26. $\begin{bmatrix} 1 & 0 & | & 1 \\ 0 & 1 & | & 1 \\ 0 & 0 & | & 0 \end{bmatrix}$

27. $\begin{bmatrix} 1 & 0 & 0 & -1 & | & 4 \\ 0 & 1 & 2 & 3 & | & 0 \end{bmatrix}$

28. $\begin{bmatrix} 1 & 0 & 2 & 4 & | & -1 \\ 0 & 1 & 3 & 5 & | & -2 \end{bmatrix}$

In Problems 29–52, solve each system of equations by finding the reduced row echelon form of the augmented matrix. If there is no solution, say the system is inconsistent.

29. $\begin{cases} x + y = 3 \\ 2x - y = 3 \end{cases}$

30. $\begin{cases} x - y = 5 \\ 2x + 3y = 15 \end{cases}$

31. $\begin{cases} 3x - 3y = 12 \\ 3x + 2y = -3 \\ 2x + y = 4 \end{cases}$

32. $\begin{cases} 6x + y = 8 \\ x - 3y = 5 \\ 2x + y = 2 \end{cases}$

33. $\begin{cases} 2x - 4y = 8 \\ x - 2y = 4 \\ -x + 2y = -4 \end{cases}$

34. $\begin{cases} 3x + y = 8 \\ 6x + 2y = 16 \\ -9x - 3y = -24 \end{cases}$

35. $\begin{cases} 2x + y + 3z = -1 \\ -x + y + 3z = 8 \\ 2x - 2y - 6z = -16 \end{cases}$

36. $\begin{cases} x + 2y + 3z = 5 \\ -2x + 6y + 4z = 0 \\ 2x + 4y + 6z = 10 \end{cases}$

37. $\begin{cases} x - y = 1 \\ y - z = 6 \\ x + z = -1 \end{cases}$

38. $\begin{cases} 2x - y + 3z = 0 \\ x + 2y - z = 5 \\ 2y + z = 1 \end{cases}$

39. $\begin{cases} x_1 + x_2 = 7 \\ x_2 - x_3 + x_4 = 5 \\ x_1 - x_2 + x_3 + x_4 = 6 \\ x_2 - x_4 = 10 \end{cases}$

40. $\begin{cases} x_1 + x_2 + x_3 + x_4 = 0 \\ 2x_1 - x_2 - x_3 + x_4 = 0 \\ x_1 - x_2 - x_3 + x_4 = 0 \\ x_1 + x_2 - x_3 - x_4 = 0 \end{cases}$

41. $\begin{cases} x_1 + 2x_2 + 3x_3 - x_4 = 0 \\ 3x_1 - x_4 = 4 \\ x_2 - x_3 - x_4 = 2 \end{cases}$

42. $\begin{cases} 2x - 3y + 4z = 7 \\ x - 2y + 3z = 2 \end{cases}$

43. $\begin{cases} x - y + z = 5 \\ 2x - 2y + 2z = 8 \end{cases}$

44. $\begin{cases} x + y + z = 3 \\ x - y + z = 7 \\ x - y - z = 1 \end{cases}$

45. $\begin{cases} 3x - y + 2z = 3 \\ 3x + 3y + z = 3 \\ 3x - 5y + 3z = 12 \end{cases}$

46. $\begin{cases} x + y - z = 12 \\ 3x - y = 1 \\ 2x - 3y + 4z = 3 \end{cases}$

47. $\begin{cases} x_1 + x_2 + x_3 + x_4 = 4 \\ 2x_1 - x_2 + x_3 = 0 \\ 3x_1 + 2x_2 + x_3 - x_4 = 6 \\ x_1 - 2x_2 - 2x_3 + 2x_4 = 1 \end{cases}$

48. $\begin{cases} x_1 + x_2 + x_3 + x_4 = 4 \\ -x_1 + 2x_2 + x_3 = 0 \\ 2x_1 + 3x_2 + x_3 - x_4 = 6 \\ -2x_1 + x_2 - 2x_3 + 2x_4 = -1 \end{cases}$

49. $\begin{cases} 2x - y - z = 0 \\ x - y - z = 1 \\ 3x - y - z = 2 \end{cases}$

50. $\begin{cases} x + y + z = 3 \\ 2x + y + z = 0 \\ 3x + y + z = 1 \end{cases}$

51. $\begin{cases} 2x - y + z = 6 \\ 3x - y + z = 6 \\ 4x - 2y + 2z = 12 \end{cases}$

52. $\begin{cases} x - y + z = 2 \\ 2x - 3y + z = 0 \\ 3x - 3y + 3z = 6 \end{cases}$

53. Investment Allocation Look again at Example 11. Suppose the couple now require $800 interest per year. Prepare a table that shows various ways this couple can achieve their goal.

54. Investment Allocation Look again at Example 11. Suppose the couple still require $500 in interest per year, but the interest rate on HH/H bonds increases to 2%. Prepare a table showing various ways the couple can achieve their goal.

55. Investment Allocation Look again at Example 11. Suppose the interest rate on I bonds goes down to 3.5%. Can the couple maintain their goal to obtain $500 per year in interest? Prepare a table that shows various investment options that the couple can use to achieve their goal.

56. Inventory Control Three species of bacteria will be kept in one test tube and will feed on three resources. Each member of the first species consumes 3 units of the first resource and 1 unit of the third. Each bacterium of the second type consumes 1 unit of the first resource and 2 units each of the second and third. Each bacterium of the third type consumes 2 units of the first resource and 4 each of the second and third. If the test tube is supplied daily with 12,000 units of the first resource, 12,000 units of the second, and 14,000 units of the third, how many of each species can coexist in equilibrium in the test tube so that all of the supplied resources are consumed? Prepare a table that shows some of the possibilities.

57. Cost of Fast Food One group of customers bought 8 deluxe hamburgers, 6 orders of large fries, and 6 large colas for $26.10. A second group ordered 10 deluxe hamburgers, 6 large fries, and 8 large colas and paid $31.60. Is there sufficient information to determine the price of each food item? If not, construct a table showing the various possibilities. Assume the hamburgers cost between $1.75 and $2.25, the fries between $0.75 and $1.00, and the colas between $0.60 and $0.90.

58. Use the information given in Problem 57 and add a third group that purchased 3 deluxe hamburgers, 2 large fries, and 4 colas for $10.95. Is there now sufficient information to determine the price of each food item?

59. Financial Planning Three retired couples each require an additional annual income of $2000 per year. As their financial consultant, you recommend that they invest some money in Treasury Bills that yield 7%, some money in corporate bonds that yield 9%, and some money in junk bonds that yield 11%. Prepare a table for each couple showing the various ways that their goals can be achieved:

(a) If the first couple has $20,000 to invest
(b) If the second couple has $25,000 to invest
(c) If the third couple has $30,000 to invest

Q (d) What advice would you give each couple regarding the amount to invest and the choices available?
[*Hint*: Higher yields generally carry more risk.]

60. Financial Planning A young couple has $25,000 to invest. As their financial consultant, you recommend that they invest some money in Treasury Bills that yield 7%, some money in corporate bonds that yield 9%, and some money in junk bonds that yield 11%. Prepare a table showing the various ways this couple can achieve the following goals:

(a) The couple want $1500 per year in income.
(b) The couple want $2000 per year in income.
(c) The couple want $2500 per year in income.

Q (d) What advice would you give this couple regarding the income that they require and the choices available?
[*Hint*: Higher yields generally carry more risk.]

61. Pharmacy A doctor's prescription calls for a daily intake of liquid containing 40 mg of vitamin C and 30 mg of vitamin D. Your pharmacy stocks three liquids that can be used: one contains 20% vitamin C and 30% vitamin D; a second, 40% vitamin C and 20% vitamin D; and a third, 30% vitamin C and 50% vitamin D. Create a table showing the possible combinations that could be used to fill the prescription.

62. Pharmacy A doctor's prescription calls for the creation of pills that contain 12 units of vitamin B_{12} and 12 units of vitamin E. Your pharmacy stocks three powders that can be used to make these pills: one contains 20% vitamin B_{12} and 30% vitamin E; a second, 40% vitamin B_{12} and 20% vitamin E; and a third, 30% vitamin B_{12} and 40% vitamin E. Create a table showing the possible combinations of each powder that could be mixed in each pill.

Q 63. Make up a system of three linear equations containing four variables that has infinitely many solutions. How many parameters will be in the solution to this system? Solve the system and create a table showing various solutions to the system of equations.

Q 64. Make up a system of two linear equations containing four variables that has infinitely many solutions. How many parameters will be in the solution to this system? Solve the system and create a table showing various solutions to the system of equations.

2.4   Matrix Algebra

PREPARING FOR THIS SECTION   *Before getting started, review the following:*

> Properties of Real Numbers (Appendix A, Section A.1, pp. 622–636)

OBJECTIVES   1   Find the dimension of a matrix
2   Find the sum of two matrices
3   Work with properties of matrices
4   Find the difference of two matrices
5   Find scalar multiples of a matrix

Matrices can be added, subtracted, and multiplied. They also possess many of the algebraic properties of real numbers. **Matrix algebra** is the study of these properties. Its importance lies in the fact that many situations in both pure and applied mathematics involve rectangular arrays of numbers. In fact, in many branches of business and the biological and social sciences, it is necessary to express and use data in a rectangular array. Let's look at an example.

Let's begin with an example that illustrates how matrices can be used to conveniently represent an array of information.

EXAMPLE 1   Arranging Data in a Matrix

In a survey of 900 people, the following information was obtained:

200	Males	Thought federal defense spending was too high
150	Males	Thought federal defense spending was too low
45	Males	Had no opinion
315	Females	Thought federal defense spending was too high
125	Females	Thought federal defense spending was too low
65	Females	Had no opinion

We can arrange the above data in a rectangular array as follows:

	Too High	Too Low	No Opinion
Male	200	150	45
Female	315	125	65

or as the matrix

$$\begin{bmatrix} 200 & 150 & 45 \\ 315 & 125 & 65 \end{bmatrix}$$

This matrix has two rows (representing males and females) and three columns (representing "too high," "too low," and "no opinion"). ▶

We now give a general definition for a matrix.

Definition of a Matrix

A **matrix** is defined as a rectangular array of the form:

$$
\begin{array}{c}
 \\
\text{Row } 1 \\
\text{Row } 2 \\
\vdots \\
\text{Row } i \\
\vdots \\
\text{Row } m
\end{array}
\begin{bmatrix}
\overset{\text{Column }1}{a_{11}} & \overset{\text{Column }2}{a_{12}} & \cdots & \overset{\text{Column }j}{a_{1j}} & \cdots & \overset{\text{Column }n}{a_{1n}} \\
a_{21} & a_{22} & \cdots & a_{2j} & \cdots & a_{2n} \\
\vdots & \vdots & & \vdots & & \vdots \\
a_{i1} & a_{i2} & \cdots & a_{ij} & \cdots & a_{in} \\
\vdots & \vdots & & \vdots & & \vdots \\
a_{m1} & a_{m2} & \cdots & a_{mj} & \cdots & a_{mn}
\end{bmatrix}
\tag{1}
$$

The symbols $a_{11}, a_{12}, \ldots$ of a matrix are referred to as the **entries** (or **elements**) of the matrix. Each entry a_{ij} of the matrix has two indices: the **row index,** i, and the **column index,** j. The symbols $a_{i1}, a_{i2}, \ldots, a_{in}$ represent the entries in the ith row, and the symbols $a_{1j}, a_{2j}, \ldots, a_{mj}$ represent the entries in the jth column.

We shall use capital letters to denote matrices. If we denote the matrix in display (1) above by A, then we can abbreviate A by

$$
A = [a_{ij}] \qquad i = 1, 2, \ldots, m \qquad j = 1, 2, \ldots, n.
$$

The matrix A has m rows and n columns.

Dimension of a Matrix

The **dimension of a matrix** A is determined by the number of rows and the number of columns in the matrix. If a matrix A has m rows and n columns, we denote the dimension of A by $m \times n$, read as "m by n."

For a 2×3 matrix, remember that the first number 2 denotes the number of rows and the second number 3 is the number of columns. A matrix with 3 rows and 2 columns is of dimension 3×2.

Square Matrix

If a matrix A has the same number of rows as it has columns, it is called a **square matrix.**

In a square matrix $A = [a_{ij}]$ the entries for which $i = j$, namely $a_{11}, a_{22}, a_{33}, a_{44}$, and so on, are the **diagonal entries** of A.

EXAMPLE 2 **Arranging Data as a Matrix**

In the recent U.S. census the following figures were obtained with regard to the city of Oak Lawn. Each year 7% of city residents move to the suburbs and 1% of the people in the suburbs move to the city. This situation can be represented by the matrix

$$P = \begin{array}{c} \\ \text{City} \\ \text{Suburbs} \end{array} \begin{array}{cc} \text{City} & \text{Suburbs} \\ \begin{bmatrix} 0.93 & 0.07 \\ 0.01 & 0.99 \end{bmatrix} \end{array}$$

Here, the entry in row 1, column 2—0.07—indicates that 7% of city residents move to the suburbs. The matrix P is a square matrix and its dimension is 2×2. The diagonal entries are 0.93 and 0.99.

A **row matrix** is a matrix with 1 row of entries. A **column matrix** is a matrix with 1 column of entries. Row matrices and column matrices are also referred to as **row vectors** and **column vectors,** respectively.

---

1 **EXAMPLE 3** **Finding the Dimension of a Matrix**

Find the dimension of each matrix. Say if the matrix is a square matrix, or a row matrix, or a column matrix.

(a) $\begin{bmatrix} 5 & -1 \\ 2 & 5 \end{bmatrix}$ **(b)** $\begin{bmatrix} 2 & 3 & 0 \\ 1 & -2 & 5 \end{bmatrix}$ **(c)** $[1 \quad 0 \quad 4]$ **(d)** $\begin{bmatrix} 2 \\ 1 \end{bmatrix}$ **(e)** $[9]$

SOLUTION
(a) 2×2, a square matrix
(b) 2×3
(c) 1×3, a row matrix
(d) 2×1, a column matrix
(e) 1×1, a square matrix, a row matrix, and a column matrix

NOW WORK PROBLEMS 3 AND 63.

Equality of Matrices

In an algebra system, it is important to know when two quantities are equal. So, we ask, "When, if at all, are two matrices equal?"

Let's try to arrive at a sound definition for equality of matrices by requiring equal matrices to have certain desirable properties. First, it would seem necessary that two equal matrices have the same dimension—that is, that they both be $m \times n$ matrices. Next, it would seem necessary that their entries be identical numbers. With these two restrictions, we define equality of matrices.

> **Equality of Matrices**
>
> Two matrices A and B are **equal** if they are of the same dimension and if corresponding entries are equal. In this case we write $A = B$, read as "matrix A is equal to matrix B."

EXAMPLE 4   **Determining Equality of Matrices**

In order for the two matrices

$$\begin{bmatrix} p & q \\ 1 & 0 \end{bmatrix} \quad \text{and} \quad \begin{bmatrix} 2 & 4 \\ n & 0 \end{bmatrix}$$

to be equal, we must have $p = 2$, $q = 4$, and $n = 1$.   ▶

EXAMPLE 5   **Determining Equality of Matrices**

Let A and B be two matrices given by

$$A = \begin{bmatrix} x + y & 6 \\ 2x - 3 & 2 - y \end{bmatrix} \quad B = \begin{bmatrix} 5 & 5x + 2 \\ y & x - y \end{bmatrix}$$

Determine if there are values of x and y so that A and B are equal.

SOLUTION   Both A and B are 2×2 matrices so $A = B$ if

$$\begin{array}{ll} x + y = 5 \quad (1) & 6 = 5x + 2 \quad (2) \\ 2x - 3 = y \quad (3) & 2 - y = x - y \quad (4) \end{array}$$

Here we have four equations containing the two variables x and y. From Equation (4) we see that $x = 2$. Using this value in Equation (1), we obtain $y = 3$. But $x = 2$, $y = 3$ do not satisfy either Equation (2) or Equation (3). Hence, there are *no* values for x and y satisfying all four equations. This means A and B can never be equal.   ▶

 NOW WORK PROBLEM 13.

Find the sum of two ② **Addition of Matrices**
matrices

In an algebra system there are operations, like adding and subtracting. Can two matrices be added? And, if so, what is the rule or law for addition of matrices?

EXAMPLE 6   **Adding Two Matrices**

Motors, Inc., produces three models of cars: a sedan, a convertible, and an SUV. If the company wishes to compare the units of raw material and the units of labor involved in 1 month's production of each of these models, the rectangular array displayed in Table 3 below may be used to present the data:

TABLE 3

	Sedan Model	Convertible Model	SUV Model
Units of Material	23	16	10
Units of Labor	7	9	11

The same information may be written concisely as the matrix

$$A = \begin{bmatrix} 23 & 16 & 10 \\ 7 & 9 & 11 \end{bmatrix}$$

Suppose the next month's production is

$$B = \begin{bmatrix} 18 & 12 & 9 \\ 14 & 6 & 8 \end{bmatrix}$$

in which the pattern of recording units and models remains the same.

The total production for the 2 months can be displayed by the matrix

$$C = \begin{bmatrix} 41 & 28 & 19 \\ 21 & 15 & 19 \end{bmatrix}$$

since the number of units of material for sedan models is $41 = 23 + 18$, the number of units of material for convertible models is $28 = 16 + 12$, and so on.   ▶

This leads to the following definition.

Addition of Matrices

We define the **sum** $A + B$ of two matrices A and B with the same dimension as the matrix consisting of the sum of corresponding entries from A and B. That is, if $A = [a_{ij}]$ and $B = [b_{ij}]$ are two $m \times n$ matrices, the **sum** $A + B$ is the $m \times n$ matrix $[a_{ij} + b_{ij}]$.

EXAMPLE 7   **Adding Two Matrices**

(a) $\begin{bmatrix} 23 & 16 & 10 \\ 7 & 9 & 11 \end{bmatrix} + \begin{bmatrix} 18 & 12 & 9 \\ 14 & 6 & 8 \end{bmatrix} = \begin{bmatrix} 23 + 18 & 16 + 12 & 10 + 9 \\ 7 + 14 & 9 + 6 & 11 + 8 \end{bmatrix}$

$$= \begin{bmatrix} 41 & 28 & 19 \\ 21 & 15 & 19 \end{bmatrix}$$

(b) $\begin{bmatrix} 0.6 & 0.4 \\ 0.1 & 0.9 \end{bmatrix} + \begin{bmatrix} 2.3 & 0.6 \\ 1.8 & 5.2 \end{bmatrix} = \begin{bmatrix} 0.6 + 2.3 & 0.4 + 0.6 \\ 0.1 + 1.8 & 0.9 + 5.2 \end{bmatrix}$

$$= \begin{bmatrix} 2.9 & 1.0 \\ 1.9 & 6.1 \end{bmatrix}$$ ▶

Notice that it is possible to add two matrices only if their dimensions are the same. Also, the dimension of the sum of two matrices is the same as that of the two original matrices.

The following pairs of matrices cannot be added since they are of different dimensions:

$$A = \begin{bmatrix} 1 & 2 \\ 7 & 2 \end{bmatrix} \qquad \text{and} \qquad B = \begin{bmatrix} 1 \\ -3 \end{bmatrix}$$

$$A = [2 \quad 3] \qquad \text{and} \qquad B = [1 \quad 1 \quad 1]$$

$$A = \begin{bmatrix} -1 & 7 & 0 \\ 2 & \frac{1}{2} & 0 \end{bmatrix} \qquad \text{and} \qquad B = \begin{bmatrix} -1 & 2 \\ 3 & 0 \\ 1 & 5 \end{bmatrix}$$

 NOW WORK PROBLEM 25.

Properties of Matrices

Work with properties of **3**
matrices

It turns out that the usual rules for the addition of real numbers (such as the commutative property and associative property) are also valid for matrix addition.

EXAMPLE 8   **Demonstrating the Commutative Property**

Let $\quad A = \begin{bmatrix} 1 & 5 \\ 7 & -3 \end{bmatrix} \quad$ and $\quad B = \begin{bmatrix} 3 & -2 \\ 4 & 1 \end{bmatrix}$

Then

$$A + B = \begin{bmatrix} 1 & 5 \\ 7 & -3 \end{bmatrix} + \begin{bmatrix} 3 & -2 \\ 4 & 1 \end{bmatrix} = \begin{bmatrix} 1 + 3 & 5 + (-2) \\ 7 + 4 & -3 + 1 \end{bmatrix}$$

$$= \begin{bmatrix} 4 & 3 \\ 11 & -2 \end{bmatrix}$$

$$B + A = \begin{bmatrix} 3 & -2 \\ 4 & 1 \end{bmatrix} + \begin{bmatrix} 1 & 5 \\ 7 & -3 \end{bmatrix} = \begin{bmatrix} 4 & 3 \\ 11 & -2 \end{bmatrix}$$

This leads us to formulate the following property for addition.

Commutative Property for Addition

If A and B are two matrices of the same dimension, then

$$\boxed{A + B = B + A}$$

The associative property for addition of matrices is also true.

Associative Property for Addition

If A, B, and C are three matrices of the same dimension, then

$$\boxed{A + (B + C) = (A + B) + C}$$

The fact that addition of matrices is associative means that the notation $A + B + C$ is *not* ambiguous, since $(A + B) + C = A + (B + C)$.

 NOW WORK PROBLEM 45.

A matrix in which all entries are 0 is called a **zero matrix.** We use the symbol **0** to represent a zero matrix of any dimension.

For real numbers, 0 has the property that $x + 0 = x$ for any x. A property of a zero matrix is that $A + \mathbf{0} = A$, provided the dimension of $\mathbf{0}$ is the same as that of A.

EXAMPLE 9   **Demonstrating a Property of the Zero Matrix**

Let
$$A = \begin{bmatrix} 3 & 4 & -\dfrac{1}{2} \\ \sqrt{2} & 0 & 3 \end{bmatrix}$$

Then

$$A + \mathbf{0} = \begin{bmatrix} 3 & 4 & -\dfrac{1}{2} \\ \sqrt{2} & 0 & 3 \end{bmatrix} + \begin{bmatrix} 0 & 0 & 0 \\ 0 & 0 & 0 \end{bmatrix}$$

$$= \begin{bmatrix} 3 + 0 & 4 + 0 & -\dfrac{1}{2} + 0 \\ \sqrt{2} + 0 & 0 + 0 & 3 + 0 \end{bmatrix} = \begin{bmatrix} 3 & 4 & -\dfrac{1}{2} \\ \sqrt{2} & 0 & 3 \end{bmatrix} = A$$ ▶

If A is any matrix, the **additive inverse** of A, denoted by $-A$, is the matrix obtained by replacing each entry in A by its negative.

EXAMPLE 10   **Finding the Additive Inverse of a Matrix**

If
$$A = \begin{bmatrix} -3 & 0 \\ 5 & -2 \\ 1 & 3 \end{bmatrix} \quad \text{then} \quad -A = \begin{bmatrix} 3 & 0 \\ -5 & 2 \\ -1 & -3 \end{bmatrix}$$ ▶

Additive Inverse Property

For any matrix A, we have the property that

$$\boxed{A + (-A) = \mathbf{0}}$$

    NOW WORK PROBLEM 47.

Subtraction of Matrices

Find the difference of two matrices 4

Now that we have defined the sum of two matrices and the additive inverse of a matrix, it is natural to ask about the *difference* of two matrices. As you will see, subtracting matrices and subtracting real numbers are much the same kind of process.

Subtraction of Matrices

We define the **difference** $A - B$ of two matrices A and B with the same dimension as the matrix consisting of the difference of corresponding entries from A and B. That is, if $A = [a_{ij}]$ and $B = [b_{ij}]$ are two $m \times n$ matrices, the **difference** $A - B$ is the $m \times n$ matrix $[a_{ij} - b_{ij}]$.

EXAMPLE 11   Subtracting Two Matrices

Let
$$A = \begin{bmatrix} 2 & 3 & 4 \\ 1 & 0 & 2 \end{bmatrix} \quad \text{and} \quad B = \begin{bmatrix} -2 & 1 & -1 \\ 3 & 0 & 3 \end{bmatrix}$$

Then

$$\begin{aligned} A - B &= \begin{bmatrix} 2 & 3 & 4 \\ 1 & 0 & 2 \end{bmatrix} - \begin{bmatrix} -2 & 1 & -1 \\ 3 & 0 & 3 \end{bmatrix} \\ &= \begin{bmatrix} 2 - (-2) & 3 - 1 & 4 - (-1) \\ 1 - 3 & 0 - 0 & 2 - 3 \end{bmatrix} = \begin{bmatrix} 4 & 2 & 5 \\ -2 & 0 & -1 \end{bmatrix} \end{aligned}$$

Notice that the difference $A - B$ is nothing more than the matrix formed by subtracting the entries in B from the corresponding entries in A.

Using the matrices A and B from Example 11, we find that

$$\begin{aligned} B - A &= \begin{bmatrix} -2 & 1 & -1 \\ 3 & 0 & 3 \end{bmatrix} - \begin{bmatrix} 2 & 3 & 4 \\ 1 & 0 & 2 \end{bmatrix} \\ &= \begin{bmatrix} -2 - 2 & 1 - 3 & -1 - 4 \\ 3 - 1 & 0 - 0 & 3 - 2 \end{bmatrix} = \begin{bmatrix} -4 & -2 & -5 \\ 2 & 0 & 1 \end{bmatrix} \end{aligned}$$

Observe that $A - B \neq B - A$, illustrating that matrix subtraction, like subtraction of real numbers, is not commutative.

EXAMPLE 12   Adding and Subtracting Matrices

Suppose that

$$A = \begin{bmatrix} 2 & 4 & 8 & -3 \\ 0 & 1 & 2 & 3 \end{bmatrix} \quad \text{and} \quad B = \begin{bmatrix} -3 & 4 & 0 & 1 \\ 6 & 8 & 2 & 0 \end{bmatrix}$$

Find: **(a)** $A + B$ **(b)** $A - B$

SOLUTION   **(a)** $A + B = \begin{bmatrix} 2 & 4 & 8 & -3 \\ 0 & 1 & 2 & 3 \end{bmatrix} + \begin{bmatrix} -3 & 4 & 0 & 1 \\ 6 & 8 & 2 & 0 \end{bmatrix}$

$= \begin{bmatrix} 2 + (-3) & 4 + 4 & 8 + 0 & -3 + 1 \\ 0 + 6 & 1 + 8 & 2 + 2 & 3 + 0 \end{bmatrix}$ Add corresponding entries.

$= \begin{bmatrix} -1 & 8 & 8 & -2 \\ 6 & 9 & 4 & 3 \end{bmatrix}$

(b) $A - B = \begin{bmatrix} 2 & 4 & 8 & -3 \\ 0 & 1 & 2 & 3 \end{bmatrix} - \begin{bmatrix} -3 & 4 & 0 & 1 \\ 6 & 8 & 2 & 0 \end{bmatrix}$

$= \begin{bmatrix} 2 - (-3) & 4 - 4 & 8 - 0 & -3 - 1 \\ 0 - 6 & 1 - 8 & 2 - 2 & 3 - 0 \end{bmatrix}$ Subtract corresponding entries.

$= \begin{bmatrix} 5 & 0 & 8 & -4 \\ -6 & -7 & 0 & 3 \end{bmatrix}$

COMMENT: Graphing utilities make the sometimes tedious process of matrix algebra easy. Let's see how a graphing utility adds and subtracts matrices by solving Example 12 using a graphing utility.

GRAPHING UTILITY SOLUTION   Enter the matrices into a graphing utility. Name them $[A]$ and $[B]$. Figure 9 shows the results of adding and subtracting $[A]$ and $[B]$.

FIGURE 9

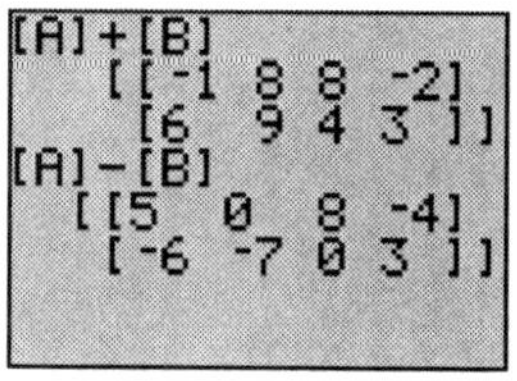

▶ ▶

NOW WORK PROBLEM 33.

Multiplying a Matrix by a Number

Let's return to the production of Motors, Inc., during the month specified in Example 6. The matrix A describing this production is

$$A = \begin{bmatrix} 23 & 16 & 10 \\ 7 & 9 & 11 \end{bmatrix}$$

Let's assume that for 3 consecutive months, the monthly production remained the same. Then the total production for the 3 months is simply the sum of the matrix A taken 3 times. If we represent the total production by the matrix T, then

$$T = \begin{bmatrix} 23 & 16 & 10 \\ 7 & 9 & 11 \end{bmatrix} + \begin{bmatrix} 23 & 16 & 10 \\ 7 & 9 & 11 \end{bmatrix} + \begin{bmatrix} 23 & 16 & 10 \\ 7 & 9 & 11 \end{bmatrix}$$

$$= \begin{bmatrix} 23 + 23 + 23 & 16 + 16 + 16 & 10 + 10 + 10 \\ 7 + 7 + 7 & 9 + 9 + 9 & 11 + 11 + 11 \end{bmatrix}$$

$$= \begin{bmatrix} 3 \cdot 23 & 3 \cdot 16 & 3 \cdot 10 \\ 3 \cdot 7 & 3 \cdot 9 & 3 \cdot 11 \end{bmatrix} = \begin{bmatrix} 69 & 48 & 30 \\ 21 & 27 & 33 \end{bmatrix}$$

In other words, when we add the matrix A 3 times, we multiply each entry of A by 3. This leads to the following definition.

Scalar Multiplication

Let A be an $m \times n$ matrix and let c be a real number, called a **scalar**. The product of the matrix A by the scalar c, called **scalar multiplication**, is the $m \times n$ matrix cA, whose entries are the product of c and the corresponding entries of A. That is, if $A = [a_{ij}]$, then $cA = [ca_{ij}]$.

When multiplying a matrix by a real number, each entry of the matrix is multiplied by the number. Notice that the dimension of A and the dimension of the product cA are the same.

5   EXAMPLE 13   Finding Scalar Multiples of a Matrix

Suppose that

$$A = \begin{bmatrix} 3 & 1 & 5 \\ -2 & 0 & 6 \end{bmatrix}, \qquad B = \begin{bmatrix} 4 & 1 & 0 \\ 8 & 1 & -3 \end{bmatrix}, \qquad C = \begin{bmatrix} 9 & 0 \\ -3 & 6 \end{bmatrix}$$

Find: **(a)** $4A$ **(b)** $\dfrac{1}{3}C$ **(c)** $3A - 2B$

SOLUTION   **(a)** $4A = 4\begin{bmatrix} 3 & 1 & 5 \\ -2 & 0 & 6 \end{bmatrix} = \begin{bmatrix} 4\cdot 3 & 4\cdot 1 & 4\cdot 5 \\ 4(-2) & 4\cdot 0 & 4\cdot 6 \end{bmatrix} = \begin{bmatrix} 12 & 4 & 20 \\ -8 & 0 & 24 \end{bmatrix}$

(b) $\dfrac{1}{3}C = \dfrac{1}{3}\begin{bmatrix} 9 & 0 \\ -3 & 6 \end{bmatrix} = \begin{bmatrix} \dfrac{1}{3}\cdot 9 & \dfrac{1}{3}\cdot 0 \\ \dfrac{1}{3}\cdot(-3) & \dfrac{1}{3}\cdot 6 \end{bmatrix} = \begin{bmatrix} 3 & 0 \\ -1 & 2 \end{bmatrix}$

(c) $3A - 2B = 3\begin{bmatrix} 3 & 1 & 5 \\ -2 & 0 & 6 \end{bmatrix} - 2\begin{bmatrix} 4 & 1 & 0 \\ 8 & 1 & -3 \end{bmatrix}$

$\qquad = \begin{bmatrix} 3\cdot 3 & 3\cdot 1 & 3\cdot 5 \\ 3(-2) & 3\cdot 0 & 3\cdot 6 \end{bmatrix} - \begin{bmatrix} 2\cdot 4 & 2\cdot 1 & 2\cdot 0 \\ 2\cdot 8 & 2\cdot 1 & 2(-3) \end{bmatrix}$

$\qquad = \begin{bmatrix} 9 & 3 & 15 \\ -6 & 0 & 18 \end{bmatrix} - \begin{bmatrix} 8 & 2 & 0 \\ 16 & 2 & -6 \end{bmatrix}$

$\qquad = \begin{bmatrix} 1 & 1 & 15 \\ -22 & -2 & 24 \end{bmatrix}$

GRAPHING UTILITY SOLUTION   Enter the matrices $[A]$, and $[B]$, and $[C]$ into a graphing utility. Figure 10 shows the required computations.

FIGURE 10

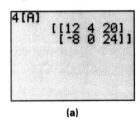
(a)

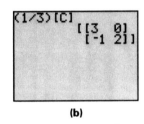
(b)

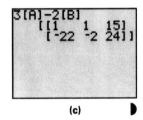
(c)

NOW WORK PROBLEM 35.

We list next some of the algebraic properties of scalar multiplication. Let h and k be real numbers, and let A and B be $m \times n$ matrices. Then

Properties of Scalar Multiplication

Let k and h be two real numbers and let A and B be two matrices of dimension $m \times n$. Then

$k(hA) = (kh)A$	(2)
$(k + h)A = kA + hA$	(3)
$k(A + B) = kA + kB$	(4)

Properties (2), (3), and (4) are illustrated in the following example.

EXAMPLE 14   **Using Properties of Scalar Multiplication**

For $A = \begin{bmatrix} 2 & -3 & -1 \\ 5 & 6 & 4 \end{bmatrix}$ and $B = \begin{bmatrix} -3 & 0 & 4 \\ 2 & -1 & 5 \end{bmatrix}$

show that

(a) $5[2A] = 10A$ **(b)** $(4 + 3)A = 4A + 3A$ **(c)** $3[A + B] = 3A + 3B$

SOLUTION   **(a)** $5[2A] = 5 \begin{bmatrix} 4 & -6 & -2 \\ 10 & 12 & 8 \end{bmatrix} = \begin{bmatrix} 20 & -30 & -10 \\ 50 & 60 & 40 \end{bmatrix}$

$10A = \begin{bmatrix} 20 & -30 & -10 \\ 50 & 60 & 40 \end{bmatrix}$

(b) $(4 + 3)A = 7A = \begin{bmatrix} 14 & -21 & -7 \\ 35 & 42 & 28 \end{bmatrix}$

$4A + 3A = \begin{bmatrix} 8 & -12 & -4 \\ 20 & 24 & 16 \end{bmatrix} + \begin{bmatrix} 6 & -9 & -3 \\ 15 & 18 & 12 \end{bmatrix}$

$= \begin{bmatrix} 14 & 21 & -7 \\ 35 & 42 & 28 \end{bmatrix}$

(c) $3[A + B] = 3 \begin{bmatrix} -1 & -3 & 3 \\ 7 & 5 & 9 \end{bmatrix} = \begin{bmatrix} -3 & -9 & 9 \\ 21 & 15 & 27 \end{bmatrix}$

$3A + 3B = \begin{bmatrix} 6 & -9 & -3 \\ 15 & 18 & 12 \end{bmatrix} + \begin{bmatrix} -9 & 0 & 12 \\ 6 & -3 & 15 \end{bmatrix}$

$= \begin{bmatrix} -3 & -9 & 9 \\ 21 & 15 & 27 \end{bmatrix}$

EXERCISE 2.4   **Answers to Odd-Numbered Problems Begin on Page AN-12.**

In Problems 1–12, find the dimension of each matrix. Say if the matrix is a square matrix, or a row matrix, or a column matrix.

1. $\begin{bmatrix} 3 & 2 \\ -1 & 3 \end{bmatrix}$

2. $\begin{bmatrix} -1 & 0 \\ 0 & 5 \end{bmatrix}$

$\begin{bmatrix} 2 & 1 & -3 \\ 1 & 0 & -1 \end{bmatrix}$

4. $\begin{bmatrix} 1 & 2 \\ 2 & 1 \\ 0 & -3 \end{bmatrix}$

5. $\begin{bmatrix} 4 & 0 \\ -1 & 2 \\ 5 & 8 \end{bmatrix}$

6. $\begin{bmatrix} 0 & -3 & 6 \\ 0 & 5 & 2 \\ 1 & 8 & 7 \end{bmatrix}$

7. $\begin{bmatrix} 1 & 4 \\ -2 & 8 \\ 0 & 0 \end{bmatrix}$

8. $\begin{bmatrix} 8 & 1 & 0 & 0 \\ 3 & -4 & 0 & 0 \end{bmatrix}$

9. $\begin{bmatrix} 4 \\ 1 \end{bmatrix}$

10. $[2 \quad 1 \quad -3]$

11. $[2]$

12. $[0]$

In Problems 13–24, determine whether the given statements are true or false. If false, tell why.

$\begin{bmatrix} 0 \\ 1 \end{bmatrix} = [0 \quad 1]$

14. $\begin{bmatrix} 3 & 2 \\ -1 & 0 \end{bmatrix} = \begin{bmatrix} 3 & 2 \\ -1 & 4 \end{bmatrix}$

15. $\begin{bmatrix} 5 & 0 \\ 0 & 1 \end{bmatrix}$ is square

16. $\begin{bmatrix} 3 & 2 & 1 \\ 4 & -1 & 0 \end{bmatrix}$ is 3×2

17. $\begin{bmatrix} x & 2 \\ 4 & 0 \end{bmatrix} = \begin{bmatrix} 3 & 2 \\ 4 & 0 \end{bmatrix}$ if $x = 3$

18. $\begin{bmatrix} x & y \\ 0 & 0 \end{bmatrix} = [x \quad y]$

19. $\begin{bmatrix} 5 & 0 \\ 1 & 1 \end{bmatrix} = \begin{bmatrix} 2+3 & 0 \\ 1 & 1 \end{bmatrix}$

20. $\begin{bmatrix} 1 & 0 \\ 0 & 1 \end{bmatrix} = \begin{bmatrix} 3-2 & 3-3 \\ 3-3 & 3-2 \end{bmatrix}$

21. $2\begin{bmatrix} 1 & 0 \\ 0 & 2 \end{bmatrix} = \begin{bmatrix} 2 & 0 \\ 0 & 4 \end{bmatrix}$

22. $-\begin{bmatrix} 8 & 0 \\ 5 & -1 \end{bmatrix} = \begin{bmatrix} -8 & 0 \\ 5 & 1 \end{bmatrix}$

23. $\begin{bmatrix} 8 \\ 1 \end{bmatrix} + \begin{bmatrix} 2 \\ 9 \end{bmatrix} = [10]$

24. $[6 \quad 0] + [1] = [7 \quad 1]$

In Problems 25–32, perform the indicated operations. Express your answer as a single matrix.

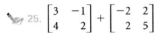 **25.** $\begin{bmatrix} 3 & -1 \\ 4 & 2 \end{bmatrix} + \begin{bmatrix} -2 & 2 \\ 2 & 5 \end{bmatrix}$

26. $\begin{bmatrix} 2 & 4 \\ 3 & -4 \end{bmatrix} - \begin{bmatrix} 8 & 9 \\ 0 & 7 \end{bmatrix}$

27. $3\begin{bmatrix} 2 & 6 & 0 \\ 4 & -2 & 1 \end{bmatrix}$

28. $-3\begin{bmatrix} 2 & 1 \\ -2 & 1 \\ 0 & 3 \end{bmatrix}$

29. $2\begin{bmatrix} 1 & -1 & 8 \\ 2 & 4 & 1 \end{bmatrix} - 3\begin{bmatrix} 0 & -2 & 8 \\ 1 & 4 & 1 \end{bmatrix}$

30. $6\begin{bmatrix} 2 & 1 \\ 3 & 1 \\ -1 & 0 \end{bmatrix} + 4\begin{bmatrix} 6 & -4 \\ -2 & -3 \\ 0 & 1 \end{bmatrix}$

31. $3\begin{bmatrix} a & 8 \\ b & 1 \\ c & -2 \end{bmatrix} + 5\begin{bmatrix} 2a & 6 \\ -b & -2 \\ -c & 0 \end{bmatrix}$

32. $2\begin{bmatrix} 2x & y & z \\ 2 & -4 & 8 \end{bmatrix} - 3\begin{bmatrix} -3x & 4y & 2z \\ 6 & -1 & 4 \end{bmatrix}$

In Problems 33–50, use the matrices below. For Problems 33–44 perform the indicated operation(s); for Problems 45–50 verify the indicated property.

$$A = \begin{bmatrix} 2 & -3 & 4 \\ 0 & 2 & 1 \end{bmatrix} \quad B = \begin{bmatrix} 1 & -2 & 0 \\ 5 & 1 & 2 \end{bmatrix} \quad C = \begin{bmatrix} -3 & 0 & 5 \\ 2 & 1 & 3 \end{bmatrix}$$

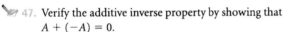

33. $A - B$

34. $B + C$

35. $2A - 3C$

36. $3C - 4B$

37. $(A + B) - 2C$

38. $4C + (A - B)$

39. $3A + 4(B + C)$

40. $(A + B) + 3C$

41. $2(A - B) - C$

42. $2A - 5(B + C)$

43. $3A - B - 6C$

44. $3A + 2B - 4C$

45. Verify the commutative property for addition by finding $A + B$ and $B + A$.

46. Verify the associative property for addition by finding $(A + B) + C$ and $A + (B + C)$.

47. Verify the additive inverse property by showing that $A + (-A) = 0$.

48. Verify Property (2) of scalar multiplication by finding $2(3A)$ and $6A$.

49. Verify Property (3) of scalar multiplication by finding $2B + 3B$ and $5B$.

50. Verify Property (4) of scalar multiplication by finding $2(A + C)$ and $2A + 2C$.

51. Find x and z so that

$$\begin{bmatrix} x \\ 4 \end{bmatrix} = \begin{bmatrix} -4 \\ z \end{bmatrix}$$

52. Find x, y, and z so that

$$\begin{bmatrix} x+y & -2 \\ 4 & 10 \end{bmatrix} = \begin{bmatrix} 6 & x-y \\ 4 & z \end{bmatrix}$$

53. Find x and y so that

$$\begin{bmatrix} x-2y & 0 \\ -2 & 6 \end{bmatrix} = \begin{bmatrix} 3 & 0 \\ -2 & x+y \end{bmatrix}$$

54. Find x, y, and z so that

$$\begin{bmatrix} x-2 & 3 & 2z \\ 6y & x & 2y \end{bmatrix} = \begin{bmatrix} y & z & 6 \\ 18z & y+2 & 6z \end{bmatrix}$$

55. Find x, y, and z so that

$$[2 \quad 3 \quad -4] + [x \quad 2y \quad z] = [6 \quad -9 \quad 2]$$

56. Find x and y so that

$$\begin{bmatrix} 3 & -2 & 2 \\ 1 & 0 & -1 \end{bmatrix} + \begin{bmatrix} x-y & 2 & -2 \\ 4 & x & 6 \end{bmatrix} = \begin{bmatrix} 6 & 0 & 0 \\ 5 & 2x+y & 5 \end{bmatrix}$$

In Problems 57–62, use a graphing utility to perform the indicated operations on the matrices given below.

$$A = \begin{bmatrix} -1 & -1 & 3 & 0 \\ 2 & 6 & 2 & 2 \\ -4 & 2 & 3 & 2 \\ 7 & 0 & 5 & -1 \end{bmatrix} \quad B = \begin{bmatrix} -1 & 2 & 4 & 5 \\ 2 & 0 & 5 & 3 \\ 0.5 & 6 & -7 & 11 \\ 5 & -1 & 2 & 7 \end{bmatrix} \quad C = \begin{bmatrix} 13 & -8 & 7 & 0 \\ 0 & 5 & 0 & -2 \\ 5 & 0 & 7 & 0 \\ 7 & 7 & 7 & 7 \end{bmatrix}$$

57. $A + B$

58. $3C - 2B$

59. $C - 3(A + B)$

60. $2(A - B) + \frac{1}{2}C$

61. $3(B + C) - A$

62. $\frac{1}{3}(A + 2C) - B$

---

63. Prison Populations In 2000, local jails and state and federal prisons contained nearly 2 million people, as follows: 613,534 prisoners were in local jails, of which 11.5% were female; 1,236,476 prisoners were in state prisons, of which 93.4% were male; and 145,416 were in federal prisons, of which 7.0% were female. Express this information using a 2 × 3 matrix. Label the rows MALE and FEMALE and the columns LOCAL, STATE, FED.

Source: United States Department of Justice, 2001.

64. Nail Production XYZ Company produces steel and aluminum nails. One week 25 gross of $\frac{1}{2}$-inch steel nails and 45 gross of 1-inch steel nails were produced. Suppose 13 gross of $\frac{1}{2}$-inch aluminum nails, 20 gross of 1-inch aluminum nails, 35 gross of 2-inch steel nails, and 23 gross of 2-inch aluminum nails were also made. Write a 2 × 3 matrix depicting this. Could you also write a 3 × 2 matrix for this situation?

65. College Degrees by Gender Post-secondary degrees include associate, bachelor's, master's, and doctoral degrees. Projections for 2003–2004 are as follows: 582,000 associate degrees, of which 218,000 will be awarded to men; 1,251,000 bachelor's degrees, of which 714,000 will be awarded to women; 442,000 master's degrees, of which 261,000 will be awarded to women; and 47,100 doctoral degrees, of which 26,700 will be awarded to men. Write a 2 × 4 matrix depicting this. Could you also write a 4 × 2 matrix for this situation?

Source: Digest of Education Statistics, National Center for Education Statistics, 2001.

66. Katy, Mike, and Danny go to the candy store. Katy buys 5 sticks of gum, 2 ice cream cones, and 20 jelly beans. Mike buys 2 sticks of gum, 15 jelly beans, and 3 candy bars. Danny buys 1 stick of gum, 1 ice cream cone, and 4 candy bars. Write a matrix depicting this situation.

67. Use a matrix to display the information given below, which was obtained in a survey of voters. Label the rows UNDER $25,000 and OVER $25,000 and label the columns

DEMOCRATS, REPUBLICANS, INDEPENDENTS.

351	Democrats earning under $25,000
271	Republicans earning under $25,000
73	Independents earning under $25,000
203	Democrats earning $25,000 or more
215	Republicans earning $25,000 or more
55	Independents earning $25,000 or more

68. Listing Stocks One day on the New York Stock Exchange, 800 issues went up and 600 went down. Of the 800 up issues, 200 went up more than $1 per share. Of the 600 down issues, 50 went down more than $1 per share. Express this information in a 2 × 2 matrix. Label the rows UP and DOWN and the columns MORE THAN $1 and LESS THAN $1.

69. Surveys In a survey of 1000 college students, the following information was obtained: 500 were liberal arts and sciences (LAS) majors, of which 50% were female; 300 were engineering (ENG) majors, of which 75% were male; and the remaining were education (EDUC) majors, of which 60% were female. Express this information using a 2 × 3 matrix. Label the rows MALE and FEMALE and the columns LAS, ENG, and EDUC.

70. Sales of Cars The sales figures for two car dealers during June showed that dealer A sold 100 compacts, 50 intermediates, and 40 full-size cars, while dealer B sold 120 compacts, 40 intermediates, and 35 full-size cars. During July, dealer A sold 80 compacts, 30 intermediates, and 10 full-size cars, while dealer B sold 70 compacts, 40 intermediates, and 20 full-size cars. Total sales over the 3 month period of June–August revealed that dealer A sold 300 compacts, 120 intermediates, and 65 full-size cars. In the same 3-month period, dealer B sold 250 compacts, 100 intermediates, and 80 full-size cars.

(a) Write 2 × 3 matrices summarizing sales data for June, July, and the 3-month period for each dealer.
(b) Use matrix addition to find the sales over the 2-month period for June and July for each dealer.
(c) Use matrix subtraction to find the sales in August for each dealer.

2.5   Multiplication of Matrices

OBJECTIVES   1   Find the product of two matrices

2   Work with properties of matrices

While addition and subtraction of matrices and the product of a scalar and a matrix are fairly straightforward, defining the *product* of two matrices requires a bit more detail.

We explain first what we mean by the product of a row matrix (a matrix with one row) with a column matrix (a matrix with one column).

Let's look at a simple example. In a given month suppose 23 units of material and 7 units of labor were required to manufacture 4-door sedans. We can represent this by the column matrix

$$\begin{bmatrix} 23 \\ 7 \end{bmatrix}$$

Also, suppose the cost per unit of material is \$450 and the cost per unit of labor is \$600. We represent these costs by the row matrix

$$[450 \quad 600]$$

The total cost of producing the sedans is calculated as follows:

$$\text{Total cost} = (\text{Cost per unit of material}) \times (\text{Units of material})$$
$$+ (\text{Cost per unit of labor}) \times (\text{Units of labor})$$
$$= 450 \times 23 + 600 \times 7 = 14{,}550$$

In terms of the matrix representations,

$$\text{Total cost} = [450 \quad 600] \begin{bmatrix} 23 \\ 7 \end{bmatrix} = 450 \times 23 + 600 \times 7 = 14{,}550$$

This leads us to formulate the following definition for multiplying a row matrix times a column matrix:

If $R = [r_1 \, r_2 \ldots r_n]$ is a **row matrix** of dimension $1 \times n$ and $C = \begin{bmatrix} c_1 \\ c_2 \\ \vdots \\ c_n \end{bmatrix}$ is a column

matrix of dimension $n \times 1$, then by the **product of R and C** we mean the number

$$RC = r_1 c_1 + r_2 c_2 + r_3 c_3 + \ldots + r_n c_n$$

EXAMPLE 1 **Finding the Product of a 1 × 3 Row Matrix and a 3 × 1 Column Matrix**

If
$$R = [1 \quad 5 \quad 3] \quad \text{and} \quad C = \begin{bmatrix} 2 \\ -1 \\ 4 \end{bmatrix}$$

then the product of R and C is

$$RC = [1 \quad 5 \quad 3] \begin{bmatrix} 2 \\ -1 \\ 4 \end{bmatrix} = 1 \cdot 2 + 5 \cdot (-1) + 3 \cdot 4 = 9$$

▶

Notice that for the product of a row matrix R and a column matrix C to be defined, if R is a $1 \times n$ row matrix, then C must have dimension $n \times 1$.

EXAMPLE 2 **Finding the Product of a 1 × 4 Row Matrix and a 4 × 1 Column Matrix**

Let
$$R = [1 \quad 0 \quad 1 \quad 5] \quad \text{and} \quad C = \begin{bmatrix} 0 \\ -11 \\ 0 \\ 8 \end{bmatrix}$$

Then the product of R and C is

$$RC = [1 \quad 0 \quad 1 \quad 5] \begin{bmatrix} 0 \\ -11 \\ 0 \\ 8 \end{bmatrix} = 1 \cdot 0 + 0 \cdot (-11) + 1 \cdot 0 + 5 \cdot 8 = 40$$

▶

NOW WORK PROBLEM 1.

Given two matrices A and B, the rows of A can be thought of as row matrices, while the columns of B can be thought of as column matrices. This observation will be used in the following definition.

Multiplication of Matrices

Let A denote an $m \times r$ matrix, and let B denote an $r \times n$ matrix. The **product** AB is defined as the $m \times n$ matrix whose entry in row i, column j is the product of the ith row of A and the jth column of B.

An example will help to clarify the definition.

1   **EXAMPLE 3**   **Finding the Product of Two Matrices**

Find the product AB if

$$A = \begin{bmatrix} 2 & 4 & -1 \\ 5 & 8 & 0 \end{bmatrix} \quad \text{and} \quad B = \begin{bmatrix} 2 & 5 & 1 & 4 \\ 4 & 8 & 0 & 6 \\ -3 & 1 & -2 & -1 \end{bmatrix}$$

SOLUTION   First, we note that A is 2×3 and B is 3×4, so the product AB is defined and will be a 2×4 matrix.

Suppose that we want the entry in row 2, column 3 of AB. To find it, we find the product of the row vector from row 2 of A and the column vector from column 3 of B.

$$\underset{\text{Row 2 of } A}{[5 \quad 8 \quad 0]} \overset{\text{Column 3 of } B}{\begin{bmatrix} 1 \\ 0 \\ -2 \end{bmatrix}} = 5 \cdot 1 + 8 \cdot 0 + 0(-2) = 5$$

So far we have

$$AB = \begin{bmatrix} \underline{} & \underline{} & \overset{\text{Column 3}}{5} & \underline{} \\ \underline{} & \underline{} & & \underline{} \end{bmatrix} \text{Row 2}$$

Now, to find the entry in row 1, column 4 of AB, we find the product of row 1 of A and column 4 of B.

$$\underset{\text{Row 1 of } A}{[2 \quad 4 \quad -1]} \overset{\text{Column 4 of } B}{\begin{bmatrix} 4 \\ 6 \\ -1 \end{bmatrix}} = 2 \cdot 4 + 4 \cdot 6 + (-1)(-1) = 33$$

Continuing in this fashion, we find AB.

$$AB = \begin{bmatrix} 2 & 4 & -1 \\ 5 & 8 & 0 \end{bmatrix} \begin{bmatrix} 2 & 5 & 1 & 4 \\ 4 & 8 & 0 & 6 \\ -3 & 1 & -2 & -1 \end{bmatrix}$$

$$= \begin{bmatrix} \text{Row 1 of } A & \text{Row 1 of } A & \text{Row 1 of } A & \text{Row 1 of } A \\ \text{times} & \text{times} & \text{times} & \text{times} \\ \text{column 1 of } B & \text{column 2 of } B & \text{column 3 of } B & \text{column 4 of } B \\ \text{Row 2 of } A & \text{Row 2 of } A & \text{Row 2 of } A & \text{Row 2 of } A \\ \text{times} & \text{times} & \text{times} & \text{times} \\ \text{column 1 of } B & \text{column 2 of } B & \text{column 3 of } B & \text{column 4 of } B \end{bmatrix}$$

$$= \begin{bmatrix} 2 \cdot 2 + 4 \cdot 4 + (-1)(-3) & 2 \cdot 5 + 4 \cdot 8 + (-1)1 & 2 \cdot 1 + 4 \cdot 0 + (-1)(-2) & 33 \text{ (from earlier)} \\ 5 \cdot 2 + 8 \cdot 4 + 0(-3) & 5 \cdot 5 + 8 \cdot 8 + 0 \cdot 1 & 5 \text{ (from earlier)} & 5 \cdot 4 + 8 \cdot 6 + 0(-1) \end{bmatrix}$$

$$= \begin{bmatrix} 23 & 41 & 4 & 33 \\ 42 & 89 & 5 & 68 \end{bmatrix}$$

A graphing utility can be used to multiply two matrices. Use a graphing utility to do Example 3.

GRAPHING UTILITY SOLUTION

Enter the matrices A and B into a graphing utility. Figure 11 shows the product AB.

FIGURE 11

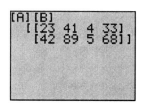

NOW WORK PROBLEM 27.

EXAMPLE 4   Manufacturing Cost

One month's production at Motors, Inc., may be given in matrix form as

$$A = \begin{bmatrix} \overset{\text{Sedan}}{23} & \overset{\text{Convertible}}{16} & \overset{\text{SUV}}{10} \\ 7 & 9 & 11 \end{bmatrix} \begin{matrix} \text{Units of material} \\ \text{Units of labor} \end{matrix}$$

Suppose that in this month's production, the cost for each unit of material is \$450 and the cost for each unit of labor is \$600. What is the total cost to manufacture the sedans, the convertibles, and the SUVs?

SOLUTION

For sedans, the cost is 23 units of material at \$450 each, plus 7 units of labor at \$600 each, for a total cost of

$$23 \cdot \$450 + 7 \cdot \$600 = 10{,}350 + 4200 = \$14{,}550$$

Similarly, for convertibles, the total cost is

$$16 \cdot \$450 + 9 \cdot \$600 = 7200 + 5400 = \$12{,}600$$

Finally, for SUVs, the total cost is

$$10 \cdot \$450 + 11 \cdot \$600 = 4500 + 6600 = \$11{,}100$$

If we represent the cost of units of material and units of labor by the 1×2 matrix

$$U = \begin{bmatrix} \overset{\text{Unit cost of}}{\underset{\text{material}}{450}} & \overset{\text{Unit cost of}}{\underset{\text{labor}}{600}} \end{bmatrix}$$

then the total cost is the product UA.

$$UA = \begin{bmatrix} 450 & 600 \end{bmatrix} \begin{bmatrix} 23 & 16 & 10 \\ 7 & 9 & 11 \end{bmatrix}$$
$$= \begin{bmatrix} 450 \cdot 23 + 600 \cdot 7 & 450 \cdot 16 + 600 \cdot 9 & 450 \cdot 10 + 600 \cdot 11 \end{bmatrix}$$
$$= \begin{bmatrix} 14{,}550 & 12{,}600 & 11{,}100 \end{bmatrix}$$

We conclude that the total cost to manufacture the sedans is \$14,550, to manufacture the convertibles is \$12,600, and to manufacture the SUVs is \$11,100.

NOW WORK PROBLEM 59.

Spreadsheets, such as Excel, can also be used to multiply matrices. Such utilities are especially useful when the dimensions of the matrices are large. Use Excel to do Example 4.

SOLUTION   **STEP 1** Enter the matrices for A and U into an Excel spreadsheet as follows:

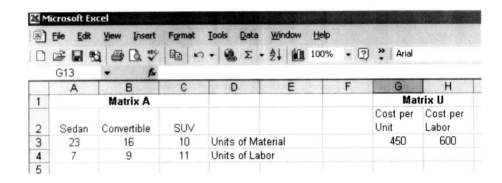

STEP 2 Highlight the cells that are to contain the product matrix. Since U is a 1 × 2 matrix and A is a 2 × 3 matrix, the product must be a 1 × 3 matrix.

STEP 3 Type: =MMULT(*highlight U, highlight A*)

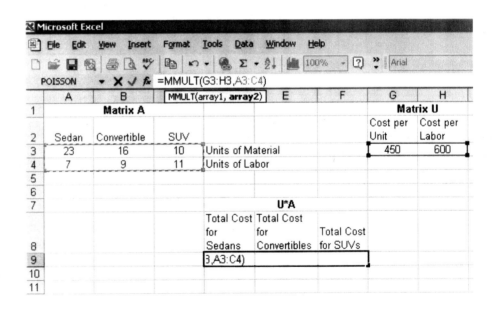

STEP 4 Press Ctrl-Shift-Enter at the same time.

	A	B	C	D	E	F	G	H
1		**Matrix A**						**Matrix U**
2	Sedan	Convertible	SUV				Cost per Unit	Cost per Labor
3	23	16	10	Units of Material			450	600
4	7	9	11	Units of Labor				
5								
6								
7					**U*A**			
8				Total Cost for Sedans	Total Cost for Convertibles	Total Cost for SUVs		
9				14550	12600	11100		
10								

Cell reference: D9, formula: {=MMULT(G3:H3,A3:C4)}

NOW WORK PROBLEM 59 USING EXCEL.

Properties of Matrix Multiplication

Work with properties of matrices **2**

If A is a matrix of dimension $m \times r$ (which has r columns) and B is a matrix of dimension $p \times n$ (which has p rows) and if $r \neq p$, the product AB is not defined.

Multiplication of matrices is possible only if the number of columns of the matrix on the left equals the number of rows of the matrix on the right.

If A is of dimension $m \times r$ and B is of dimension $r \times n$, then the product AB is of dimension $m \times n$.

FIGURE 12

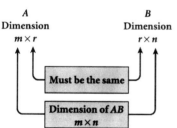

Figure 12 illustrates the result stated above.

For example, for the matrices given in Example 3 the product BA is not defined, because B is 3×4 and A is 2×3.

Another result of matrix multiplication is illustrated in the next example.

EXAMPLE 5   Multiplying Two Matrices

If

$$A = \begin{bmatrix} 2 & 1 & 3 \\ 1 & -1 & 0 \end{bmatrix} \quad \text{and} \quad B = \begin{bmatrix} 1 & 0 \\ 2 & 1 \\ 3 & 2 \end{bmatrix}$$

find: **(a)** AB **(b)** BA

475

SOLUTION   (a) $AB = \begin{bmatrix} 2 & 1 & 3 \\ 1 & -1 & 0 \end{bmatrix} \begin{bmatrix} 1 & 0 \\ 2 & 1 \\ 3 & 2 \end{bmatrix} = \begin{bmatrix} 13 & 7 \\ -1 & -1 \end{bmatrix}$

$\qquad\qquad\quad$ 2 × 3 $\qquad$ 3 × 2 $\qquad$ 2 × 2

(b) $BA = \begin{bmatrix} 1 & 0 \\ 2 & 1 \\ 3 & 2 \end{bmatrix} \begin{bmatrix} 2 & 1 & 3 \\ 1 & -1 & 0 \end{bmatrix} = \begin{bmatrix} 2 & 1 & 3 \\ 5 & 1 & 6 \\ 8 & 1 & 9 \end{bmatrix}$

$\qquad\qquad\quad$ 3 × 2 $\qquad$ 2 × 3 $\qquad$ 3 × 3

Notice in Example 5 that AB is 2 × 2 and BA is 3 × 3. It is possible for both AB and BA to be defined yet be unequal. In fact, even if A and B are both $n \times n$ matrices, so that AB and BA are each defined and $n \times n$, AB and BA will usually be unequal.

EXAMPLE 6   **Multiplying Two Square Matrices**

If

$$A = \begin{bmatrix} 2 & 1 \\ 0 & 4 \end{bmatrix} \quad \text{and} \quad B = \begin{bmatrix} -3 & 1 \\ 1 & 2 \end{bmatrix}$$

find: **(a)** AB **(b)** BA

SOLUTION   **(a)** $AB = \begin{bmatrix} 2 & 1 \\ 0 & 4 \end{bmatrix} \begin{bmatrix} -3 & 1 \\ 1 & 2 \end{bmatrix} = \begin{bmatrix} -5 & 4 \\ 4 & 8 \end{bmatrix}$

(b) $BA = \begin{bmatrix} -3 & 1 \\ 1 & 2 \end{bmatrix} \begin{bmatrix} 2 & 1 \\ 0 & 4 \end{bmatrix} = \begin{bmatrix} -6 & 1 \\ 2 & 9 \end{bmatrix}$

The preceding examples demonstrate that an important property of real numbers, the commutative property of multiplication, is not shared by matrices.

Matrix multiplication is not commutative. That is, in general,

$$\boxed{AB \text{ is not equal to } BA}$$

Matrix multiplication is associative.

Associative Property of Multiplication

Let A be a matrix of dimension $m \times r$, let B be a matrix of dimension $r \times p$, and let C be a matrix of dimension $p \times n$. Then matrix multiplication is **associative.** That is,

$$\boxed{A(BC) = (AB)C}$$

The resulting matrix ABC is of dimension $m \times n$.

Notice the limitations that are placed on the dimensions of the matrices in order for multiplication to be associative.

Distributive Property

Let A be a matrix of dimension $m \times r$. Let B and C be matrices of dimension $r \times n$. Then the **distributive property** states that

$$A(B + C) = AB + AC$$

The resulting matrix $AB + AC$ is of dimension $m \times n$.

NOW WORK PROBLEM 41.

The Identity Matrix

For an $n \times n$ square matrix, the entries located in row i, column i, $1 \le i \le n$, are called the **diagonal entries.** An $n \times n$ square matrix whose diagonal entries are 1s, while all other entries are 0s, is called the **identity matrix** I_n. For example,

$$I_2 = \begin{bmatrix} 1 & 0 \\ 0 & 1 \end{bmatrix}, \quad I_3 = \begin{bmatrix} 1 & 0 & 0 \\ 0 & 1 & 0 \\ 0 & 0 & 1 \end{bmatrix}$$

and so on.

EXAMPLE 7   **Finding the Product of a Matrix and the Identity Matrix I_2**

For
$$A = \begin{bmatrix} 3 & 2 \\ -4 & 5 \end{bmatrix}$$

compute   **(a)** AI_2 **(b)** I_2A

SOLUTION   **(a)** $AI_2 = \begin{bmatrix} 3 & 2 \\ -4 & 5 \end{bmatrix} \begin{bmatrix} 1 & 0 \\ 0 & 1 \end{bmatrix} = \begin{bmatrix} 3 & 2 \\ -4 & 5 \end{bmatrix} = A$

(b) $I_2A = \begin{bmatrix} 1 & 0 \\ 0 & 1 \end{bmatrix} \begin{bmatrix} 3 & 2 \\ -4 & 5 \end{bmatrix} = \begin{bmatrix} 3 & 2 \\ -4 & 5 \end{bmatrix} = A$

Example 7 demonstrates a more general result.

If A is a square matrix of dimension $n \times n$, then $AI_n = I_nA = A$.

For square matrices the identity matrix plays the role that the number 1 plays for multiplication in the set of real numbers.

When the matrix A is not square, care must be taken when forming the products AI and IA. For example, if

$$A = \begin{bmatrix} 1 & 2 \\ 3 & 2 \\ 1 & 1 \end{bmatrix}$$

then A is of dimension 3×2 and

$$AI_2 = A\begin{bmatrix} 1 & 0 \\ 0 & 1 \end{bmatrix} = \begin{bmatrix} 1 & 2 \\ 3 & 2 \\ 1 & 1 \end{bmatrix}\begin{bmatrix} 1 & 0 \\ 0 & 1 \end{bmatrix} = \begin{bmatrix} 1 & 2 \\ 3 & 2 \\ 1 & 1 \end{bmatrix} = A$$

Although the product I_2A is not defined, we can calculate the product I_3A as follows:

$$I_3A = \begin{bmatrix} 1 & 0 & 0 \\ 0 & 1 & 0 \\ 0 & 0 & 1 \end{bmatrix}\begin{bmatrix} 1 & 2 \\ 3 & 2 \\ 1 & 1 \end{bmatrix} = \begin{bmatrix} 1 & 2 \\ 3 & 2 \\ 1 & 1 \end{bmatrix} = A$$

The above observations can be generalized as follows:

Identity Property

If A is a matrix of dimension $m \times n$ and if I_n denotes the identity matrix of dimension $n \times n$, and I_m denotes the identity matrix of dimension $m \times m$, then

$$\boxed{I_mA = A \qquad \text{and} \qquad AI_n = A}$$

EXERCISE 2.5   Answers to Odd-Numbered Problems Begin on Page AN-13.

In Problems 1–16, find the product.

1. $[1 \quad 3]\begin{bmatrix} 2 \\ 4 \end{bmatrix}$

2. $[-1 \quad 4]\begin{bmatrix} 5 \\ 2 \end{bmatrix}$

3. $[1 \quad -2 \quad 3]\begin{bmatrix} 0 \\ 1 \\ 2 \end{bmatrix}$

4. $[-1 \quad 1 \quad 0]\begin{bmatrix} 1 \\ -1 \\ 1 \end{bmatrix}$

5. $[1 \quad 4]\begin{bmatrix} 2 & 0 \\ 4 & -2 \end{bmatrix}$

6. $\begin{bmatrix} 2 & 0 \\ 4 & -2 \end{bmatrix}\begin{bmatrix} 2 \\ 4 \end{bmatrix}$

7. $\begin{bmatrix} 2 & 0 \\ 4 & -2 \end{bmatrix}\begin{bmatrix} 2 & 1 \\ 3 & -2 \end{bmatrix}$

8. $\begin{bmatrix} 1 & 4 \\ -1 & 2 \end{bmatrix}\begin{bmatrix} 3 & 0 \\ -2 & 2 \end{bmatrix}$

9. $[1 \quad -2 \quad 3]\begin{bmatrix} 0 & 1 \\ 1 & 2 \\ 2 & 3 \end{bmatrix}$

10. $\begin{bmatrix} 1 & -2 & 3 \\ 4 & 0 & 6 \end{bmatrix}\begin{bmatrix} 0 \\ 1 \\ 2 \end{bmatrix}$

11. $\begin{bmatrix} 1 & -2 & 3 \\ 4 & 0 & 6 \end{bmatrix}\begin{bmatrix} 0 & -2 \\ 1 & 0 \\ 2 & -4 \end{bmatrix}$

12. $\begin{bmatrix} 1 & -2 & 3 \\ 4 & 0 & 6 \end{bmatrix}\begin{bmatrix} -1 & 2 & 1 \\ 1 & 3 & 0 \\ 0 & 4 & -2 \end{bmatrix}$

13. $\begin{bmatrix} 2 & 0 \\ 4 & -2 \\ 6 & -1 \end{bmatrix}\begin{bmatrix} 2 & 1 \\ 3 & -2 \end{bmatrix}$ **14.** $\begin{bmatrix} 1 & 4 \\ -1 & 2 \end{bmatrix}\begin{bmatrix} 2 & 0 & 6 \\ -1 & 4 & 1 \end{bmatrix}$ **15.** $\begin{bmatrix} 1 & -1 & 6 \\ 2 & 0 & -1 \\ 3 & 1 & 2 \end{bmatrix}\begin{bmatrix} 3 & 2 \\ 0 & 1 \\ 1 & 0 \end{bmatrix}$ **16.** $\begin{bmatrix} 2 & 0 & 1 \\ -1 & 1 & 1 \end{bmatrix}\begin{bmatrix} 1 & 0 & 0 \\ 2 & 1 & 2 \\ 3 & 2 & 4 \end{bmatrix}$

In Problems 17–26, let A be of dimension 3 × 4, B be of dimension 3 × 3, C be of dimension 2 × 3, and D be of dimension 3 × 2. Determine which of the following expressions are defined and, for those that are, give the dimension.

17. BA **18.** CD **19.** AB **20.** DC **21.** $(BA)C$

22. $A(CD)$ **23.** $BA + A$ **24.** $CD + BA$ **25.** $DC + B$ **26.** $CB - A$

In Problems 27–42, use the matrices given below. For Problems 27–40 perform the indicated operation(s); for Problems 41 and 42 verify the indicated property.

$$A = \begin{bmatrix} 1 & 2 \\ 0 & 4 \end{bmatrix} \quad B = \begin{bmatrix} 1 & 2 & 3 \\ -1 & 4 & -2 \end{bmatrix} \quad C = \begin{bmatrix} 3 & 1 \\ 4 & -1 \\ 0 & 2 \end{bmatrix} \quad D = \begin{bmatrix} 1 & 0 & 4 \\ 0 & 1 & 2 \\ 0 & -1 & 1 \end{bmatrix} \quad E = \begin{bmatrix} 3 & -1 \\ 4 & 2 \end{bmatrix}$$

27. AB **28.** DC **29.** BC **30.** AA **31.** $(D + I_3)C$

32. $DC + C$ **33.** EI_2 **34.** I_3D **35.** $(2E)B$ **36.** $E(2B)$

37. $-5E + A$ **38.** $3A + 2E$ **39.** $3CB + 4D$ **40.** $2EA - 3BC$

41. Verify the associative property of matrix multiplication by finding $D(CB)$ and $(DC)B$.

42. Verify the distributive property by finding $(A + E)B$ and $AB + EB$.

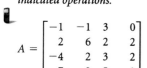

In Problems 43–50, use a graphing utility or EXCEL and the matrices given below to perform the indicated operations.

$$A = \begin{bmatrix} -1 & -1 & 3 & 0 \\ 2 & 6 & 2 & 2 \\ -4 & 2 & 3 & 2 \\ 7 & 0 & 5 & -1 \end{bmatrix} \quad B = \begin{bmatrix} -1 & 2 & 4 & 5 \\ 2 & 0 & 5 & 3 \\ 0.5 & 6 & -7 & 11 \\ 5 & -1 & 2 & 7 \end{bmatrix} \quad C = \begin{bmatrix} 13 & -8 & 7 & 0 \\ 0 & 5 & 0 & -2 \\ 5 & 0 & 7 & 0 \\ 7 & 7 & 7 & 7 \end{bmatrix}$$

43. AB **44.** BA **45.** $(AB)C$ **46.** $A(BC)$

47. $B(A + C)$ **48.** $(A + C)B$ **49.** $A(2B - 3C)$ **50.** $(A + B)(A - B)$

51. For
$$A = \begin{bmatrix} 1 & -1 \\ 2 & 0 \end{bmatrix} \quad \text{and} \quad B = \begin{bmatrix} 3 & 2 \\ -2 & 4 \end{bmatrix}$$
find AB and BA. Notice that $AB \neq BA$.

52. Show that, for all values a, b, c, and d, the matrices
$$A = \begin{bmatrix} a & b \\ -b & a \end{bmatrix} \quad \text{and} \quad B = \begin{bmatrix} c & d \\ -d & c \end{bmatrix}$$
are commutative; that is, $AB = BA$.

53. For what numbers x will the following be true?

$$[x \quad 4 \quad 1]\begin{bmatrix} 2 & 1 & 0 \\ 1 & 0 & 2 \\ 0 & 2 & 4 \end{bmatrix}\begin{bmatrix} x \\ -7 \\ \frac{5}{4} \end{bmatrix} = 0$$

54. Let

$$A = \begin{bmatrix} 1 & 2 & 5 \\ 2 & 4 & 10 \\ -1 & -2 & -5 \end{bmatrix}$$

Show that $A \cdot A = A^2 = \mathbf{0}$. Thus the rule in the real number system that if $a^2 = 0$, then $a = 0$ does not hold for matrices.

55. What must be true about a, b, c, and d if we demand that $AB = BA$ for the following matrices?

$$A = \begin{bmatrix} a & b \\ c & d \end{bmatrix} \qquad B = \begin{bmatrix} 1 & 1 \\ -1 & 1 \end{bmatrix}$$

Assume that

$$\begin{bmatrix} a & b \\ c & d \end{bmatrix} \neq \begin{bmatrix} 1 & 0 \\ 0 & 1 \end{bmatrix}$$

56. Let

$$A = \begin{bmatrix} a & b \\ b & a \end{bmatrix}$$

Find a and b such that $A^2 + A = \mathbf{0}$, where $A^2 = A \cdot A$.

57. For the matrix

$$A = \begin{bmatrix} a & 1-a \\ 1+a & -a \end{bmatrix}$$

show that $A^2 = A \cdot A = I_2$.

58. Find the row vector $[x_1 \quad x_2]$ for which

$$[x_1 \quad x_2]\begin{bmatrix} \frac{1}{2} & \frac{1}{2} \\ \frac{1}{4} & \frac{3}{4} \end{bmatrix} = [x_1 \quad x_2]$$

under the condition that $x_1 + x_2 = 1$. Here, the row vector $[x_1 \quad x_2]$ is called a **fixed vector** of the matrix

$$\begin{bmatrix} \frac{1}{2} & \frac{1}{2} \\ \frac{1}{4} & \frac{3}{4} \end{bmatrix}$$

59. Department Store Purchases Lee went to a department store and purchased 6 pairs of pants, 8 shirts, and 2 jackets. Chan purchased 2 pairs of pants, 5 shirts, and 3 jackets. If pants are \$25 each, shirts are \$18 each, and jackets are \$39 each, use matrix multiplication to find the amounts spent by Lee and Chan.

60. Factory Production Suppose a factory is asked to produce three types of products, which we will call P_1, P_2, P_3. Suppose the following purchase order was received: $P_1 = 7$, $P_2 = 12$, $P_3 = 5$. Represent this order by a row vector and call it P:

$$P = [7 \quad 12 \quad 5]$$

To produce each of the products, raw material of four kinds is needed. Call the raw material M_1, M_2, M_3, and M_4. The matrix below gives the amount of material needed for each product:

$$\begin{array}{c} \\ P_1 \\ Q = P_2 \\ P_3 \end{array}\begin{array}{cccc} M_1 & M_2 & M_3 & M_4 \\ \begin{bmatrix} 2 & 3 & 1 & 12 \\ 7 & 9 & 5 & 20 \\ 8 & 12 & 6 & 15 \end{bmatrix} \end{array}$$

Suppose the cost for each of the materials M_1, M_2, M_3, and M_4 is \$10, \$12, \$15, and \$20, respectively. The cost vector is

$$C = \begin{bmatrix} 10 \\ 12 \\ 15 \\ 20 \end{bmatrix}$$

Compute each of the following and interpret each one:

(a) PQ (b) QC (c) PQC

Problems 61–69 require the following definition:

Powers of Matrices For a square matrix A it is possible to find $A \cdot A = A^2$. We can also compute

$$A^n = \underbrace{A \cdot A \cdot \ldots \cdot A}_{n \text{ factors}}$$

In Problems 61–64, find A^2, A^3, and A^4.

61. $A = \begin{bmatrix} 1 & 0 \\ 3 & 2 \end{bmatrix}$

62. $A = \begin{bmatrix} 3 & 1 \\ -2 & -1 \end{bmatrix}$

63. $A = \begin{bmatrix} 1 & 0 \\ 0 & 1 \end{bmatrix}$

64. $A = \begin{bmatrix} \frac{1}{2} & \frac{1}{2} \\ \frac{1}{4} & \frac{3}{4} \end{bmatrix}$

65. Can you guess what A^n looks like for the matrix given in Problem 63?

66. Can you guess what A^n looks like for the matrix given in Problem 64?

In Problems, 67–69 use a graphing utility to compute A^2, A^{10}, and A^{15}.

67. $A = \begin{bmatrix} -0.5 & -1 & 0.3 & 0 & 0.3 \\ 2 & 1.6 & 1 & -1 & 0.4 \\ -4 & 2 & 0.7 & 2 & 0.2 \\ 1 & 0 & 0 & -1 & 0 \\ 0 & 0 & -0.9 & 0 & 0 \end{bmatrix}$

68. $A = \begin{bmatrix} -1 & 0.02 & 0.24 & 0 \\ 2 & 0 & 0 & 1.3 \\ 0.5 & 6 & -0.7 & 1.1 \\ 2.5 & -1 & 0.02 & 0.7 \end{bmatrix}$

69. $A = \begin{bmatrix} 0 & 1 & 0 \\ 1 & 0 & 1 \\ 1 & 1 & 1 \end{bmatrix}$

70. Make up two matrices A and B for which $AB = BA$.

71. Make up two square matrices A and B for which AB and BA are both defined, but $AB \neq BA$.

72. Make up two matrices A and B for which AB is defined but BA is not.

2.6   The Inverse of a Matrix

OBJECTIVES   **1** Find the inverse of a matrix
              2 Use the inverse of a matrix to solve a system of equations

The *inverse* of a matrix, if it exists, plays the role in matrix algebra that the reciprocal of a number plays in the set of real numbers. For example, the product of 2 and its reciprocal, $\frac{1}{2}$, equals 1, the multiplicative identity. The product of a matrix and its inverse equals the identity matrix.

Inverse of a Matrix

Let A be a matrix of dimension $n \times n$. A matrix B of dimension $n \times n$ is called the **inverse** of A if $AB = BA = I_n$. We denote the inverse of a matrix A, if it exists, by A^{-1}.

EXAMPLE 1   **Verifying That One Matrix Is the Inverse of Another Matrix**

Show that $\begin{bmatrix} \frac{1}{2} & -\frac{1}{2} \\ 0 & 1 \end{bmatrix}$ is the inverse of $\begin{bmatrix} 2 & 1 \\ 0 & 1 \end{bmatrix}$.

SOLUTION   Since

$$\begin{bmatrix} 2 & 1 \\ 0 & 1 \end{bmatrix}\begin{bmatrix} \frac{1}{2} & -\frac{1}{2} \\ 0 & 1 \end{bmatrix} = \begin{bmatrix} 1 & 0 \\ 0 & 1 \end{bmatrix}$$

and

$$\begin{bmatrix} \frac{1}{2} & -\frac{1}{2} \\ 0 & 1 \end{bmatrix}\begin{bmatrix} 2 & 1 \\ 0 & 1 \end{bmatrix} = \begin{bmatrix} 1 & 0 \\ 0 & 1 \end{bmatrix}$$

the required condition is met.

    NOW WORK PROBLEM 1.

The next example provides a technique for finding the inverse of a matrix. Although this technique is not the one we shall ultimately use, it is illustrative.

EXAMPLE 2   Finding the Inverse of a Matrix

Find the inverse of the matrix $A = \begin{bmatrix} 2 & 1 \\ 0 & 1 \end{bmatrix}$.

SOLUTION   We begin by assuming that this matrix has an inverse of the form

$$A^{-1} = \begin{bmatrix} a & b \\ c & d \end{bmatrix}$$

Then the product of A and A^{-1} must be the identity matrix:

$$\begin{bmatrix} 2 & 1 \\ 0 & 1 \end{bmatrix} \begin{bmatrix} a & b \\ c & d \end{bmatrix} = \begin{bmatrix} 1 & 0 \\ 0 & 1 \end{bmatrix}$$

Multiplying the matrices on the left side, we get

$$\begin{bmatrix} 2a + c & 2b + d \\ c & d \end{bmatrix} = \begin{bmatrix} 1 & 0 \\ 0 & 1 \end{bmatrix}$$

The condition for equality requires that

$$2a + c = 1 \qquad 2b + d = 0 \qquad c = 0 \qquad d = 1$$

Using $c = 0$ and $d = 1$ in the first two equations, we find

$$a = \tfrac{1}{2} \qquad b = -\tfrac{1}{2} \qquad c = 0 \qquad d = 1$$

The inverse of

$$A = \begin{bmatrix} 2 & 1 \\ 0 & 1 \end{bmatrix} \qquad \text{is} \qquad A^{-1} = \begin{bmatrix} a & b \\ c & d \end{bmatrix} = \begin{bmatrix} \tfrac{1}{2} & -\tfrac{1}{2} \\ 0 & 1 \end{bmatrix}$$

Sometimes a square matrix does not have an inverse.

EXAMPLE 3   Showing a Matrix Has No Inverse

Show that the matrix below does not have an inverse.

$$A = \begin{bmatrix} 0 & 1 \\ 0 & 0 \end{bmatrix}$$

SOLUTION   We proceed as in Example 2 by assuming that A does have an inverse. It will be of the form

$$A^{-1} = \begin{bmatrix} a & b \\ c & d \end{bmatrix}$$

The product of A and A^{-1} must be the identity matrix.

$$\begin{bmatrix} 0 & 1 \\ 0 & 0 \end{bmatrix} \begin{bmatrix} a & b \\ c & d \end{bmatrix} = \begin{bmatrix} 1 & 0 \\ 0 & 1 \end{bmatrix}$$

Performing the multiplication on the left side, we have

$$\begin{bmatrix} c & d \\ 0 & 0 \end{bmatrix} = \begin{bmatrix} 1 & 0 \\ 0 & 1 \end{bmatrix}$$

But these two matrices can never be equal (look at row 2, column 2: $0 \neq 1$). We conclude that our assumption that A has an inverse is false. That is, A does not have an inverse. ▶

NOW WORK PROBLEM 21.

So far, we have shown that a square matrix may or may not have an inverse. What about nonsquare matrices? Can they have inverses? The answer is "No." By definition, whenever a matrix has an inverse, it will commute with its inverse under multiplication. So if the nonsquare matrix A had the alleged inverse B, then AB would have to be equal to BA. But the fact that A is not square causes AB and BA to have different dimensions and prevents them from being equal. So such a B could not exist.

A nonsquare matrix has no inverse.

The procedure used in Example 2 to find the inverse, if it exists, of a square matrix becomes quite involved as the dimension of the matrix gets larger. A more efficient method that uses the reduced row-echelon form of a matrix is provided next.

Reduced Row-Echelon Technique for Finding Inverses

We will introduce this technique by looking at an example.

EXAMPLE 4   Finding the Inverse of a Matrix

Find the inverse of the matrix

$$A = \begin{bmatrix} 4 & 2 \\ 3 & 1 \end{bmatrix}$$

SOLUTION   Assuming A has an inverse, we will denote it by

$$X = \begin{bmatrix} x_1 & x_2 \\ x_3 & x_4 \end{bmatrix}$$

Then the product of A and X is the identity matrix of dimension 2×2. That is,

$$AX = I_2$$

$$\begin{bmatrix} 4 & 2 \\ 3 & 1 \end{bmatrix} \begin{bmatrix} x_1 & x_2 \\ x_3 & x_4 \end{bmatrix} = \begin{bmatrix} 1 & 0 \\ 0 & 1 \end{bmatrix}$$

Performing the multiplication on the left yields

$$\begin{bmatrix} 4x_1 + 2x_3 & 4x_2 + 2x_4 \\ 3x_1 + x_3 & 3x_2 + x_4 \end{bmatrix} = \begin{bmatrix} 1 & 0 \\ 0 & 1 \end{bmatrix}$$

This matrix equation can be written as the following system of four equations containing four variables:

$$\begin{cases} 4x_1 + 2x_3 = 1 & \quad 4x_2 + 2x_4 = 0 \\ 3x_1 + x_3 = 0 & \quad 3x_2 + x_4 = 1 \end{cases}$$

We find the solution to be

$$x_1 = -\tfrac{1}{2} \qquad x_2 = 1 \qquad x_3 = \tfrac{3}{2} \qquad x_4 = -2$$

The inverse of A is

$$A^{-1} = \begin{bmatrix} -\tfrac{1}{2} & 1 \\ \tfrac{3}{2} & -2 \end{bmatrix}$$

Let's look at this example more closely. The system of four equations containing four variables can be written as two systems:

(a) $\begin{cases} 4x_1 + 2x_3 = 1 \\ 3x_1 + x_3 = 0 \end{cases}$ (b) $\begin{cases} 4x_2 + 2x_4 = 0 \\ 3x_2 + x_4 = 1 \end{cases}$

Their augmented matrices are

(a) $\begin{bmatrix} 4 & 2 & | & 1 \\ 3 & 1 & | & 0 \end{bmatrix}$ (b) $\begin{bmatrix} 4 & 2 & | & 0 \\ 3 & 1 & | & 1 \end{bmatrix}$

Since the matrix A appears in both (a) and (b), any row operation we perform on (a) and (b) can be performed on the single augmented matrix that combines the two right-hand columns. We denote this matrix by $A|I_2$ and write

$$[A|I_2] = \begin{bmatrix} 4 & 2 & | & 1 & 0 \\ 3 & 1 & | & 0 & 1 \end{bmatrix}$$

If we perform the row operations on $[A|I_2]$ needed to obtain the reduced row echelon form of A, we get

$$\begin{bmatrix} 1 & 0 & | & -\tfrac{1}{2} & 1 \\ 0 & 1 & | & \tfrac{3}{2} & -2 \end{bmatrix}$$

The 2×2 matrix on the right-hand side of the vertical bar is A^{-1}.

This example illustrates the general procedure:

To find the inverse, if it exists, of a square matrix of dimension $n \times n$, follow these steps:

> **Steps for Finding the Inverse of a Matrix**
>
> **STEP 1** Form the matrix $[A|I_n]$.
> **STEP 2** Using row operations, write $[A|I_n]$ in reduced row echelon form.
> **STEP 3** If the resulting matrix is of the form $[I_n|B]$, that is, if the identity matrix appears on the left side of the bar, then B is the inverse of A. Otherwise, A has no inverse.

Let's work another example.

1 **EXAMPLE 5** **Finding the Inverse of a Matrix**

Find the inverse of

$$A = \begin{bmatrix} 1 & 1 & 2 \\ 2 & 1 & 0 \\ 1 & 2 & 2 \end{bmatrix}$$

SOLUTION **STEP 1** Since A is of dimension 3×3, we use the identity matrix I_3. The matrix $[A|I_3]$ is

$$\begin{bmatrix} 1 & 1 & 2 & | & 1 & 0 & 0 \\ 2 & 1 & 0 & | & 0 & 1 & 0 \\ 1 & 2 & 2 & | & 0 & 0 & 1 \end{bmatrix}$$

STEP 2 We proceed to obtain the reduced row echelon form of this matrix:

Use $\begin{array}{l} R_2 = -2r_1 + r_2 \\ R_3 = -1r_1 + r_3 \end{array}$ to obtain $\begin{bmatrix} 1 & 1 & 2 & | & 1 & 0 & 0 \\ 0 & -1 & -4 & | & -2 & 1 & 0 \\ 0 & 1 & 0 & | & -1 & 0 & 1 \end{bmatrix}$

Use $R_2 = -1r_2$ to obtain $\begin{bmatrix} 1 & 1 & 2 & | & 1 & 0 & 0 \\ 0 & 1 & 4 & | & 2 & -1 & 0 \\ 0 & 1 & 0 & | & -1 & 0 & 1 \end{bmatrix}$

Use $\begin{array}{l} R_1 = -1r_2 + r_1 \\ R_3 = -1r_2 + r_3 \end{array}$ to obtain $\begin{bmatrix} 1 & 0 & -2 & | & -1 & 1 & 0 \\ 0 & 1 & 4 & | & 2 & -1 & 0 \\ 0 & 0 & -4 & | & -3 & 1 & 1 \end{bmatrix}$

Use $R_3 = -\frac{1}{4}r_3$ to obtain $\begin{bmatrix} 1 & 0 & -2 & | & -1 & 1 & 0 \\ 0 & 1 & 4 & | & 2 & -1 & 0 \\ 0 & 0 & 1 & | & \frac{3}{4} & -\frac{1}{4} & -\frac{1}{4} \end{bmatrix}$

Use $\begin{array}{l} R_1 = 2r_3 + r_1 \\ R_2 = -4r_3 + r_2 \end{array}$ to obtain $\begin{bmatrix} 1 & 0 & 0 & | & \frac{1}{2} & \frac{1}{2} & -\frac{1}{2} \\ 0 & 1 & 0 & | & -1 & 0 & 1 \\ 0 & 0 & 1 & | & \frac{3}{4} & -\frac{1}{4} & -\frac{1}{4} \end{bmatrix}$

The matrix $[A|I_3]$ is in reduced row echelon form.

STEP 3 Since the identity matrix I_3 appears on the left side, the matrix appearing on the right is the inverse. That is,

$$A^{-1} = \begin{bmatrix} \frac{1}{2} & \frac{1}{2} & -\frac{1}{2} \\ -1 & 0 & 1 \\ \frac{3}{4} & -\frac{1}{4} & -\frac{1}{4} \end{bmatrix}$$

▶

You should verify that $AA^{-1} = A^{-1}A = I_3$.

 NOW WORK PROBLEM 15.

COMMENT: A graphing utility can be used to find the inverse of a matrix. Use a graphing utility to do Example 5.

GRAPHING UTILITY   Enter the matrix A into a graphing utility and use the $\boxed{x^{-1}}$ key to obtain $[A]^{-1}$.
SOLUTION   Figure 13 shows A^{-1}.

FIGURE 13

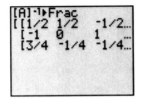

NOW WORK PROBLEM 51 USING A GRAPHING UTILITY.

Use EXCEL to do Example 5.

SOLUTION   **STEP 1** Enter the matrix A.
STEP 2 Highlight the cells that will contain A^{-1}; A^{-1} is a 3×3 matrix.
STEP 3 Type: **=MINVERSE** (*highlight matrix A*). The Excel screen should look like the one below.

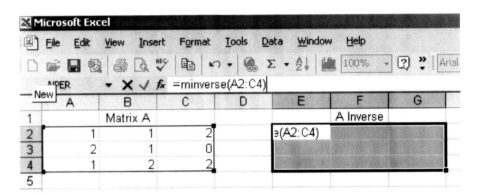

STEP 4 Press **Ctrl-Shift-Enter** all at the same time. The inverse will be displayed in the highlighted cells, as in the screen below.

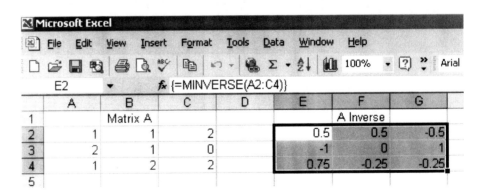

✓**Check:** Multiply $A*A^{-1}$ using the steps below. This product should be the identity matrix.

1. Highlight the cells that are to contain the product matrix.
2. Type: **=MMULT**(*highlight A, highlight* A^{-1})
3. Press **Ctrl-Shift-Enter** at the same time.

	Microsoft Excel						
File	Edit	View	Insert	Format	Tools	Data	Window	Help

C7 f_x {=MMULT(A2:C4,E2:G4)}

	A	B	C	D	E	F	G
1		Matrix A				A Inverse	
2	1	1	2		0.5	0.5	-0.5
3	2	1	0		-1	0	1
4	1	2	2		0.75	-0.25	-0.25
5							
6				A*A Inverse			
7			1	0	1.11E 16		
8			0	1	0		
9			0	0	1		
10							

Notice that $a_{13} = 1.11\text{E-}16$ is in scientific notation. In decimal notation it is 0.000000000000000111, very close to zero. So, $A*A^{-1} = \begin{bmatrix} 1 & 0 & 0 \\ 0 & 1 & 0 \\ 0 & 0 & 1 \end{bmatrix}$, as required.

 NOW WORK PROBLEM 51 USING EXCEL.

EXAMPLE 6   **Showing that a Matrix Has No Inverse**

Show that the matrix given below has no inverse.

$$\begin{bmatrix} 3 & 2 \\ 6 & 4 \end{bmatrix}$$

SOLUTION   We set up the matrix

$$\begin{bmatrix} 3 & 2 & | & 1 & 0 \\ 6 & 4 & | & 0 & 1 \end{bmatrix}$$

Use $R_1 = \frac{1}{3} r_1$ to obtain $\begin{bmatrix} 1 & \frac{2}{3} & | & \frac{1}{3} & 0 \\ 6 & 4 & | & 0 & 1 \end{bmatrix}$

Use $R_2 = -6r_1 + r_2$ to obtain $\begin{bmatrix} 1 & \frac{2}{3} & | & \frac{1}{3} & 0 \\ 0 & 0 & | & -2 & 1 \end{bmatrix}$

The 0s in row 2 tell us we cannot get the identity matrix. This, in turn, tells us the original matrix has no inverse.

COMMENT: If a matrix has no inverse, a graphing utility will display an Error message. See Figure 14 for the result of doing Example 6.

FIGURE 14

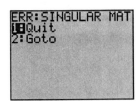

Solving a System of *n* Linear Equations Containing *n* Variables Using Inverses

Use the inverse of a matrix to solve a system of equations

The inverse of a matrix can be used to solve a system of *n* linear equations containing *n* variables. We begin with a system of three linear equations containing three variables:

$$\begin{cases} x + y + 2z = 1 \\ 2x + y = 2 \\ x + 2y + 2z = 3 \end{cases}$$

If we let

$$A = \begin{bmatrix} 1 & 1 & 2 \\ 2 & 1 & 0 \\ 1 & 2 & 2 \end{bmatrix} \qquad X = \begin{bmatrix} x \\ y \\ z \end{bmatrix} \qquad B = \begin{bmatrix} 1 \\ 2 \\ 3 \end{bmatrix}$$

the above system of equations can be written compactly as the matrix equation

$$AX = B$$

In general, any system of *n* linear equations containing *n* variables $x_1, x_2, \ldots, x_n$, can be written in the form

$$AX = B$$

where *A* is the $n \times n$ matrix of the coefficients of the unknowns, *B* is an $n \times 1$ column matrix whose entries are the numbers appearing to the right of each equal sign in the system, and *X* is an $n \times 1$ column matrix containing the *n* variables.

To find *X*, we start with the matrix equation $AX = B$ and use properties of matrices. Our assumption is that the $n \times n$ matrix *A* has an inverse A^{-1}.

$AX = B$	*A* has an inverse A^{-1}
$A^{-1}(AX) = A^{-1}B$	Multiply both sides by A^{-1}
$(A^{-1}A)X = A^{-1}B$	Apply the Associative Property on the left side
$I_n X = A^{-1}B$	Apply the Inverse Property: $A^{-1}A = I_n$
$X = A^{-1}B$	Apply the Identity Property: $I_n X = X$

This leads us to formulate the following result:

A system of n linear equations containing n variables

$$AX = B$$

for which A is a square matrix and A^{-1} exists, always has a unique solution that is given by

$$X = A^{-1}B$$

We use the above result in the next example.

EXAMPLE 7   **Solving a System of Equations Using Inverses**

Solve the system of equations:
$$\begin{cases} x + y + 2z = 1 \\ 2x + y = 2 \\ x + 2y + 2z = 3 \end{cases}$$

SOLUTION   Here

$$A = \begin{bmatrix} 1 & 1 & 2 \\ 2 & 1 & 0 \\ 1 & 2 & 2 \end{bmatrix} \quad X = \begin{bmatrix} x \\ y \\ z \end{bmatrix} \quad B = \begin{bmatrix} 1 \\ 2 \\ 3 \end{bmatrix}$$

From Example 5 we know A has the inverse.

$$A^{-1} = \begin{bmatrix} \frac{1}{2} & \frac{1}{2} & -\frac{1}{2} \\ -1 & 0 & 1 \\ \frac{3}{4} & -\frac{1}{4} & -\frac{1}{4} \end{bmatrix}$$

Based on the result just stated, the solution X of the system is

$$X = A^{-1}B$$

$$\begin{bmatrix} x \\ y \\ z \end{bmatrix} = \begin{bmatrix} \frac{1}{2} & \frac{1}{2} & -\frac{1}{2} \\ -1 & 0 & 1 \\ \frac{3}{4} & -\frac{1}{4} & -\frac{1}{4} \end{bmatrix} \begin{bmatrix} 1 \\ 2 \\ 3 \end{bmatrix} = \begin{bmatrix} 0 \\ 2 \\ -\frac{1}{2} \end{bmatrix}$$

The solution is $x = 0, y = 2, z = -\frac{1}{2}$.

NOW WORK PROBLEMS 39 AND 41.

The method used to solve the system in Example 7 requires that A have an inverse. If, for a system of equations $AX = B$, the matrix A has no inverse, then the system must be analyzed using the methods discussed in Section 2.3.

The method used in Example 7 for solving a system of equations is particularly useful for applications in which the constants appearing to the right of the equal sign change while the coefficients of the variables on the left side do not. See Problems 39–50 for an illustration. See also the discussion of Leontief models in Section 2.7.

Use EXCEL to do Example 7.

SOLUTION   From Example 5, we know the inverse of A is $A^{-1} = \begin{bmatrix} 0.5 & 0.5 & -0.5 \\ -1 & 0 & 1 \\ 0.75 & -0.25 & -0.25 \end{bmatrix}$.

To find the solution to the system of equations, multiply $A^{-1} * B$, where $B = \begin{bmatrix} 1 \\ 2 \\ 3 \end{bmatrix}$.

STEP 1 Highlight the cells that are to contain the product matrix. This will be a 3×1 matrix.

STEP 2 Type: **=MMULT** (*highlight A^{-1}, highlight B*)

STEP 3 Press **Ctrl-Shift-Enter** at the same time, to get $A^{-1}B$.

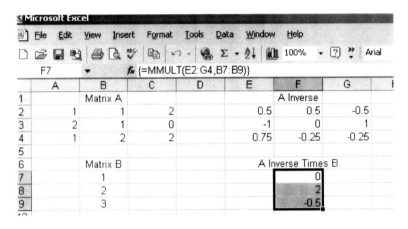

So, $x = 0, y = 2,$ and $z = -0.5$.

 NOW WORK PROBLEMS 39 AND 41 USING EXCEL.

EXERCISE 2.6   Answers to Odd-Numbered Problems Begin on Page AN-14.

In Problems 1–6, show that the given matrices are inverses of each other.

1. $\begin{bmatrix} 1 & 2 \\ 2 & 3 \end{bmatrix}\begin{bmatrix} -3 & 2 \\ 2 & -1 \end{bmatrix}$

2. $\begin{bmatrix} 1 & 5 \\ 2 & 0 \end{bmatrix}\begin{bmatrix} 0 & \frac{1}{2} \\ \frac{1}{5} & -\frac{1}{10} \end{bmatrix}$

3. $\begin{bmatrix} -1 & -2 \\ 3 & 4 \end{bmatrix}\begin{bmatrix} 2 & 1 \\ -\frac{3}{2} & -\frac{1}{2} \end{bmatrix}$

4. $\begin{bmatrix} 1 & 3 \\ 2 & -1 \end{bmatrix}\begin{bmatrix} \frac{1}{7} & \frac{3}{7} \\ \frac{2}{7} & -\frac{1}{7} \end{bmatrix}$

5. $\begin{bmatrix} 1 & 2 & 3 \\ 2 & 3 & 4 \\ 1 & 2 & 1 \end{bmatrix}\begin{bmatrix} -\frac{5}{2} & 2 & -\frac{1}{2} \\ 1 & -1 & 1 \\ \frac{1}{2} & 0 & -\frac{1}{2} \end{bmatrix}$

6. $\begin{bmatrix} 1 & 3 & 3 \\ 1 & 4 & 3 \\ 1 & 3 & 4 \end{bmatrix}\begin{bmatrix} 7 & -3 & -3 \\ -1 & 1 & 0 \\ -1 & 0 & 1 \end{bmatrix}$

In Problems 7–20, find the inverse of each matrix using the reduced row-echelon technique.

7. $\begin{bmatrix} 3 & 7 \\ 2 & 5 \end{bmatrix}$

8. $\begin{bmatrix} 4 & 1 \\ 3 & 1 \end{bmatrix}$

9. $\begin{bmatrix} 1 & -1 \\ 3 & -4 \end{bmatrix}$

10. $\begin{bmatrix} 5 & 3 \\ 3 & 2 \end{bmatrix}$

11. $\begin{bmatrix} 2 & 1 \\ 4 & 3 \end{bmatrix}$

12. $\begin{bmatrix} 2 & 3 \\ 2 & -2 \end{bmatrix}$

13. $\begin{bmatrix} 0 & 0 & 1 \\ 0 & 1 & 0 \\ 1 & 0 & 0 \end{bmatrix}$

14. $\begin{bmatrix} -1 & 1 & 0 \\ 1 & 0 & 2 \\ 3 & 1 & 0 \end{bmatrix}$

15. $\begin{bmatrix} 1 & 1 & -1 \\ 3 & -1 & 0 \\ 2 & -3 & 4 \end{bmatrix}$

16. $\begin{bmatrix} 1 & 1 & 1 \\ 2 & 1 & 1 \\ 1 & 1 & 2 \end{bmatrix}$

17. $\begin{bmatrix} 1 & 1 & -1 \\ 2 & 1 & 1 \\ 1 & 0 & 1 \end{bmatrix}$

18. $\begin{bmatrix} 2 & 3 & -1 \\ 1 & 1 & 1 \\ 0 & 2 & -1 \end{bmatrix}$

19. $\begin{bmatrix} 1 & 1 & 0 & 0 \\ 0 & 1 & -1 & 1 \\ 1 & -1 & 1 & 1 \\ 0 & 1 & 0 & -1 \end{bmatrix}$

20. $\begin{bmatrix} 1 & 2 & -3 & -2 \\ 0 & 1 & 4 & -2 \\ 3 & -1 & 4 & 0 \\ 2 & 1 & 0 & 3 \end{bmatrix}$

In Problems 21–26, show that each matrix has no inverse.

21. $\begin{bmatrix} 4 & 6 \\ 2 & 3 \end{bmatrix}$ 22. $\begin{bmatrix} -1 & 2 \\ 3 & -6 \end{bmatrix}$ 23. $\begin{bmatrix} -8 & 4 \\ -4 & 2 \end{bmatrix}$ 24. $\begin{bmatrix} 2 & 10 \\ 1 & 5 \end{bmatrix}$ 25. $\begin{bmatrix} 1 & 1 & 1 \\ 3 & -4 & 2 \\ 0 & 0 & 0 \end{bmatrix}$ 26. $\begin{bmatrix} -1 & 2 & 3 \\ 5 & 2 & 0 \\ 2 & -4 & -6 \end{bmatrix}$

In Problems 27–34, find the inverse, if it exists, of each matrix.

27. $\begin{bmatrix} 1 & 1 \\ 1 & 2 \end{bmatrix}$ 28. $\begin{bmatrix} 2 & 1 \\ 1 & 1 \end{bmatrix}$ 29. $\begin{bmatrix} 3 & -2 \\ 0 & 4 \end{bmatrix}$ 30. $\begin{bmatrix} 4 & -1 \\ -2 & 0 \end{bmatrix}$

31. $\begin{bmatrix} 3 & 2 \\ 6 & 4 \end{bmatrix}$ 32. $\begin{bmatrix} 4 & 2 \\ 2 & 1 \end{bmatrix}$ 33. $\begin{bmatrix} 1 & -2 & -1 \\ -2 & 5 & 4 \\ 3 & -8 & -5 \end{bmatrix}$ 34. $\begin{bmatrix} 1 & 1 & -1 \\ -2 & -1 & 4 \\ 3 & 2 & -8 \end{bmatrix}$

35. Find the inverse of both

$$A = \begin{bmatrix} 1 & 2 \\ 2 & -1 \end{bmatrix} \quad \text{and} \quad B = \begin{bmatrix} 1 & 3 \\ 2 & 1 \end{bmatrix}$$

to determine $A^{-1} - B^{-1}$.

36. Find the inverse of

$$A = \begin{bmatrix} 1 & -4 \\ 2 & -3 \end{bmatrix} \quad \text{and} \quad B = \begin{bmatrix} 2 & -2 \\ 3 & 2 \end{bmatrix}$$

Determine $A^{-1} - B^{-1}$.

37. Write the matrix product $A^{-1}B$ used to find the solution to the system $AX = B$.

$$\begin{cases} x + 3y + 2z = 2 \\ 2x + 7y + 3z = 1 \\ x \qquad + 6z = 3 \end{cases}$$

38. Write the matrix product $A^{-1}B$ used to find the solution to the system $AX = B$.

$$\begin{cases} x + 2y + 2z = 3 \\ 2x + 5y + 7z = 2 \\ 2x + y - 4z = 4 \end{cases}$$

In Problems 39–50, solve each system of equations by the method of Example 7. For Problems 39–44, use the inverse found in Problem 7. For Problems 45–50, use the inverse found in Problem 15.

39. $\begin{cases} 3x + 7y = 10 \\ 2x + 5y = 2 \end{cases}$ 40. $\begin{cases} 3x + 7y = -4 \\ 2x + 5y = -3 \end{cases}$ 41. $\begin{cases} 3x + 7y = 13 \\ 2x + 5y = 9 \end{cases}$

42. $\begin{cases} 3x + 7y = 0 \\ 2x + 5y = 14 \end{cases}$ 43. $\begin{cases} 3x + 7y = 12 \\ 2x + 5y = -4 \end{cases}$ 44. $\begin{cases} 3x + 7y = -2 \\ 2x + 5y = 10 \end{cases}$

45. $\begin{cases} x + y - z = 3 \\ 3x - y = -4 \\ 2x - 3y + 4z = 6 \end{cases}$ 46. $\begin{cases} x + y - z = 6 \\ 3x - y = 8 \\ 2x - 3y + 4z = -3 \end{cases}$ 47. $\begin{cases} x + y - z = 12 \\ 3x - y = -4 \\ 2x - 3y + 4z = 16 \end{cases}$

48. $\begin{cases} x + y - z = -8 \\ 3x - y = 12 \\ 2x - 3y + 4z = -2 \end{cases}$ 49. $\begin{cases} x + y - z = 0 \\ 3x - y = -8 \\ 2x - 3y + 4z = -6 \end{cases}$ 50. $\begin{cases} x + y - z = 21 \\ 3x - y = 12 \\ 2x - 3y + 4z = 14 \end{cases}$

In Problems 51–56, use a graphing utility or EXCEL to find the inverse, if it exists, of each matrix.

51. $\begin{bmatrix} 25 & 61 & -12 \\ 18 & -2 & 4 \\ 8 & 35 & 21 \end{bmatrix}$ 52. $\begin{bmatrix} 18 & -3 & 4 \\ 6 & -20 & 14 \\ 10 & 25 & -15 \end{bmatrix}$ 53. $\begin{bmatrix} 44 & 21 & 18 & 6 \\ -2 & 10 & 15 & 5 \\ 21 & 12 & -12 & 4 \\ -8 & -16 & 4 & 9 \end{bmatrix}$

54. $\begin{bmatrix} 16 & 22 & -3 & 5 \\ 21 & -17 & 4 & 8 \\ 2 & 8 & 27 & 20 \\ 5 & 15 & -3 & -10 \end{bmatrix}$ 55. $A = \begin{bmatrix} 3 & 0 & 2 & -1 & 3 \\ -2 & 1 & 2 & 3 & 0 \\ 2 & 2 & 1 & 1 & -1 \\ 1 & 2 & 0 & 2 & -3 \\ 4 & 0 & -1 & 1 & -1 \end{bmatrix}$ 56. $A = \begin{bmatrix} 0 & 0 & 2 & -1 & 3 \\ -2 & 0 & 2 & 3 & 0 \\ 2 & 2 & 0 & 0 & -1 \\ 1 & 2 & 0 & 2 & -3 \\ 4 & 4 & 0 & 0 & -2 \end{bmatrix}$

In Problems 57–60, use the idea behind Example 7 and either a graphing utility or EXCEL to solve each system of equations.

57. $\begin{cases} 25x + 61y - 12z = 10 \\ 18x - 12y + 7z = -9 \\ 3x + 4y - z = 12 \end{cases}$

58. $\begin{cases} 25x + 61y - 12z = 5 \\ 18x - 12y + 7z = -3 \\ 3x + 4y - z = 12 \end{cases}$

59. $\begin{cases} 25x + 61y - 12z = 21 \\ 18x - 12y + 7z = 7 \\ 3x + 4y - z = -2 \end{cases}$

60. $\begin{cases} 25x + 61y - 12z = 25 \\ 18x - 12y + 7z = 10 \\ 3x + 4y - z = -4 \end{cases}$

61. Show that the inverse of $A = \begin{bmatrix} a & b \\ c & d \end{bmatrix}$ is given by the formula $A^{-1} = \begin{bmatrix} \dfrac{d}{\Delta} & \dfrac{-b}{\Delta} \\ \dfrac{-c}{\Delta} & \dfrac{a}{\Delta} \end{bmatrix}$

where $\Delta = ad - bc \neq 0$. The number Δ is called the **determinant** of A.

In Problems 62–65, use the result of Problem 61 to find the inverse of each matrix.

62. $\begin{bmatrix} 1 & 2 \\ 2 & 3 \end{bmatrix}$

63. $\begin{bmatrix} 1 & 5 \\ 2 & 0 \end{bmatrix}$

64. $\begin{bmatrix} -1 & -2 \\ 3 & 4 \end{bmatrix}$

65. $\begin{bmatrix} 1 & 2 \\ 8 & 15 \end{bmatrix}$

2.7   Applications: Leontief Model; Cryptography; Accounting; The Method of Least Squares*

OBJECTIVES

1 Determine relative income using a closed Leontief model
2 Determine production levels necessary to meet forecasted demand using an open Leontief model
3 Use matrices in cryptography
4 Determine the full cost of manufactured products
5 Find the transpose of a matrix
6 Use the method of least squares to find the line of best fit

Application 1: Leontief Models

The Leontief models in economics are named after Wassily Leontief, who received the Nobel Prize in economics in 1973. These models can be characterized as a description of an economy in which input equals output or, in other words, consumption equals production. That is, the models assume that whatever is produced is always consumed.

Leontief models are of two types: closed, in which the entire production is consumed by those participating in the production; and open, in which some of the production is consumed by those who produce it and the rest of the production is consumed by external bodies.

In the *closed model* we seek the relative income of each participant in the system. In the *open model* we seek the amount of production needed to achieve a forecast demand, when the amount of production needed to achieve current demand is known.

THE CLOSED MODEL

We begin with an example to illustrate the idea.

Each application is optional, and the applications given are independent of one another.

1   **EXAMPLE 1   Determining Relative Income Using a Closed Leontief Model**

Three homeowners, Juan, Luis, and Carlos, each with certain skills, agreed to pool their talents to make repairs on their houses. As it turned out, Juan spent 20% of his time on his own house, 40% of his time on Luis' house, and 40% on Carlos' house. Luis spent 10% of his time on Juan's house, 50% of his time on his own house, and 40% on Carlos' house. Of Carlos' time, 60% was spent on Juan's house, 10% on Luis', and 30% on his own. Now that the projects are finished, they need to figure out how much money each should get for his work, including the work performed on his own house, so that the amount paid by each person equals the amount received by each one. They agreed in advance that the payment to each one should be approximately $3000.00.

SOLUTION   We place the information given in the problem in a 3 × 3 matrix, as follows:

$$
\begin{array}{c}
\text{Work done by} \\
\begin{array}{ccc} \text{Juan} & \text{Luis} & \text{Carlos} \end{array}
\end{array}
$$

$$
\begin{array}{r}
\text{Proportion of work done on Juan's house} \\
\text{Proportion of work done on Luis' house} \\
\text{Proportion of work done on Carlos' house}
\end{array}
\begin{bmatrix}
0.2 & 0.1 & 0.6 \\
0.4 & 0.5 & 0.1 \\
0.4 & 0.4 & 0.3
\end{bmatrix}
$$

Next, we define the variables:

$$x = \text{Juan's wages}$$
$$y = \text{Luis' wages}$$
$$z = \text{Carlos' wages}$$

We require that the amount paid out by each one equals the amount received by each one. Let's analyze this requirement, by looking just at the work done on Juan's house. Juan's wages are x. Juan's expenditures for work done on his house are $0.2x + 0.1y + 0.6z$. Juan's wages and expenditures are required to be equal, so

$$x = 0.2x + 0.1y + 0.6z$$

Similarly,

$$y = 0.4x + 0.5y + 0.1z$$
$$z = 0.4x + 0.4y + 0.3z$$

These three equations can be written compactly as

$$
\begin{bmatrix} x \\ y \\ z \end{bmatrix} =
\begin{bmatrix}
0.2 & 0.1 & 0.6 \\
0.4 & 0.5 & 0.1 \\
0.4 & 0.4 & 0.3
\end{bmatrix}
\begin{bmatrix} x \\ y \\ z \end{bmatrix}
$$

Some computation reduces the system to

$$
\begin{cases}
0.8x - 0.1y - 0.6z = 0 \\
-0.4x + 0.5y - 0.1z = 0 \\
-0.4x - 0.4y + 0.7z = 0
\end{cases}
$$

493

Solving for x, y, z, we find that

$$x = \tfrac{31}{36}z \qquad y = \tfrac{32}{36}z$$

where z is the parameter. To get solutions that fall close to \$3000, we set $z = 3600$.* The wages to be paid out are therefore

$$x = \$3100 \qquad y = \$3200 \qquad z = \$3600$$

The matrix in Example 1, namely,

$$\begin{bmatrix} 0.2 & 0.1 & 0.6 \\ 0.4 & 0.5 & 0.1 \\ 0.4 & 0.4 & 0.3 \end{bmatrix}$$

is called an **input–output matrix.**

In the general closed model we have an economy consisting of n components. Each component produces an *output* of some goods or services, which, in turn, is completely used up by the n components. The proportionate use of each component's output by the economy makes up the input–output matrix of the economy. The problem is to find suitable pricing levels for each component so that income equals expenditure.

Closed Leontief Model

In general, an input–output matrix for a closed Leontief model is of the form

$$\boxed{A = [a_{ij}] \quad i, j = 1, 2, \ldots, n}$$

where the a_{ij} represent the fractional amount of goods or services used by i and produced by j. For a closed model the sum of each column equals 1 (this is the condition that all production is consumed internally) and $0 \le a_{ij} \le 1$ for all entries (this is the restriction that each entry is a fraction).

If A is the input–output matrix of a closed system with n components and X is a column vector representing the price of each output of the system, then

$$\boxed{X = AX}$$

represents the requirement that income equal expenditure.

For example, the first entry of the matrix equality $X = AX$ requires that

$$x_1 = a_{11}x_1 + a_{12}x_2 + \ldots + a_{1n}x_n$$

*Other choices for z are, of course, possible. The choice of which value to use is up to the homeowners. No matter what choice is made, the amount paid by each one equals the amount received by each one.

The right side represents the price paid by component 1 for the goods it uses, while x_1 represents the income of component 1; we are requiring they be equal.

We can rewrite the equation $X = AX$ as

$$X - AX = \mathbf{0}$$
$$I_n X - AX = \mathbf{0}$$
$$(I_n - A)X = \mathbf{0}$$

This matrix equation, which represents a system of equations in which the right-hand side is always $\mathbf{0}$, is called a **homogeneous system of equations.** It can be shown that if the entries in the input–output matrix A are positive and if the sum of each column of A equals 1, then this system has a one-parameter solution; that is, we can solve for $n - 1$ of the variables in terms of the remaining one, which serves as the parameter. This parameter serves as a "scale factor."

NOW WORK PROBLEM 1.

THE OPEN MODEL

Determine production levels **2**
necessary to meet fore-
casted demand using an
open Leontief model

For the open model, in addition to internal consumption of goods produced, there is an outside demand for the goods produced. This outside demand may take the form of exportation of goods or may be the goods needed to support consumer demand. Again, however, we make the assumption that whatever is produced is also consumed.

For example, suppose an economy consists of three industries R, S, and T, and suppose each one produces a single product. We assume that a portion of R's production is used by each of the three industries, while the remainder is used by consumers.

The same is true of the production of S and T. To organize our thoughts, we construct a table that describes the interaction of the use of R, S, and T's production over some fixed period of time. See Table 4.

TABLE 4

Amounts Consumed

		R	S	T	Consumer	Total
Amounts Produced	R	50	20	40	70	180
	S	20	30	20	90	160
	T	30	20	20	50	120

All entries in the table are in appropriate units, say, in dollars. The first row (row R) represents the production in dollars of industry R (input). Out of the total of $180 worth of goods produced, R, S, and T use $50, $20, and $40, respectively, for the production of their goods, while consumers purchase the remaining $70 for their consumption (output). Observe that input equals output since everything produced by R is used up by R, S, T, and consumers.

The second and third rows are interpreted in the same way.

An important observation is that the goal of R's production is to produce $70 worth of goods, since this is the demand of consumers. In order to meet this demand, R must produce a total of $180, since the difference, $110, is required internally by R, S, and T.

Suppose, however, that consumer demand is expected to change. To effect this change, how much should each industry now produce? For example, in Table 4, current demand for R, S, and T can be represented by a **demand vector:**

$$D_0 = \begin{bmatrix} 70 \\ 90 \\ 50 \end{bmatrix}$$

But suppose marketing forecasts predict that in 3 years the demand vector will be

$$D_3 = \begin{bmatrix} 60 \\ 110 \\ 60 \end{bmatrix}$$

Here, the demand for item R has decreased; the demand for item S has significantly increased, and the demand for item T is higher. Given the current total output of R, S, and T at 180, 160, and 120, respectively, what must it be in 3 years to meet this projected demand?

The solution of this type of forecasting problem is derived from the *open Leontief model* in input–output analysis. In using input–output analysis to obtain a solution to such a forecasting problem, we take into account the fact that the output of any one of these industries is affected by changes in the other two, since the total demand for, say, R in 3 years depends not only on consumer demand for R, but also on consumer demand for S and T. That is, the industries are interrelated.

To obtain the solution, we need to determine how much of each of the three products R, S, and T is required to produce 1 unit of R. For example, to obtain 180 units of R requires the use of 50 units of R, 20 units of S, and 30 units of T (the entries in column 1). Forming the ratios, we find that to produce 1 unit of R requires $\frac{50}{180} = 0.278$ of R, $\frac{20}{180} = 0.111$ of S, and $\frac{30}{180} = 0.167$ of T. If we want, say, x units of R, we will require $0.278x$ units of R, $0.111x$ units of S, and $0.167x$ units of T.

Continuing in this way, we can construct the matrix

$$
A = \begin{array}{c} \\ R \\ S \\ T \end{array}
\begin{array}{c}
\begin{array}{ccc} R & S & T \end{array} \\
\begin{bmatrix} 0.278 & 0.125 & 0.333 \\ 0.111 & 0.188 & 0.167 \\ 0.167 & 0.125 & 0.167 \end{bmatrix}
\end{array}
$$

Observe that column 1 represents the amounts of R, S, T required for 1 unit of R; column 2 represents the amounts of R, S, and T required for 1 unit of S; and column 3 represents the amounts of R, S, and T required for 1 unit of T. For example, the entry in row 3, column 2 (0.125), represents the amount of T needed to produce 1 unit of S.

As a result of placing the entries this way, if

$$X = \begin{bmatrix} x \\ y \\ z \end{bmatrix}$$

represents the total output required to obtain a given demand, the product AX represents the amount of R, S, and T required for internal consumption. The condition that production = consumption requires that

Internal consumption + Consumer demand = Total output

In terms of the matrix A, the total output X, and the demand vector D, this requirement is equivalent to the equation

$$AX + D = X$$

In this equation we seek to find X for a prescribed demand D. The matrix A is calculated as above for some initial production process.*

EXAMPLE 2   **Finding Production Levels to Meet Future Demand: The Open Leontief Model**

For the data given in Table 4, find the total output X required to achieve a future demand of

$$D_3 = \begin{bmatrix} 60 \\ 110 \\ 60 \end{bmatrix}$$

SOLUTION   We need to solve for X in

$$AX + D_3 = X$$

Simplifying, we have

$$[I_3 - A]X = D_3$$

Solving for X, we have

$$X = [I_3 - A]^{-1} \cdot D_3$$

$$= \begin{bmatrix} 0.722 & -0.125 & -0.333 \\ -0.111 & 0.812 & -0.167 \\ -0.167 & -0.125 & 0.833 \end{bmatrix}^{-1} \begin{bmatrix} 60 \\ 110 \\ 60 \end{bmatrix}$$

$$= \begin{bmatrix} 1.6048 & 0.3568 & 0.7131 \\ 0.2946 & 1.3363 & 0.3857 \\ 0.3660 & 0.2721 & 1.4013 \end{bmatrix} \begin{bmatrix} 60 \\ 110 \\ 60 \end{bmatrix}$$

$$= \begin{bmatrix} 178.33 \\ 187.81 \\ 135.96 \end{bmatrix}$$

The total output of R, S, and T required for the forecast demand D_3 is

$$x = 178.322 \qquad y = 187.811 \qquad z = 135.969$$

NOW WORK PROBLEM 9.

*The entries in A can be checked by using the requirement that $AX + D_0 = X$, for $D_0 = $ Initial demand and $X = $ Total output. For our example, it must happen that

| Internal consumption | + Consumer demand | = Total output |

$$\begin{matrix} AX & + & D_0 & = & X \end{matrix}$$

$$\begin{bmatrix} 0.278 & 0.125 & 0.333 \\ 0.111 & 0.188 & 0.167 \\ 0.167 & 0.125 & 0.167 \end{bmatrix} \begin{bmatrix} 180 \\ 160 \\ 120 \end{bmatrix} + \begin{bmatrix} 70 \\ 90 \\ 50 \end{bmatrix} = \begin{bmatrix} 180 \\ 160 \\ 120 \end{bmatrix}$$

Use EXCEL to do Example 2.

SOLUTION   We need to solve $AX + D_3 = X$ for X. The solution to this equation is
$$X = (I_3 - A)^{-1} \cdot D_3$$
We begin by using EXCEL to compute $(I - A)^{-1} \cdot D$.

STEP 1   Enter matrices A, I, and D_3 into an EXCEL spreadsheet.

	File Edit View Insert Format Tools Data Window Help

	A	B	C	D	E	F	G
1			**Matrix A**				**D_3**
2				Output			
3			R	S	T		
4	Input	R	0.278	0.125	0.333		60
5		S	0.111	0.188	0.167		110
6		T	0.167	0.125	0.167		60
7							
8		**Matrix I**					
9							
10							
11	1	0	0				
12	0	1	0				
13	0	0	1				
14							

STEP 2   Compute $I - A$.
- **(a)** Highlight the cells where the difference is to go. This must be a 3×3 matrix.
- **(b)** Type $= (highlight\ I) - (highlight\ A)$ and press **Ctrl-Shift-Enter.**

Microsoft Excel
File Edit View Insert Format Tools Data Window Help Acrobat

AVERAGE   =C4:E6-A9:C11

	A	B	C	D	E	F	G	H
1			**Matrix A**				**D_3**	
2				Output				
3			R	S	T			
4	Input	R	0.278	0.278	0.278		60	
5		S	0.278	0.278	0.278		110	
6		T	0.278	0.278	0.278		60	
7								
8		**Matrix I**				**I - A**		
9	1	0	0		A9:C11			
10	0	1	0					
11	0	0	1					
12								
13								

STEP 3 Compute $(I - A)^{-1}$
   (a) Highlight the cells that will contain $(I - A)^{-1}$; $(I - A)^{-1}$ is a 3×3 matrix.
   (b) Type: **=MINVERSE**(*highlight matrix* $(I - A)$) in the formula bar, and press **Ctrl-Shift-Enter.**

STEP 4 Compute $(I - A)^{-1}$ D.
   (a) Highlight the cells that are to contain the product matrix. This will be a 3×1 matrix.
   (b) Type: **=MMULT**(*highlight*$(I - A)^{-1}$, *highlight* D), and press **Ctrl-Shift-Enter.**

The results are given below.

	File	Edit	View	Insert	Format	Tools	Data	Window	Help		

F16   ▼ f_x {=MMULT(A16:C18,G4:G6)}

	A	B	C	D	E	F	G
1			**Matrix A**				**D₃**
2				Output			
3			R	S	T		
4	Input	R	0.278	0.125	0.333		60
5		S	0.111	0.188	0.167		110
6		T	0.167	0.125	0.167		60
7							
8		**Matrix I**				**I - A**	
9							
10	1	0	0		0.722	-0.125	-0.333
11	0	1	0		-0.111	0.812	-0.167
12	0	0	1		-0.167	-0.125	0.833
13							
14		**(I - A) Inverse**				**(I - A)⁻¹*D₃**	
15							
16	1.604837	0.356823	0.713086			178.3259	
17	0.294644	1.336257	0.38568			187.8077	
18	0.365952	0.272055	1.401315			135.9621	
19							

So $x = 178.33, y = 187.81, z = 135.96.$

NOW WORK PROBLEM 9 USING EXCEL.

The general open model can be described as follows: Suppose there are n industries in the economy. Each industry produces some goods or services, which are partially consumed by the n industries, while the rest are used to meet a prescribed current demand. Given the output required of each industry to meet current demand, what should the output of each industry be to meet some different future demand?

Open Leontief Model

The matrix $A = [a_{ij}]$, $i, j = 1, \ldots, n$ of the open model is defined to consist of entries a_{ij}, where a_{ij} is the amount of output of industry j required for one unit of output of industry i. If X is a column vector representing the production of each industry in the system and D is a column vector representing future demand for goods produced in the system, then

$$X = AX + D$$

From the equation above we find

$$[I_n - A]X = D$$

It can be shown that the matrix $I_n - A$ has an inverse, provided each entry in A is positive and the sum of each column in A is less than 1. Under these conditions we may solve for X to get

$$X = [I_n - A]^{-1} \cdot D$$

This form of the solution is particularly useful since it allows us to find X for a variety of demands D by doing one calculation: $[I_n - A]^{-1}$.

We conclude by noting that the use of an input–output matrix to solve forecasting problems assumes that each industry produces a single commodity and that no technological advances take place in the period of time under investigation (in other words, the proportions found in the matrix A are fixed).

EXERCISE 2.7   APPLICATION 1   Answers to Odd-Numbered Problems Begin on Page AN-15.

In Problems 1–4, find the relative wages of each person for the given closed input–output matrix. In each case take the wages of C to be the parameter and use $z = C$'s wages $= \$30,000$.

1.
$$\begin{array}{c} \\ A \\ B \\ C \end{array} \begin{array}{ccc} A & B & C \\ \begin{bmatrix} \dfrac{1}{2} & \dfrac{1}{3} & \dfrac{1}{4} \\[6pt] \dfrac{1}{4} & \dfrac{1}{3} & \dfrac{1}{4} \\[6pt] \dfrac{1}{4} & \dfrac{1}{3} & \dfrac{1}{2} \end{bmatrix} \end{array}$$

2.
$$\begin{array}{c} \\ A \\ B \\ C \end{array} \begin{array}{ccc} A & B & C \\ \begin{bmatrix} \dfrac{1}{4} & \dfrac{2}{3} & \dfrac{1}{2} \\[6pt] \dfrac{1}{2} & \dfrac{1}{6} & \dfrac{1}{4} \\[6pt] \dfrac{1}{4} & \dfrac{1}{6} & \dfrac{1}{4} \end{bmatrix} \end{array}$$

3.
$$\begin{array}{c} \\ A \\ B \\ C \end{array} \begin{array}{ccc} A & B & C \\ \begin{bmatrix} 0.2 & 0.3 & 0.1 \\ 0.6 & 0.4 & 0.2 \\ 0.2 & 0.3 & 0.7 \end{bmatrix} \end{array}$$

4.
$$\begin{array}{c} \\ A \\ B \\ C \end{array} \begin{array}{ccc} A & B & C \\ \begin{bmatrix} 0.4 & 0.3 & 0.2 \\ 0.2 & 0.3 & 0.3 \\ 0.4 & 0.4 & 0.5 \end{bmatrix} \end{array}$$

5. For the three industries R, S, and T in the open Leontief model of Table 4 on page 131, compute the total output vector X if the forecast demand vector is

$$D_2 = \begin{bmatrix} 80 \\ 90 \\ 60 \end{bmatrix}$$

6. Rework Problem 5 if the forecast demand vector is

$$D_4 = \begin{bmatrix} 100 \\ 80 \\ 60 \end{bmatrix}$$

7. **Closed Leontief Model** A society consists of four individuals: a farmer, a builder, a tailor, and a rancher (who produces meat products). Of the food produced by the farmer, $\frac{3}{10}$ is used by the farmer, $\frac{2}{10}$ by the builder, $\frac{2}{10}$ by the tailor, and $\frac{3}{10}$ by the rancher. The builder's production is utilized 30% by the farmer, 30% by the builder, 10% by the tailor, and 30% by the rancher. The tailor's production is used in the ratios $\frac{3}{10}, \frac{3}{10}, \frac{1}{10}$, and $\frac{3}{10}$ by the farmer, builder, tailor, and rancher, respectively. Finally, meat products are used 20% by each of the farmer, builder, and tailor, and 40% by the rancher. What is the relative income of each if the rancher's income is scaled at $\$25,000$?

8. If in Problem 7 the meat production utilization changes so that it is used equally by all four individuals, while everyone else's production utilization remains the same, what are the relative incomes?

9. Open Leontief Model Suppose the interrelationships between the production of two industries R and S in a given year are given in the table:

	R	S	Current Consumer Demand	Total Output
R	30	40	60	130
S	20	10	40	70

If the forecast demand in 2 years is

$$D_2 = \begin{bmatrix} 80 \\ 40 \end{bmatrix}$$

what should the total output X be?

10. Open Leontief Model Suppose the interrelationships between the production of five industries (manufacturing, electric power, petroleum, transportation, and textiles) in a given year are as listed in the table:

	Manu.	E.P.	Petr.	Tran.	Text.
Manu.	0.2	0.12	0.15	0.18	0.1
E.P.	0.17	0.11	0	0.19	0.28
Petr.	0.11	0.11	0.12	0.46	0.12
Tran.	0.1	0.14	0.18	0.17	0.19
Text.	0.16	0.18	0.02	0.1	0.3

If the demand vector is given by

$$D = \begin{bmatrix} 100 \\ 100 \\ 200 \\ 100 \\ 100 \end{bmatrix}$$

what should the production vector X be?

Application 2: Cryptography

Use matrices in ③ cryptography

Our second application is to *cryptography*, the art of writing or deciphering secret codes. We begin by giving examples of elementary codes.

EXAMPLE 1   Encoding a Message

A message can be encoded by associating each letter of the alphabet with some other letter of the alphabet according to a prescribed pattern. For example, we might have

A B C D E F G H I J K L M N O P Q R S T U V W X Y Z
↓ ↓
C D E F G H I J K L M N O P Q R S T U V W X Y Z A B

With the above code the word MESSAGE would become OGUUCIG.

501

EXAMPLE 2   Encoding a Message

Another code may associate numbers with the letters of the alphabet. For example, we might have

A B C D E F G H I J K L M N O P Q R S T U V W X Y Z

↓ ↓

26 25 24 23 22 21 20 19 18 17 16 15 14 13 12 11 10 9 8 7 6 5 4 3 2 1

In this code the word PEACE looks like 11 22 26 24 22.   ▶

Both the above codes have one important feature in common. The association of letters with the coding symbols is made using a one-to-one correspondence so that no possible ambiguities can arise.

Suppose we want to encode the following message:

$$\text{TOP SECURITY CLEARANCE}$$

If we decide to divide the message into pairs of letters, the message becomes

$$\text{TO PS EC UR IT YC LE AR AN CE}$$

(If there is a letter left over, we arbitrarily assign Z to the last position.) Using the correspondence of letters to numbers given in Example 2, and writing each pair of letters as a column vector, we obtain

$$\begin{bmatrix} T \\ O \end{bmatrix} = \begin{bmatrix} 7 \\ 12 \end{bmatrix} \quad \begin{bmatrix} P \\ S \end{bmatrix} = \begin{bmatrix} 11 \\ 8 \end{bmatrix} \quad \begin{bmatrix} E \\ C \end{bmatrix} = \begin{bmatrix} 22 \\ 24 \end{bmatrix} \quad \begin{bmatrix} U \\ R \end{bmatrix} = \begin{bmatrix} 6 \\ 9 \end{bmatrix} \quad \begin{bmatrix} I \\ T \end{bmatrix} = \begin{bmatrix} 18 \\ 7 \end{bmatrix}$$

$$\begin{bmatrix} Y \\ C \end{bmatrix} = \begin{bmatrix} 2 \\ 24 \end{bmatrix} \quad \begin{bmatrix} L \\ E \end{bmatrix} = \begin{bmatrix} 15 \\ 22 \end{bmatrix} \quad \begin{bmatrix} A \\ R \end{bmatrix} = \begin{bmatrix} 26 \\ 9 \end{bmatrix} \quad \begin{bmatrix} A \\ N \end{bmatrix} = \begin{bmatrix} 26 \\ 13 \end{bmatrix} \quad \begin{bmatrix} C \\ E \end{bmatrix} = \begin{bmatrix} 24 \\ 22 \end{bmatrix}$$

Next, we arbitrarily choose a 2 × 2 matrix A, which we know has an inverse A^{-1} (we'll see why later). Let's choose

$$A = \begin{bmatrix} 2 & 3 \\ 1 & 2 \end{bmatrix}$$

The inverse A^{-1} of A is

$$A^{-1} = \begin{bmatrix} 2 & -3 \\ -1 & 2 \end{bmatrix}$$

Now we transform the column vectors representing the message by multiplying each of them on the left by matrix A:

$$A\begin{bmatrix} T \\ O \end{bmatrix} = \begin{bmatrix} 2 & 3 \\ 1 & 2 \end{bmatrix} \begin{bmatrix} 7 \\ 12 \end{bmatrix} = \begin{bmatrix} 50 \\ 31 \end{bmatrix}$$

$$A\begin{bmatrix} P \\ S \end{bmatrix} = \begin{bmatrix} 2 & 3 \\ 1 & 2 \end{bmatrix} \begin{bmatrix} 11 \\ 8 \end{bmatrix} = \begin{bmatrix} 46 \\ 27 \end{bmatrix}$$

$$A\begin{bmatrix} E \\ C \end{bmatrix} = \begin{bmatrix} 2 & 3 \\ 1 & 2 \end{bmatrix} \begin{bmatrix} 22 \\ 24 \end{bmatrix} = \begin{bmatrix} 116 \\ 70 \end{bmatrix}$$

$$A\begin{bmatrix} U \\ R \end{bmatrix} = \begin{bmatrix} 2 & 3 \\ 1 & 2 \end{bmatrix} \begin{bmatrix} 6 \\ 9 \end{bmatrix} = \begin{bmatrix} 39 \\ 24 \end{bmatrix}$$

$$A\begin{bmatrix} I \\ T \end{bmatrix} = \begin{bmatrix} 2 & 3 \\ 1 & 2 \end{bmatrix}\begin{bmatrix} 18 \\ 7 \end{bmatrix} = \begin{bmatrix} 57 \\ 32 \end{bmatrix}$$

$$A\begin{bmatrix} Y \\ C \end{bmatrix} = \begin{bmatrix} 2 & 3 \\ 1 & 2 \end{bmatrix}\begin{bmatrix} 2 \\ 24 \end{bmatrix} = \begin{bmatrix} 76 \\ 50 \end{bmatrix}$$

$$A\begin{bmatrix} L \\ E \end{bmatrix} = \begin{bmatrix} 2 & 3 \\ 1 & 2 \end{bmatrix}\begin{bmatrix} 15 \\ 22 \end{bmatrix} = \begin{bmatrix} 96 \\ 59 \end{bmatrix}$$

$$A\begin{bmatrix} A \\ R \end{bmatrix} = \begin{bmatrix} 2 & 3 \\ 1 & 2 \end{bmatrix}\begin{bmatrix} 26 \\ 9 \end{bmatrix} = \begin{bmatrix} 79 \\ 24 \end{bmatrix}$$

$$A\begin{bmatrix} A \\ N \end{bmatrix} = \begin{bmatrix} 2 & 3 \\ 1 & 2 \end{bmatrix}\begin{bmatrix} 26 \\ 13 \end{bmatrix} = \begin{bmatrix} 79 \\ 52 \end{bmatrix}$$

$$A\begin{bmatrix} C \\ E \end{bmatrix} = \begin{bmatrix} 2 & 3 \\ 1 & 2 \end{bmatrix}\begin{bmatrix} 24 \\ 22 \end{bmatrix} = \begin{bmatrix} 76 \\ 116 \end{bmatrix}$$

The coded message is

50 31 46 27 116 70 39 24 57 32 76 50 96 59 79 24 79 52 76 116

To decode or unscramble the above message, pair the numbers in 2×1 column vectors. Then on the left multiply each column vector by A^{-1}. For example, the first two column vectors then become

$$A^{-1}\begin{bmatrix} 50 \\ 31 \end{bmatrix} = \begin{bmatrix} 2 & -3 \\ -1 & 2 \end{bmatrix}\begin{bmatrix} 50 \\ 31 \end{bmatrix} = \begin{bmatrix} 7 \\ 12 \end{bmatrix} = \begin{bmatrix} T \\ O \end{bmatrix}$$

$$A^{-1}\begin{bmatrix} 46 \\ 17 \end{bmatrix} = \begin{bmatrix} 2 & -3 \\ -1 & 2 \end{bmatrix}\begin{bmatrix} 46 \\ 27 \end{bmatrix} = \begin{bmatrix} 11 \\ 8 \end{bmatrix} = \begin{bmatrix} P \\ S \end{bmatrix} \text{ etc.}$$

Continuing in this way, the original message is obtained.

 NOW WORK PROBLEM 1.

EXAMPLE 3   **Encoding a Message**

The message to be encoded is

SEPTEMBER IS OKAY

The encoded message is to be formed using triplets of numbers and the 3×3 matrix

$$A = \begin{bmatrix} 1 & 0 & 0 \\ 3 & 1 & 5 \\ -2 & 0 & 1 \end{bmatrix} \text{ whose inverse is } A^{-1} = \begin{bmatrix} 1 & 0 & 0 \\ -13 & 1 & -5 \\ 2 & 0 & 1 \end{bmatrix}.$$

Use the following association of letters to numbers:

A B C D E F G H I J K L M N O P Q R S T U V W X Y Z
↓ ↓
1 2 3 4 5 6 7 8 9 10 11 12 13 14 15 16 17 18 19 20 21 22 23 24 25 26

SOLUTION   When we divide the message into triplets of letters, we obtain

SEP TEM BER ISO KAY

Now multiply the matrix A times each column vector of the message.

[*Note*: If additional letters had been required to complete a triplet, we would have used Z or YZ.]

$$A\begin{bmatrix} S \\ E \\ P \end{bmatrix} = \begin{bmatrix} 1 & 0 & 0 \\ 3 & 1 & 5 \\ -2 & 0 & 1 \end{bmatrix}\begin{bmatrix} 19 \\ 5 \\ 16 \end{bmatrix} = \begin{bmatrix} 19 \\ 142 \\ -22 \end{bmatrix}$$

$$A\begin{bmatrix} T \\ E \\ M \end{bmatrix} = \begin{bmatrix} 1 & 0 & 0 \\ 3 & 1 & 5 \\ -2 & 0 & 1 \end{bmatrix}\begin{bmatrix} 20 \\ 5 \\ 13 \end{bmatrix} = \begin{bmatrix} 20 \\ 130 \\ -27 \end{bmatrix}$$

$$A\begin{bmatrix} B \\ E \\ R \end{bmatrix} = \begin{bmatrix} 1 & 0 & 0 \\ 3 & 1 & 5 \\ -2 & 0 & 1 \end{bmatrix}\begin{bmatrix} 2 \\ 5 \\ 18 \end{bmatrix} = \begin{bmatrix} 2 \\ 101 \\ 14 \end{bmatrix}$$

$$A\begin{bmatrix} I \\ S \\ O \end{bmatrix} = \begin{bmatrix} 1 & 0 & 0 \\ 3 & 1 & 5 \\ -2 & 0 & 1 \end{bmatrix}\begin{bmatrix} 9 \\ 19 \\ 15 \end{bmatrix} = \begin{bmatrix} 9 \\ 121 \\ -3 \end{bmatrix}$$

$$A\begin{bmatrix} K \\ A \\ Y \end{bmatrix} = \begin{bmatrix} 1 & 0 & 0 \\ 3 & 1 & 5 \\ -2 & 0 & 1 \end{bmatrix}\begin{bmatrix} 11 \\ 1 \\ 25 \end{bmatrix} = \begin{bmatrix} 11 \\ 109 \\ 3 \end{bmatrix}$$

The coded message is

19 142 −22 20 130 −27 2 101 14 9 121 −3 11 109 3

To decode the message in Example 3, form 3×1 column vectors of the numbers in the coded message and multiply on the left by A^{-1}.

The above are elementary examples of encoding and decoding. Modern-day cryptography uses sophisticated computer-implemented codes that depend on higher-level mathematics.

EXERCISE 2.7   APPLICATION 2   Answers to Odd-Numbered Problems Begin on Page AN-15.

In Problems 1–5, use the correspondence

A B C D E F G H I J K L M N O P Q R S T U V W X Y Z
↓ ↓
1 2 3 4 5 6 7 8 9 10 11 12 13 14 15 16 17 18 19 20 21 22 23 24 25 26

1. The matrix

$$A = \begin{bmatrix} 2 & 3 \\ 1 & 2 \end{bmatrix}$$

was used to code the following messages:

(a) 64 36 75 47 75 47 49 29 60 36 60 37 53 33 71
    39 35 18
(b) 76 49 64 41 95 55 43 24 69 42 59 32 77 45 27
    15 37 21 128 77

Use the inverse to decode each message.

2. Use the inverse of matrix

$$A = \begin{bmatrix} 1 & 0 & 0 \\ 3 & 1 & 5 \\ -2 & 0 & 1 \end{bmatrix}$$

to decode the message

4 152 17 15 106 −22 1 50 3 1 52 7 19 195 −12

3. The matrix A used to encode a message has as its inverse the matrix

$$A^{-1} = \begin{bmatrix} 2 & -3 \\ -1 & 2 \end{bmatrix}$$

Decode the message

70 39 62 41 113 69 93 57 51 28 29 15 54 33 14 9 62
40 67 43 116 71

4. Use the matrix

$$A = \begin{bmatrix} 1 & 0 & 0 \\ 3 & 1 & 5 \\ -2 & 0 & 1 \end{bmatrix}$$

to encode the message: I HAVE TICKETS TO THE FINAL
FOUR.

5. Use the matrix $A = \begin{bmatrix} 2 & 3 \\ 1 & 2 \end{bmatrix}$

to encode each of the following messages:

(a) I AM GOING TO DISNEY
(b) WE SURF THE NET
(c) LETS GO CLUBBING

6. Use the matrix

$$A = \begin{bmatrix} 1 & 0 & 0 \\ 3 & 1 & 5 \\ -2 & 0 & 1 \end{bmatrix}$$

to encode the messages given in Problem 5.

Application 3: Accounting

Determine the full cost of manufactured products

4

Consider a firm that has two types of departments, production and service. The production departments produce goods that can be sold in the market, and the service departments provide services to the production departments. A major objective of the cost accounting process is the determination of the full cost of manufactured products on a per unit basis. This requires an allocation of indirect costs, first, from the service department (where they are incurred) to the producing department in which the goods are manufactured and, second, to the specific goods themselves. For example, an accounting department usually provides accounting services for service departments as well as for the production departments. The indirect costs of service rendered by a service department must be determined in order to correctly assess the production departments. The total costs of a service department consist of its direct costs (salaries, wages, and materials) and its indirect costs (charges for the services it receives from other service departments). The nature of the problem and its solution are illustrated by the following example.

EXAMPLE 1   **Finding Total Monthly Costs**

Consider a firm with two production departments, P_1 and P_2, and three service departments, S_1, S_2, and S_3. These five departments are listed in the leftmost column of Table 5.

TABLE 5

Department	Total Costs	Direct Costs, Dollars	Indirect Costs for Services from Departments		
			S_1	S_2	S_3
S_1	x_1	600	$0.25x_1$	$0.15x_2$	$0.15x_3$
S_2	x_2	1100	$0.35x_1$	$0.20x_2$	$0.25x_3$
S_3	x_3	600	$0.10x_1$	$0.10x_2$	$0.35x_3$
P_1	x_4	2100	$0.15x_1$	$0.25x_2$	$0.15x_3$
P_2	x_5	1500	$0.15x_1$	$0.30x_2$	$0.10x_3$
Totals		5900	x_1	x_2	x_3

The total monthly costs of these departments are unknown and are denoted by x_1, x_2, x_3, x_4, x_5. The direct monthly costs of the five departments are shown in the third column of the table. The fourth, fifth, and sixth columns show the allocation of charges for the services of S_1, S_2, and S_3 to the various departments. Since the total cost for each department is its direct costs plus its indirect costs, the first three rows of the table yield the total costs for the three service departments:

$$x_1 = 600 + 0.25x_1 + 0.15x_2 + 0.15x_3$$
$$x_2 = 1100 + 0.35x_1 + 0.20x_2 + 0.25x_3$$
$$x_3 = 600 + 0.10x_1 + 0.10x_2 + 0.35x_3$$

Let X, C, and D denote the following matrices:

$$X = \begin{bmatrix} x_1 \\ x_2 \\ x_3 \end{bmatrix} \qquad C = \begin{bmatrix} 0.25 & 0.15 & 0.15 \\ 0.35 & 0.20 & 0.25 \\ 0.10 & 0.10 & 0.35 \end{bmatrix} \qquad D = \begin{bmatrix} 600 \\ 1100 \\ 600 \end{bmatrix}$$

Then the system of equations above can be written in matrix notation as

$$X = D + CX$$

which is equivalent to

$$[I_3 - C]X = D$$

The total costs of the three service departments can be obtained by solving this matrix equation for X:

$$X = [I_3 - C]^{-1}D$$

Now

$$[I_3 - C] = \begin{bmatrix} 0.75 & -0.15 & -0.15 \\ -0.35 & 0.80 & -0.25 \\ -0.10 & -0.10 & 0.65 \end{bmatrix}$$

from which it can be shown that

$$[I_3 - C]^{-1} = \begin{bmatrix} 1.57 & 0.36 & 0.50 \\ 0.79 & 1.49 & 0.76 \\ 0.36 & 0.28 & 1.73 \end{bmatrix}$$

It is significant that the inverse of $[I_3 - C]$ exists, and that all of its entries are nonnegative. Because of this and the fact that the matrix D contains only nonnegative entries, the matrix X will also have only nonnegative entries. This means there is a meaningful solution to the accounting problem:

$$X = [I_3 - C]^{-1}D = \begin{bmatrix} 1638.00 \\ 2569.00 \\ 1562.00 \end{bmatrix}$$

Thus, $x_1 = \$1638.00$, $x_2 = \$2569.00$, and $x_3 = \$1562.00$. All direct and indirect costs can now be determined by substituting these values in Table 5, as shown in Table 6.

From Table 6 we learn that department P_1 pays \$1122.25 for the services it receives from S_1, S_2, S_3, and P_2 pays \$1172.60 for the services it receives from these departments. The procedure we have followed charges the direct costs of the service departments to the production departments, and each production department is charged according to the services it utilizes. Furthermore, the total cost for P_1 and P_2 is \$5894.85, and this figure approximates the sum of the direct costs of the three service departments and the two production departments. The results are consistent with conventional accounting procedure. Discrepancies that occur are due to rounding.

TABLE 6

Department	Total Costs, Dollars	Direct Costs, Dollars	Indirect Costs for Services from Departments, Dollars		
			S_1	S_2	S_3
S_1	1629.15	600	409.50	385.35	234.30
S_2	2577.60	1100	573.30	513.80	390.50
S_3	1567.40	600	163.80	256.90	546.70
P_1	3222.25	2100	245.70	642.25	234.30
P_2	2672.60	1500	245.70	770.70	156.20

Finally, a comment should be made about the allocation of charges for services as shown in Table 5. How is it determined that 25% of the total cost x_1 of S_1 should be charged to S_1, 35% to S_2, 10% to S_3, 15% to P_1, and 15% to P_2? The services of each department can be measured in some suitable unit, and each department can be charged according to the number of these units of service it receives. If 20% of the accounting items concern a given department, that department is charged 20% of the total cost of the accounting department. When services are not readily measurable, the allocation basis is subjectively determined. ▸

EXERCISE 2.7   APPLICATION 3   Answers to Odd-Numbered Problems Begin on Page AN-15.

1. Consider the accounting problem described by the data in the table:

Department	Total Costs	Direct Costs, Dollars	Indirect Costs for Services from Departments	
			S_1	S_2
S_1	x_1	2,000	$\frac{1}{9}x_1$	$\frac{3}{9}x_2$
S_2	x_2	1,000	$\frac{3}{9}x_1$	$\frac{1}{9}x_2$
P_1	x_3	2,500	$\frac{1}{9}x_1$	$\frac{2}{9}x_2$
P_2	x_4	1,500	$\frac{3}{9}x_1$	$\frac{1}{9}x_2$
P_3	x_5	3,000	$\frac{1}{9}x_1$	$\frac{2}{9}x_2$
Totals		10,000	x_1	x_2

Determine whether this accounting problem has a solution. If it does, find the total costs. Prepare a table similar to

Table 6. Show that the total of the service charges allocated to P_1, P_2, and P_3 is equal to the sum of the direct costs of the service departments S_1 and S_2.

2. Follow the directions of Problem 1 for the accounting problem described by the following data:

Department	Total Costs	Direct Costs, Dollars	Indirect Costs for Services from Departments		
			S_1	S_2	S_3
S_1	x_1	500	$0.20x_1$	$0.10x_2$	$0.10x_3$
S_2	x_2	1000	$0.40x_1$	$0.15x_2$	$0.30x_3$
S_3	x_3	500	$0.10x_1$	$0.05x_2$	$0.30x_3$
P_1	x_4	2000	$0.20x_1$	$0.35x_2$	$0.20x_3$
P_2	x_5	1500	$0.10x_1$	$0.35x_2$	$0.10x_3$
Totals		5500	x_1	x_2	x_3

Application 4: The Method of Least Squares

In Section 1.4 we were given a set of data points (x, y) and we drew the scatter diagram representing the relation between x and y. We analyzed the scatter diagram to determine if the relation between the variables was linear. When it was linear, we selected two points and found an equation of the line containing the points. We then compared our line to the *line of best fit*, found using a graphing utility. In this application we define the *line of best fit* and learn how matrices are used to find the *line of best fit*.

The method of least squares refers to a technique that is often used in data analysis to find the "best" linear equation that fits a given collection of experimental data. (We shall see a little later just what we mean by the word *best*.) It is a technique employed by statisticians, economists, business forecasters, and those who try to interpret and analyze data. Before we begin our discussion, we will need to define the *transpose* of a matrix.

Transpose

Let A be a matrix of dimension $m \times n$. The **transpose** of A, written A^T, is the $n \times m$ matrix obtained from A by interchanging the rows and columns of A.

The first row of A^T is the first column of A; the second row of A^T is the second column of A; and so on.

5   **EXAMPLE 1**   **Finding the Transpose of a Matrix**

If

$$A = \begin{bmatrix} 1 & 2 & 3 \\ 0 & -1 & 2 \end{bmatrix} \qquad B = \begin{bmatrix} 1 & 1 \\ 0 & 1 \\ 2 & 3 \end{bmatrix} \qquad C = [1 \quad 0 \quad -1]$$

then

$$A^T = \begin{bmatrix} 1 & 0 \\ 2 & -1 \\ 3 & 2 \end{bmatrix} \qquad B^T = \begin{bmatrix} 1 & 0 & 2 \\ 1 & 1 & 3 \end{bmatrix} \qquad C^T = \begin{bmatrix} 1 \\ 0 \\ -1 \end{bmatrix} \qquad \blacktriangleright$$

Note how dimensions are reversed when computing A^T: A has the dimension 2×3, while A^T has the dimension 3×2, and so on.

We begin our discussion of the method of least squares with an example.

Suppose a product has been sold over time at various prices and that we have some data that show the demand for the product (in thousands of units) in terms of its price (in dollars). If we use x to represent price and y to represent demand, the data might look like the information in Table 7.

Suppose also that we have reason to assume that y and x are *linearly related*. That is, we assume we can write

$$y = mx + b$$

TABLE 7

Price, x	4	5	9	12
Demand, y	9	8	6	3

for some, as yet unknown, m and b. In other words, we are assuming that when demand is graphed against price, the resulting graph will be a line. Our belief that y and x are linearly related might be based on past experience or economic theory.

If we plot the data from Table 7 in a scatter diagram (see Figure 15), it seems pretty clear that no single line passes through the plotted points. Though no line will pass through all the points, we might still ask: "Is there a line that 'best fits' the points?" That is, we seek a line of the sort in Figure 16.

FIGURE 15

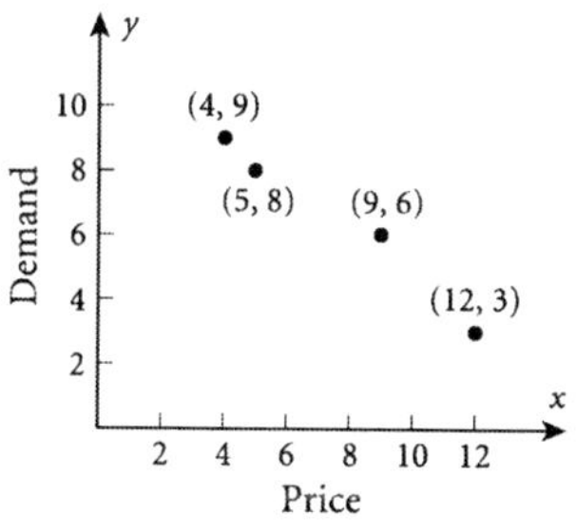

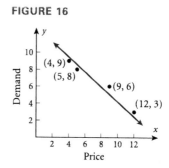

FIGURE 16

Statement of the Problem

Given a set of noncollinear points, find the line that "best fits" these points.

We still need to clarify our use of the phrase "best fits." First, though, suppose there are four data points, labeled from left to right as (x_1, y_1), (x_2, y_2), and so on. See Figure 17. Now suppose the equation of the line L in Figure 17 is

$$y = mx + b$$

where we do not know m and b. If (x_1, y_1) is actually on the line, it would be true that

$$y_1 = mx_1 + b$$

FIGURE 17

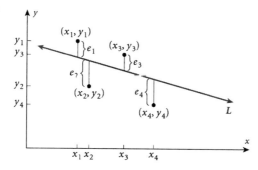

But, since we can't be sure (x_1, y_1) lies on L, the above equation may not be true. That is, there may be a nonzero difference between y_1 and $mx_1 + b$. We designate this difference by e_1, where e_1 represents the 'error'. Then

$$e_1 = y_1 - (mx_1 + b)$$

so, solving for y, we have

$$y_1 = mx_1 + b + e_1$$

What we have just said about (x_1, y_1) we can repeat for the remaining data points (x_i, y_i), $i = 2, 3$, and 4, obtaining

$$e_i = y_i - (mx_i + b)$$

or, equivalently,

$$y_i = mx_i + b + e_i \qquad i = 1, 2, 3, 4 \tag{1}$$

Geometrically, $|e_i|$ measures the vertical distance between the sought-after line L and the data point (x_i, y_i).

Recall that our problem is to find a line L that "best fits" the data points (x_i, y_i). Intuitively, we would seek a line L for which the e_i's are all small. Since some e_i's may be positive and some may be negative, it will not do to just add them up. To eliminate the signs, we square each e_i. The method of least squares consists of finding a line L for which the sum of the squares of the e_i's is as small as possible. Since finding the line $y = mx + b$ is the same as finding m and b, we restate the least squares problem for this example as follows:

The Least Squares Problem

Given the data points (x_i, y_i), $i = 1, 2, 3, 4$, find m and b satisfying

$$\boxed{y_i = mx_i + b + e_i \qquad i = 1, 2, 3, 4} \tag{2}$$

so that $e_1^2 + e_2^2 + e_3^2 + e_4^2$ is minimized.

A solution to the problem can be compactly expressed using matrix language. Since a derivation requires either calculus or more advanced linear algebra, we omit the proof and simply present the solution:

Solution to the Least Squares Problem

The line of best fit

$$\boxed{y = mx + b} \tag{3}$$

to a set of four points (x_1, y_1), (x_2, y_2), (x_3, y_3), (x_4, y_4) is obtained by solving the system of two equations in two unknowns

$$\boxed{A^T A X = A^T Y} \tag{4}$$

for m and b, where

$$A = \begin{bmatrix} x_1 & 1 \\ x_2 & 1 \\ x_3 & 1 \\ x_4 & 1 \end{bmatrix} \qquad X = \begin{bmatrix} m \\ b \end{bmatrix} \qquad Y = \begin{bmatrix} y_1 \\ y_2 \\ y_3 \\ y_4 \end{bmatrix} \tag{5}$$

6 **EXAMPLE 2**   **Finding the Line of Best Fit**

Use least squares to find the line of best fit for the points

$$(4, \ 9), \ (5, \ 8), \ (9, \ 6), \ (12, \ 3)$$

SOLUTION   We use Equations (5) to set up the matrices A and Y:

$$A = \begin{bmatrix} 4 & 1 \\ 5 & 1 \\ 9 & 1 \\ 12 & 1 \end{bmatrix} \qquad Y = \begin{bmatrix} 9 \\ 8 \\ 6 \\ 3 \end{bmatrix}$$

Equation (4), $A^T A X = A^T Y$, becomes

$$\begin{bmatrix} 4 & 5 & 9 & 12 \\ 1 & 1 & 1 & 1 \end{bmatrix} \begin{bmatrix} 4 & 1 \\ 5 & 1 \\ 9 & 1 \\ 12 & 1 \end{bmatrix} \begin{bmatrix} m \\ b \end{bmatrix} = \begin{bmatrix} 4 & 5 & 9 & 12 \\ 1 & 1 & 1 & 1 \end{bmatrix} \begin{bmatrix} 9 \\ 8 \\ 6 \\ 3 \end{bmatrix}$$

This reduces to a system of two equations in two unknowns:

$$\begin{cases} 266m + 30b = 166 \\ 30m + \ 4b = \ \ 26 \end{cases}$$

The solution, which you can verify, is given by

$$m = -\tfrac{29}{41} \qquad \text{and} \qquad b = \tfrac{484}{41}$$

The least squares solution to the problem is given by the line

$$y = -\tfrac{29}{41}x + \tfrac{484}{41}$$

Note, for example, that corresponding to the value $x = 5$, the above line of "best fit" has $y = -\tfrac{29}{41} \cdot 5 + \tfrac{484}{41} = 8.27$, while the experimentally observed demand had value 8. Were we to use our computed line to approximate the connection between price and demand, we would predict that a selling price of $x = 6$ would yield a demand of $y = -\tfrac{29}{41} \cdot 6 + \tfrac{484}{41} = 7.56$.

COMMENT: The line of best fit obtained in Example 2 is

$$y = -\tfrac{29}{41}x + \tfrac{484}{41} = -0.7073170732x + 11.80487805$$

In Figure 18, we compare the line of best fit obtained above to that generated by a graphing utility and show the graph of the line of best fit on the scatter diagram.

FIGURE 18

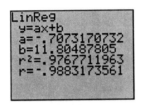   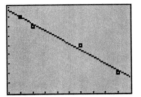

THE GENERAL LEAST SQUARES PROBLEM

We presented the method of least squares using a particular example with four data points. It is easy to extend this analysis.

The general least squares problem consists of finding the line of best fit to a set of n data points:

$$(x_1, y_1), (x_2, y_2), \ldots, (x_n, y_n) \tag{6}$$

If we let

$$A = \begin{bmatrix} x_1 & 1 \\ x_2 & 1 \\ \cdot & \cdot \\ \cdot & \cdot \\ \cdot & \cdot \\ x_n & 1 \end{bmatrix} \qquad X = \begin{bmatrix} m \\ b \end{bmatrix} \qquad Y = \begin{bmatrix} y_1 \\ y_2 \\ \cdot \\ \cdot \\ \cdot \\ y_n \end{bmatrix}$$

then the line of best fit to the data points in Equation (6) is

$$y = mx + b$$

where $X = \begin{bmatrix} m \\ b \end{bmatrix}$ is found by solving the system of two equations in two unknowns,

$$A^T A X = A^T Y \tag{7}$$

It can be shown that the system given by Equation (7) always has a unique solution provided the data points do not all lie on the same vertical line.

The least squares techniques presented here along with more elaborate variations are frequently used today by statisticians and researchers.

In Problems 1–6, compute A^T.

1. $A = \begin{bmatrix} 4 & 1 & 2 \\ 3 & 1 & 0 \end{bmatrix}$

2. $A = \begin{bmatrix} 5 & 2 & -1 \\ 1 & 3 & 6 \\ 1 & -1 & 2 \end{bmatrix}$

3. $A = \begin{bmatrix} 1 & 11 \\ 0 & 12 \\ 1 & 4 \end{bmatrix}$

4. $A = \begin{bmatrix} -1 & 6 & 4 \end{bmatrix}$

5. $A = \begin{bmatrix} 8 \\ 6 \\ 3 \end{bmatrix}$

6. $A = \begin{bmatrix} 5 & 3 \\ 3 & 7 \end{bmatrix}$

7. Supply Equation The following table shows the supply (in thousands of units) of a product at various prices (in dollars):

Price, x	3	5	6	7
Supply, y	10	13	15	16

(a) Find the least squares line of best fit to the above data.
(b) Use the equation of this line to estimate the supply of the product at a price of $8.

8. Study Time vs. Performance Data giving the number of hours a person studied compared to his or her performance on an exam are given in the table:

Hours Studied x	Exam Score y
0	50
2	74
4	85
6	90
8	92

(a) Find a least squares line of best fit to the above data.
(b) What prediction would this line make for a student who studied 9 hours?

9. Advertising vs. Sales A business would like to determine the relationship between the amount of money spent on advertising and its total weekly sales. Over a period of 5 weeks it gathers the following data:

Amount Spent on Advertising (in thousands) x	Weekly Sales Volume (in thousands) y
10	50
17	61
11	55
18	60
21	70

Find a least squares line of best fit to the above data.

10. Drug Concentration vs. Time The following data show the connection between the number of hours a drug has been in a person's body and its concentration in the body.

Number of Hours	Drug Concentration (parts per million)
2	2.1
4	1.6
6	1.4
8	1.0

(a) Fit a least squares line to the above data.
(b) Use the equation of the line to estimate the drug concentration after 5 hours.

11. A matrix is **symmetric** if $A^T = A$. Which of the following matrices are symmetric?

(a) $\begin{bmatrix} 1 & 1 & 2 \\ 1 & 0 & 1 \\ 3 & 2 & 3 \end{bmatrix}$ (b) $\begin{bmatrix} 0 & 1 & 3 \\ 1 & 4 & 7 \\ 3 & 7 & 5 \end{bmatrix}$ (c) $\begin{bmatrix} 1 & 2 & 3 & 0 \\ 2 & 4 & 5 & 0 \\ 3 & 5 & 1 & 0 \end{bmatrix}$

Need a symmetric matrix be square?

12. Show that the matrix $A^T A$ is always symmetric.

Chapter 2   Review

OBJECTIVES

THINGS TO KNOW

Systems of Linear Equations
Systems with a solution are consistent and either have a unique solution or have infinitely many solutions
Systems with no solution are inconsistent

Solving Systems of Linear Equations
Using substitution (p. 53)
Using elimination (p. 55)
Using matrices: row echelon method (pp. 70–71)
Using matrices: reduced row echelon method (p. 73)

Matrix (pp. 65 and 94)
Augmented matrix of a system of linear equations (p. 65)
Row operations (p. 67)
Dimension of a matrix (p. 94)
Square matrix (p. 94)
Zero matrix (p. 98)
Identity matrix (p. 113)
Inverse of a matrix (p. 117)

TRUE–FALSE ITEMS   Answers are on page AN-15.

T F **1.** Matrices of the same dimension can always be added.

T F **2.** Matrices of the same dimension can always be multiplied.

T F **3.** A square matrix will always have an inverse.

T F **4.** The reduced row echelon form of a matrix A is unique.

T F **5.** If A and B are each of dimension 4×4, then $AB = BA$.

T F **6.** Matrix addition is always defined.

T F **7.** Matrix multiplication is commutative.

FILL IN THE BLANKS   Answers are on page AN-15.

1. If matrix A is of dimension 3×4 and matrix B is of dimension 4×2, then AB is of dimension _____.

2. A system of three linear equations containing three variables has either _____ solution, or no solution, or _____ _____ solutions.

3. If A is a matrix of dimension 3×4, the 3 tells the number of _____ and the 4 tells the number of _____ .

4. If $AB = I$, the identity matrix, then B is called the _____ of A.

5. If B is a 2×3 matrix and BA^2 is defined, then A is a matrix of dimension _____.

6. If A is a 4×5 matrix and AB^3 is defined, then B is a matrix of dimension _____.

REVIEW EXERCISES   Answers to odd-numbered problems begin on page AN-15.
Blue problem numbers represent the author's suggestions for a Practice Test.

In Problems 1–10, solve each system of equations algebraically using the method of substitution or the method of elimination. If the system has no solution, say it is inconsistent.

1. $\begin{cases} 2x - y = 5 \\ 5x + 2y = 8 \end{cases}$

2. $\begin{cases} 2x + 3y = 2 \\ 7x - y = 3 \end{cases}$

3. $\begin{cases} x - 2y - 4 = 0 \\ 3x + 2y - 4 = 0 \end{cases}$

4. $\begin{cases} x - 3y + 4 = 0 \\ \dfrac{1}{2}x - \dfrac{3}{2}y + \dfrac{4}{3} = 0 \end{cases}$

5. $\begin{cases} 3x - 2y = 8 \\ x - \dfrac{2}{3}y = 12 \end{cases}$

6. $\begin{cases} 2x + 5y = 10 \\ 4x + 10y = 20 \end{cases}$

7. $\begin{cases} x + 2y - z = 6 \\ 2x - y + 3z = -13 \\ 3x - 2y + 3z = -16 \end{cases}$

8. $\begin{cases} x + 5y - z = 2 \\ 2x + y + z = 7 \\ x - y + 2z = 11 \end{cases}$

9. $\begin{cases} 2x - 4y + z = -15 \\ x + 2y - 4z = 27 \\ 5x - 6y - 2z = -3 \end{cases}$

10. $\begin{cases} x - 4y + 3z = 15 \\ -3x + y - 5z = -5 \\ -7x - 5y - 9z = 10 \end{cases}$

In Problems 11–14, write the system of equations corresponding to the given augmented matrix. In Problems 13–14, analyze the solution.

11. $\begin{bmatrix} 3 & 2 & | & 8 \\ 1 & 4 & | & -1 \end{bmatrix}$

12. $\begin{bmatrix} 1 & 2 & 5 & | & -2 \\ 5 & 0 & -3 & | & 8 \\ 2 & -1 & 0 & | & 0 \end{bmatrix}$

13. $\begin{bmatrix} 1 & 0 & 0 & | & 4 \\ 0 & 1 & 0 & | & 6 \\ 0 & 0 & 1 & | & -1 \end{bmatrix}$

14. $\begin{bmatrix} 1 & 0 & 3 & | & 5 \\ 0 & 1 & 2 & | & 8 \end{bmatrix}$

In Problems 15–32, use matrices to find the solution, if it exists, of each system of linear equations. If the system has infinitely many solutions, write the solution using parameters and then list at least three solutions. If the system has no solution, say it is inconsistent.

15. $\begin{cases} -5x + 2y = -2 \\ -3x + 3y = 4 \end{cases}$

16. $\begin{cases} -3x + 2y = 3 \\ -3x + 4y = 4 \end{cases}$

17. $\begin{cases} x + 2y + 5z = 6 \\ 3x + 7y + 12z = 23 \\ x + 4y = 25 \end{cases}$

18. $\begin{cases} x + 2y - z = -4 \\ 3x + 7y - z = -21 \\ x + 4y - 6z = -17 \end{cases}$

19. $\begin{cases} x + 2y + 7z = 2 \\ 3x + 7y + 18z = -1 \\ x + 4y + 2z = -13 \end{cases}$

20. $\begin{cases} x + 2y - 7z = -1 \\ 3x + 7y - 24z = 6 \\ x + 4y - 12z = 26 \end{cases}$

21. $\begin{cases} 2x - y + z = 1 \\ x + y - z = 2 \\ 3x - y + z = 0 \end{cases}$

22. $\begin{cases} 2x + 3y - z = 5 \\ x - y + z = 1 \\ 3x - 3y + 3z = 3 \end{cases}$

23. $\begin{cases} y - 2z = 6 \\ 3x + 2y - z = 2 \\ 4x + 3z = -1 \end{cases}$

24. $\begin{cases} 2x - y + 3z = 5 \\ x + 2z = 0 \\ 3x + 2y + z = -3 \end{cases}$

25. $\begin{cases} x - 3y = 5 \\ 3y + z = 0 \\ 2x - y + 2z = 2 \end{cases}$

26. $\begin{cases} x - z = 2 \\ 2x - y = 4 \\ x + y + z = 6 \end{cases}$

27. $\begin{cases} 3x + y - 2z = 3 \\ x - 2y + z = 4 \end{cases}$

28. $\begin{cases} 2x - y - 3z = 0 \\ x - 2y + z = 4 \end{cases}$

29. $\begin{cases} x + 2y - z = 5 \\ 2x - y + 2z = 0 \end{cases}$

30. $\begin{cases} x - y + 2z = 6 \\ 2x + 2y - z = -1 \end{cases}$

31. $\begin{cases} 2x - y = 6 \\ x - 2y = 0 \\ 3x - y = 6 \end{cases}$

32. $\begin{cases} x - 2y = 0 \\ 2x + y = 5 \\ x - 3y = 6 \end{cases}$

In Problems 33–36, analyze the solution, if any, of the system of equations having the given augmented matrix.

33. $\begin{bmatrix} 1 & 4 & 3 & | & 4 \\ 0 & 1 & 0 & | & -1 \\ 0 & 0 & 1 & | & 1 \end{bmatrix}$

34. $\begin{bmatrix} 1 & 0 & -3 & | & 4 \\ 0 & 1 & 0 & | & 8 \\ 0 & 0 & 1 & | & 0 \end{bmatrix}$

35. $\begin{bmatrix} 1 & 0 & 0 & 2 & | & 1 \\ 0 & 1 & 1 & 2 & | & 2 \\ 0 & 0 & 1 & 0 & | & 3 \end{bmatrix}$

36. $\begin{bmatrix} 1 & 0 & 3 & 1 & | & 2 \\ 0 & 1 & 4 & 2 & | & 1 \end{bmatrix}$

In Problems 37–50, use the following matrices to compute each expression. State the dimension of the resulting matrix.

$$A = \begin{bmatrix} 1 & 0 \\ 2 & 4 \\ -1 & 2 \end{bmatrix} \quad B = \begin{bmatrix} 4 & -3 & 0 \\ 1 & 1 & -2 \end{bmatrix} \quad C = \begin{bmatrix} 3 & -4 \\ 1 & 5 \\ 5 & -2 \end{bmatrix}$$

37. $A + C$

38. $A - C$

39. $6A$

40. $-4B$

41. AB

42. BA

43. CB

44. BC

45. $(A + C)B$

46. $B(2C - A)$

47. A^T

48. B^T

49. $C + 0$

50. $A + (-A)$

In Problems 51–56, find the inverse, if it exists, of each matrix.

51. $\begin{bmatrix} 3 & 0 \\ -2 & 1 \end{bmatrix}$

52. $\begin{bmatrix} 4 & 1 \\ 3 & 1 \end{bmatrix}$

53. $\begin{bmatrix} 4 & 2 \\ 6 & 3 \end{bmatrix}$

54. $\begin{bmatrix} 1 & 2 & 3 \\ 2 & 4 & 5 \\ 3 & 5 & 6 \end{bmatrix}$

55. $\begin{bmatrix} 4 & 3 & -1 \\ 0 & 2 & 2 \\ 3 & -1 & 0 \end{bmatrix}$

56. $\begin{bmatrix} 1 & 2 & -3 \\ 4 & 6 & 2 \\ -1 & -6 & 9 \end{bmatrix}$

57. What must be true about x, y, z, w, if we require $AB = BA$ for the matrices

$$A = \begin{bmatrix} x & y \\ z & w \end{bmatrix} \quad \text{and} \quad B = \begin{bmatrix} 1 & 1 \\ -1 & 1 \end{bmatrix}$$

58. Let $t = [t_1 \ t_2]$, with $t_1 + t_2 = 1$, and let $A = \begin{bmatrix} \dfrac{1}{4} & \dfrac{3}{4} \\ \dfrac{2}{3} & \dfrac{1}{3} \end{bmatrix}$.

Find t such that $tA = t$.

59. Mixture Sweet Delight Candies, Inc., sells boxes of candy consisting of creams and caramels. Each box sells for $4 and holds 50 pieces of candy (all pieces are the same size). If the caramels cost $0.05 to produce and the creams cost $0.10 to produce, how many caramels and creams should be in each box for no profit and no loss? Would you increase or decrease the number of caramels in order to obtain a profit?

60. Cookie Orders A cookie company makes three kinds of cookies—oatmeal raisin, chocolate chip, and shortbread—packaged in small, medium, and large boxes. The small box contains 1 dozen oatmeal raisin and 1 dozen chocolate chip; the medium box has 2 dozen oatmeal raisin, 1 dozen chocolate chip, and 1 dozen shortbread; the large box contains 2 dozen oatmeal raisin, 2 dozen chocolate chip, and 3 dozen shortbread. If you require exactly 15 dozen oatmeal raisin, 10 dozen chocolate chip, and 11 dozen shortbread cookies, how many of each size box should you buy?

61. Mixture Problem A store sells almonds for $6 per pound, cashews for $5 per pound, and peanuts for $2 per pound. One week the manager decides to prepare 100 16-ounce packages of nuts by mixing the peanuts, almonds, and cashews. Each package will be sold for $4. The mixture is to produce the same revenue as selling the nuts separately. Prepare a table that shows some of the possible ways the manager can prepare the mixture.

62. Financial Planning Three retired couples each require an additional annual income of $1800 per year. As their financial consultant, you recommend they invest some money in Treasury Bills that yield 6%, some money in corporate bonds that yield 8%, and some money in junk

bonds that yield 10%. Prepare a table for each couple showing the various ways their goal can be achieved

(a) If the first couple has $20,000 to invest
(b) If the second couple has $25,000 to invest
(c) If the third couple has $30,000 to invest

63. Financial Planning A retired couple have $40,000 to invest. As their financial consultant, you recommend they invest some money in Treasury Bills that yield 6%, some money in corporate bonds that yield 8%, and some money in junk bonds that yield 10%. Prepare a table showing the various ways this couple can achieve the following goals:

(a) They want $2500 per year in income.
(b) They want $3000 per year in income.
(c) They want $3500 per year in income.

64. Cost of College

The National Center for Education Statistics collects data on the costs of college. The data in the table show the average cost of tuition, fees, room, and board paid by full-time undergraduate students at degree-granting institutions for select years.

Year	Average Cost
1976–77	$2,275
1981–82	3,489
1987–88	5,494
1991–92	7,077
1995–96	8,800
2000–01	10,876

(a) Draw a scatter diagram of the data, and comment on its apparent shape. [*Hint*: Represent 1976–77 by 1, 1981–82 by 6, 1987–88 by 12, and so on.]

(b) Find the least squares line of best fit relating years to average cost.

(c) Interpret the slope of the line found in part (b).

(d) Use the line of best fit to predict the cost of tuition, fees, room, and board for the school year 2003–04.

(e) Use the least squares line to predict the cost of college in 2005–06.

(f) Compute the line of best fit using a graphing utility and compare it to the line obtained in part (b).

Source: Digest of Education Statistics, National Center for Education Statistics, 2001.

65. The U.S. Bureau of the Census publishes estimates of emigration from the United States to other countries. The following table lists the estimated emigration rates for selected years from 1991–2001.

Year	Emigrates
1991	252,000
1993	258,000
1996	267,000
1998	278,000
2000	287,000
2001	293,000

Use these data in the following questions.

(a) Draw a scatter diagram, and comment on its apparent shape. [*Hint*: Use 1 for 1991, 3 for 1993, 6 for 1996 and so on.]

(b) Find the least squares line of best fit relating years to emigration numbers.

(c) Interpret the slope of the line found in part (b).

(d) Use the line of best fit to predict the emigration from the United States in 2005.

(e) Use the least squares line to estimate the number of people who emigrated from the United States in 1995.

(f) Compute the line of best fit using a graphing utility and compare it to the line obtained in part (b).

Source: U.S. Bureau of the Census, Internet release, 2003.

66. In a professional cooperative a physician, an attorney, and a financial planner trade services. They agree that the value of each one's work is approximately $20,000. At the end of the period the physician had spent 30% of his time on his own care, 40% on the attorney's care, and 30% on the financial planner's care. The attorney had

spent 50% of her time on the physician's affairs, 20% on her own affairs, and 30% on the financial planner's affairs. Finally, the financial planner spent 20% of his time managing the physician's portfolio, 20% managing the attorney's portfolio, and 60% managing his own portfolio. How should each of these three professionals be paid for their work?

67. Suppose three corporations each manufacture a different product, although each needs the others' goods to produce its own. Moreover, suppose some of each manufacturer's production is consumed by individuals outside the system. Current internal and external consumptions are given in the table below, but it has been forecast that consumer demand will change in the next several years and that in 5 years consumers will demand 80 units of product *A*, 40 units of product *B*, and 80 units of product *C*. Find the total output needed to be made by each corporation to meet the predicted demand.

	A	*B*	*C*	Consumer	Total
A	100	50	40	60	250
B	20	10	30	40	100
C	30	40	30	100	200

68. Use the correspondence

A B C D E F G H I J K L M N O P
26 25 24 23 22 21 20 19 18 17 16 15 14 13 12 11

Q R S T U V W X Y Z
10 9 8 7 6 5 4 3 2 1

and the matrices (I) $A = \begin{bmatrix} 3 & 1 \\ 2 & 2 \end{bmatrix}$ and (II) $B = \begin{bmatrix} 1 & 4 & 2 \\ 2 & 0 & 2 \\ 0 & 0 & 4 \end{bmatrix}$

to encode the following messages:

(a) HEY DUDE WHATS UP
(b) CALL ME ON MY CELL

Then use (I) to decode the message: 75 78 45 42 72 64 35 42 17 26 67 62.

69. A local firm has two service departments, purchasing and legal, and three production departments, P_1, P_2, and P_3. The total costs of each department are unknown and are denoted x_1, x_2, x_3, x_4, and x_5. The direct costs of manufacture and the indirect costs of service are listed in the table on page 155.

Department	Total Costs	Direct Cost in Dollars	Indirect Costs for Services	
S_1	x_1	800	$0.20x_1$	$0.10x_2$
S_2	x_2	4000	$0.10x_1$	$0.30x_2$
P_1	x_3	1500	$0.20x_1$	$0.10x_2$
P_2	x_4	500	$0.30x_1$	$0.20x_2$
P_3	x_5	1200	$0.20x_1$	$0.30x_2$
Totals		8000	x_1	x_2

Find the total costs. Prepare a table similar to Table 6 on page 143. Show that the total of the service charges allocated to P_1, P_2, and P_3 is equal to the sum of the direct costs of the service departments S_1 and S_2.

Chapter 2 Project

THE OPEN LEONTIEF MODEL AND THE AMERICAN ECONOMY

The open Leontief model, introduced in Section 2.7, is used to model the economies of many countries throughout the world, as well as the global economy itself. The first example Leontief himself used was the American economy in 1947. He divided the economy into 42 industries (or **sectors**) for his original model. For purposes of this example, the data from the 42 sectors used by Leontief has been collected into just 3: agriculture, manufacturing, and services (shown in Table 1). Of course, the open sector (which consumes only goods and services) or demand vector is also present.

TABLE 1   EXCHANGE OF GOODS AND SERVICES IN THE U.S. FOR 1947 (IN BILLIONS OF 1947 DOLLARS)

	Agriculture	Manufacturing	Services	Open Sector	Total Gross Output
Agriculture	34.69	4.92	5.62	39.24	84.56
Manufacturing	5.28	61.82	22.99	60.02	163.43
Services	10.45	25.95	42.03	130.65	219.03

1. Create the matrix A and the demand vector D_0 for the open Leontief model from Table 1.

2. Calculate the production vector X_0 for the open Leontief model from Table 1.

3. Now suppose that the demand vector is changed to

$$D_1 = \begin{bmatrix} 40.24 \\ 60.02 \\ 130.65 \end{bmatrix}$$

Notice that the only change from the original demand vector D_0 to the new demand vector D_1 is the addition of 1 unit of consumer demand in the agricultural sector. Calculate a new production vector X_1 for this new demand vector D_1.

4. Compute $X_1 - X_0$.

You should find that the difference in the production vectors X_0 and X_1 is the first column of the matrix $(I - A)^{-1}$. This is not a coincidence. It can be shown that the difference in the old and new production vectors must be the first column of $(I - A)^{-1}$. This observation leads to an important interpretation of the entries of $(I - A)^{-1}$:

Interpretation: The (i, j) entry in the matrix $(I - A)^{-1}$ is the amount by which industry i must change its production level to satisfy an increase of one unit in the demand from industry j.

5. How much would the service production level need to increase if demand from the agriculture sector is increased by 1 unit?

6. How much would the manufacturing production level need to increase if demand from the agricultural sector is increased by 1 unit?

The most up-to-date available input–output table for the American economy is the 1998 table. The entire table (which is officially called an I-O Use table) may be found in the publication "Annual Input–Output Accounts of the U.S. Economy, 1998," which may be downloaded from the Bureau of Economic Analysis website: www.bea.gov/bea/an2.htm.

This document provides a good overview of different types of tables, as well as some applications of the 1998 table. The table for 1998 divides the economy into nearly 500 sectors, which are then consolidated into over 90 sectors. There is also a table that further consolidates the economy into just 9 sectors: This table is given as Table 2.

TABLE 2   EXCHANGE OF GOODS AND SERVICES IN THE U.S. FOR 1998 (IN MILLIONS OF 1998 DOLLARS)

Industry	Agric.	Mining	Const.	Manuf.	T.C.U.	Trade	F.I.R.	Services	Other	Demand	Total Output
Agriculture	68682	78	5860	144622	154	1816	11476	12310	567	34940	283290
Mining	368	31478	7368	81722	52354	31	6	32	3061	−39241	147738
Construction	3369	4693	895	28756	47369	12694	66515	28785	25895	787208	1006179
Manufacturing	49395	14510	299429	1380590	70485	68005	19318	340944	17593	1611520	3953584
Transportation, Communication, Utilities	12625	12652	24847	179922	200933	68214	52626	120762	22872	586248	1250307
Trade	13948	3498	81671	230668	15091	32765	4925	68036	2646	1103110	1552311
Finance, Insurance, Real Estate	20647	33253	16485	71167	40283	108418	445679	243750	7945	1520718	2547553
Services	8998	5851	103708	240141	144495	219223	191363	530971	13585	2214382	3479631
Other	166	29	1076	13826	3306	11226	28196	24713	3034	1032052	1211707

7. Find the consumption matrix A and demand vector D_0 for this table.

8. Find the production vector X for the consumption matrix A and demand vector D_0.

9. If the demand for construction increases by 1 million, how much will the production of the transportation, communication, and utilities sector have to increase to compensate?

10. Which three sectors are most affected by an increase of $1 million of demand for construction?

11. If the entire jth column of the matrix $(I - A)^{-1}$ is added, the result is the amount of new production in the **entire economy** caused by an increase of $1 million in demand for the jth sector. How much new production is produced in the entire economy by an increase of $1 million in demand for construction? This amount is referred to in economic literature as a **backward linkage.**

12. What would it mean for the (i, j) entry in $(I - A)^{-1}$ to be zero?

MATHEMATICAL QUESTIONS FROM PROFESSIONAL EXAMS*

Use the following information to answer Problems 1–4:
Akron, Inc. owns 80% of the capital stock of Benson Company and 70% of the capital stock of Cashin, Inc. Benson Company owns 15% of the capital stock of Cashin, Inc. Cashin, Inc., in turn, owns 25% of the capital stock of Akron, Inc. These ownership interrelationships are illustrated in the diagram.

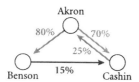

Net income before adjusting for interests in intercompany net income for each corporation follows:

Akron, Inc.	$190,000
Benson Co.	$170,000
Cashin, Inc.	$230,000

Ignore all income tax considerations.
A_e = *Akron's consolidated net income;*
that is, its net income plus its share of the consolidated net income of Benson and Cashin
B_e = *Benson's consolidated net income;*
that is, its net income plus its share of the consolidated net income of Cashin
C_e = *Cashin's consolidated net income;*
that is, its net income plus its share of the consolidated net income of Akron

1. **CPA Exam** The equation, in a set of simultaneous equations, which computes A_e is

   (a) $A_e = .75(190,000 + .8B_e + .7C_e)$
   (b) $A_e = 190,000 + .8B_e + .7C_e$
   (c) $A_e = .75(190,000) + .8(170,000) + .7(230,000)$
   (d) $A_e = .75(190,000) + .8B_e + .7C_e$

2. **CPA Exam** The equation, in a set of simultaneous equations, which computes B_e is

   (a) $B_e = 170,000 + .15C_e - .75A_e$
   (b) $B_e = 170,000 + .15C_e$
   (c) $B_e = .2(170,000) + .15(230,000)$
   (d) $B_e = .2(170,000) + .15C_e$

3. **CPA Exam** Cashin's minority interest in consolidated net income is

   (a) $.15(230,000)$ (b) $230,000 + .25A_e$
   (c) $.15(230,000) + .25A_e$ (d) $.15C_e$

4. **CPA Exam** Benson's minority interest in consolidated net income is

   (a) $34,316   (b) $25,500   (c) $45,755   (d) $30,675

CPSIA information can be obtained at www.ICGtesting.com
Printed in the USA
241224LV00001B/1/P

9 780470 891797